AA007222

44th Riso International Symposium on Materials Science (RISO 2024)

IOP Conference Series: Materials Science and Engineering
Volume 1310

Roskilde, Denmark
2-6 September 2024

ISBN: 979-8-3313-0267-2
ISSN: 1757-8981

Printed from e-media with permission by:

Curran Associates, Inc.
57 Morehouse Lane
Red Hook, NY 12571

Some format issues inherent in the e-media version may also appear in this print version.

Institute of Physics
Dirac House, Temple Back
Bristol BS1 6BE UK

Phone: 44 1 17 929 7481
Fax: 44 1 17 920 0979

techtracking@iop.org

Additional copies of this publication are available from:

Curran Associates, Inc.
57 Morehouse Lane
Red Hook, NY 12571 USA
Phone: 845-758-0400
Fax: 845-758-2633
Email: curran@proceedings.com
Web: www.proceedings.com

TABLE OF CONTENTS

IOP Conf. Series: Materials Science and Engineering **1310** (2024) 011001 doi:10.1088/1757-899X/1310/1/011001

This Symposium focusses on the microstructures of metallic samples and components prepared by additive manufacturing (AM), but covers also comparative aspects of microstructural evolution during conventional thermomechanical processing, where this is informative for understanding microstructures resulting from AM.

AM of metals has become very popular as a next generation digitally-driven manufacturing technique, and huge successes have already been achieved in obtaining high material densities, as well as in managing stress relief to allow precise control of component shapes. However, proper characterization of metal microstructures formed through AM still lags behind these successes. In particular, the potential for property optimization through microstructural engineering has hardly been explored yet for metal AM. As a reflection of this, the 44th Risø International Symposium focus on 2D and 3D characterization of additively manufactured metal microstructures, the in-service microstructural evolution of components made using AM, relationships between microstructure and properties, and the possibilities for optimizing metal microstructures during the AM process, as well as by use of post-AM treatments, as studied through either experiments or simulations. Contributions regarding all types of metal AM processes are included, as are those relating microstructures to a wide range of properties, from mechanical to physical and chemical. Contributions based both on experimental as well as on theoretical work are embraced. The Symposium Proceedings contains 7 keynote papers, 7 invited papers, and 31 contributed papers, covering a wide range of topics including:

- Characterization of AM microstructures
- Simulation of AM microstructures
- Relations between AM processing and AM microstructures
- Relations between AM microstructures and properties
- Comparison between AM microstructures and those evolving during conventional manufacturing (e.g. during rolling, and annealing)
- Design of defect-tolerant microstructures in additively manufactured metals
- Microstructural enhancements using advanced AM processing and metal combinations for better properties
- Possibilities in use of post-AM microstructural engineering treatments
- Microstructural evolution during in-service operation of AM components
- Challenges related to microstructural optimization for practical applications of AM components

Local Organizing Committee

A. T. Brandt (Technical University of Denmark, Denmark)
A. Godfrey (Tsinghua University, China)
D. Juul Jensen (Technical University of Denmark, Denmark)
J. V. Laursen (Technical University of Denmark, Denmark)
S. Mohanty (Technical University of Denmark, Denmark)
V. K. Nadimpalli (Technical University of Denmark, Denmark)
W. Pantleon (Technical University of Denmark, Denmark)
T. Yu (Technical University of Denmark, Denmark)
Y. Zhang (Technical University of Denmark, Denmark)
X. Zhang (Technical University of Denmark, Denmark)

International Advisory Committee

B. Boyce (Sandia National Laboratory, USA)
X. Huang (Chongqing University, China)
L. Lu (Shenyang National Laboratory for Materials Science, China)
I. Raffeis (RWTH Aachen University, Germany)
N. Tsuji (Kyoto University, Japan)
P. Withers (University of Manchester, UK)

IOP Conf. Series: Materials Science and Engineering **1310** (2024) 011001 doi:10.1088/1757-899X/1310/1/011001

Editorial Committee

IOP Conf. Series: Materials Science and Engineering **1310** (2024) 011002 doi:10.1088/1757-899X/1310/1/011002

Peer Review Statement

All papers published in this volume have been reviewed through processes administered by the Editors. Reviews were conducted by expert referees to the professional and scientific standards expected of a proceedings journal published by IOP Publishing.

- **Type of peer review:** Single Anonymous
- **Conference submission management system:** Morressier
- **Number of submissions received:** 49
- **Number of submissions sent for review:** 49
- **Number of submissions accepted:** 45
- **Acceptance Rate (Submissions Accepted / Submissions Received × 100):** 91.8
- **Average number of reviews per paper:** 2
- **Total number of reviewers involved:** 9
- **Contact person for queries:**
  Name: Jytte Laursen
  Email: risoesymp@construct.dtu.dk
  Affiliation: Technical University of Denmark

In-Situ EBSD Study of Austenitisation in a Wire-Arc Additively Manufactured High-Strength Steel

M Taylor[1*], Y H Mozumder[1], A D Smith[2], A E Davis[1], F Scenini[1], P B Prangnell[1] and E J Pickering[1,3]

[1] Department of Materials, University of Manchester, M13 9PL, UK
[2] TESCAN-UK, Cambridge, UK
[3] Henry Royce Institute, Oxford Road, Manchester M13 9PL, UK

[*]E-mail: mark.taylor-5@manchester.ac.uk

Abstract. As-built high-strength steel wire-arc additive manufactured (WAAM) components often contain coarse columnar prior austenite grain (PAG) structures. These microstructures can result in poor mechanical properties, hence post-build re-austenitisation treatments are of interest to improve microstructures. Here, the potential for engineering austenite refinement was investigated using high-temperature in-situ EBSD with high temporal resolution. It was found that 300M first austenitised with a memory effect, reforming the columnar PAG structures in both morphology and crystallographic orientation. This was observed to occur by the coalescence of acicular austenite and happened earlier in the depleted dendritic regions. On further heating, austenite underwent recrystallisation without the application of external deformation via a discontinuous mechanism, involving the bulging of the prior high angle boundaries. Recrystallisation twins were also seen to form and grow alongside regular recrystallised grains. The predominance of discontinuous recrystallisation meant that a coarse grain structure was still retained after austenitisation, owing to the influence of a grain 'swapping' effect.

1. Introduction

Wire-arc additive manufacturing (WAAM) may be used to produce near-net-shape high-strength steel components, thereby reducing waste and costly machining compared to traditional forging and machining manufacturing routes [1]. However, in martensitic steels, such as 300M, WAAM may produce coarse, textured columnar prior austenite grain (PAG) structures [2], which can result in poor mechanical properties and mechanical anisotropy if not dealt with in process or by a post-build heat treatment. Here, we investigate the latter.

In some cases of austenitisation in ferritic steels, an austenite memory effect [3–5] occurs, wherein PAGs in a martensitic or bainitic steel are reformed in both morphology and crystallographic orientation during austenitisation. In a few of these steel alloys, autogenous recrystallisation, *i.e.* without the application of external deformation (sometimes called 'spontaneous recrystallisation'), has been shown to follow the memory effect during austenitisation heat treatments, which leads to a refined grain structure [4–11]. If the memory effect occurs in 300M, the success of a post-build heat treatment to optimise the microstructure may depend upon the promotion of the post-memory recrystallisation effect.

Recently, the memory effect and subsequent austenite recrystallisation have been observed through *in-situ* high temperature EBSD [8–12]. The memory effect has been reported to progress via the growth of acicular austenite that nucleates between martensite laths, with the new austenite having the same orientation of the previous PAG [4,9,13]. The PAG's morphology is reproduced when similarly orientated acicular austenite grains coalesce. Several authors [9,11,14] have also noted that acicular austenite has a high dislocation density and, under certain conditions, this may provide sufficient driving force for subsequent austenite recrystallisation [8–12]. There are several theories [3,7,15,16] concerning the mechanism behind the nucleation and growth of memory orientation acicular austenite. Importantly, the memory effect has been reported in steels that contain no detectable retained austenite [8,12,16,17], suggesting that it does not necessarily originate from the regrowth of retained austenite. It is also unclear whether the high dislocation density seen in acicular austenite is inherited from the martensite into which it grows [5,12,14], is generated by the phase transformation [15], or results from slight misorientation differences between similarly-orientated austenite grains on coalescence [8].

In this work, high-temperature *in-situ* EBSD was used to directly observe the evolution of a 300M as-built WAAM microstructure during austenitisation. The memory effect and subsequent recrystallisation during further heating were observed with higher temporal resolution than has been reported previously in the literature [8–12], thus providing new insights into the process.

2. Method

A 2 mm-thick slice was cut from a WAAM 300M wall in the as-built condition, the process parameters had been optimised to deliberately produce coarse columnar PAGs. This was further sectioned into a 5×5×2 mm^3 sample that was ground and polished to a 0.04 μm colloidal silica finish, before final polishing using a broad ion beam. *In-situ* EBSD was performed using a TESCAN and NewTec In-Situ Testing (TANIST) system, which consisted of a NewTec FurnaSEM heating stage in a TESCAN Clara, fitted with an Oxford Instruments Symmetry 3 EBSD detector and Ultim Max 170 EDS detector. The sample was fixed to the heating crucible using conductive carbon paste with the WAAM build direction pointing upwards in the collected maps and the welding direction out of page. After curing and out gassing, an EBSD map with simultaneous EDS was captured before the heating experiment started. NewTec FurnaSEM software interfaced with the heating stage, microscope and EBSD detector allowed the experiment to be automated and run according to a pre-determined temperature and imaging program. Brightness, contrast, focusing, region tracking and data collection were all automated within the program, with region tracking performed by template matching of the secondary electron (SE) images.

The sample was heated according to the temperature profile shown in figure 1(a). Data analysis was performed using MATLAB scripts written using the MTEX 5.10.2 toolbox [18]. Further details concerning the material composition, experimental set up, EBSD data collection parameters and data analysis can be found alongside the raw data and gifs in reference [19].

3. Results

Figure 1(b) displays the room temperature pre-test EBSD, alongside 1(c) a PAG reconstruction of this map performed using the MTEX 5.10.2 variant graph approach [20,21] and 1(d) an EDS map captured before the start of heating. The initial microstructure was predominately martensitic with some bainite and contained compositional microsegregation typical of WAAM. Interdendritic

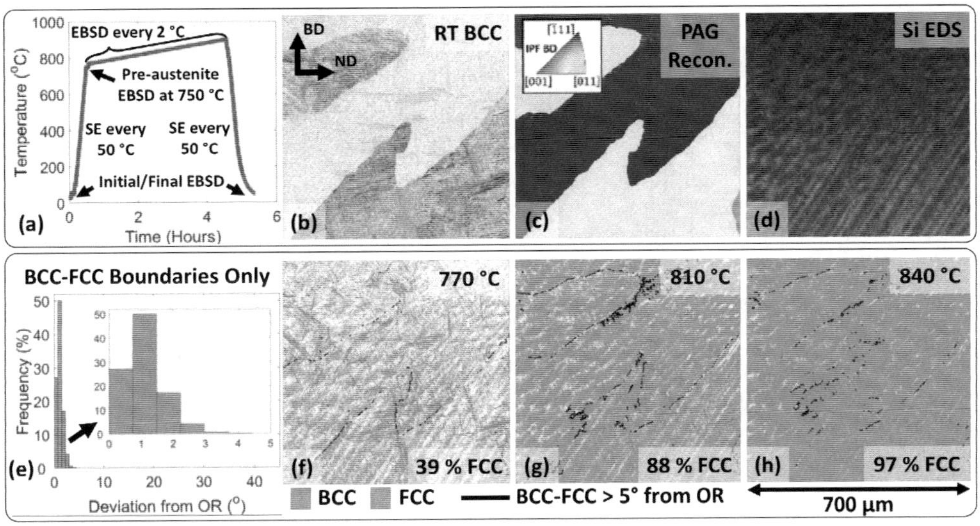

Figure 1. (a) *In-situ* EBSD temperature profile. (b) A room-temperature EBSD map (before heating) showing the BCC phase only (martensite). IPF colours are aligned with the build direction (BD). (c) An MTEX PAG reconstruction of (b). (d) An EDS map collected simultaneously displaying the Si Kα, it highlights the regions that are also enriched in Cr, Mn, Mo and Ni by microsegregation. (e) A histogram showing the deviation angles of each BCC-FCC boundary in the 770 °C EBSD map, from the OR refined during the PAG reconstruction. (f-h) Selected EBSD frames from the *in-situ* experiment coloured by phase, white pixels are unindexed. Boundaries that were more than 5 ° from the OR are in black.

regions were enriched in Si, Cr, Mn, Mo and Ni, as has been reported previously in the literature [2].

The real martensite-austenite orientation relationship (OR) is irrational [22], usually varying by a few degrees from a known OR such as the Kurdjumov-Sachs (KS) OR. Here, the MTEX reconstruction algorithm [23] was used to calculate a refined OR for this material, from the pre-test map (figure 1(b)). The KS OR was used as an initial starting point and the refined OR (through 3 iterations) was 2.79 ° from the KS OR and 2.58 ° from the Nishiyama-Wassermann (NW). The deviation from this refined OR was calculated for every martensite-austenite boundary, in each frame collected, during the transformation to austenite.

The deviation angles for the 770 °C frame are shown in the histogram in figure 1(e). Almost all the martensite-austenite phase boundaries were less than 5 ° from the refined OR. The few exceptions are highlighted in black on the phase maps in figure 1(f-h), with those boundaries corresponding to memory effect austenite grains from either side of high-angle PAG boundaries. This clearly demonstrates that the transformation progresses with an orientation relationship between austenite and martensite on almost all transformation fronts, as has been reported previously for the memory effect [3,16]. Notably, figure 1(f-h) also shows that the interdendritic regions enriched in alloying elements austenitised over 100 °C later than the depleted regions (compare EDS in figure 1(d) with 1(g-h)), the Ac_1 temperature (start of transformation) raised locally. Usually, these regions still austenitise with a memory orientation.

Selected frames from heating displaying the austenite phase only are shown in figure 2, 2(a-c) are in IPF colours (aligned to the build direction). Comparison of figure 2(a) with the PAG reconstruction of the pre-austenitisation map in figure 1(c) demonstrates a memory effect, as the PAG grains are reformed both morphologically and crystallographically via the coalescence of

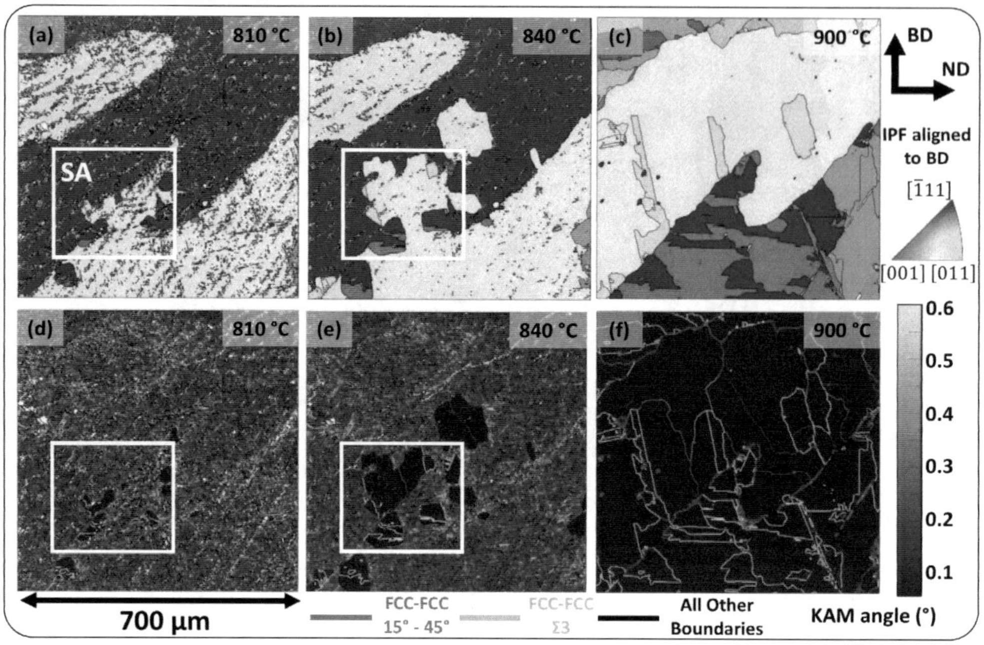

Figure 2. (a-c) Selected ESBD frames from the *in-situ* heating in IPF colours aligned to the build direction (BD). (d-f) The denoised KAM maps from the same frames, a misorientation threshold between 0.06 ° – 0.6 ° was used to better highlight recrystallised grains, these thresholds were based on the histograms in figure 3. High angle (15 ° – 45 °) boundaries and twin boundaries (Σ3) were highlighted in red and gold respectively. A selected area (SA) that is enlarged in figure 3(a-d) is outlined in white. In all maps only the FCC phase is shown, BCC or unindexed pixels appearing white.

acicular austenite. Interestingly, the post-memory effect recrystallisation reproduced an approximation of the columnar PAG morphology (figure 2(c)). When comparing figures 2(a) and 2(c) (near the start of recrystallisation, and after recrystallisation, respectively), it appears that the two neighbouring columnar grains have swapped orientation and developed recrystallisation twins – the blue-coloured grain at 840°C becomes yellow at 900°C, and vice versa. The boundary between them was retained. This effect is even more evident when comparing figure 1(c) (the reconstructed PAG structure before any austenitisation) to figure 2(c) (after austenitisation and recrystallisation).

Figure 2(d-f) displays KAM maps taken from the same frames in 2(a-c), with high angle (15° - 45°) and twin (within 3° of an FCC-FCC Σ3 boundary) boundaries highlighted in red and gold respectively. Figure 3(a-d) magnifies a selected area (SA) from the IPF and KAM maps of the 810 °C and 840 °C frames in figure 2. The KAM maps in figure 2(d-f) show that the memory austenite grains contain considerable internal misorientation, a dense sub-grain structure with misorientations typically < 0.6 ° can be viewed in figure 3(c-d). This value was used as the upper threshold for all KAM maps. Upon further heating above 800 °C, the material underwent recrystallisation even before austenitisation has fully completed. This can be seen clearly by the development of a bimodal distribution when the KAM values of individual frames are plotted on histograms in figure 3(e-g). The misorientation angle of the trough between the two peaks seen in the FCC frequency distribution (0.06 °) in figure 3(f) was used as the lower threshold for the figure 2(d-f) and figure 3(c-d) KAM maps, in order to better highlight the recrystallised grains.

Grains 1 and 2 labelled in figure 3 'swap' orientations through the bulging of the high-angle grain boundary between them into the neighbouring grain, sweeping up strained regions. This is the mechanism that leads to the grain 'swapping' described above. The complete datasets showing all frames composed as a gif can be downloaded from the following link in reference [19].

4. Discussion

The results clearly demonstrate that the austenitisation of WAAM'd 300M was a two-step process: first, the growth of acicular austenite occurred with a memory effect, and then this austenite recrystallised. In agreement with the literature [6,9,10,13,24], the orientation relationship between acicular austenite and the martensite was near to the KS and NW ORs. The effect of compositional microsegregation was to raise the A_3 temperature (end of transformation) in the alloying element enriched interdendritic regions, which also austenitised with a memory effect, but nearly 100 °C later than the dendritic regions.

The recrystallisation observed was predominantly discontinuous, wherein pre-existing high angle boundaries (15°-45°) bulged out following a gradient in dislocation density (as shown clearly in figure 3(a)), which is implied from the distinct difference in KAM between recrystallised and acicular austenite. In this case, the existing boundaries are between acicular austenite grains with memory orientations belonging to either side of a PAG boundary, as exemplified by figure 3(a). The mechanism is similar to that described by Tsuchiyama *et al.* [24] for recrystallisation in martensite. It is hard to determine what causes the presumed difference in dislocation density across either side of the boundary. It is assumed to be random local difference in recovery rates since the same high angle boundaries can be seen to bulge in opposite directions along its length – indeed, it is this that led to the grain 'swapping' seen most clearly comparing figures 1(c) and 2(c).

This grain 'swapping' is an intriguing result, both in terms of its mechanism and its effect on the new austenite grain structure. The mechanism involves the motion of the high-angle grain boundary (HAGB) between the two columnar grains (*i.e.*, the blue- and yellow-coloured grains in figure 1(c) that have 'swapped' places by figure 2(c)). To maintain a high mobility, a HAGB requires a disparity in dislocation density across the boundary locally. When the HAGB between the columnar grains bulges out (see the example regions labelled Grain 1 and Grain 2 in figures 3(c) and 3(d)), the new grain area formed has a comparatively low dislocation density, but remains connected to the dislocation-dense segment of the grain from which it grew. The new dislocation-sparse segment of the grain cannot consume the dislocation-dense segment of the same orientation because they share no HAGB. Therefore, the dislocation-dense segment may only be consumed by a dislocation-sparse grain with a different orientation, one sharing a high angle misorientation with the first (*e.g.*, see the dislocation-sparse regions of Grain 2 consuming the dislocation-dense area of Grain 1, and vice versa, in figure 3). The dislocation-sparse grain areas cease to grow when they meet another dislocation-sparse grain. This happens at the original HAGB, as indicated by the "impingement" label in figure 3(d). This is because the driving force for HAGB migration is reduced, as there is no longer a significant disparity in dislocation density on either side of the boundary. Thus, the original PAG is retained. This intriguing result means that the prior columnar grain structure is largely retained, even after a recrystallisation step that might normally be expected to refine the grain structure generated by the memory effect.

Note, it would seem intuitive to suggest that the retention of the HAGB boundary between the two 'swapped' grains is related to the presence of some boundary pinning particles (Mo or V carbides). However, in this case HAGB migration only appears to have been obstructed by phase

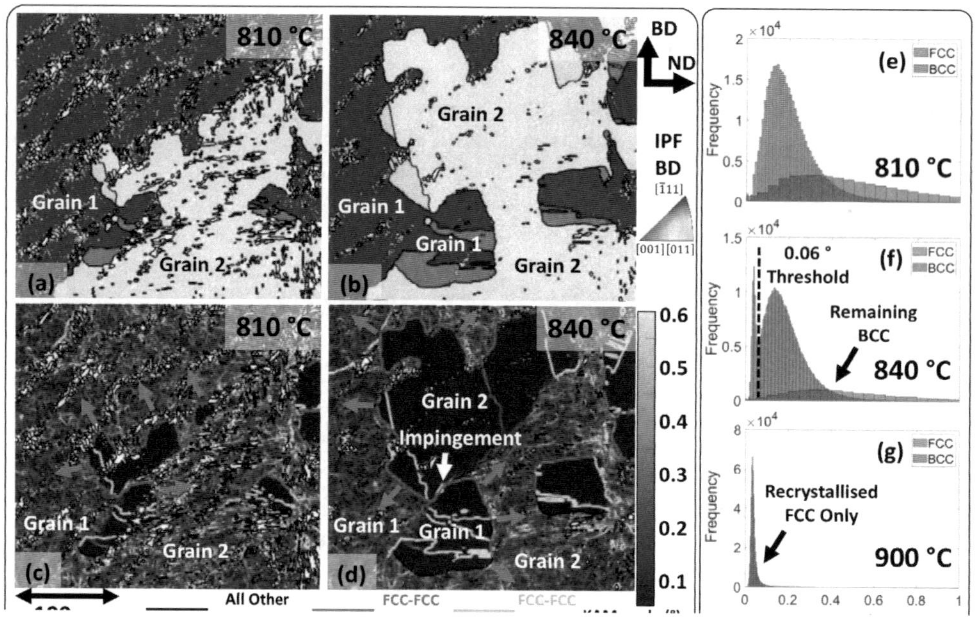

Figure 3. (a-d) Magnified areas of selected ESBD frames, highlighted in figure 2(a-b) and (d-e) respectively. (a-b) are in IPF colours aligned to the build direction, grain boundaries are highlighted in black. (c-d) are denoised KAM maps with a threshold between 0.06 ° – 0.6 °. High angle (15 ° – 45 °) boundaries and twin boundaries (Σ3) were highlighted in red and gold respectively. Red arrows indicated the direction of high angle boundary migration and a white arrow indicates where two recrystallised grains impinged. Only the FCC phase is displayed, BCC or unindexed pixels appearing white. (e-g) Histograms of the BCC and FCC phase denoised KAM values from the selected ESBD frames also used in figures 1-2. Average KAM values decrease with heating and a distinct bimodal distribution can be found in (f) where the material is mid-recrystallisation. The threshold marked on (f) was used in figure 2(d-f) to better highlight recrystallised grains.

boundaries or impingement on other recrystallised grains during discontinuous recrystallisation, without the need for the effect of pinning particles. Although, their presence cannot be ruled out here. The irregular growth front of HAGB during the migration (figure 3(d)) was clearly caused by retained martensite in the interdendritic regions obstructing growth.

It is also interesting to note that if the material had a finer equiaxed grain structure to start with, each grain would have multiple neighbours with which it shares HAGB. During discontinuous recrystallisation, these HAGB would bulge out and then impinge to produce a new grain structure that was more refined that the initial structure [25]. Conversely, in the WAAM'd 300M considered here, just two neighbouring austenite grains simultaneously bulged out and impinged on each other. The aspect ratio and size of the columnar grains allowed this to occur before any other HAGB 'carrying a different orientation could migrate to the same boundary and disrupt this.

Recrystallisation twins (Σ3) formed during the recrystallisation step, providing some small level of grain refinement. It is likely that these twins are normal recrystallisation twins, which formed due to stacking faults developed during the migration of high angle grain boundaries through the dislocation dense material [26]. Non-memory variants of austenite have been reported [13,16] forming with an OR to the prior martensite phase in steels that display a memory effect. Due to the nature of the OR, these grains are usually twin orientations to the memory orientation [16]. However, the twins observed here were probably not related to this effect since they were not observed until the start of recrystallisation. Some authors [12,15] have found examples in other steels where recrystallisation following a memory effect does not appear to follow a discontinuous mechanism, and instead the new recrystallised grains nucleate within PAGs or at PAG boundaries. Potentially, grain refinement might be improved by stimulating such intra-PAG nucleation through the use of faster heating rates [4–6,10,13] or multiple heating cycles [27], and this should be investigated in further work.

With respect to the coarse columnar prior-austenite microstructures found in WAAM'd ferritic steels, the results discussed here suggest that poor heat treatment design could fail to remove the coarse columnar PAG structures despite the steel undergoing austenitisation. This can occur because there is a memory effect and, importantly, because the subsequent recrystallisation is predominantly discontinuous.

5. Conclusions

In this work *in-situ* high temperature EBSD was used to observe the grain structure evolution during the austenitisation of a WAAM'd a low- alloy steel (300M). The following conclusions have been drawn:

- 300M austenitises with a memory effect and then undergoes recrystallisation upon further heating, without the application of external deformation.
- Memory austenite shares an OR near the KS and NW with neighbouring martensite laths, save for at PAG boundaries.
- Compositionally enriched areas (interdendritic) austenitised >100 °C later than depleted regions, though the mechanisms of austenitisation and recrystallisation were the same.
- Under the conditions investigated, recrystallisation is discontinuous, occurring through the movement of high angle grain boundaries between prior-austenite grains, which were retained by the memory effect.
- The motion of the high-angle grain boundaries can lead to the 'swapping' of orientations locally, whilst preserving the original boundary, meaning columnar grain structures are not broken up.
- Recrystallisation twins were observed forming alongside recrystallised austenite grains.

Acknowledgements

The authors gratefully acknowledge financial support from Airbus Operations ltd and Innovate UK through the I-Break project (Landing Gear Industrial Breakthroughs; #10003486). The authors acknowledge the use of equipment associated with the Advanced Metals Processing and Characterisation themes of the Henry Royce Institute for Advanced Materials, funded through EPSRC grants EP/R00661X/1, EP/S019367/1, EP/P025021/1 and EP/P025498/1. Many thanks to Jun Wang at Cranfield Universty for building the WAAM component.

References

[1] Williams S W, Martina F, Addison A C, Ding J, Pardal G and Colegrove P 2016 *Materials Science and Technology* **32** 641–7
[2] Wang J, Diao C, Taylor M, Wang C, Pickering E, Ding J, Pimentel M and Williams S 2023 *Int J Adv Manuf Technol* **129** 3751–67

[3] Watanabe S and Kunitake T 1976 On the Formation of Austenite Grains from Prior Martensitic Structure *ISIJ Int.* **16** 28–35
[4] Kimmins S T and Gooch D J 1983 *Metal Science* **17** 519–32
[5] D'yachenko S S 2000 *Met Sci Heat Treat* **42** 122–7
[6] Matsuda S and Okamura Y 1974 *ISIJ Int.* **14** 444–9
[7] Hara T, Maruyama N, Shinohara Y, Asahi H, Shigesato G, Sugiyama M and Koseki T 2009 *ISIJ Int.* **49** 1792–800
[8] Shinozaki T, Tomota Y, Fukino T and Suzuki T 2017 *ISIJ International* **57** 533–9
[9] Tomota Y, Gong W, Harjo S and Shinozaki T 2017 *Scripta Materialia* **133** 79–82
[10] Ressel G, Brandl D, Hönigmann T, Lukas M, Stark A, Gruber C, Lukas S, Stockinger M and Kozeschnik E 2023 *Materials & Design* **234** 112370
[11] Rosenauer A et al. 2024 *Materials Characterization* **208** 113554
[12] Shirazi H, Miyamoto G, Hossein Nedjad S, Chiba T, Nili Ahmadabadi M and Furuhara T 2018 *Acta Materialia* **144** 269–80
[13] Zhang X, Miyamoto G, Toji Y, Nambu S, Koseki T and Furuhara T 2018 *Acta Materialia* **144** 601–12
[14] Nakada N, Fukagawa R, Tsuchiyama T, Takaki S, Ponge D and Raabe D 2013 *ISIJ International* **53** 1286–8
[15] Liu L, Yang Z-G, Zhang C and Liu W-B 2010 *Materials Science and Engineering: A* **527** 7204–9
[16] Nakada N, Tsuchiyama T, Takaki S and Hashizume S 2007 *ISIJ Int.* **47** 1527–32
[17] Brandl D, Lukas M, Stockinger M, Ploberger S and Ressel G 2019 *Materials & Design* **176** 107841
[18] Bachmann F, Hielscher R and Schaeben H 2010 *SSP* **160** 63–8
[19] Taylor M 2024 *Data for "In-Situ EBSD Study of Austenitisation in a Wire-Arc Additively Manufactured High-Strength Steel"* last viewed on 31/05/2024 available online: https://doi.org/10.5281/zenodo.11098666
[20] Hielscher R, Nyyssönen T, Niessen F and Gazder A A 2022 *Materialia* **22** 101399
[21] Niessen F, Nyyssönen T, Gazder A A and Hielscher R 2022 *J Appl Crystallogr* **55** 180–94
[22] Christian J W 1990 *Metallurgical Transactions A* **21** 799–803
[23] Nyyssönen T, Isakov M, Peura P and Kuokkala V-T 2016 *Metall Mater Trans A* **47** 2587–90
[24] Tsuchiyama T, Natori M, Nakada N and Takaki S 2010 *ISIJ Int.* **50** 771–3
[25] Humphreys F J and Hatherly M 2004 *Recrystallization and Related Annealing Phenomena* (Elsevier)
[26] Mahajan S 1997 *Acta Materialia* **45** 2633–8
[27] Nasiri Z, Ghaemifar S, Naghizadeh M and Mirzadeh H 2021 *Metals and Materials International* **27** 2078–94

Additive manufacturing to design a new material TiB_2-reinforced SUS316L using TiB_2-Fe composite powder

R. Otsuka[1*], T. Kikuchi[1], T. Yamamoto[2] and S. Katayama[1*]

[1]R&D Laboratory, Nippon Denko Co., Ltd., Tokyo, Japan
[2] Toyama Industrial Technology R&D Center, Takaoka, Toyama, Japan

*E-mail: r_otsuka@nippondenko.co.jp, s-kataya@nippondenko.co.jp

Abstract. The metal additive manufacturing (AM) process allows for the production of three-dimensional (3D) parts with unique geometries that could not be formed by conventional methods. However, our aim is to utilize the AM process to create never-before-seen new materials because we believe it holds potential not only for new processing but also for creating a new material. The TiB_2-Fe composite powder developed as a new raw material of the AM process, which was used as an additive powder in additive manufactured (AMed) SUS316L for effectively improving the mechanical properties of AMed SUS316L. The TiB_2-reinfoced SUS316L synthesized by a Powder Bed Fusion (PBF) method using the TiB_2-Fe composite powder had an extremely high Vickers hardness and strength of about 400 HV and 1250 MPa, respectively, being close to twice those of AMed SUS316L.

1. Introduction

The AM process has attracted significant attention for its flexible process capability, allowing for the customized manufacturing of three-dimensionally (3D) complex parts without forming by conventional methods such as cutting process, powder metallurgy process, and liquid metallurgy process. This has led to the widespread development of AMed parts in various industries such as aerospace, medical, energy, and automotive[1-9]. The AM process is also recognized for combing the Digital Transformation (DX) with the AM process due to AMed parts guided by a digital model of 3D CAD (Computer Aided Design), influencing the manufacturing, telecommunications, and information industries. Thus, many efforts have been made in design of creatively shaped parts by AM process using conventional materials to minimize the number of parts and reduce assembly work.

In the metal AM process, a variety of alloys such as Fe-, Ni-, Cu-, Ti-, Al-, and Mg-based alloys have been successfully AMed[9-11]. When focusing on Fe-based alloys, it is common to use well-known steels such as SUS316L, SUS410, SUS420, SUS630, maraging steel (18Ni300), SKD61, and 15-5PH for developing highly complex parts by the AM process. However, it is considered that the AM process does not only have a potential for designing a new structure but also for creating an advanced material. By utilizing the rapid heating-melting and rapid cooling-solidification process in the AM process, a new material could be designed. Some attempts have been made to study a new material based on a novel microstructure design. For example, a hierarchical anisotropic structure was designed by studying laser and scanning conditions of the laser powder bed fusion method, leading to improving properties of alloys[12-14].

In this study, we focused on dispersion-reinforced alloys for another new material design using the AM process. Titanium diboride (TiB_2) was chosen as a dispersion material for designing dispersion-reinforced alloys because of its high values of hardness (34.0 GPa), elastic modulus (565 GPa), and melting point (about 3,225 °C) [15]. TiB_2 particles cannot be dispersed in steels by the conventional casting method because of the phase separation resulted from the large difference in density between steel (Fe: 7.87 g/cm^3) and TiB_2 (4.52 g/cm^3). The powder metallurgy method has been used to fabricate TiB_2-reinforced steels by sintering ferric powders and TiB_2 powders[16-19]. However, TiB_2 particles have a tendency to undergo grain growth and agglomeration during sintering. The powder metallurgy method is thought to have a limit to achieving further high dispersion of TiB_2 powders although the high dispersion of TiB_2 particles provides the effective reinforcement in steels. The AM process was applied to the fabrication of TiB_2-reinfored steels using TiB_2 powders and SUS316L powders[20-22]. The higher dispersion of TiB_2 particles was obtained by the AM process because of its rapid heating-melting and rapid cooling-solidification. A fine TiB_2 powder is needed for it, but it is very difficult to prepare fine particles of TiB_2 by means of crushing and milling extremely hard TiB_2. For example, the pre-treatment of high-energy ball milling was carried out to homogenize the mixture of matrix and reinforcement materials and to refine the TiB_2 particle size.

We have newly developed TiB_2-Fe composite powder suitable for TiB_2 source of TiB_2-reinforced alloys fabricated by the AM process. The developed TiB_2-Fe composite powder has a specific structure where fine TiB_2 particles are dispersed in the Fe matrix. In addition, the developed TiB_2-Fe composite powder is synthesized by a new simple method, which uses a reaction of Ferroboron Fe-B and Ferrotitanium Fe-Ti to form TiB_2 in Fe matrix. The developed TiB_2-Fe composite powder has high fluidity, being effectively used as an additive powder for the fabrication of TiB_2-reinfored SUS316L by the AM process.

In this study, we investigated the application of the TiB_2-Fe composite powder in the PBF method with selective laser melting for fabricating TiB_2-reinforced SUS316L.

2. Experimental Procedure

2.1 Preparation of TiB_2-Fe composite powders from Fe-Ti and Fe-B

Fe-Ti powder (OFT-40, 40 wt%-Ti, size: 0.25 mm or below, OSAKA Titanium technologies Co., Ltd.) and Fe-B powder (FBH S1L 17 wt%-B, size: 1-30 mm, NIPPONDENKO Co., Ltd.) were used as starting materials. The Fe-B powder was milled and classified to give a Fe-B powder with a size of 75μm or below. Fe-Ti and Fe-B powders in a molar ratio of B/Ti = 1.9 were mixed by a vibration ball mill filled with Ar gas, resulting in the TiB_2-content of 28 wt%. The mixed powders were fired at 1,200°C for 10h in a flow of Ar. After grinding the surface of fired bulky samples, they were crushed by a jaw crusher. The crushed samples were milled and classified to fabricate a TiB_2-Fe powder with a size of 25-75μm.

2.2 Additive manufacturing of PBF method with laser melting

The TiB_2-Fe composite powder was additively mixed with atomized SUS316L powder(size: 25-53μm, Höganäs) to obtain mixed powders with TiB_2-contents of 2.0, 3.0, 3.5, 4.0 and 5.0 wt%, respectively. The mixed powders were then used as a starting powder for the AM process of PBF method with laser melting (M280,EOS GmbH) under high-purity Ar gas.

A laser with a spot size of 0.1 mm was applied to produce cylindrical samples with diameters of 10 mm and heights of 5 mm with preheating at a temperature of 80 °C. The scan strategy with a rotation of 67° between layers was conducted. The powder layer thickness was set to 0.02 mm.

The laser power and scanning speed were 100-250 W and 750-1500 mm/s, respectively, at hatch spacing of 0.09 mm (the distance between two adjacent scanning tracks) to study the condition being suitable for only the SUS316L powder and the mixed powders.

2.3 Characterisations

The assignment of TiB_2 phase in the TiB_2-Fe composite powder was perfomed by the X-ray diffraction (XRD) method. The density measurement of AMed samples was performed using the Archimedes method. The hardness of AMed samples was determined using the Vickers hardness testing method. The tensile strength was measured using AMed samples shown in figure 1.

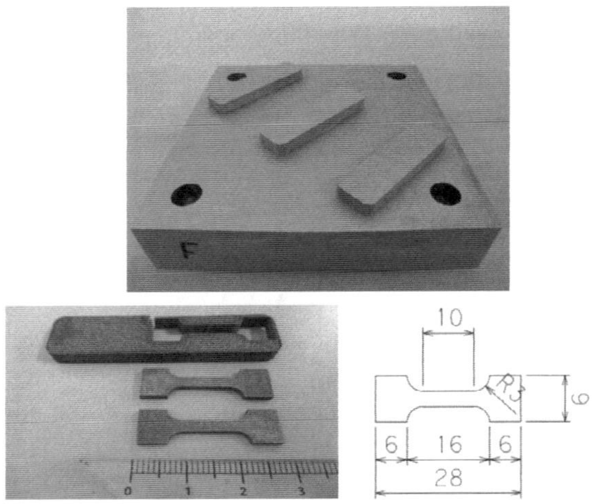

Figure 1. Representative samples and size for measuring tensile strength.

3. Results and Discussion

3.1 TiB_2-Fe composite powder prepared from Fe-Ti and Fe-B

The particle morphology of TiB_2-Fe composite powder is not spherical as shown in figure 2, but it is cut into small particles to ensure flowability. The cross-sectional observation of TiB_2-Fe composite particle showed the dispersion structure of fine TiB_2 with several microns in size in an

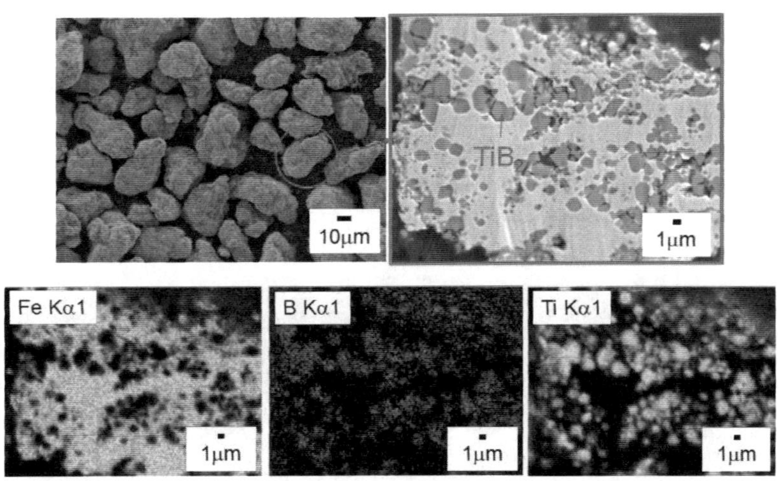

Figure 2. SEM image and cross-sectional SEM-EDS mapping of TiB_2-Fe composite powder.

Fe matrix, which was formed by a well-controlled solid-state reaction method using Fe-Ti and Fe-B. The heat-treatment condition is to prevent significant grain growth after the formation of TiB_2 nuclei through mutual diffusion of Ti and B, while ensuring complete reaction between Ti and B. The formation of TiB_2 phase was also confirmed by XRD.

3.2 AM condition of PBF method using SUS316L and TiB_2-Fe composite powders

First, the basic condition was determined in the use of only the SUS316L powder. Samples were prepared with the laser powers of 100, 150, 200 and 250 W and the scanning speeds of 750, 1000, 1250 and 1500 mm/s. The relative densities of the samples are shown in figure 3. The laser power of 200 W or above was found to achieve a high relative density over 98% at any scanning speed. The scanning speed of 750 mm/s was also found to achive a high relative density over 98% at any laser power. Looking at the energy density, a high relative density was obtained over 750 J/mm^3 as shown in figure 3, which was consistent with the previous report[23]. The cross-section of these samples was observed under a metallurgical microscope, and no defects were found.

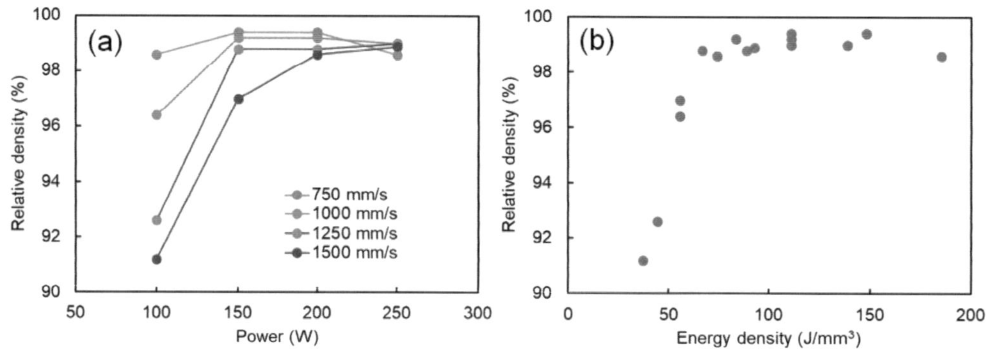

Figure 3. Relative densities of cylindrical samples depending on laser power (a), scanning speed (a), and energy density (b).

Based on the above results, the next step of 3D fabrication was proceeded using SUS316L and TiB_2-Fe composite powders. Samples with a variety of TiB_2-content were prepared using SUS316L and TiB_2-Fe composite powders. As shown in fugure 4, a high relative density was obtained over 100 J/mm^3, which was slightly higher than that of samples without TiB_2. This may be the result of the influence of TiB_2, which has different thermal capacity and thermal conductivity from the matrix, being included. But it is not clear because of the data including imperfect samples that have defects and/or cracks.

Although crack-free cylindrical samples were fomed up to the TiB_2-content of 3.5 wt%, a significant number of cracks occurred in samples with the TiB_2-content of 4.0 wt% or above. This is due to the high hardness of samples with high TiB_2-contents. The rapid expansion and rapid cooling caused by laser irradiation generate residual stress in 3D-fabricated samples, leading to crack formation especially in samples with higher hardness.

The formation of cracks in the horizontal direction were qualitatively determined by visually observing the external surface of the cylinder. The cracks tended to decrease as the laser power decreased and the scanning speed increased, but the minimum necessary energy density is needed to obtain a high dense sample. The cracks were formed in the horizontal direction on the side surfaces and often found in the lower to middle layers of cylindrical samples. The cracks propagated from the outer surface towards the interior. Based on these results, it can be considered that the cracks occur when the residual stress, which accumulates and builds up during the solidification and layered-manufacturing process, reaches a critical point. Thus, it is considered that the lower scan speed with keeping the minimum necessary energy density would be effective to reduce the accumulation of the residual stress.

Thus, a typical appropriate condition of the PBF method for AMed TiB_2-reinfoced SUS316L using TiB_2-Fe composite powder was determined to be a laser power of 100 W and a scanning speed of 310 mm/s with a laser spot diameter of 0.1 mm at hatch spacing of 0.09 mm and powder layer thickness of 0.02 mm.

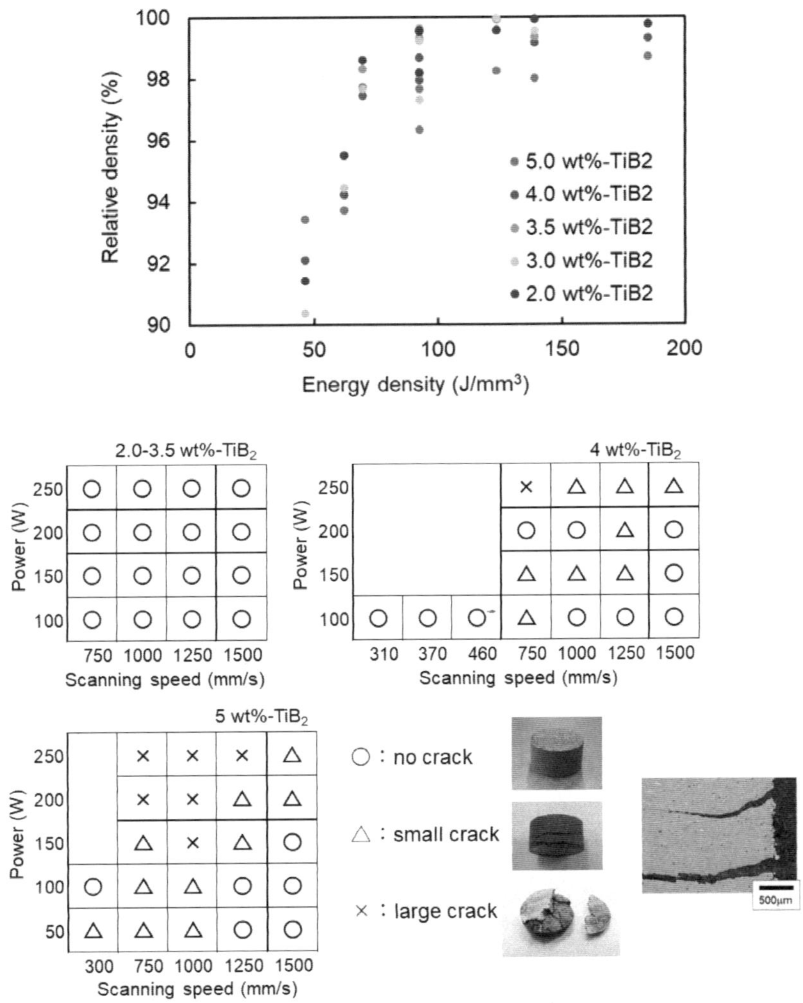

Figure 4. Effect of energy density on relative density of cylindrical samples and effect of laser power and scanning speed on formation of cylindrical samples.

3.3 Properties of TiB_2-reinfoced SUS316L samples fabricated by PBF method using SUS316L and TiB_2-Fe composite powders

The Vickers hardness linearly increased with increasing the TiB_2-content as shown in figure 5. It is clearly known that the TiB_2-dispersion is effective in enhancing the hardness of SUS316L. The Vickers hardness of SUS316L was improved to more than double through the dispersion of TiB_2

at only the small TiB_2-content of 3-4 wt%, which was extremely smalle in comparison to TiB_2-dispersion strengthening achieved by the powder metallurgy process[16-19, 24].

The tensile strength also linearly increased with increasing the TiB_2-content as shown in fugure 5. This also shows that the TiB_2-dispersion is effective in enhancing the strength of SUS316L. The tesile strength was also improved to nearly double that of SUS316L by the dispartion of TiB_2 with the TiB_2-content of 4 wt% althogth there is a significant variation in strength due to the occurrence of cracks at the TiB_2-content of 4 wt%. Figure 6 shows the stress-strain curves of some samples whose tensile strength was measured. It is clear that the higher content of TiB_2 results in better tensile strength but less elongation.

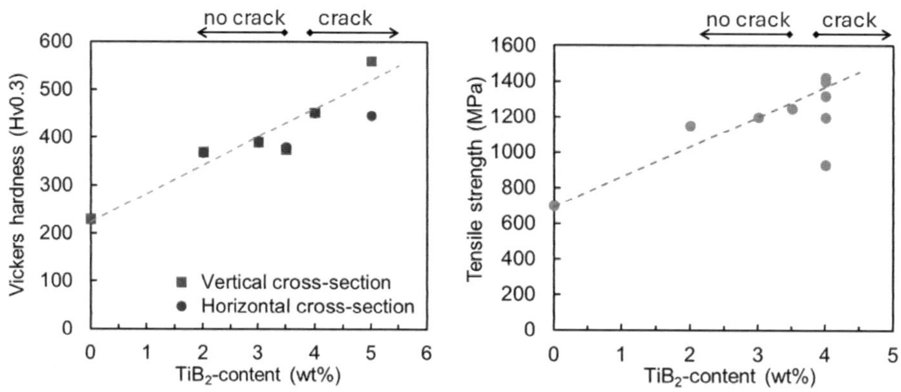

Figure 5. Effects of TiB_2 -content on Vickers hardness and tensile strength of TiB_2-reinfoced SUS316L fabricated by PBF method using SUS316L and TiB_2-Fe composite powders.

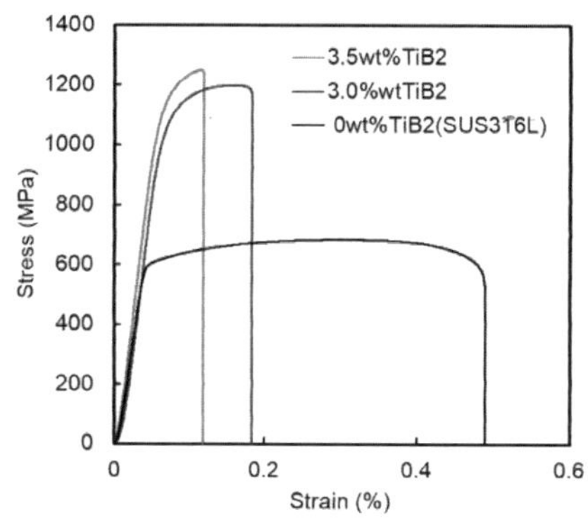

Figure 6. Representative stress-strain curves of TiB_2-reinfoced SUS316L fabricated by PBF method using SUS316L and TiB_2-Fe composite powders.

3.4 Morphology and distribution of TiB_2 inTiB_2-reinfoced SUS316L

The SEM observation of the TiB_2-reinfoced SUS316L revealed the dispersion sate of TiB_2 particles in SUS316L matrix as shown in figure 7. The hierarchical distribution of TiB_2 particles was observed, where the clusters consisting of several tens of TiB_2 particles with several micrometre-size were scattered in SUS316L matrix. The size and morphology of TiB_2 particles in the AMed TiB_2-reinfoced SUS316L appear to be the same as those of the starting TiB_2-Fe composite powder shown in figure 2, possibly meaning that the dispersion of TiB_2 without melting and promoting

IOP Conf. Series: Materials Science and Engineering **1310** (2024) 012002 doi:10.1088/1757-899X/1310/1/012002

particle growth of TiB_2 could be accomplished by rapid laser melting and solidification of only matrix as expected.

The following formation of TiB_2-dispersed SUS316L could be assumed as shown in figure 8. The mixed powder is rapidly melted by a laser beam, where at that time, TiB_2 particles in TiB_2-Fe composite powder is rapidly dispersed and spread within the molten pool by fluid flow of melted metal[25-27] and/or concentration gradient of TiB_2-perticles. Subsequently, rapid solidification of the molten pool occurs, resulting in the TiB_2-dispersion structure.

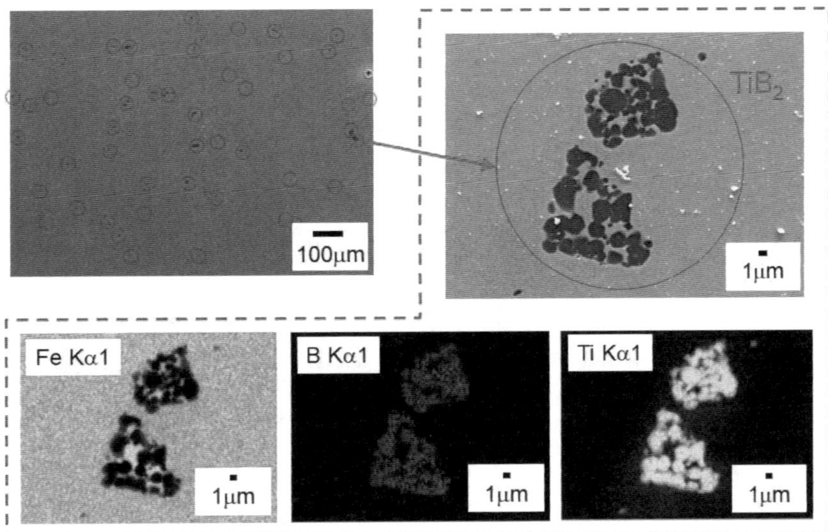

Figure 7. SEM-EDS mapping of TiB_2-reinfoced SUS316L fabricated by PBF method using SUS316L and TiB_2-Fe composite powders.

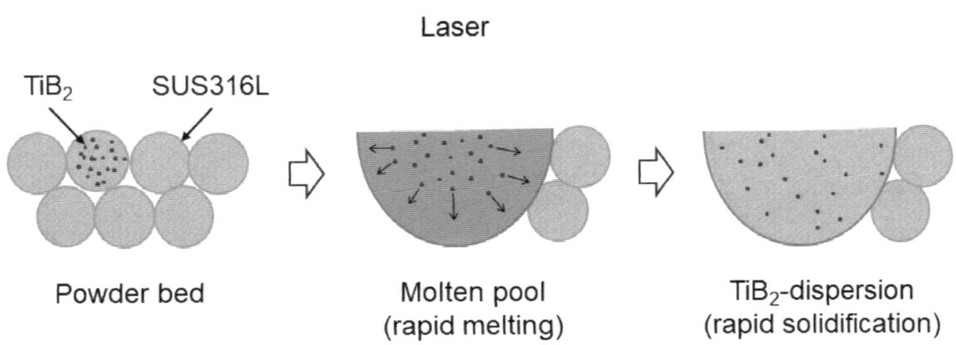

Figure 8. Simplified schematic image for forming TiB_2-dispersed SUS316L by PBF method using SUS316L and TiB_2-Fe composite powders.

4. Conclusions

Using the TiB_2-Fe composite powder prepared by the simple method with a reaction of Fe-Ti and Fe-B, TiB_2-reinfoced SUS316L has successfully fabricated by the PBF method with selective laser melting. Fine TiB_2 particles with several micrometre-size were dispersed in SUS316L matrix. The appropriate condition of the PBF method for AMed TiB_2-reinfoced SUS316L using TiB_2-Fe composite powder was determined to fabricate perfect AMed samples up to the TiB_2 -content of 3.5 wt%. Cracks occurred in samples with the TiB_2-content of 4.0 wt% or above. The Vickers hardness and tesile strength of in TiB_2-dspersed SUS316 increased with increasing the TiB_2 content. At the TiB_2 -content of 3.5 wt%, the Vickers hardness and strength were improved to reach the high values of about 400 HV and 1250 MPa, respectively, being closed to twice those of AMed SUS316L.

This study revealed that the AM processing was effective for designing a new material that was not achieved by conventional methods. Our further works are the development of AMed new materials, as well as the more improvement of AMed TiB_2-reinfoced SUS316L with higher contents of TiB_2.

Acknowledgements

The authors wish to acknowledge Y. Koyama and N. Matsumoto for preparing TiB_2-Fe composite powders.

References

[1] Levy G N, Schindel R and Kruth J 2003 CIRP *Ann. Manuf. Technol.* **52**(2) 589-609
[2] DebRoy T, Wei H L, Zuback J S, Mukherjee T, Elmer J W, Milewski J O, Beese A M, Wilson-Heid A, De A and Zhang W 2018 *Prog. Mater. Sci.* **92** 112 -224
[3] Tebianian M, Aghaie S, Sadat Razavi Jafari N, Elmi Hosseine S R, Pereira A B, Fernandes F A O, Farbakhti M, Chen C and Huo Y 2023 *Mater.* **16** 7514
[4] Frazier W E 2014 *J. Mater. Eng. Perf.* **23** 1917-1928
[5] Nakano T, Ishimoto T, Ozasa R and Matsugaki A 2023 *Mater. Trans.* **64** 2361-2367
[6] Matsugaki A, Matsuzaka T and Nakano T 2023 *Mater. Trans.* **64** 25
[7] Liu H, Yu H, Guo C, Chen X, Zhong S, Zhou L, Osman A and Lu J 2023 *Adv. Mater.* **36** 2306570
[8] Wong K V and Hernandez A 2012 *ISRN Mech. Eng.* **2012** 208760
[9] Gu D, Meiners W, Wissenbach K and Poprawe R. 2012 *Int. Mater. Rev.* **57**(3) 133-164
[10] Zeng Z, Salehi M, Kopp A, Xu S, Esmaily M and Birbilis N 2022 *J. Magnes. Alloy.* **10**(6) 1511-154
[11] Ghoussoub J N Utada S Pedraza F Dick-Cleland W J B Tang Y T Reed R 2023 *Metall. Mater. Trans.* A **54A** 1721
[12] Gokcekaya O et al 2023 *Mater. Res. Lett.* **11** 274-280
[13] Ekubaru Y, Gokcekaya O, Ishimoto T, Sato K, Manabe K, Wang P and Nakano T 2022 *Mater. Design* **221** 110976
[14] Gokcekaya G, Ishimoto T, Hibino S, Yasutomi J, Narushima T and Nakano T 2021 *Acta Mater.* **212** 116876
[15] Munro R G 2000 *J. Res. Natl. Inst. Stand. Technol.* **105**(5) 709-720
[16] Tanaka K, Oshima T and Saito T, 1998 *Tetsu-to-Hagane* **84**(8) 586-591
[17] Tanaka K, Oshima T and Saito T 1998 *Tetsu-to-Hagane* **84**(10) 747-754
[18] Tanaka K and Saito T 1999 *J. Phase Equilib.* **20** 207-214
[19] Matsumoto N, Miyake T, Kondoh M, Oishi Y and Toyama K 2018 *J. Jpn. Soc. Powder Powder Metallurgy* **65** 227-232
[20] AlMangour B, Grzesiak D and Yang J -M 2016 *J. Alloys Compd.* **680** 480-493
[21] AlMangour B, Grzesiak D and Yang J -M 2017 *Powder Technol.* **309** 37-48
[22] AlMangour B, Kim Y -K, Grzesiak D and Lee K -A, 2019 *Compos. B Eng.* **156** 51-63
[23] Ahmed N, Barsoum I, Haidemenopoulos G and Abu Al-Rub R K, 2022 *J. Manuf. Process* **75** 415-434
[24] Ke Y, Matsugi K, Xu Z, Choi Y, Wang M and Yu J 2020 *Mater. Trans.* **61** 548
[25] Kyogoku H and Ikeshoji T -T 2020 *Mech. Eng. Rev.* **7** 19-00182
[26] Khairallah S A, Anderson A T, Rubenchik A and King W E 2016 *Acta Mater.* **108** 36-45
[27] Sun Z, Tan X, Tor S B and Chua C K 2018 *NPG Asia Mater.* **10** 127-136

Characterisation of T_1 (Al_2CuLi) Precipitates in Conventional (2099-T83) and Laser Additive Manufactured (PBF-LB/M, DED-LB/M) Microstructures

F Adjei-Kyeremeh[1*], C K Akuata[2], N Bansal[1], S Zischke[3], I Raffeis[1] and A B Polaczek[1]

[1] Foundry Institute, RWTH-Aachen University, 52072 Aachen, Germany.
[2] Chair of Corrosion and Corrosion Protection, RWTH Aachen University, 52072 Aachen, Germany.
[3] Central Facility for Electron Microscopy, RWTH Aachen University, 52074 Aachen, Germany.

[*] E-mail: f.kyeremeh@gi.rwth-aachen.de

Abstract. The high strength and lightweight benefits of Al-Cu-Li alloys make them attractive for aerospace applications. The major strength contributing phase is the T_1(Al_2CuLi) phase. Although extensive reporting is known on its nucleation behaviour and structure in conventional T_1 microstructures, little is known for additive manufacturing (AM). This work focusses on the comparative characterisation of T_1 phase (structure, nucleation behaviour) in conventional (2099-T83), Powder Bed Fusion (PBF-LB/M) and Direct Energy Deposition (DED-LB/M) AA2099 microstructures. It is shown that with different AM processes and heat treatments, different T_1 microstructures are achievable.

1. Introduction

Al-Li alloys are of immense benefit to aerospace and other allied industries owing to, among others, their lightweight, high specific strength and stiffness properties [1]. With maximum solubility of 4.2 wt.% in Al, alloying of each 1 wt.% Li leads to a reduction in the density of the alloy by 3 % and an increase in the elastic modulus between 5 – 6 wt.% [2]. Consequently, 1st and 2nd generation Al-Li alloys with Li > 2wt.% deployed in commercial application suffered setbacks and were withdrawn due to, among other factors, deterioration in fracture toughness properties because of planar slip and ease of shearability of Al_3Li (δ') precipitates by dislocations [3]. The Al_3Li (δ') is considered as the dominant phase in the high lithium containing Al-Li alloys [4].

Third generation Al-Li alloys such as the Al-Cu-Li series with Li < 2wt.% were therefore developed [4]. With Cu as the principal alloying element, along with the presence of solute elements such as Mg and Mn, including dispersoids forming elements like Zr, the precipitation microstructure becomes complex [5]. The most notable phases which form in 3rd generation Al-Cu-Li alloys are the T_1 (Al_2CuLi), T_B ($Al_{7.5}Cu_4Li$), θ (Al_2Cu) and S (Al_2CuMg), with T_1 regarded as the major strengthening phase [6]. To attain high strength and ductility, a high volume fraction of finely distributed T_1 needs to be achieved [7]. Although the crystallographic parameters of T_1 are well documented [8], the length and thickness are significantly influenced by the extent of cold working, aging time and temperature [2]. From nucleation perspective, conventional manufacturing reports Cu/Li ratio, presence of Mg and Ag and thermomechanical processing as influencing factors on kinetics [9]. Several T_1 nucleation theories have been proposed for conventional microstructures and these are well documented in the literature [10, 11]. Additive Manufacturing (AM) of Al-Cu-Li alloys have been limited to published work by [12-17], the author's work [2, 7, 8] for PBF-LB/M and [18-21] for DED-LB/M. Despite the use of in-situ preheating by [13] and [16], T_1 was either not found or its nucleation mechanism was not discussed. After T6 heat treatment on PBF-LB/M 2195 samples, matrix (intragranular) T_1 of varying length (~50 – 160 nm) was reported by [10]. For DED-LB/M samples, at room temperature (RT), thermal cycling effect was described as facilitating T_1 nucleation [15]. Although no mention was made of where they nucleated from the microstructure, T_1 with length 200 – 940 nm could be observed from bright field (BF) TEM images. DED-LB/M cladded 2A97 was found to have small amounts of T_1 of up to 1 µm in the matrix when viewed in the [100] zone axis [19].

Considering the complexities in T_1 nucleation behaviour across different manufacturing processes and with differently applied heat treatments, the objective of this work is to characterise and compare T_1 microstructures of AA2099 Al-Cu-Li alloy across different manufacturing processes, such as conventional (AA2099-T83), and Laser Additive Manufacturing (PBF-LB/M and DED-LB/M). Focus is given to the understanding of the nucleation behaviour of T_1 across the different processes and the comparative characterisation of their morphology and structure. The goal is to provide the understanding which could be tailored towards achieving the optimum microstructure that delivers the required mechanical properties in AM built Al-Li alloys.

2. Material and Methods

Commercial AA2099 –T83 bar was procured from Smith Metals UK for microscopic investigations. The composition of the alloy as supplied by the manufacturer is detailed in table 1. The AA2099 metal powder utilized for the Laser Additive Manufacturing (LAM) process: PBF-LB/M and DED-LB/M was gas atomized by Nanoval GMBH, Berlin Germany using Argon (Ar). Particle Size Distribution (PSD) was performed using the CAMSIZER X2, from Microtrac Retsch GMBH, Düsseldorf Germany. D10, D50 and D90 values of 16.1, 32.6and 62.4 µm respectively were measured. The particles were highly symmetrical (0.903), with good sphericity and aspect ratio of 0.756 and 0.795 respectively. Flowability of the gas atomized powder was measured with a revolution powder analyser (RPA) from Mercury Scientific using a 100 cm^3 powder drum size, with settings of 3 revolutions per minute (RPM) and avalanche threshold of 120 avalanches. Average avalanche energy, time and angle of 7.8 MJ/kg, 3.6 Sec and 47.6° respectively were recorded.

Table 1. Chemical composition of the gas atomized AA2099 measured with Inductively Coupled Plasma Optical Emission Spectroscopy (ICP-OES) and the commercial 2099-T83.

Elements	Al	Cu	Li	Zn	Mg	Mn	Zr	Si
AA2099	Bal.	2.63	1.56	0.67	0.28	0.17	0.09	0.02
AA2099 – T83	Bal.	2.40	1.60	0.04	0.10	0.10	0.05	0.05

DED cladding in this work was carried out with the Lunovu iLMD with laser spot size of 1.5 mm, from Lunovu GmbH, Herzogenrath Germany. The rotation disc of the powder hopper was

3 RPM, while powder mass flow rate of 8 g/min was set for cladding. Argon was used both as shielding and carrier gas with a flow rate of 6l/min. Single tracks of 15 mm length and coupons of 7.5x15x7.5 mm^3 were cladded both in room temperature (RT) and in-situ – preheating conditions (350°C, 400°C, 450°C and 500°C). Single-track Linear Energy Densities (LED) were 43.62 J/mm (RT), 19.98 J/mm (400°C, 450°C) and 14.44 J/mm (500°C). Preheated coupons were cladded with LED of single-tracks but with hatching and layer thickness of 0.75 µm and 1 mm respectively. RT single-track samples were subjected to post heat treatment (solution treatment 520°C, 2hours + water quenching and artificial aging at 450°C and 500°C for 5 hours) using the Nabertherm N120/85HA oven.

The Aconity 3D Mini PBF system with an 80 µm spot diameter, equipped with IPG YLR 400, F-Theta lens with focus of 420 mm, and wavelength between 1030–1080 nm was used for printing of PBF-LB/M 5x5x10 mm^3 cuboid samples under in-situ preheating conditions (500°C, Volume Energy Density; 44.1 J/mm^3). In both LAM printings of cuboids and coupons, simple bidirectional scanning strategy was used. Cross-sections of the samples were metallographically prepared for Scanning Electron Microscopy (SEM -Zeiss – SUPRA 55 VP FEG SEM with Oxford Instrument Inca EDS/EBSD Detectors). For Transmission Electron Microscopy (TEM - JEOL JEM-F200), 3 mm diameter cylinders with thickness of ~80 µm were prepared from the AA 2099-T83 samples, followed by thinning in a solution of 75 % CH_3OH + 25 % HNO_3 at – 30 °C by double jet electropolishing using a Streuers TenuPol 5 device. For the AM built samples, electron transparent lamellae were prepared using a Helios 5 Plasma Focussed Ion Beam (P-FIB) device..

3. Results and Discussion

3.1 Conventional Microstructure: AA2099-T83

The as-received commercial AA2099 was subjected to a T-83 heat treatment with pre-aging (150°C) stretching of 3 %. The pre-ageing stretching is purposed to increase the dislocation density in the material. Increase in dislocation density increases the number of nucleation sites. As dislocations, in an attempt to minimize total energy split into Schockley partials, they become preferential nucleation sites for T_1 [10]. From TEM of the as-received T-83, as shown below in figure 1, high volume fraction of matrix (intragranular) T_1 can be observed (figure 1a), with lengths varying from under 100 nm to ~300 nm. The observed length is consistent with reported lengths observed in conventionally manufactured Al-Cu-Li alloys [22, 23]. Viewed from the $[\bar{1}12]_{Al}$ zone axis, the reflections from the Selected Area Diffraction (SAD) image in figure 1b, not only confirms the presence of the matrix T_1. As it is known from literature, these T_1 plates can be described as nucleated and grown from dislocation sites. The presence of other intermetallics such as β, θ or δ is also confirmed as reflexes at ½ $\{220\}_{Al}$ position by the SAD reflections (figure 1b).

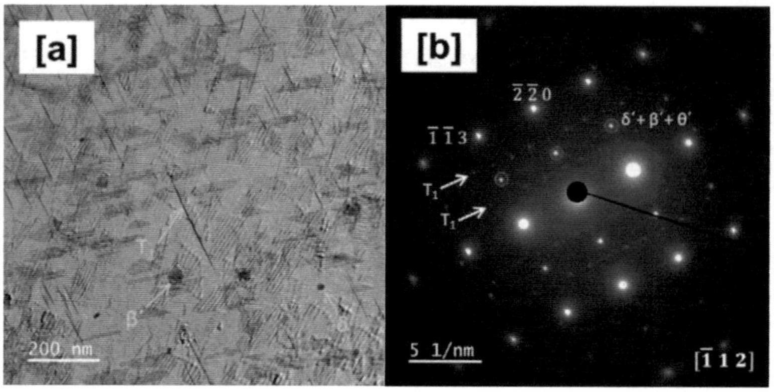

Figure 1. (a) BF TEM image showing the presence of T_1 and other intermetallic particles in the AA2099-T83 (b) SAD taken from the $[\bar{1}12]_{Al}$ zone axis confirming the presence of T_1.

3.2 DED Microstructure: Single-Tracks (Preheating, T6) and Coupons (Preheating)

In order to circumvent post heat treatment steps which not only consumes time and energy but also prolongs production steps, in-situ preheat treatment are under continuous exploration by researchers to exploit and enhance material properties. In-situ preheat treatment is typically beneficial not only in the minimization of solidification cracks, residual stresses, cooling rate and temperature gradient, but also enhances precipitation. In precipitation hardenable alloys such as AA2099, in-situ preheating at optimum nucleation temperatures favours the nucleation and growth of targeted precipitates. From work done by the authors [2, 7, 8], based on Scheil calculations, the nucleation temperature of T_1 in AA2099 is predicted to start at ~520 °C. Single tracks and coupons were therefore cladded under in-situ preheating conditions from 350 °C to 500 °C with a step size of 50°C. SEM micrographs of the characterised samples are shown in figure 2. At 350°C (figure 2a), the microstructure is characterized by intragranular sub-grain structures, enriched with Cu and no observed T_1. Between 400°C - 500°C (figure 2b-d), the nucleation and growth of T_1 can be observed. At 500°C preheating temperature, not many of T_1 were observed at the middle-top section of the sample. Compared to the intragranular T_1 plates observed in the conventional T-83 (figure 1a), only grain boundary (intergranular) T_1 is observed in the DED-LB/M. DED-LB/M compared to other AM processes like PBF-LB/M has slower cooling and solidification rate and as a result allows more time for segregation and solute atoms diffusion to the grain boundaries. This is exacerbated by the activation of in-situ preheating which further prolongs the melt pool lifetime. Nucleation and growth of precipitate at grain boundaries compared to the matrix is favoured by the comparatively lower free energy barrier to the matrix, grain boundary segregation and pronounce solute diffusivity [24]. [20] explained that intragranular T_1 nucleation requires high activation energy compared to that of intergranular T_1. As a result, at high aging temperatures where undercooling is low and activation energy barrier is high, high nucleation rate shifts to regions which require low activation energy like grain boundaries. A critical T_1 nucleus formed at the grain boundary is further clustered by solute atoms into thin layers of T_1 plates [24].

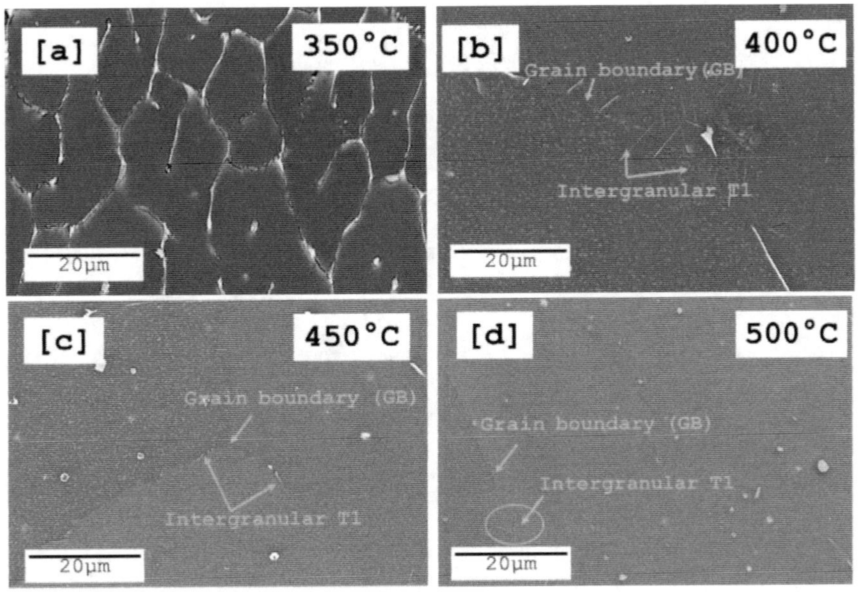

Figure 2. SEM micrograph of DED cladded single-track (preheating: 350°C - 500 °C): (a) 350°C, showing no T_1 presence (b)-(c) 400-450°C, observed intergranular T_1 (d) 500°C, no intergranular T_1 observed at middle section.

Prior to this work, the only report known to the authors of intergranular T_1 for additively manufactured microstructures was PBF-LB/M T6 treated AA2195 by [25]. It was argued that the high dislocation density in the as-built state prior to T6 heat treatment provided nucleation sites for both intragranular and intergranular T_1. It must be mentioned that work done by the previous authors on PBF-LB/M T6 treated AA2195 did not report of intergranular T_1. No work is currently known of DED-intergranular T_1, especially with preheating. From conventional manufacturing however, after 200°C aging with 10% rolling and without rolling, T_1 phase was reported by [26]. It was inferred that intergranular T_1 also nucleated from nucleation sites promoted by the pre-aging stretching. This cannot be the case for this current work as no pre-aging was performed for the nucleation of the intergranular T_1. In any case, no explanation was given to the grain boundary T_1 in the unrolled state. Since dislocations form in LAM microstructures, it is plausible to suggest that the intergranular T_1 could also have nucleated at grain boundaries on existing dislocations at the grain boundaries. In the authors [26] conclusion, intergranular T_1 was found to have more propensity to coarsen than intragranular T_1. In this work, intergranular T_1 was further characterized by TEM as shown in figure 3a-d below, showing both thin and coarse intergranular T_1 plates. The coarse grain boundary T_1 plate shown in figure 3d and confirmed by the SAD in figure 3e has a measured thickness of 22 (+/-1) nm slightly higher than the 19 nm reported by [26]. After performing T6 heat treatment (520°C for 2 hours + artificial ageing at 450 °C and 500 °C for 5 hours) on room temperature cladded DED single-tracks, the samples were characterised by SEM as shown in figure 4. Contrary to the nucleation and growth of intergranular T_1 with preheating (figure 3b, c), intragranular T_1 can be observed after T6 post heat treatment similar to previous findings [10]. The observed T_1 plates at 450 °C are finer than the T_1 plates at 500 °C ageing temperature which may be described as an overaged condition. Thermal heating and cooling cycles associated with LAM process especially under RT built conditions generate thermal stresses which induce the formation of dislocations [2].

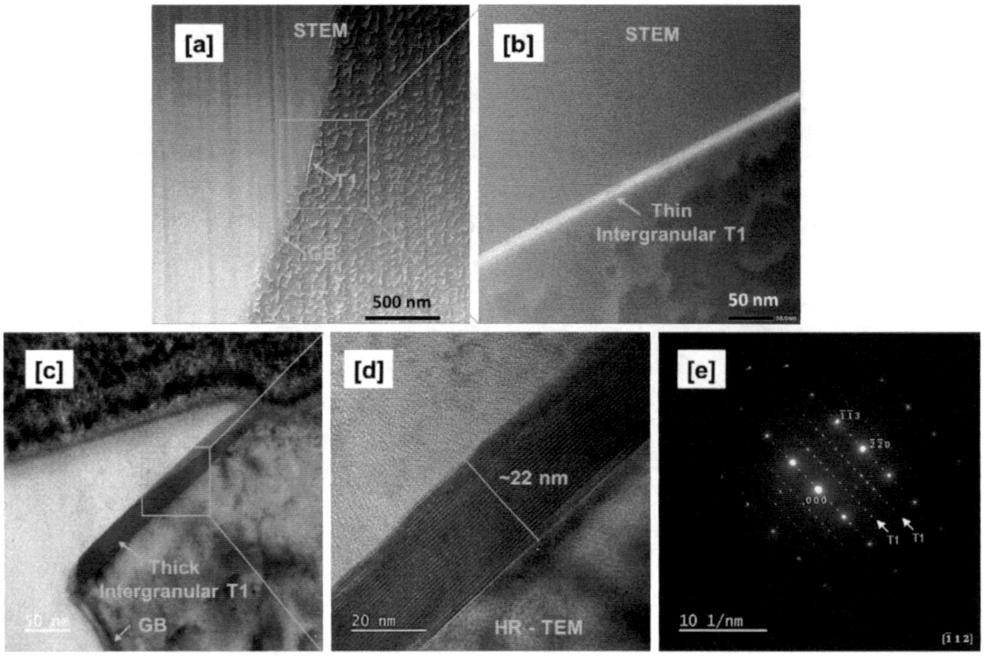

Figure 3. DED cladded single-track (preheating at 450 °C): STEM images (a), (b) and TEM image (c) showing thin and thick intergranular T_1, respectively (d) HR TEM image of thick intergranular T_1 (e) SAD diffraction pattern taken from $[\bar{1}12]_{Al}$ zone axis showing T_1 streaks (inclined variant).

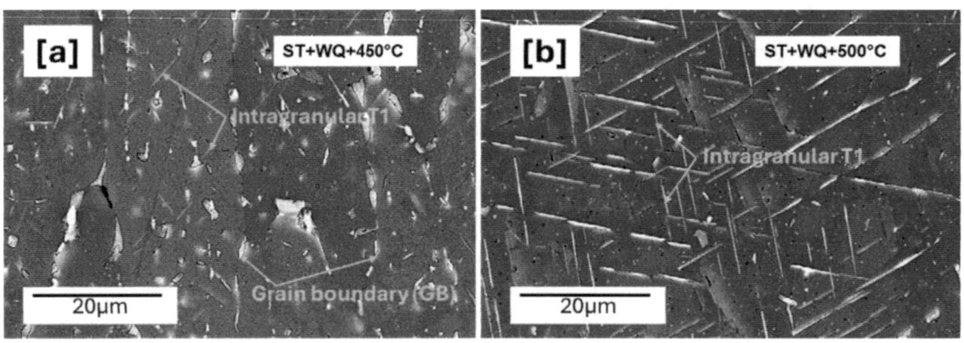

Figure 4. SEM micrograph: DED cladded single-track (a) solution treated (ST) + water quenching (WQ) with artificial aging (450°C, 5hours) (b) ST + WQ with artificial aging (500°C, 5hours).

The dislocation densities quantified in LAM microstructures are reported to be in the same order of magnitude with that of the conventional cold worked samples [27]. Dislocation density is therefore high enough to provide more nucleation sites for the intragranular T_1 nucleation and growth when subjected to appropriate aging temperature and time. The intragranular T_1 observed in the T6 heat treated DED single-tracks can be deemed to have nucleated on matrix dislocations as opposed to the grain boundary nucleation mechanism via preheating (intergranular T_1). The scaling up of the DED cladded single-tracks (figure 2,3,4) to coupons (figure 5) leads to different microstructures. Lamella-like structures are observed in the matrix at the middle section of the coupons (preheating: 400 – 500°C) in figure 5. The lamella-like structures appear to be fine at lower temperature (400°C), but coarsen as the temperature increases to 500°C. They also appear to be dissolving at 500°C where there's a longer exposure time to heat from the build substrate as well as thermal cycling. The formation of T_1 phase is not observed at this magnification. The complete evolution of single-track microstructures to coupon microstructures can be attributed to thermal heterogeneities from pronounce thermal cycling introduced not only from the addition of further hatching tracks, multiple layers and contours, but also from the scan strategy as the build volume increased. Therefore, based on the build volume, T_1 microstructure must be tailored in DED-LB/M.

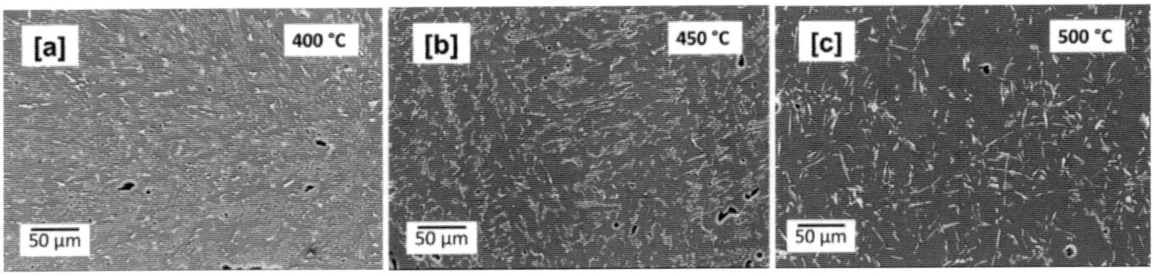

Figure 5. SEM micrograph of DED cladded coupons: (preheating): (a) 400°C (b) 450°C (c) 500°C.

3.3 PBF-LB/M Microstructure: Cuboids (Preheating)

Previous publications [2, 7, 8] investigated and reported the influence of in-situ preheat treatment on T_1 nucleation in PBF-LB/M and concluded that pre-ageing stretching was not a pre-requisite for T_1 nucleation as the observed intragranular T_1 nucleated from the melt during solidification. From TEM imaging shown in figure 6a-d of 500°C in-situ PBF-LB/M preheating, the intragranular T_1 confirmed from the SAD reflections taken from the $[\bar{1}22]_{Al}$ zone axis, can be said to have rather nucleated and grown on an intermetallic phase (figure 6a, b). Point EDS measurement showed the presence of Al, Cu, Mn and Fe in the intermetallic phase.

There is no reported work of T_1 nucleation on intermetallic phases in LAM microstructures but such phenomenon is not uncommon for conventionally manufactured Al-Cu-Li alloys [2, 28]. T_1 thickening is known to be relatively stable but grows quicker along the edges to form thin plates [26]. It can therefore be presumed that once T_1 nucleated on the intermetallic phase during solidification, they grew along the edges in atom layers into the observed thin plates. From the HR-TEM shown in figure 6c, the resolved intragranular T_1 can also be seen to consist of ~ 5 atom layers similar to a previous report [7].

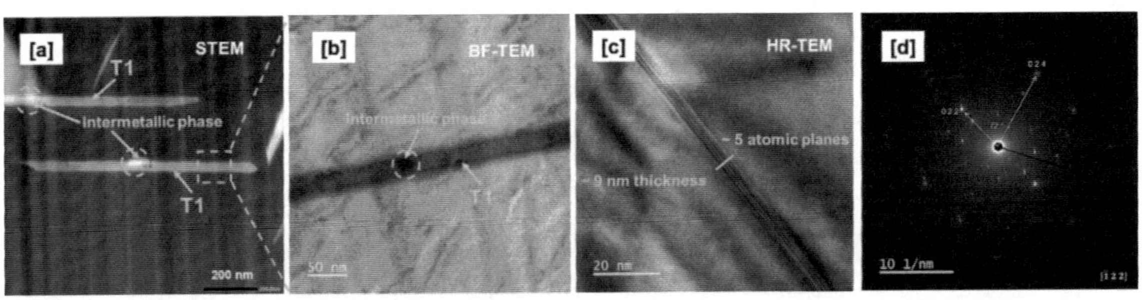

Figure 6. PBF-LB/M cuboid printed at 450°C preheating temperature: (a) STEM Intragranular T_1 with intermetallic phase at the centre (b) BF TEM image (c) HR image showing the thin plates consists of ~ 5 atom (d) SAD diffraction pattern taken from the $[\bar{1}22]_{Al}$ zone axis.

4. Conclusion

In conclusion, the focus of this work has been on characterising and comparing T_1 plate and its nucleation behaviour across different processes such as conventional manufacturing, and LAM (DED-LB/M and PBF-LB/M). Different manufacturing routes and heat treatment methods produce different T_1 microstructures. After 3% cold work and artificial aging (150°C), intragranular T_1 plates that nucleated on dislocations were characterised in AA2099-T83. In DED-LB/M single track with in-situ preheating, characterised microstructures rather show grain boundary T_1. In room temperature cladded DED-LB/M single tracks, intragranular T_1 plates were formed after the application of T6 heat treatment. For In-situ-preheated (500°C) PBF-LB/M, the T_1 plates were found to nucleate on intermetallic phase.

References

[1] Jagadeesh C, Shivananda Nayaka H, Sekar P, Ramesh S and Anne G 2024 *Mater. Res. Express.* **11** 36509
[2] Adjei-Kyeremeh F, Pratesa Y, Shen X, Song W, Raffeis I, Vroomen U et al. 2023 *Materials.* **16** 4916
[3] Abd El-Aty A, Xu Y, Guo X, Zhang S-H, Ma Y and Chen D 2017 *J Adv Res.* 2018 **10** 49–67
[4] Li Y, Shi Z and Lin J 2019 *Mater. Des.* **183** 108121
[5] Decreus B, Deschamps A, Geuser F de and Sigli C 2013 *Adv Eng Mater.* **15** 1082–5
[6] Khushaim M, Boll T, Seibert J, Haider F and Al-Kassab T 2015 *Adv Cond Matter Phys.* **1** 647468
[7] Raffeis I, Adjei-Kyeremeh F, Vroomen U, Suwanpinij P, Ewald S and Bührig-Polazcek A 2019 *JOM* **71** 1543–53
[8] Raffeis I, Adjei-Kyeremeh F, Vroomen U, Richter S and Bührig-Polaczek A 2020 *Materials* **13** 5188
[9] Araullo-Peters V, Gault B, Geuser F de, Deschamps A and Cairney JM 2014 *Acta Mater.* **66** 199–208
[10] Yang Y, He G, Liu Y, Li K, Wu W and Huang C 2021 J *Mater Sci.* **56** 18368–90
[11] Tsivoulas D 2015 *Metall Mater Trans A.* **46** 2342–6
[12] Qi Y, Hu Z, Zhang H, Nie X, Zhang C and Zhu H 2021 *Addit. Manuf.* **47** 102249
[13] Wu S, Lei Z, Li B, Liang J and Chen Y 2022 *Addit. Manuf.* **54** 102762
[14] Qi Y, Zhang H, Nie X, Hu Z, Zhu H and Zeng X 2020 *Addit. Manuf.* **35** 101346
[15] Chen R, Wang H, He B, Tu Y and Tang H 2021 *Vacuum* **190** 110299
[16] Shao S, Liang Z, Yin P, Li X and Zhang Y 2024 *Materials* **17** 657
[17] Baehr S, Matheson G, Ammann T, Mayr P and Zaeh MF 2024 *Adv. Ind. Manuf. Eng.* **8** 100142
[18] Zhong H, Qi B, Cong B, Qi Z and Sun H 2019 *Chin. J. Mech. Eng.* **32** 1-7
[19] Arbo SM, Tomovic-Petrovic S, Aunemo J, Dahle N and Jensrud O 2022 *J. Adv. Join. Process.* **5** 100096
[20] Chen PS, Kuruvilla AK, Malone TW and Stanton WP 1998 *J. Mater. Eng. Perform.* **7** 682–90
[21] Jiao S, Cheng X, Shen S, Wang X, He B, Liu D and Wang H 2020 *J. Alloys Compd.* **821** 153125
[22] Balducci E, Ceschini L, Messieri S, Wenner S and Holmestad R 2017 *Mater. Des.* **119** 54–64
[23] Decreus B, Deschamps A, Geuser F de, Donnadieu P, Sigli C and Weyland M 2013 *Acta Mater.* **61** 2207–18
[24] Butler EP and Swann PR. 1976 *Acta Metall.* **24** 343–52
[25] Wang C, Wu S, Lei Z, Zhang X, Fu W, Li X, et al. 2024 *J. Mater. Res. Technol.* **30** 6520–35
[26] Duan S, Wang Y, Liu Z, Guo F, Matsuda K and Zou Y 2023 *J. Mater. Res. Technol.* **24** 7782–7
[27] Hengsbach F, Koppa P, Duschik K, Holzweissig MJ, Burns M, Nellesen J, et al. 2017 *Mater. Des.* **133** 136–42
[28] Huang B-P and Zheng Z-Q 1998 *Acta Mater.* **46** 4381–93

Microstructure of additive manufactured materials for plasma-facing components of future fusion reactors

D A H Wartacz[1], H Becker[1,2], S Antusch[3], N Ordás[4], C Gundlach[5], O V Mishin[1] and W Pantleon[1*]

[1] Department of Civil and Mechanical Engineering, Technical University of Denmark, Kongens Lyngby, Denmark
[2] Institute of Materials and Joining Technology, Otto-von-Guericke-University, Magdeburg, Germany
[3] Institute for Applied Materials, Karlsruhe Institute of Technology, Karlsruhe, Germany
[4] Ceit-BRTA, Donostia – San Sebastian, Spain
[5] Department of Physics, Technical University of Denmark, Kongens Lyngby, Denmark

*E-mail: pawo@dtu.dk

Abstract. Two metallic materials considered for the divertor in fusion reactors are manufactured by powder bed fusion through electron beam melting: tungsten as armor of plasma-facing components and an age-hardenable CuCrZr alloy as heat sink material for the divertor. Cuboids are additively manufactured from both materials, and cross sections containing the build direction are characterized by electron backscatter diffraction. A peculiar heterogeneity is observed in the microstructure of tungsten and traced to the scanning strategy. Large columnar grains along the building direction with slight outward inclination are seen on both sides of the cross section i.e. where grains are observed in viewing planes perpendicular to the printing direction. Grains appear only slightly elongated in the center; neither their entire length nor their inclination is detected when the plane of view contains the printing direction. Many incidental twin boundaries are identified in the CuCrZr alloy; their occurrence is rationalized by the presence of an almost perfect <110> fiber texture. Additionally, X-ray computed tomography confirmed the low porosity of the CuCrZr specimen.

1. Introduction

Future fusion reactors will expose plasma-facing components to high heat fluxes and only few materials will be able to withstand them [1]. Tungsten (W) is the material of choice for the armor of all plasma-facing components, while copper chromium zirconium (CuCrZr) alloys will act as heat sink material for high heat flux components and plasma-facing components in the divertor of EU-DEMO [2]. Both materials experience challenges when operating at high temperatures as deformed tungsten may recrystallize [1,3] and age-hardened CuCrZr may overage [1,4]. The present work is concerned with additively manufactured parts of both materials and peculiarities of the microstructure resulting from the manufacturing process which has to be optimized to mitigate some of the risks. Despite the advantage of additive manufacturing to produce parts with complicated geometries, only cuboidal specimens with simple geometry manufactured by powder bed fusion through electron beam melting are investigated. Electron backscatter diffraction (EBSD) and X-ray computed tomography are used to characterize the microstructure.

IOP Conf. Series: Materials Science and Engineering **1310** (2024) 012004 doi:10.1088/1757-899X/1310/1/012004

2. Materials and experimental techniques

2.1 Tungsten

Specimens of pure tungsten are additively manufactured by powder bed fusion via electron beam melting using a GE Additive Arcam A2X Electron Beam Melting machine at the Karlsruhe Institute of Technology [5]. Figure 1 shows the cuboidal specimen and illustrates its geometrical sizes. In each printed layer, the electron beam is scanned along nested, concentric squares as sketched in figure 1(c); the first square at the perimeter, the last one in the center of the layer. The printing directions defined in this manner can be recognized on the top surface in figure 1(a).

A 2.5 mm thick plate containing the (vertical) build direction (BD) is cut at a distance of about 3.5 mm from the outer surface (blue in figures 1(b) and 1(c)). The surface closest to the back surface is mechanically ground, polished and finally electro-polished at room temperature using an aqueous solution containing 3 wt.% NaOH with an applied voltage of 12 V and a current of approximately 2 A. Orientation data are acquired by EBSD on the entire prepared surface.

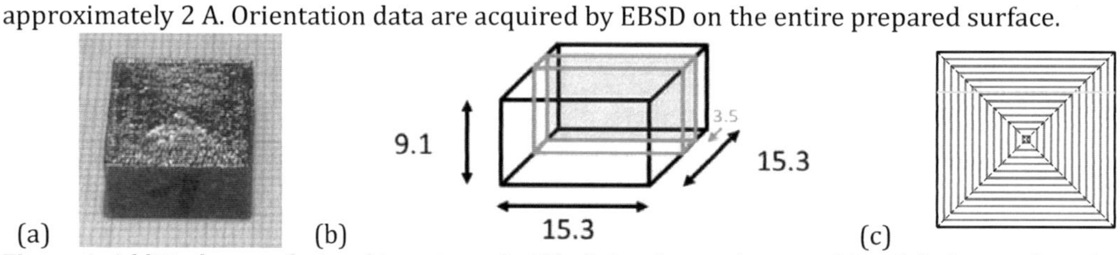

Figure 1: Additively manufactured tungsten cuboid built by electron beam melting: (a) picture of specimen on millimeter paper, (b) geometry with sketch of cutting scheme (the investigated cross section is highlighted in light blue, sizes are in millimeter), and (c) sketch of electron beam path in each layer in red.

2.2 CuCrZr

Specimens of CuCrZr are additively manufactured by powder bed fusion via electron beam melting using an adapted GE Additive Arcam Electron Beam Melting A2X device at AIDIMME Instituto Tecnológica in collaboration with Ceit [4]. Cubes with sizes 13×13×13 mm^3 are printed from an alloy powder containing 0.8 wt.% Cr and 0.11 wt.% Zr obtained by gas atomization at Ceit. In each layer, the electron beam is scanned uni-directionally, i.e. along parallel lines in the same direction. The printing direction is rotated by 90° between successive layers. The powder bed temperature is set to approximately 380 °C, slightly lower than the standard ageing temperature of CuCrZr (450 °C to 475 °C). The printed cubes have a very low oxygen content in their as-build condition. Hot isostatic pressing at 400 °C and 150 MPa for 3 h reduces their porosity while the Cr precipitates are prevented to coarsen [4]. A cross section containing the build direction from the center of the cube is investigated by EBSD. Preparation of the soft CuCrZr alloy without surface damage turns out to be challenging. Following grinding and polishing, the surface is gently polished chemo-mechanically (OP-S) for 4 h to obtain a surface suitable for EBSD.

2.3 Electron backscatter diffraction

Orientation data are gathered on the prepared cross sections by electron back scatter diffraction at an accelerating voltage of 20 kV using different steps sizes. Two different scanning electron microscopes (SEMs) are used: a ZEISS Supra 35 SEM with a Nordlys EBSD detector and a ZEISS Sigma SEM using a C-Nano EBSD detector from Oxford instruments (only for a detailed map on tungsten). To gain an overview over the entire cross section of the tungsten sample, several data sets acquired from 30 smaller regions after sequential stage movements are stitched together to obtain a large orientation map. The data are analyzed using the Matlab plugin toolbox MTEX Version 5.8.2 [6] and evaluated further by own purposely developed routines.

IOP Conf. Series: Materials Science and Engineering **1310** (2024) 012004 doi:10.1088/1757-899X/1310/1/012004

2.4 Computed tomography

X-ray computed tomography is used to reveal internal pores in the printed CuCrZr alloy. A ZEISS Xradia 410 Versa X-ray microscope with an X-ray energy of 150 keV is utilized. For computed tomography, a slender bar-shaped specimen is cut to a size of about 1×4×12 mm^3 with its long axis along the build direction and its short axis pointing from the previous outer surface into the bulk. The X-ray microscope is operated at 150 kV and 10 W using a HE1 filter and 4× objective. 2400 projections with an exposure time of 6 s per projection are acquired. Reconstruction with a voxel size of 3.54 μm^3 is achieved using a 1024×1024 camera resolution (using 2×2 binning) covering the full alloy piece within the measured volume. Image reconstruction is performed with an inbuilt acquisition and reconstruction software package (ZEISS) based on a Feldkamp, Davis and Kress algorithm with filtered back-projection [7]. The data are further analyzed and visualized using the software Avizo 3D from Thermo Fisher Scientific.

In the reconstructed volume, pores are distinguishable by their lower attenuation compared to the CuCrZr alloy. (No intensity variation attributable to precipitates are recognized.) Internal pores are identified through an intensity threshold analysis. Position, volume, and equivalent spherical diameter are determined for each individual pore.

3. Results and discussion

3.1 Armor material tungsten

Figure 2 shows a detailed orientation map obtained with a small step size of 1 μm from the center of the cross section. The two maps illustrate the presence of high angle boundaries (HABs) with disorientation angles above 15° and low angle boundaries (LABs) with disorientation angles above a certain threshold angle, but not larger than 15°. Grains can be distinguished by their different colors designating the crystallographic directions along BD or the high angle boundaries separating them. The grains appear elongated along BD and non-convex. The observed waviness of the high angle boundaries is attributed to printing layer-by-layer. The dominance of reddish and bluish colors indicates a preference for <100> and <111> directions to align with BD.

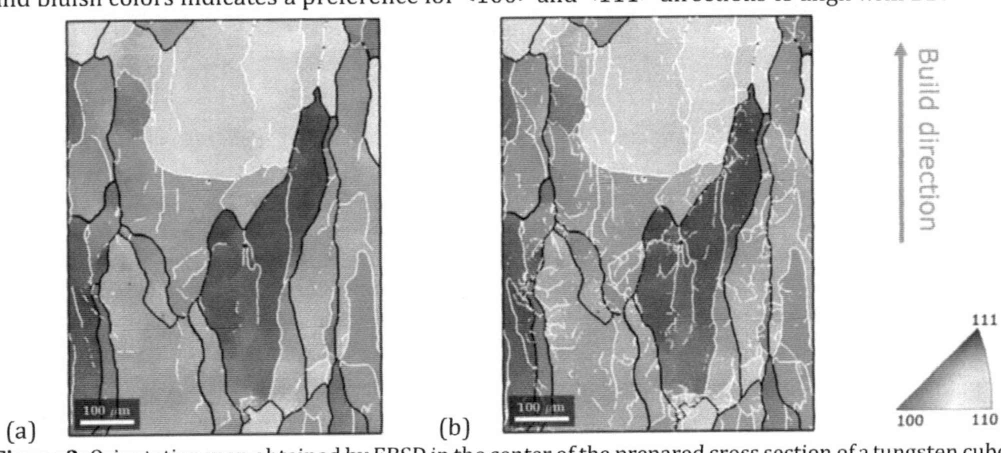

Figure 2: Orientation map obtained by EBSD in the center of the prepared cross section of a tungsten cuboid additively manufactured by electron beam melting. The colors represent the crystallographic direction along BD according to the inverse pole figure (IPF) on the right. High angle boundaries (with disorientation angles above 15°) are shown in black; low angle boundaries are marked in white for different threshold values for the disorientation angle: (a) 2° and (b) 1°.

IOP Conf. Series: Materials Science and Engineering **1310** (2024) 012004 doi:10.1088/1757-899X/1310/1/012004

The grains do not have a unique orientation as seen from the color variations and the presence of low angle boundaries within them. A much finer subdivision in the hierarchical structure becomes obvious in figure 2(b) when low angle boundaries with disorientation angles above 1° are shown compared to low angle boundaries above 2° in figure 2(a). From sudden changes in color apparent in the yellow-orange grain on the top of the map or the purple grain stretching almost from the bottom to the center, it becomes obvious that distinct orientation differences exist, which are resolved neither by low angle boundaries above 2°, nor such above 1°. Lowering the threshold disorientation angle for low angle boundaries even further is not advisable, as spurious artefacts appear due to the limited angular resolution of the ESBD system when acquiring orientations. Consequently, more advanced evaluation procedures (analogous to e.g. [8,9]) must be devised for characterizing the substructure within the grains in future.

The orientation map gathered from the entire cross section with a large step size of 20 µm is presented in figure 3. It illustrates the crystallographic direction along BD as well as the presence of high angle boundaries with disorientation angles above 15° and low angle boundaries with disorientation angles above 2°. All orientations of pixels indexed as tungsten are presented as collected without any filtering or removal of unindexed points (about 9 %; a higher indexing rate is preferable, but deemed unnecessary for the present purpose). The dominating red and blue colors in the entire map confirm indeed the alignment of <100> and<111> directions with BD.

Figure 3: Orientation map obtained by EBSD on an entire cross section through a tungsten cuboid additively manufactured by electron beam melting. The colors represent the crystallographic direction along the vertical build direction according to the IPF on the right. High angle boundaries with disorientation angles larger than 15° are shown in black, low angle boundaries with disorientation angles above 2° in white. The overview map bases on 30 individual maps stitched together.

Figure 3 clearly reveals macroscopic heterogeneity in the microstructure. The entire map resembles a triptych with a central panel and two side wings. On both sides, left and right, columnar grains are inclined about 15° outwards in an opposite sense with respect to the build direction; the grains in the middle appear more equiaxed with a less, but notable elongation along the build direction. The observed outward inclination of the grains on both sides is presumably a direct consequence of the applied printing strategy within each printed layer (discussed below).

IOP Conf. Series: Materials Science and Engineering **1310** (2024) 012004 doi:10.1088/1757-899X/1310/1/012004

For further analysis of microstructure and texture, the entire orientation map is subdivided into three separate regions (left, middle, and right, in the manner of a triptych). For each of the three regions, the texture is analyzed individually; figure 4 shows the 100 and 111 pole figures. They confirm the preferred alignment of <100> and <111> directions with the build direction for each of the three regions, with a dominance of aligning of <100> directions over <111> directions and even similar maximal pole densities for the three regions. No fiber textures are observed; in all cases, the dominant orientation has one of its <100> directions close to the build direction and a <110> direction perpendicular to BD in the cross section as well as normal to the cross section.

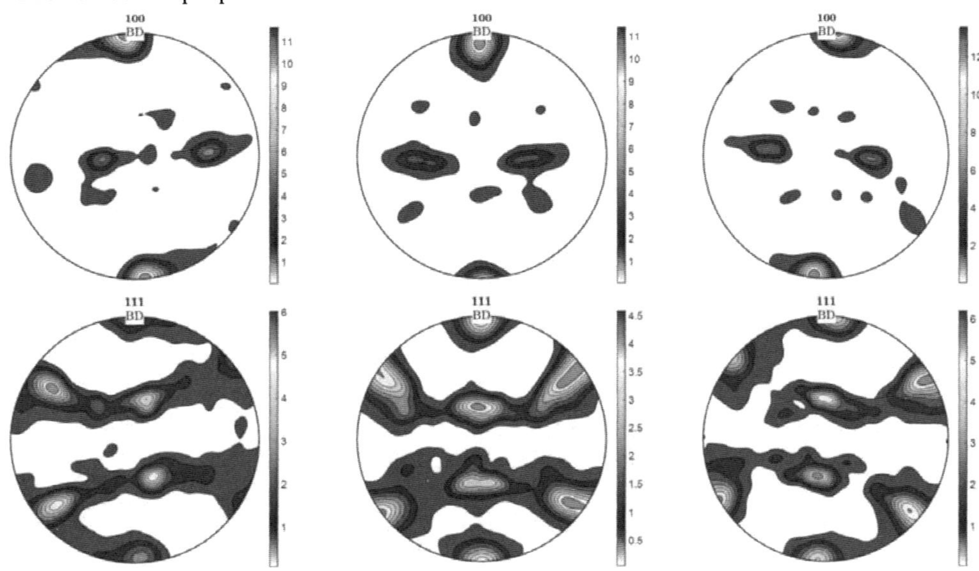

(a) left (b) middle (c) right

Figure 4: 100 (top row) and 111 (bottom row) pole figures for three different regions on the cross section shown in figure 3 through a tungsten cuboid additively manufactured by electron beam melting.

Obviously, the textures in the three different regions are not identical and systematic tilts occur. This is most easily seen for the 100 pole figures. The 100 pole figure from the left hand side (figure 4(a)) is rotated anti-clockwise by about 6°, while the one from the right hand side (figure 4(c)) is rotated clockwise by the same amount. The 100 pole figure of the middle region in figure 4(b) is tilted 6° out of the plane of observation. All tilts and rotations are consistent with each other and the squared path of the electron beam used for printing each layer. This particular scanning scheme leads to four distinct regions in the printed cuboid separated by diagonals (see figure 1(c)). Within each region the electron beam scans on parallel paths. When printing the next layer, solidifying grains pick up an existing orientation from the layer beneath and continue to grow homo-epitaxially. As the outermost squares are printed first followed by the nested inner squares, the columnar grains become inclined perpendicular to the electron beam path. Simultaneously, the effective build direction of each columnar grain becomes slanted causing an inclination of the axis of preferred crystallographic direction. The inclination of the columnar grains, however, appears larger than that of the poles.

Scanning along nested squares causes the apparently different microstructure in the middle region of the section as well. Consequently, the grains there are also presumed to be elongated and inclined in a direction perpendicular to the mapped surface (i.e. out of plane as the 100 pole in figure 4(b)). The middle region where grains do not show any obvious inclination in the observation plane and have only a slight elongation along the build direction is about 7.7 mm wide, in good agreement with the expectation (8.3 mm) considering the section being cut at about 3.5 mm from the outer surface and subjected to some width reduction during surface preparation.

Grains are identified based on the disorientation angle between neighboring pixels as contiguous regions separated by high angle boundaries from other grains. If a grain is required to contain at least 6 pixels, the resolved grains in the entire map consist in average of 132 pixels. Only slight differences are observed for the average grain areas with some smaller grains in the middle region (129 pixels in average) compared to the left and the right (136 pixels in average). With the step size of 20 µm used for the overview map in figure 3, the average grain area becomes 52.800 µm² amounting to 18.9 grains per mm² corresponding to an ASTM grain size number [10] of 1.3 which is by far smaller than the ASTM grain size number of 3 or larger required for tungsten plates in the ITER materials specification [11].

3.2 Heat sink material CuCrZr

As sample preparation without surface damage is difficult for the CuCrZr alloy, the orientation data acquired on a cross section with a step size of 4 µm still contain many unindexed points (about 20%). They are removed by filtering of the data in the orientation map in figure 5(a). Columnar grains elongated along the build direction with corrugated grain boundaries dominate the microstructure. Their zigzag appearance is attributed to the layered deposition and the rotation of the scanning direction of the electron beam by 90° between the layers, cf. [4]. Both, the greenish appearance of the orientation map and the 110 pole figure in figure 5(b) reveal a strong alignment of one of the crystallographic <110> directions with the build direction; an almost perfect <110> fiber texture with only minor variations perpendicular to the build direction exists.

Several twin boundaries (as defined by their Σ3 orientation relationship) are recognized as blue lines in the orientation map of figure 5(a), in particular on the right hand side. They are neither particularly straight, nor associated with twin lamellae typical for face-centered cubic materials [12]. The occurrence of such incoherent twin boundaries between neighboring grains is purely incidental, but a direct consequence of the strong <110> fiber texture along the build direction. The Σ3 orientation relationship between twins is determined by a 60° rotation about a common <111> axis. In case of cubic symmetry, the same orientation relation can be achieved by a 72° rotation about a <110> axis. Due to the strong <110> fiber texture, most of the grains have a common <110> axis, such that the statistical occurrence of a Σ3 relationship becomes large. For an ideal <110> fiber texture with cylindrical symmetry and rotations up to 90° about the <110> axis, a rather high fraction of twin relations is expected depending on the inclination between the closest <110> direction and the fiber axis. For an average inclination angle of 5°, about 10 % of the boundaries are classified as twin boundaries according to the Brandon criterion [13], i.e. their deviation from the ideal twin relationship 60°<111> is less than 8.7°. From the distribution of disorientation angles of high angle boundaries presented in figure 5(c), a significant fraction of boundaries with disorientation angles less than 30° are observed, despite the strong <110> fiber texture. On the other hand, 2.8 % of all high angle boundaries are twin boundaries complying with the Brandon criterion; only slightly more than the 1.8 % expected for a random distribution of orientations. However, 20.6 % of all boundaries with disorientation angles above 51.3°, which possibly could be classified as twin boundaries by the Brandon criterion, are twin boundaries, compared to 9.6 % for a random orientation distribution. If only deviations less than 3° from the ideal 60°<111> rotation orientation relationship are considered as suggested in [14], even 37.8 % of all HABs with disorientation angles above 57° are classified as twin boundaries.

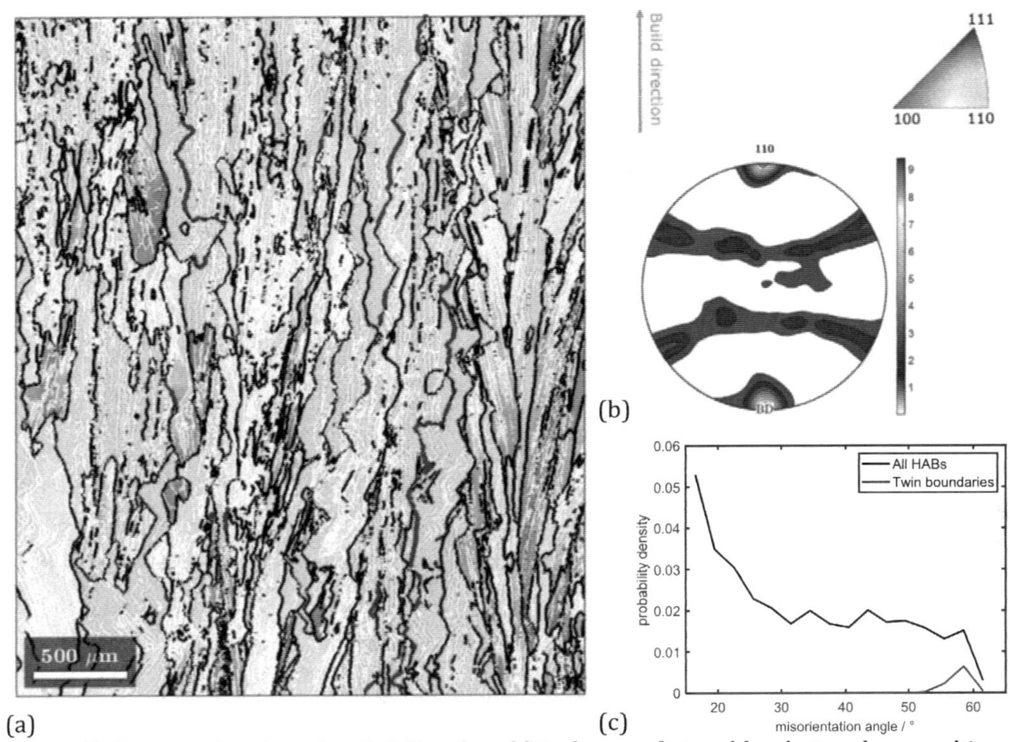

Figure 5: Cross section through a CuCrZr cube additively manufactured by electron beam melting: (a) orientation map obtained by ESBD with build direction vertically. The colors represent the crystallographic directions along BD according to the IPF on the right. High angle boundaries with disorientation angles larger than 15° are delineated in black, low angle boundaries above 2° in white and twin boundaries are shown as blue lines. (b) 110 pole figure with the building direction vertically. (c) boundary disorientation distribution function for all high angle boundaries, i.e. with disorientation angles larger than 15°. The contribution of twin boundaries identified according to the Brandon criterion is indicated in blue.

Reconstruction of the X-ray computed tomography data with a pixel resolution of 3.5 μm revealed 103 pores with a minimum equivalent sphere diameter of at least 4.4 μm. Figure 6(a) presents the reconstructed volume with clearly discernible pores occurring more frequently close to the former outer surface (on the left hand side in the image). The porosity amounts to a volume fraction smaller than 0.01 %. If the pore size distribution in figure 6(b) is approximated by an exponential distribution, an average pore diameter of 6.3 μm is deduced (including even the unresolved pores with equivalent sphere diameter less than 4.4 μm, i.e. with volumes smaller than the pixel size).

4. Conclusions

The microstructure and texture of pure tungsten and a CuCrZr alloy both additively manufactured by electron beam melting is characterized. In pure tungsten, a peculiar heterogeneity is observed in the microstructure of tungsten and traced to the scanning strategy of the electron beam. Large columnar grains inclined outwards are observed when grains are observed in planes perpendicular to the printing direction. In viewing planes parallel to the printing direction, grains appear only slightly elongated and neither their entire length, nor their inclination can be observed. Many incidental twin boundaries are identified in the CuCrZr alloy which are rationalized by the development on an almost perfect <110> fiber texture. X-ray computed tomography confirms the low porosity of the CuCrZr specimen.

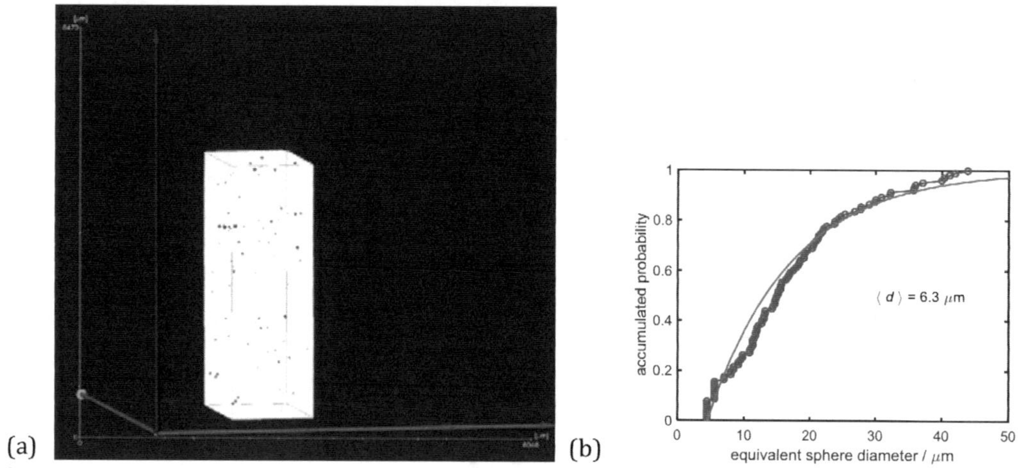

Figure 6: Results of X-ray micro computed tomography on additively manufactured CuCrZr sample: (a) reconstructed volume illustrating the spatial distribution of pores; the blue axis points along the build direction, the red axis into the specimen; the volume appears shorter along the green axis due to the used projection; the surface on the left is close to the former outer surface and (b) accumulated probability distribution of the equivalent sphere diameter of the 103 resolved pores.

Acknowledgment

This work has been carried out within the framework of the EUROfusion Consortium, funded by the European Union via the Euratom Research and Training Programme (Grant Agreement No 101052200 – EUROfusion). Views and opinions expressed are however those of the authors only and do not necessarily reflect those of the European Union or the European Commission. Neither the European Union nor the European Commission can be held responsible for them. The authors would also like to acknowledge AIDIMME for providing the CuCrZr cubes.

References

[1] Pintsuk G, Aiello G, Dudarev S L, Gorley M, Henry J, Richou M, Rieth M, Terentyev D and Vila R 2022 *Fus. Eng. Des.* **174** 112994:1-21
[2] You J H et al. 2016 Fusion Eng. Des. **109-111** 1598-1603
[3] Pantleon W 2021 *Phys. Scr.* **96** 124036:1-9
[4] Ordás N, Portolés L, Azpeleta M, Gómez A, Blasco J R, Martinez M, Ureña J and Iturriza I 2021 *J. Nucl. Mater.* **548** 152841:1-9
[5] Antusch S et al. 2024 *Nucl. Mater. Energy* **39** 101683:1-24
[6] Bachmann F, Hielscher R and Schaeben H 2010 *Solid State Phenom.* **160** 63-68
[7] Feldkamp L A, Davis L C and Kress J W 1984 *J. Opt. Soc. Am.* **1** 612-619
[8] Pantleon W 2005 *Mater. Sci. Techn.* **21** 1392-1396
[9] Pantleon W, He W, Johansson T P and Gundlach C 2008 *Mater. Sci. Eng. A* **483-484** 668-671
[10] ASTM E112-13 2013 Standard Test Methods for Determining Average Grain Size
[11] Hirai T et al. 2016 *Nucl. Mater. Energy* **9** 616-622
[12] Alimadadi H, Fanta A B, Somers M A J and Pantleon K 2014 *Surf. Coat. Techn.* **254** 207-216
[13] Brandon D G 1966 *Acta Metall.* **14** 1479-1484
[14] Mishin O V and Huang X 1999 *Mater. Sci. Forum* **294-296** 401-404

Large-scale microstructure modelling of an additively manufactured part using cellular automata

O Zinovieva[1*], A Zinoviev[1], O Gokcekaya[2] and Y Tang[3]

[1] School of Engineering and Technology, University of New South Wales Canberra at the Australian Defence Force Academy, Canberra, Australia
[2] Department of Materials Processing, Osaka University, Osaka, Japan
[3] Faculty of Engineering, Monash University, Clayton, Australia

*E-mail: o.zinovieva@unsw.edu.au

Abstract. This study represents the first attempt, to our knowledge, to simulate the grain structure of a laser powder bed fusion printed metal part exceeding 1,000 mm^3 in volume using a workstation. Utilising our in-house cellular automata finite difference code with OpenMP for parallelisation, the model predictions closely align with experimental observations. The implemented model shows promising capabilities for estimating microstructures with reasonable accuracy and could serve as a powerful tool for digital manufacturing.

1. Introduction

Metal additive manufacturing (MAM) holds the potential to transform the production landscape by enabling the creation of complex components with superior properties achieved by microstructure engineering [1]. Manufacturing near-net shape components with minimal post-processing remains a goal of MAM practitioners. Even after post-processing, the microstructure and mechanical properties of MAM parts are influenced by the as-built microstructure, making a deep understanding of the latter critical for predicting the performance of MAM components.

The field of simulating as-built AM microstructures is rapidly evolving. Geometrically based methods [2] and image reconstruction can be efficient but do not provide physics-based insights into the microstructure formation. Physically based microstructure modelling covers this gap [3]. Macroscale empirical approaches [4] focus on homogenised estimations. At finer scales, attention is directed towards understanding the effects of process conditions on cellular-dendritic solidification, employing sophisticated models like phase field [5] or micro-cellular automata (CA) [6]. These models provide deeper insights into solidification process but are limited by high computational cost, restricting simulations to small regions containing a few dendrites. Grain structure simulations during MAM are commonly performed using the kinetic Monte Carlo (kMC) or CA methods [3]. However, the kMC method does not consider crystallographic orientations of grains, does not explicitly integrate physical time and is not directly related to the temperature evolution during MAM [7], while the CA method does not have the limitations of kMC [8,9].

While both kMC and CA approaches are considered efficient enough to enable microstructure simulations of large domains, few of those have been reported so far. Considering direct energy deposition with kMC, Rodgers et al. [7] simulated the microstructure evolution in a 30 mm tall cylinder consisting of 100 build layers printed with two scan passes per layer (~ more than 35 mln cells). Using CA integrated with finite differences (FD), Zinovieva et al. [8] described the microstructure formation in a 316L steel cuboid sample measuring ~ 2.06 × 2.06 × 4.75 mm^3,

IOP Conf. Series: Materials Science and Engineering **1310** (2024) 012005 doi:10.1088/1757-899X/1310/1/012005

which consisted of more than 150 layers virtually printed with powder bed fusion (PBF) (~ more than 160 mln cells). Coupling CA with the finite element method, Koepf et al. [9] predicted the grain structure of a PBF-produced CMSX-4 sample of ~ 3 × 3× 8 mm³ and 160 layers.

To address this gap, this paper reports on the large-scale simulation of IN718 nickel-based superalloy using the CAFD approach. The simulation covers a multilayer build process, integrating thermal and microstructural dynamics to predict the grain structure in a simple PBF-produced part. Large-scale microstructure modelling tools are of key importance for the emerging idea of holistic computational design of MAM parts [10]. As the macroscale properties of a MAM part are determined by processes and structural features at smaller scales, the computational design approach for MAM requires fine control over different spatial scales involved in both the part design and manufacturing process. Therefore, large-scale CA represent a critical tool for optimising manufacturing parameters and improving the mechanical performance of the final components, in particular when involved in integrated computational materials engineering.

2. Methodology

2.1 Experimental

Gas-atomised IN718 spherical powder, with particle sizes characterised by a distribution where D10, D50, and D90 measure 18.6, 32.3, and 57.9 μm respectively, was used for sample manufacturing. The laser beam PBF (PBF-LB) was carried out using an EOS M290 printer equipped with a 400 W Yb-fibre laser with a Gaussian profile. The following process parameters were applied: the laser power, P, of 180 W; the scan speed, v_{scan}, of 1400 mm/s; the layer thickness, d, of 40 μm; the hatch distance, h, of 80 μm, and the volumetric energy density, $E = P/(hdv_{scan})$, of 40 J/mm³. The base plate was preheated to 80°C to ensure stable fabrication conditions. A bidirectional scan strategy with no rotations or shifts between layers was applied.

After printing, electrical discharge machining was employed to remove the samples from the base plate. For materials characterisation, we considered the BD-TD plane perpendicular to the laser scan direction, where BD and TD stand for the build direction and transverse direction, respectively. The sample was mechanically polished using emery paper up to a 4000 grade and then chemically polished with colloidal silica to achieve a mirror finish. Electron backscatter diffraction (EBSD) analysis was performed using a field-emission scanning electron microscope (FE-SEM; JEOL JIB-4610F, Japan) equipped with an EBSD detector (Aztec HKL, Oxford Instruments, UK). The parameters for data acquisition in EBSD are a 20 kV accelerating voltage and a 2 μm step. The EBSD data were processed using HKL Channel5 software from Oxford Instruments, UK, to produce inverse pole figure (IPF) maps and corresponding pole figures. The melt pool shape was evaluated on the mirror-polished and etched surfaces using a mixture of 31% HNO3, 6% HF, and 63% H2O. After a 20-minute etch, the samples were washed with water and ethanol and thoroughly dried. The melt pool shape was then examined using an FE-SEM.

2.2 Thermal simulations

The transient heat transfer during PBF is modelled by solving the local form of the energy conservation law within a 3D volume V_H bounded by surface S and subjected to a volumetric moving heat source with an input Q. At the scale of the melt pool, the conservation of energy is governed by Fourier's law of heat conduction and can be mathematically expressed as

$$\rho c_H \frac{\partial T(\vec{r},t)}{\partial t} = \nabla[\lambda \nabla T(\vec{r},t)] + Q(\vec{r},t) \qquad \text{for } \vec{r} \in V_H. \qquad (1)$$

Here ρ (kg/m³), c_H (J kg⁻¹ K⁻¹) and λ (W m⁻¹ K⁻¹) are the density, specific heat capacity and the thermal conductivity as functions of temperature T (K), as shown in figure 1(a-c), [11,12]; $\vec{r}(\mathrm{X}, \mathrm{Y}, \mathrm{Z})$ is the coordinate vector within the global coordinate system (figure 1(d)), and t (s) is the time.

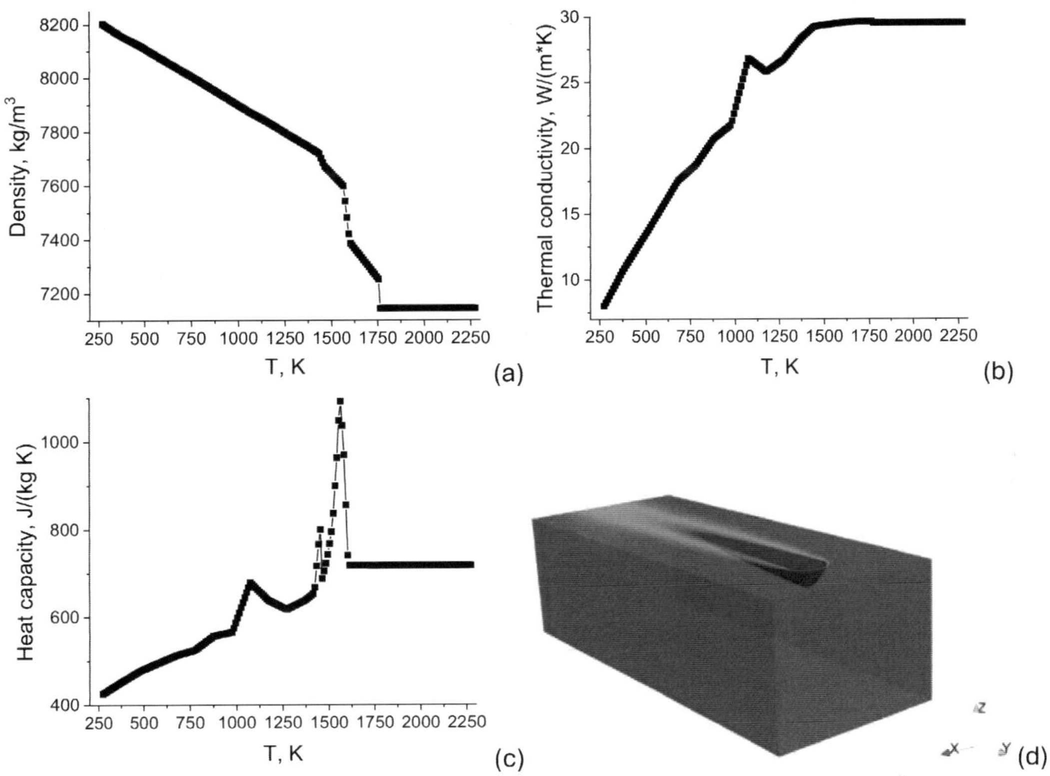

Figure 1. Temperature-dependent material properties of IN718 alloy [11,12]: (a) density; (b) thermal conductivity; (c) heat capacity; (d) solution of the heat conduction problem describing a single track, with the temperature varying from 1600 K (red) to 300 K (dark blue).

The volumetric heat source is described by Goldak's double ellipsoid model [13]:

$$Q(\vec{r}_{\mathrm{hs}}, t) = \frac{6\sqrt{3} f \eta P}{a b \gamma \pi \sqrt{\pi}} \exp\left(-\frac{3x_{\mathrm{hs}}^2}{\alpha^2} - \frac{3z_{\mathrm{hs}}^2}{\beta^2} - \frac{3y_{\mathrm{hs}}^2}{\gamma^2}\right) \tag{2}$$

The vector $\vec{r}_{\mathrm{hs}} = (\mathrm{x_{hs}}, \mathrm{y_{hs}}, \mathrm{z_{hs}})$ is defined in a heat source coordinate system with the origin in the melt pool centre; η is the absorption coefficient; a, b, and γ (all in μm) denote the half-width, depth, and length of semi-ellipsoids. While semi-ellipsoids share the width and depth, the length is different for the front and rear semi-ellipsoids, $\gamma = \gamma_f$ for $y_{\mathrm{hs}} \geq 0$ and $\gamma = \gamma_r$ for $y_{\mathrm{hs}} < 0$. The heat fractions deposited in the front and rear semi-ellipsoids are calculated as $f = f_{\mathrm{f}} = 2/(1 + \gamma_{\mathrm{r}}/\gamma_{\mathrm{f}})$ for $y_{\mathrm{hs}} \geq 0$ and $f = \gamma_{\mathrm{r}} = 2/(1 + \gamma_{\mathrm{f}}/\gamma_{\mathrm{r}})$ for $y_{\mathrm{hs}} < 0$.

Boundary conditions are imposed as follows

$$\partial T(\vec{r}, t)/\partial n = 0 \qquad \text{for } \vec{r} \in \mathrm{S}_{\mathrm{X}_k} \cup \mathrm{S}_{\mathrm{Y}_k}, \tag{3}$$

$$\lambda \frac{\partial T(\vec{r}, t)}{\partial t} = \kappa_1 [T^e - T(\vec{r},\ t)] \qquad \text{for } \vec{r} \in \mathrm{S}_{\mathrm{Z}_1}, \tag{4}$$

$$\lambda \frac{\partial T(\vec{r}, t)}{\partial t} = \kappa_2 [T^e - T(\vec{r},\ t)] + \sigma_B \varepsilon [T^{e4} - T^4(\vec{r},\ t)] \qquad \text{for } \vec{r} \in \mathrm{S}_{\mathrm{Z}_2}. \tag{5}$$

Here $\mathrm{S}_{\mathrm{X}_k}$ are the lateral surfaces, $\mathrm{S}_{\mathrm{Y}_k}$ are the back and front surfaces, $\mathrm{S}_{\mathrm{Z}_k}$ are bottom and top surfaces, respectively, $k = 1{,}2$; n is the outward normal to the surface, κ (W m⁻² K⁻¹) is the heat transfer coefficient, σ_B (kg s⁻³ K⁻⁴) is the Stefan–Boltzmann constant, and ε is the emissivity.

Equations 1–5 are solved using an explicit FD Forward-Time Central-Space (FTCS) scheme, which is conditionally stable. The time step is chosen to satisfy $\tau = k\rho c h_{\rm h}^2/\lambda$, $k \leq 0.5$, h_h (μm) is the FD voxel size. To reduce the computational costs associated with thermal simulations, we utilise the concept of a local steady-state temperature field [14] to describe multi-track multi-layer printing. The laser beam irradiation is modelled as a steady-state temperature field that traverses a larger computational domain at a predefined scan speed, following the defined scan strategy. This steady-state region includes the melt pool and surrounding area, as shown in figure 1(d). The size of this adjacent area is defined based on outcomes from test simulations. Earlier we successfully utilised this algorithm for PBF of Ti-6Al-4V alloy and 316L steel [14,15].

2.3 Microstructure modelling

The MAM-induced microstructure is modelled using grain-scale CA [8,15], which integrates the liquid-solid phase transformation, epitaxial solidification, and the orientation-dependent crystal growth in molten metal. The grain structure evolution is described through the state variables of CA cells (voxels) and the local conditions of these cells (e.g., temperature, crystallographic orientation). The numerical model employs a regular voxel mesh $h_{CA} = h_h$. The transient temperature field is derived from thermal analysis as detailed in Section 2.2.

Each CA cell has a predefined neighbourhood (first-order Moore, considering the 26 nearest voxels) and can exist in one of four states: 'liquid' (molten CA voxel), 'undercooled liquid' (CA voxel with a temperature $T < T_L$, where T_L is the liquidus temperature), 'growing' (solidification front), or 'solid'. Cell state transformations occur through nucleation or dendritic growth processes, as detailed in our earlier publications (refer to table 1 in [15]). A CA voxel can transition to a 'growing' state either through nucleation – realised in the base plate in this model – or by being engulfed by a growing grain. A 'growing' CA cell is initialised with a crystallographic orientation defined by a rotation matrix, either randomly assigned from a set of orientations generated with steps of 0.05 on (−1; 1) and 1° on (0°; 360°) for the first and second basis vectors (nucleation) or is inherited from a parent cell (epitaxial solidification or grain growth). These 'growing' CA cells demarcate the boundary between 'liquid' and 'solid' cells, effectively representing the solidification front. For epitaxial solidification to occur, a 'solid' cell must be adjacent to at least one 'liquid' cell. Each 'growing' CA voxel contains a so-called grain envelope, shaped as a regular octahedron in 3D cubic alloys. These octahedrons grow along the diagonals which align with preferred growth directions (⟨100⟩ for cubic alloys), with the growth rate according to the grain growth law. Once there are no 'liquid' cells in the neighbourhood of the parent cell, it stops growing. Our CA model derives the growth law from the analytical KGT model proposed by Kurz et al. [16], linking the growth rate along preferred directions, v (m/s), to the local undercooling of octahedron vertices, ΔT_t (K). We seek a solution of KGT model formulated for rapid dendritic/cellular growth as

$$v^2 \frac{\pi^2 \Gamma}{P^2 D^2} + v \frac{m_{\rm L} C_0 (k-1)\xi}{D[1-(1-k){\rm Iv}(P)]} + G = 0, \qquad (6)$$

$$\Delta T_t = \Delta T_c + \Delta T_r = m_{\rm L} C_0 \left(1 - \frac{1}{1-(1-k){\rm Iv}(P)}\right) + \frac{v\Gamma}{PD}. \qquad (7)$$

Here $P = \frac{Rv}{2D}$ is the dimensionless solutal Peclet number; R (m) is the dendrite tip radius; ${\rm Iv}(P) = P \cdot \exp(P) \cdot E(P)$ is the Ivantsov function; $E(P) = \int_P^\infty \frac{\exp(-x)}{x} dx$ is the exponential integral function; the function $\xi = 1 - \frac{2k}{\sqrt{1+(2\pi/P)^2} - 1 + 2k}$; G (K/m) is the temperature gradient; ΔT_c (K) and ΔT_r (K) are the solutal and curvature undercooling, respectively. Other variables are listed in table 1. We do not consider the contribution of thermal undercooling in equation 7, as it is related to latent heat release at the dendrite tip [17] and our temperature field is imposed from the exterior. Solving equations 6–7 for different growth rates, we add tabular data for v and ΔT_t as the input data for CAFD simulations. To eliminate the CA mesh-induced anisotropy, we employ the modified decentred octahedron algorithm [18].

2.4 Simulation parameters

Initially, the CA model is used to create a base plate with non-textured, equiaxed grains. For this purpose, the approach we proposed earlier [8] is employed. The average grain size in the base plate is determined from experimental observations and informs the number of nuclei to be distributed randomly in the base plate. The latter is assumed to solidify at a constant temperature, that is, all the nuclei grow at the same rate. PBF-LB process parameters mimic the experiment (Section 2.1). The parameters used in the KGT model are summarised in table 1. The temperature gradient considered in the KGT model was estimated as $G = 10^6$ K/m from the transient heat transfer results (Section 2.2).

Table 1. Parameters used in the KGT model [6]

Description	Symbol	Value	Unit
Gibbs-Thomson coefficient	Γ	$3.65 \cdot 10^{-7}$	K·m
Liquidus slope	m_L	-10.5	K/wt%
Initial composition	C_0	5.0	wt%
Partition coefficient	k	0.48	-
Diffusion coefficient	D	$3.0 \cdot 10^{-9}$	m^2/s

Figure 2. (a) Melt pool structure in the BD-TD cross-section of the PBF-LB printed IN718 part obtained using FE-SEM, with the melt pool boundaries, width and depth indicated in orange; (b) comparison of the width and depth of experimental and simulated melt pools

The heat source model was calibrated using experimental results from PBF-LB IN718, see figure 2. The calibration followed a standard approach, which involves measuring the melt pool dimensions in a BD-TD cross-section, as shown in figure 2(a), and adjusting the geometrical parameters a, b, and γ in the FD FTCS solution of equations 1–5 until the melt pool size calculated by the FD model matches the experimental data. The computer-aided design (CAD) model of the part was generated with the following dimensions $14.0 \times 5.2 \times 16.0$ mm³ and the hole centred in the part centre (7.0×8.0 mm²) with a diameter of 10 mm, as schematically shown in figure 3(a).

3. Results and Discussion

Figure 3 represents the CAD model of the virtually printed part (figure 3(a)), microstructure maps taken in the central BD-TD and horizontal SD-TD cross sections (figure 3(b)) as indicated on the CAD schematic, and an orientation map in inverse pole figure (IPF) colours taken in a part of the central BD-TD cross-section (figure 3(c)). Here, SD stands for scan direction. Large microstructure maps are defined by the misorientation angle. The latter is defined as an angle between the BD axis in the global coordinate system and the closest ⟨100⟩ direction of the crystal coordinate system. The grains observed in the central hole represent the unmelted powder that is omitted from consideration on the IPF-coloured orientation map. The ATEX software [19] was used to generate IPF-coloured orientation maps, aligning the crystallographic direction parallel to the BD, following the experiment. Misorientation-angle-based microstructure maps are plotted in Paraview [20]. The virtually printed part consisted of 400 layers and 18.2 bn cells. The calculation using the OpenMP-based in-house CAFD software took 12 days on a workstation with Intel® Core™ i7-10700F.

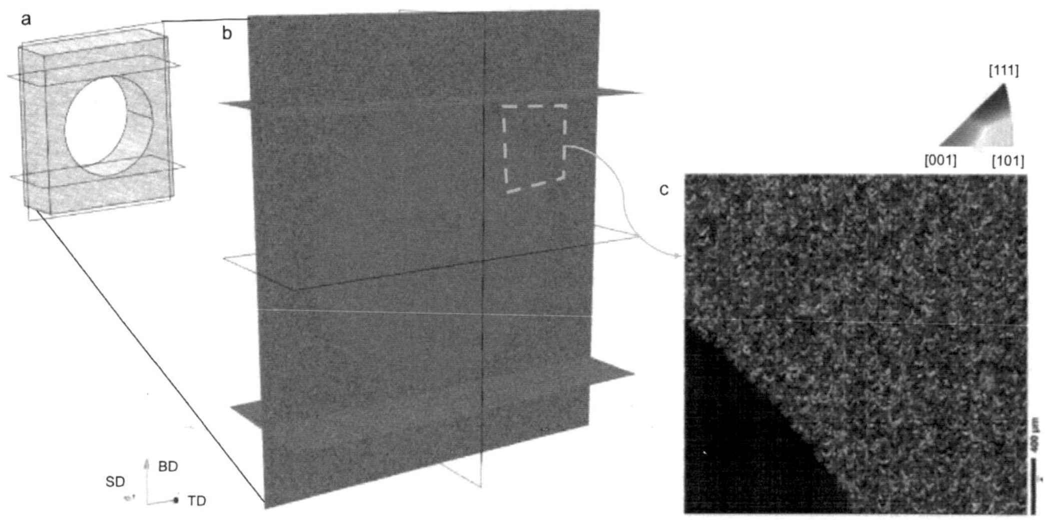

Figure 3. The virtually 3D printed IN718 part and its microstructure: (a) the CAD model; (b) microstructure maps taken in the central BD-TD cross section and the horizontal SD-TD cross sections as indicated on the CAD schematic (figure 3a); (c) the orientation map in inverse pole figure (IPF) colours taken in a part of the central BD-TD cross-section, see figure 3(b).

Figure 4 compares a part of the model-predicted central BD-TD cross-section (figure 4(b)) with the experimentally observed microstructure (figure 4(a)) as well as the respective {100} pole figures. The experimentally observed and model-predicted microstructures agree well. In both cases, thin columnar grains and relatively fine v-shaped grains of observably randomised orientations can be seen. Both types of grains grow epitaxially from the powder. Slender grains grow along the centreline of melt pools. Whereas v-shaped grains form by first growing to the left, to the top centre of the melt pool, normal to the liquidus isotherm, and with the next scan track growing to the right, to the top centre of the neighbouring melt pool. In doing so, they occupy occupying a part of each of the neighbouring melt pools. The mechanism of the v-shaped grain formation is discussed in detail in [8].

A weak texture is present, with the intensity maxima in {001}⟨100⟩ and {001}⟨110⟩. Similar microstructures, in terms of both grain morphology and crystallographic orientations, were reported for PBF-LB 316L steel in [21]. Note that the observed microstructure – weakly textured, characterised by rather fine grains – differs from a typical PBF-LB-induced microstructure characterised by large columnar grains and a strong texture. This is related to a rather shallow melt pool and relatively small hatch overlap, see figure 2(a). In [22], different laser powers and scan speeds were considered in PBF-LB manufacturing of IN718, and it was shown that higher volumetric energy densities lead to the formation of a pronounced crystallographic texture and columnar grain structure typically observed in 3D printed metals. For example, the same process parameters and two times higher laser power led to a single-crystal-like microstructure [22].

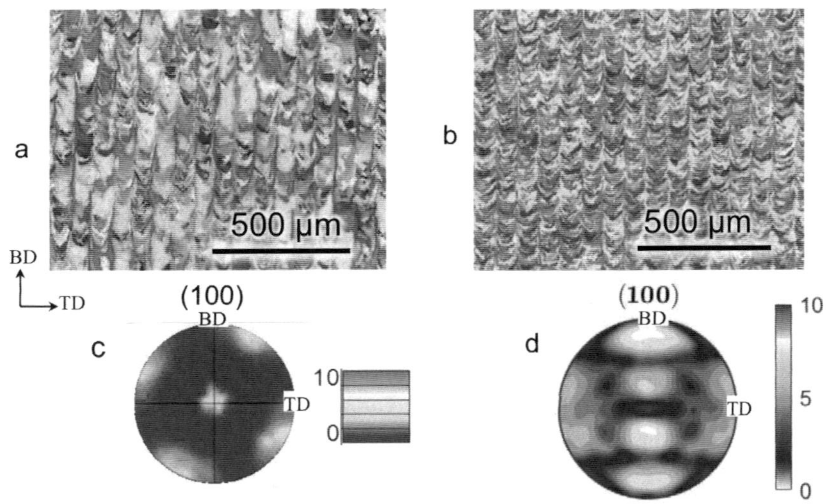

Figure 4. As-built (a) experimental and (b) model-predicted microstructures of PBF-LB-produced IN718 and (c, d) respective (100) pole figures.

4. Conclusions

This work employed the in-house CAFD code that uses OpenMP paradigm for parallelisation to simulate the microstructure of PBF-LB-printed metal part of more than 1,000 mm^3 in volume on a workstation. The model-predicted microstructure agrees well with the experimental data. The chosen process parameters yielding the relatively low volumetric energy density of 40 J/mm^3 resulted in a polycrystalline-like microstructure characterised by a weak texture and relatively fine grains. Two types of grains were observed, namely thin columnar grains growing along the melt pool centrelines and v-shaped grains spread within two hatches. The crystallographic orientations of grains were observably random, yet grains exhibited a weak degree of texture with the intensity maxima in $\langle 100 \rangle \parallel$ SD and in $\langle 110 \rangle \parallel$ BD. This is related to a relatively shallow melt pool and a relatively small overlap between neighbouring tracks.

Future workflow improvements include the integration of mechanical calculations to estimate the resulting mechanical behaviour and the steps for improved model accuracy. For instance, in the current state the model does not account for the macroscopic thermal variations due to build height and geometrical features that could lead to overheating regions. A steady-state moving thermal field concept based on the geometrical heat source model and transient heat conduction problem solution does not capture perturbations or complex melt pool shapes, including keyholing. However, implementing a more complex computational fluid dynamic model that could solve these issues is non-trivial as it would not compromise the existing framework's capacity for quick, computationally efficient, and large-scale predictions of MAM microstructures. Saying that, we believe that the model implemented offers significant capability for estimating microstructures in small parts with reasonable accuracy. It could serve as a powerful tool for digital manufacturing, enhancing the potential of BAM to tailor microstructures through process parameters.

Acknowledgements. OZ acknowledges the UNSW Canberra start-up grant [PS66347]. AZ and OZ contributed equally to this work.

References

[1] Hagihara K and Nakano T 2022 Control of anisotropic crystallographic texture in powder bed fusion additive manufacturing of metals and ceramics—a review *JOM* **74** 1760–73

[2] van Nuland T F W, van Dommelen J A W and Geers M G D 2021 An anisotropic Voronoi algorithm for generating polycrystalline microstructures with preferred growth directions *Comput Mater Sci* **186** 109947

[3] Zinovieva O, Romanova V and Balokhonov R 2023 Predictive simulation of microstructural pattern in additively manufactured metallic materials *Quality Analysis of Additively Manufactured Metals* ed J Kadkhodapour, S Schmauder and F Sajadi (Elsevier) pp 547–82

[4] Charles Murgau C, Pederson R and Lindgren L E 2012 A model for Ti6Al4V microstructure evolution for arbitrary temperature changes *Model Simul Mat Sci Eng* **20** 055006

[5] O'Toole P I, Patel M J, Tang C, Gunasegaram D, Murphy A B and Cole I S 2021 Multiscale simulation of rapid solidification of an aluminium–silicon alloy under additive manufacturing conditions *Addit Manuf* **48** 102353

[6] Nie P, Ojo O A and Li Z 2014 Numerical modeling of microstructure evolution during laser additive manufacturing of a nickel-based superalloy *Acta Mater* **77** 85–95

[7] Rodgers T M, Bishop J E and Madison J D 2018 Direct numerical simulation of mechanical response in synthetic additively manufactured microstructures *Model Simul Mat Sci Eng* **26** 055010

[8] Zinovieva O, Zinoviev A, Romanova V and Balokhonov R 2020 Three-dimensional analysis of grain structure and texture of additively manufactured 316L austenitic stainless steel *Addit Manuf* **36** 101521

[9] Koepf J A, Soldner D, Ramsperger M, Mergheim J, Markl M and Körner C 2019 Numerical microstructure prediction by a coupled finite element cellular automaton model for selective electron beam melting *Comput Mater Sci* **162** 148–55

[10] Bayat M, Zinovieva O, Ferrari F, Ayas C, Langelaar M, Spangenberg J, Salajeghe R, Poulios K, Mohanty S, Sigmund O and Hattel J 2023 Holistic computational design within additive manufacturing through topology optimization combined with multiphysics multi-scale materials and process modelling *Prog Mater Sci* **138** 101129

[11] Zafar M Q, Wang J, Zhang Z, Wu C, Zhao H, Hussain G and Ma N 2023 Thermomechanical process simulation and experimental verification for laser additive manufacturing of Inconel®718 *Materials* **16** 2595

[12] Le T N, Lo Y L and Lin Z H 2020 Numerical simulation and experimental validation of melting and solidification process in selective laser melting of IN718 alloy *Addit Manuf* **36** 101519

[13] Goldak J, Chakravarti A and Bibby M 1984 A new finite element model for welding heat sources *Metallurgical and Materials Transactions B* **15** 299–305

[14] Zinovieva O, Zinoviev A and Ploshikhin V 2019 Fast simulations of the grain structure evolution in additively manufactured materials *Proceedings of the Second International Conference on Simulation for Additive Manufacturing SIM-AM* (Pavia)

[15] Zinovieva O, Romanova V, Zinoviev A, Nekhorosheva O and Balokhonov R 2023 Elastic properties of additively manufactured steel produced with different scan strategies *Int J Mech Sci* **244** 108089

[16] Kurz W, Giovanola B and Trivedi R 1986 Theory of microstructural development during rapid solidification *Acta Metallurgica* **34** 823–30

[17] Kurz W and Fisher D J 1992 *Fundamentals of Solidification* (Trans Tech Publications)

[18] Wang W, Lee P D and McLean M 2003 A model of solidification microstructures in nickel-based superalloys: Predicting primary dendrite spacing selection *Acta Mater* **51** 2971–87

[19] Beausir B and Fundenberger J-J 2017 Analysis Tools for Electron and X-ray diffraction, ATEX - software

[20] Ahrens J, Geveci B and Law C 2005 ParaView: An End-User Tool for Large-Data Visualization *Visualization Handbook* (Elsevier) pp 717–31

[21] Bahshwan M, Myant C W, Reddyhoff T and Pham M S 2020 The role of microstructure on wear mechanisms and anisotropy of additively manufactured 316L stainless steel in dry sliding *Mater Des* **196** 109076

[22] Gokcekaya O, Ishimoto T, Hibino S, Yasutomi J, Narushima T and Nakano T 2021 Unique crystallographic texture formation in Inconel 718 by laser powder bed fusion and its effect on mechanical anisotropy *Acta Mater* **212** 116876

Understanding heterogeneous nucleation and predicting grain size in wire and arc additive manufacturing of steels with inoculation

Y Xu[1] , G. Tranell[2] and Y Li[2*]

[1] Department of Sustainable Energy Technology, SINTEF Industry, Trondheim, Norway
[2] Department of Materials Science and Engineering, Norwegian University of Science and Technology (NTNU), Trondheim, Norway

*E-mail: yanjun.li@ntnu.no

Abstract. Inoculation has become a common practice to achieve grain refinement in metal additive manufacturing, and its application in steels has been extensively investigated in recent years. However, the nucleation behavior of equiaxed grains under high temperature gradient and fast cooling rate solidification conditions in additive manufacturing is still not well understood. And grain size prediction models for additive manufacturing of steel is lacking. In this work, a numerical microstructure simulation model is developed to simulate the growth and nucleation behaviors of grains in additive manufacturing, and to predict the grain size of inoculated steels with equiaxed grain structure. Here a wire and arc additive manufacturing (WAAM) of duplex stainless steel is selected for case study. The simulation results are in a good agreement with the experimental results.

1. Introduction

Metal additive manufacturing (AM) has achieved significant development in the past decade and has become one of the most exciting technologies in advanced metal manufacturing [1, 2]. Due to its special process involving partial remelting of previous layer and subsequent epitaxial growth from the melt pool boundary, it is quite often to form columnar grains. Such columnar grain structure could cause hot tearing (cracking), anisotropy of mechanical properties etc. Therefore, numerous research has been carried out to solve these issues, and among them, inoculation by particles or certain alloying elements are widely investigated and applied now in AM.

In additive manufacturing of steels, one of the most common elements that has been added to steels is Ti [3-5]. And TiN could be formed in the melt, which helps to promote heterogeneous nucleation of equiaxed grains [6]. Micro/nano carbide and boride particles are also found to have the grain refinement effect. AlMangour et al. [7] added TiB_2 in 316L stainless steel during selective laser melting. Rodrigues et al. added SiC particles to the molten pool during wire and arc additive manufacturing (WAAM) of a high strength low alloy steel to achieve grain refinement. Recently, TiC/TiB_2 nano-inoculants have introduced in PH 13-8Mo martensitic stainless steel during WAAM [8]. Despite some successful addition of particles for grain refinement, there is still challenges to obtain fully equiaxed grains in AM of steels. One of the reasons could be due to the superhigh temperature gradient during solidification in AM. Deep understanding of the nucleation of new grains under such solidification conditions has been limited.

There are a few transient models based on cellular automaton and phase field [9] approaches to simulate the epitaxial growth and bulk nucleation in AM. Li et al.[10] simulated the solidification behavior during direct energy deposition of 304 steel, and revealed the influence of nucleation number density on the final grain structure. Yang et al. [11, 12] simulated grain evolutions with 3D phase field model in a process scale for multi-layer powder bed fusion of 316L stainless steel refined by TiB_2 nanoparticles and gave a qualitative reproduction of experimental results. As discussed by the authors, many simplifications are made, especially the liquid to solid transformation is just controlled by the temperature field without considering any undercooling and growth kinetics.

In this work, a transient growth model for grain growth is developed to simulate the growth behavior of a single grain in AM and the corresponding undercooling in the liquid. Non-equilibrium effects are taken into account. Together with a nucleation model, a multi-grains model is further extended to predict the grain size in AM of inoculated steels.

2. Model description

The present numerical model for grain size prediction in WAAM, is based on a further development of the grain sized prediction model used for normal directional solidification of aluminum alloys [13]. A schematic drawing of the 2D cross section of solidification domain used in the model is shown in figure 1. In the model the nucleation and growth of grains are modelled in 3D. Constant cooling rate and temperature gradient is applied in such local domains, while the local solidification parameters of WAAM process were determined by finite element simulation. The interfacial non-equilibrium solidification effect on the dendrite tip growth is taken into account by implementing growth rate dependent kinetic partition coefficient k_v and kinetic liquidus slope m_v. The temperature gradient effect on the tip radius of dendrite was also included.

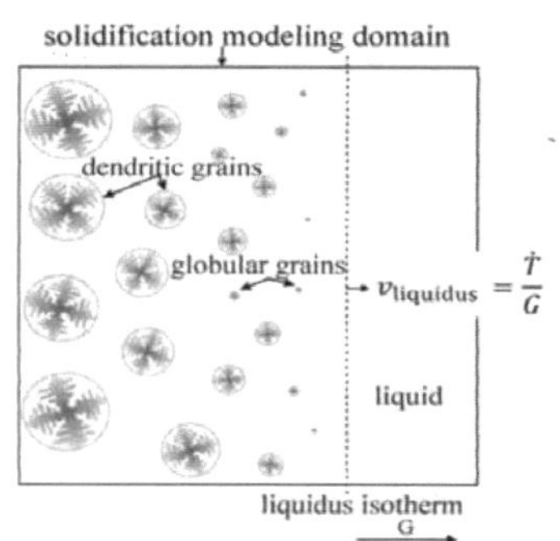

Figure 1. Schematic drawing of the 2D cross section of solidification modelling domain along the temperature gradient G direction. Nucleation and grain growth propagates from the left colder side towards the right hotter side.

The heterogeneous nucleation of grains on inoculant particles is based on the free growth criterion. The governing equations to determine the free growth undercooling and local undercooling experienced by inoculant particles, the size of inhibited nucleation zone around growing grains, and grain growth velocity were described in a previous model, and could be found in [13]. Here only some of the new equations are described.

During rapid solidification, the solute partition coefficient k_v is dependent of the moving velocity of solid/liquid interface V, and is estimated here based on [14]:

$$k_v = \frac{k + a_0 V/D_i}{1 + a_0 V/D_i} \tag{1}$$

where k is the equilibrium partition coefficient (could be assumed as constant or temperature dependent), a_0 is a characteristic length scale for solute trapping (of the order of interatomic distance), V is the growth velocity. D_i is the solute interface diffusion coefficient, which is often approximated to be the diffusion coefficient in the liquid D_l. And the velocity dependent liquidus slope m_v is calculated by the following equation [15]:

$$m_v = m\frac{1 - k_v[1 - \ln\left(\frac{k_v}{k}\right)]}{1 - k} \tag{2}$$

where m is the equilibrium liquidus slope.

For the growing grain, the total undercooling at the growth tip (solid-liquid interface) ΔT is consisted of [16, 17]:

$$\Delta T = T_{l\_ne} - T^* = \Delta T_s + \Delta T_r + \Delta T_k \tag{3}$$

where, T^* is the interface temperature. ΔT_s is the solute undercooling, ΔT_r is the curvature undercooling and ΔT_k is the kinetic undercooling for atom attachment.

For the spherical grain growth, the growth velocity is nearly the same as normal solidification condition, however, for dendritic growth, the temperature gradient will have a direct effect on the tip radius and growth rate. Here, the marginal stability criteria and Ivantsov solution combining with the tip undercooling description are used to calculate the growth velocity, tip radius and tip concentration in the liquid.

Based on marginal stability criteria, the tip radius R_{tip} is [15]:

$$R_{tip} = [\frac{\Gamma}{\sigma^*(m_v G_c \xi_c - G)}]^{0.5} \tag{4}$$

Γ is the Gibbs-Thomson coefficient and σ^* is a stability constant given by $1/4\pi^2$, G is the temperature gradient. G_c is the solute concentration gradient in the liquid at the interface to solid phase, and ξ_c is a stability parameter.

For parabolic tip of dendrites, the relation between supersaturation Ω and solute Peclet number P_c is described by Ivantsov solution [15]:

$$\Omega = Iv(P_c), \tag{5}$$

$$\Omega = \frac{C_l^* - C_0}{C_l^*(1-k)} \tag{6}$$

Iv is the Ivantsov function [16] $Iv(P_s) = P_c * exp(P_c)E1(P_c)$, and for computational convenience, the inverse Ivantsov function proposed by Wang and Berkmann [18] has been used:

$$P_c = 0.4567\left(\frac{\Omega}{1-\Omega}\right)^{1.195} \tag{7}$$

And P_c is the Peclet number for solute diffusion at the dendrite tip:

$$P_c = \frac{R_{tip}V}{2D_l} \tag{8}$$

Local undercooling in the liquid ahead of the growing solid-liquid interface is determined by the local equilibrium liquidus temperature and local melt temperature distribution. The solute concentration distribution is taken from reference [19].

3. Results and discussion

WAAM of DSS from reference [5] was selected for case study. Ti was added in the wire to refine the grain size of ferrite. The chemical compositions of the deposition metal of the base wire are shown in table 1. Different amount of Ti was added to the wire, including DSS- 0.01Ti, DSS-0.05Ti, DSS-0.1Ti, DSS-0.5Ti and DSS-1.0Ti (wt.%). The average voltage and current of the arc for the deposition process is 30 V and 145 A, respectively. And the deposition speed is 5.4 mm/s.

Table 1. Chemical composition of the deposition DSS metal of the base wire [5].

Element	C	Si	Mn	Cr	Ni	Mo	N	Fe
Concentration (wt.%)	0.07	0.43	1.52	29.89	6.54	3.47	0.30	Ba.

Finite element modelling of WAAM process using COMOSL Multiphysics 6.1 was done to get the cooling rate and temperature gradient at different locations in the melt pool. Part of the results are shown in figure 2, where the left figure shows the temperature field in the cross section parallel to moving direction of arc, and the right figure shows the temperature history curves at the location 1 and location 2 marked in Figure 2(a). From the temperature curves during cooling stage close to the liquidus temperature of the alloy, the cooling rate and temperature gradient at location 1 are determined, and they are 340 K/s, and 187.6 K/mm. These are used as main solidification parameters in the microstructure model. According to the thermodynamic calculation, the primary phase forming during the solidification is ferrite. The steel is simplified as a binary Fe-12Ni (wt.%) alloy which has the same equivalent grain growth restriction factor as the experimental alloy. The thermodynamic and physical parameters of this steel [20] are listed in table 2. Besides, using the concentration of N and Ti, the volume faction of TiN particles is calculated by Thermo-Calc as 1.92 vol.% for the alloy with 1wt.% Ti.

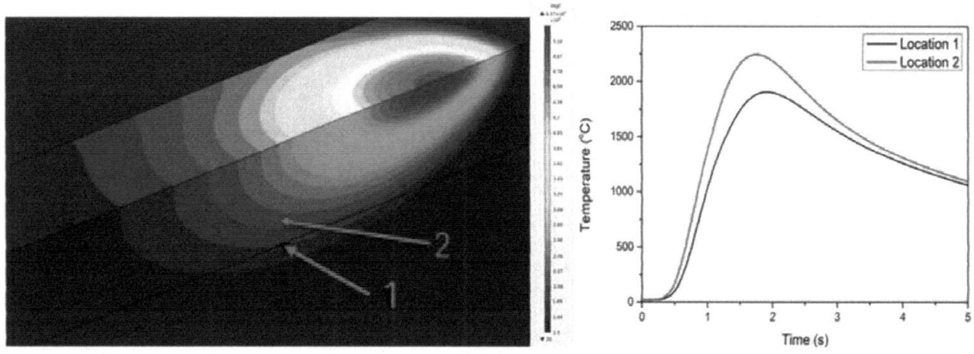

Figure 2. Temperature field simulated by the COMSOL model at the cross section along the AM direction and the temperature curves at location 1 and location 2 in the melt pool.

Table 2. Physical and thermodynamic parameters used in the microstructure model for the DSS steel.

Quantity	Symbol	Units	Value
Gibbs-Thomson coefficient	Γ	m K	2.56×10^{-7}
Pre-exponentiel coefficient for Ni diffusion	D_0	$m^2 s^{-1}$	4.92×10^{-7}
Activation energy for diffusion of Ni	Q	kJ/mol	67.7
Equilibrium Liquidus slope	m	K /wt.%	-5.5

First, a one-grain model was used to simulate the dendritic growth of one grain after nucleation, where no other potential heterogeneous nucleation site was considered. Figure 3 shows the growth velocity evolution of the single grain, and the propagation of the growth front and the liquidus isotherm, as a function of solidification time for the steel during WAAM. Here the interaction with other grains that may nucleate in front of the growth front is neglected. As can be seen, the growth velocity increases from very small value and quickly reaches the steady state growth after 0.08 s, with the growth velocity equals to 1.8 mm/s. This growth velocity is exactly the same as the moving velocity of liquidus isotherm. After that, the growth velocity of dendrite tip will not change. From figure 3(b), we can see that in the initial growth stage, the distance between the growth front and the liquidus isotherm is very small. As the solidification time increases, the distance becomes larger. When the grain growth reaches the steady stage, the growth front has propagated 90 μm, and the liquidus isotherm is 50 μm ahead of the growth front.

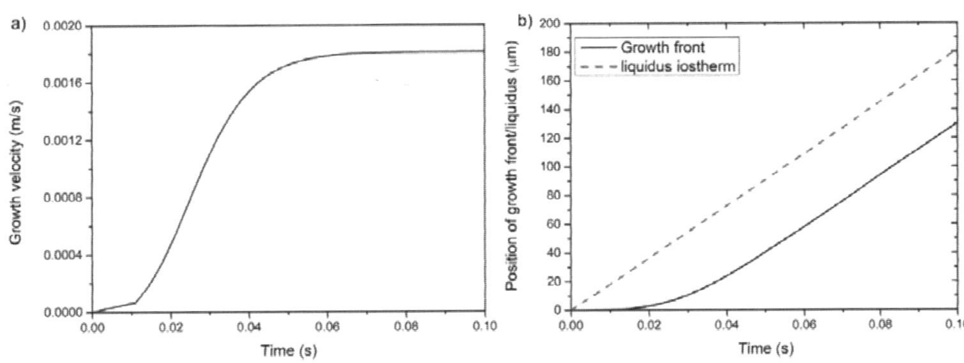

Figure 3. (a) Growth velocity evolution, (b) the propagation of the growth front and the liquidus isotherm, of a single grain as a function of solidification time for DSS steel during WAAM ($G=1.88*10^5$ K/m, $\dot{T}$ =340 K/s).

Figure 4 shows the equilibrium liquidus temperature and local temperature profile in the liquid ahead of the growth front (solid-liquid interface) at 2 different solidification time (t=0.012 s, and 0.1 s), where the gap between them is the local effective undercooling in the liquid availabe for nucleation of new grains, which is further shown in figure 4(c). In the beginning of the dendritic growth, the maximum effective undercooling in the liquid is only 1 K. As time increases and grain growth continues, the tip temperature further decreases and the local undercooling in the liquid increases. At t>=0.1 s, the maximum undercooling in the liquid is around 8.8 K and the length of the undercooled zone is 51 μm. Under this condition, based on the criteria proposed by Kurz et al. [21], that the particle spacing l_{ps} should fulfill $l_{ps} < 0.6l_{uz}$, where l_{uz} is the length of undercooled zone, the minimum amount of inoculation particles necessary to promote equiaxed grains formation is 3.49 $*10^{12}$ m^{-3}, if the critical nucleation undercooling of these particles are below 8.8 K. An estimation based on the TiN volume fraction gives the particle number density around 10^{15}-10^{16} for 0.5 and 1 wt.% Ti. Therefore, it is possible to form equiaxed grains in those samples. However, if less particle are added, then columnar grains could be the dominant solidification structure. And if no extra particles are added in the alloy, the grain structure is fully columnar in the final AM component.

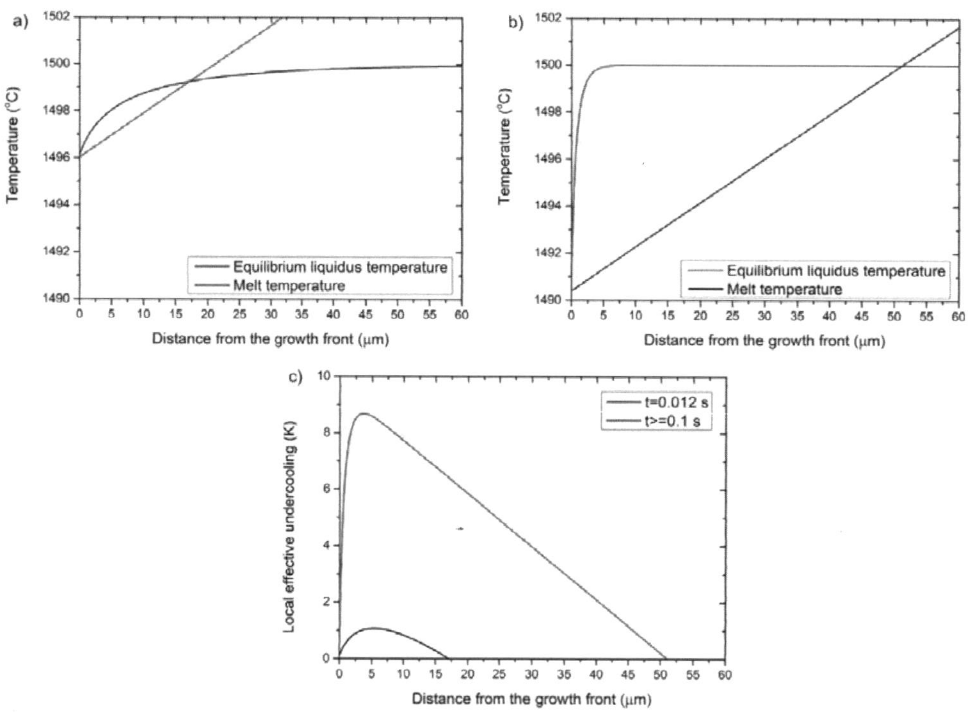

Figure 4. Equilibrium liquidus temperature and local temperature profile in the liquid ahead of the growth front (solid-liquid interface), at 2 different solidification time a) 0.012s b) 0.1 s, c) the local effective undercooling in the liquid, for DSS steel during WAAM, ($G=1.88*10^5$ K/m, $\dot{T}$ =340 K/s).

By the single grain model, the primary dendrite arm spacing is in the range of 16.8-47.3 μm, which is close to the experimental alloy with a low Ti content with a columnar dendrite structure. In the high Ti-content alloy, a smaller primary dendrite arm spacing can be expected due to the epitaxial growth from previous layer with equiaxed grain structure and nucleation of more new grains.

For the equiaxed grain solidification during WAAM of th experimental alloys with high Ti contents, grain size has been simulated by using the multi-grains model where the heterogeneous nucleation of grains on TiN particles formed in liquid metal are considered. The results are shown in figure 5. For 1 wt.% Ti addition, where the grain structure is fully equiaxed in the experiment, the predicted grain size by the present model is in good agreement with the experimental results.

However, for 0.5 wt.% Ti alloy, according to the theoretical number density of TiN particles, the predicted grain size is slightly smaller than the measured one. The reason could be due to the uncertainty of TiN particles that form during cooling and solidification process. Some dedicated experimental characterization of the TiN particles would help to clarify the reasons behind. In addition, further simulation of experimental cases using more stable inoculant particles which does not dissolve in the melt or form during cooling would be interesting. However, such experimental data is still lacking.

In addition, we can see that the cooling rate in WAAM process is still not very high in comparison to the powder bed fusion process. In the WAAM process, the cooling rate and temperature gradient is not enough to trigger rapid solidification far from equilibrium. It would be interesting to further test the current model in the powder bed fusion type AM of steels.

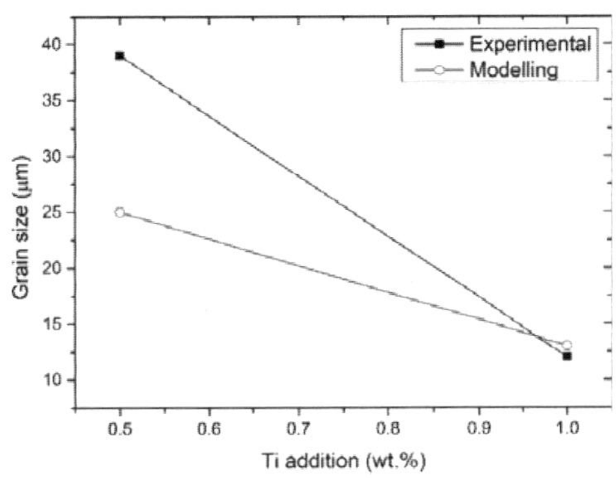

Figure 5. Grain size prediction for DSS steel added with different amount of Ti, in comparison to the experimental results reported in reference [5].

4. Conclusion

A solidification microstructure model considering the effects of cooling rate, temperature gradient, non-equilibrium solidification kinetics etc. has been developed to simulate the grain structure development in the WAAM of steels. Based on the single grain growth model, the model could predict the solidification grain structure in terms of columnar or equiaxed structure under different addition amount of inoculation particles. While the multi-grains model could predict the grain size of the inoculated samples showing equiaxed grain structure. Besides, the primary dendrite arm spacing of columnar grains could be simulated for the non-inoculated alloy. The modelling results are in a good agreement with the experimental results.

Acknowledgements

This research work has received funding from the European Institute of Innovation and Technology (EIT), a body of the European Union, under the Horizon 2020, the EU framework Programme for Research and Innovation, via Smart WAAM project. It has been co-funded by the SFI PhysMet, (Centre for Research-based Innovation, 309584). The authors gratefully acknowledge the financial support from the Research Council of Norway and the partners of the SFI PhysMet.

IOP Conf. Series: Materials Science and Engineering **1310** (2024) 012006 doi:10.1088/1757-899X/1310/1/012006

References

[1] Sames W J, List F A, Pannala S, Dehoff R R and Babu S S 2016 The metallurgy and processing science of metal additive manufacturing *International Materials Reviews* **61** 315-60
[2] Mukherjee T, Elmer J W, Wei H L, Lienert T J, Zhang W, Kou S and DebRoy T 2023 Control of grain structure, phases, and defects in additive manufacturing of high-performance metallic components *Progress in Materials Science* **138** 101153
[3] Durga A, Pettersson N H, Malladi S B A, Chen Z, Guo S, Nyborg L and Lindwall G 2021 Grain refinement in additively manufactured ferritic stainless steel by in situ inoculation using pre-alloyed powder *Scripta Materialia* **194** 113690
[4] Zhai W, Zhou W and Nai S M L 2022 Grain refinement of 316L stainless steel through in-situ alloying with Ti in additive manufacturing *Materials Science and Engineering: A* **840** 142912
[5] Wu S, Zhang Y, Xue M and Cheng F 2024 Grain refinement of flux-cored wire arc additively manufactured duplex stainless steel through in-situ alloying of Ti *Materials Characterization* **209** 113749
[6] Tan Q, Chang H, Lindwall G, Li E, Durga A, Liang G, Yin Y, Wang G and Zhang M-X 2024 Unravelling the roles of TiN-nanoparticle inoculant in additively manufactured 316 stainless steel *Journal of Materials Science & Technology* **175** 153-69
[7] AlMangour B, Kim Y-K, Grzesiak D and Lee K-A 2019 Novel TiB2-reinforced 316L stainless steel nanocomposites with excellent room- and high-temperature yield strength developed by additive manufacturing *Composites Part B: Engineering* **156** 51-63
[8] Ghaffari M, Vahedi Nemani A, Shakerin S, Mohammadi M and Nasiri A 2023 Grain refinement and strengthening of PH 13-8Mo martensitic stainless steel through TiC/TiB2 inoculation during wire arc additive manufacturing *Materialia* **28** 101721
[9] Azizi H, Ebrahimi A, Ofori-Opoku N, Greenwood M, Provatas N and Mohammadi M 2021 Characterizing the microstructural effect of build direction during solidification of laser-powder bed fusion of Al-Si alloys in the dilute limit: A phase-field study *Acta Mater.* **214** 116983
[10] Li X and Tan W 2018 Numerical investigation of effects of nucleation mechanisms on grain structure in metal additive manufacturing *Comput. Mater. Sci.* **153** 159-69
[11] Yang M, Wang L and Yan W 2021 Phase-field modeling of grain evolutions in additive manufacturing from nucleation, growth, to coarsening *npj Computational Materials* **7** 56
[12] Yang M, Wang L and Yan W 2021 Phase-field modeling of grain evolution in additive manufacturing with addition of reinforcing particles *Additive Manufacturing* **47** 102286
[13] Xu Y, Casari D, Mathiesen R H and Li Y 2018 Revealing the heterogeneous nucleation behavior of equiaxed grains of inoculated Al alloys during directional solidification *Acta Mater.* **149** 312-25
[14] Aziz M J 1982 Model for solute redistribution during rapid solidification *J. Appl. Phys.* **53** 1158-68
[15] Kurz W, Giovanola B and Trivedi R 1986 Theory of microstructural development during rapid solidification *Acta Metallurgica* **34** 823-30
[16] Kurz W and Fisher D J 1992 *Fundamentals of solidification*: Trans Tech Publications.
[17] Gäumann M, Trivedi R and Kurz W 1997 Nucleation ahead of the advancing interface in directional solidification *Materials Science and Engineering: A* **226-228** 763-9
[18] Wang C Y and Beckermann C 1994 Prediction of Columnar to Equiaxed Transition during Diffusion-Controlled Dendritic Alloy Solidification *Metall and Mat Trans A* **25** 1081-93
[19] Gäumann M, Trivedi R and Kurz W 1997 Nucleation ahead of the advancing interface in directional solidification *Materials Science and Engineering: A* **226–228** 763-9
[20] Mizukami H, Suzuki T, Umeda T and Kurz W 1993 Initial stage of rapid solidification of 18-8 stainless steel *Materials Science and Engineering: A* **173** 361-4
[21] Kurz W, Bezençon C and Gäumann M 2001 Columnar to equiaxed transition in solidification processing *Science and Technology of Advanced Materials* **2** 185

IOP Conf. Series: Materials Science and Engineering **1310** (2024) 012007
doi:10.1088/1757-899X/1310/1/012007

Influence of print-chamber oxygen content on the microstructure and properties of 3D-printed 316L

WY Wang, W Liu and A Godfrey*

Laboratory of Advanced Materials (MOE), School of Materials Science and Engineering, Tsinghua University, Beijing, 100084, China

*E-mail: awgodfrey@mail.tsinghua.edu.cn

Abstract. Samples of 316L stainless steel have been prepared using laser-powder bed fusion from the same batch of powder using different print-chamber oxygen levels, ranging from 50 ppm to 1500 ppm. The oxide particle density is found to increase with oxygen content, while the cell structure is invariant with oxygen level and the grain size shows a relatively sharp transition for measured oxygen levels of above 450 ppm. Based on the microstructural observations it is suggested that the increasing oxygen levels leads to a transition in the solidification pattern. Samples printed at the higher oxygen level show higher strength and lower mechanical anisotropy than samples with a coarser grains structure printed at lower oxygen levels. The main influence of the higher oxide particle content on thermal stability is on the kinetics of recrystallization during isothermal annealing at 1000 °C.

1. Introduction

Metal additive manufacturing by laser-powder bed fusion (L-PBF) is of increasing interest for industrial applications [1,2]. The L-PBF process is characterized by rapid cooling during the layer-by-layer build process, which leads to some unique and complex microstructures in as-printed metal samples that are associated with enhanced mechanical properties [3,4]. Understanding the microstructural evolution during L-PBF, and using this information to achieve microstructures with optimal mechanical properties is therefore a topic of current interest. One specific problem in this regard is how to reduce, or even eliminate, the microstructural anisotropy associated with the grain structure in as-printed metal samples [5-7]. During rapid solidification in the L-PBF process, grains tend to grow epitaxially from grains in the previous layer along the direction of maximum temperature gradient, resulting in a columnar grain morphology from the bottom to top of the sample when observed from the side view (i.e., any cross section containing the build direction). Such microstructural anisotropy causes anisotropy in mechanical properties, resulting in a lower yield strength (and higher uniform elongation) for tension loading parallel to build direction (BD) compared to loading perpendicular to the BD [8]. This anisotropy is in general undesirable and restricts the wider application of L-PBF for component manufacture.

One of the ways to reduce microstructural anisotropy in additively manufactured metals is via grain refinement, i.e., by transforming the large columnar grain morphology to a fine equiaxial grain morphology [9,10]. Grain refinement in L-PBF can be achieved by several ways, including addition of particles in the metal powder, such as TiC particles [11], to create melt pool nuclei, or by modifying the composition to include solutes (e.g., Ti) that increase the growth restriction

IOP Conf. Series: Materials Science and Engineering **1310** (2024) 012007 doi:10.1088/1757-899X/1310/1/012007

factor, and thus promote constitutional supercooling, thereby leading to a finer solidification structure [12]. These methods all require the addition of extrinsic particles or solute elements, which may cause problems in steels regarding the austenite-to-ferrite transition, or more widely in the formation of intermetallic compounds leading to loss of ductility and toughness. According to many previous studies, as-printed L-PBF 316L stainless steel samples contain a significant number of oxide particles, which are believed to be formed in-situ in the melt pool due to the reaction between alloying elements (mainly Si and Mn) in 316L SS powder and free chamber oxygen [13-15]. Furthermore, differences in oxygen content can also affect melt pool dynamics during the L-PBF process [16], which may also influence the solidification microstructure.

To explore this further in this work, 316L SS samples have been prepared from a single batch of powder under different oxygen chamber content, ranging from 50 ppm to 1500 ppm, with all other L-PBF parameters kept the same, with the resulting microstructures characterized over several length scales and compared with mechanical properties and thermal stability.

Table 1. Composition of the 316L powder used in this investigation.

Element	C	Si	Mn	P	S	Cr	Ni	Mo	N	Fe
Weight %	0.015	0.06	0.72	0.009	0.004	17.70	12.60	2.38	0.10	bal.

2. Experimental methods

Samples for this investigation were prepared using a SLM 280 instrument (Nikon SLM Solutions). All samples were prepared from the same batch of sample powder, prepared by gas atomization with 78% of powder particle sizes in the range from 10 to 45 μm. The nominal composition of the initial powder (supplied by the powder manufacturer) is given in table 1. Prior to this experimental study a processing map was used to determine a set of optimal printing parameters, based on measurements of sample density. Accordingly, all specimens in this study were prepared using L-PBF printing conditions of 180W laser power and scanning speed of 600 mm s^{-1} (corresponding to a sample density of >99.9%). Samples were printed using a 67° rotation between layers with a nominal layer thickness of 30 μm using Ar as the shielding gas. The oxygen level was measured using a sensor located at the top of the print chamber. The oxygen level was varied by adjustment of a vacuum pressure release valve, and values are quoted in this study to the nearest 50 ppm (based on the typical observed fluctuations in oxygen level over the sample print duration). The oxygen sensor reading was allowed to stabilize before starting each print run using different settings of the pressure release valve, typically taking 1 – 2 hours.

Samples were printed as rectangular blocks from which smaller samples were cut using electro-discharge machining for microstructural investigations. For observations in the scanning electron microscope (SEM; TESCAN MIRA 3 LMH) samples were mechanically ground then electro-polished using a 10% perchloric acid and 90% alcohol solution. Orientation mapping in the SEM was carried out using an Oxford Instruments electron back-scatter diffraction (EBSD) system equipped with a CMOS-based detector operated at 20 kV. Thin-foils for observation in the transmission electron microscope were prepared by mechanical slicing and griding, followed by mechanical polishing and twin-jet polishing in an electrolyte of 10% perchloric acid and 90% alcohol. For mechanical testing dog-bone shape tensile samples were cut from the bulk sample by electro-discharge machining with gauge length of 5 mm, and thickness (width) of 1.8 (1) mm. Tensile tests were carried out until failure using an AG-X tensile machine under constant displacement rate of 0.5mm min^{-1} (corresponding to an initial strain rate of about 1.6×10^{-3} s^{-1}).

3. Results and discussion

3.1 Microstructural observations

Figure 1 shows the microstructure in a cross-section containing the build direction for samples printed with measured oxygen chamber levels of 150 ppm (figure 1a,b) and 1500 ppm (figure 1 c,d) under SEM observation using back-scatter electron imaging optimized for orientation contrast. At lower magnification a finer scale microstructure is clearly visible in the samples printed under an oxygen content of 1500 ppm. At higher magnification the cell structure that is typical for L-PBF-printed 316L is also visible, seen as features that are either elongated or equiaxed depending on the relationship between the sectioning plane and the (columnar) cells. Interestingly for both samples, printed under low (150 ppm) and high (1500 ppm) oxygen levels the length scale of the cell structure is similar, with a mean diameter (estimated from locations where the cells as seen as equiaxed features) of around 0.5 μm. Both samples also contain a large number of fine particles - seen either as bright white features (where the particles are still present) or black features (where the particles have been incidentally removed during samples preparation).

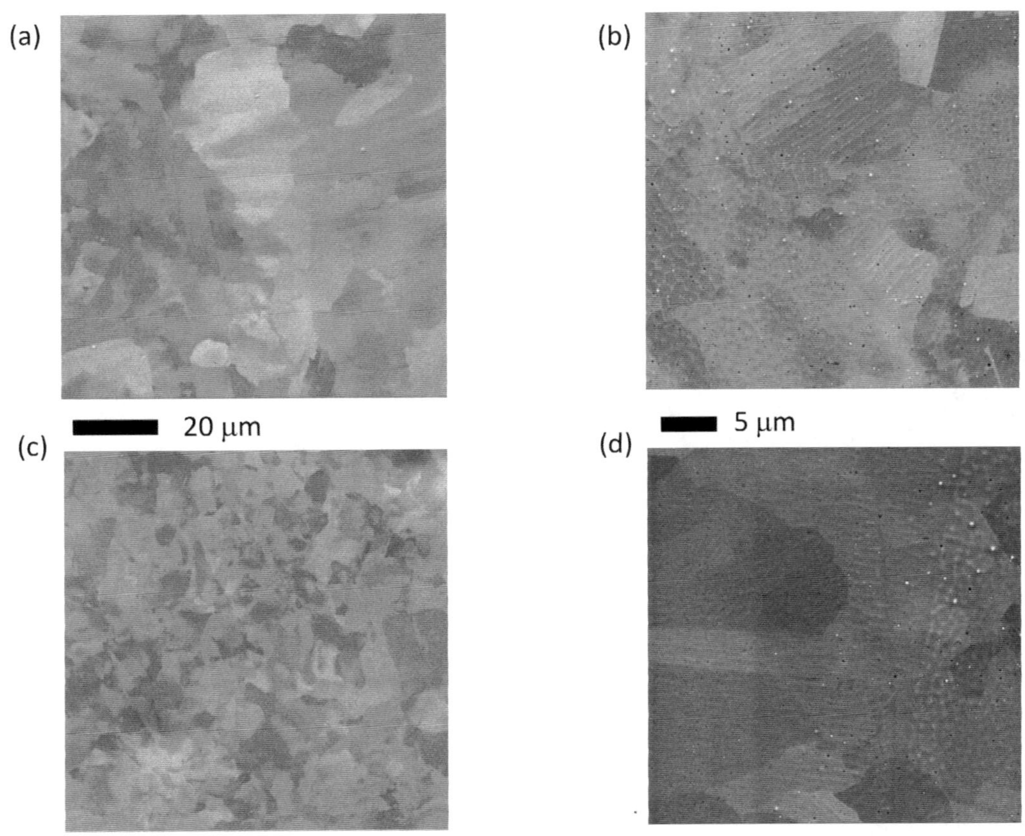

Figure 1. SEM micrographs showing the grain structure (a,b) and cell structure (c,d) of samples printed under 150 ppm oxygen (a,c) and 1500 ppm oxygen (b,d). Oxide particles can also be seen in the higher magnification images as black or white speckles.

The differences in microstructure between the samples printed under low (150 ppm) and high (1500 ppm) oxygen levels are revealed more clearly in orientation maps constructed from the EBSD data, as seen in figure 2, which show the microstructure of the samples as viewed both in plane parallel to the build direction (in the plane with the build direction as the surface normal) and in a plane containing the build direction (in this case with the build direction parallel to the page vertical axis). In both cases an inverse pole-figure (IPF) coloring is used such corresponding to the build direction. Accordingly, it can be readily seen that for the low oxygen case a strong fiber texture is present, with a large fraction of the material volume with the build direction parallel to <110>. For the sample printed under high oxygen level conditions a preference for a <110> fiber texture is also seen, though to a significantly lesser extent compared to the low oxygen case. In both viewing section the microstructural scale is much finer for the sample printed under high oxygen level conditions, so much so that the melt pools are only weakly visible in the IPF orientation maps.

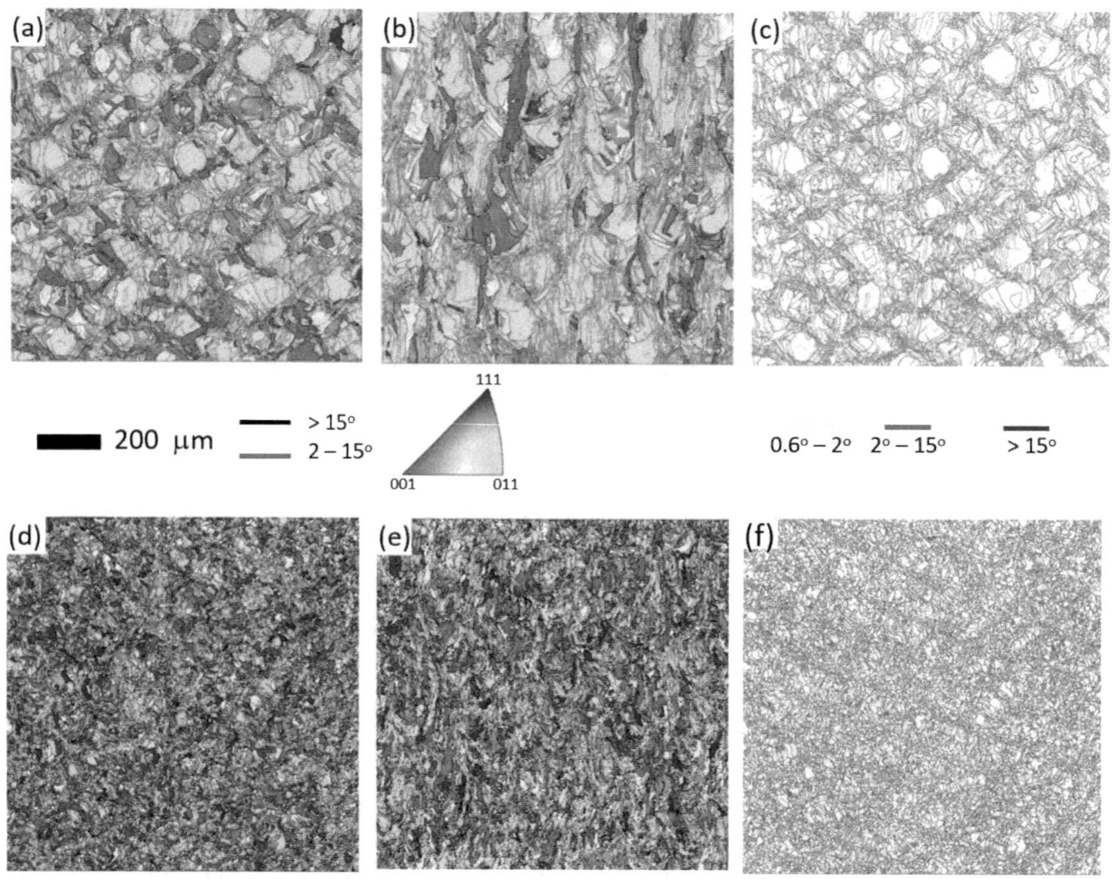

Figure 2. EBSD maps for samples printed under oxygen levels of 150 ppm (a-c) and 1500 ppm (d-f); (a, d) IPF coloring (build direction) viewed along the build direction, (b,e) IPF coloring (build direction) in a cross-section plane containing the build direction; (c,f) misorientation maps corresponding to (a,c) showing misorientations as indicated in the legend.

Figure 2c and 2f are boundary misorientation maps (corresponding to the IPF maps of figure 2b and 2e respectively) where misorientations 0.6° – 2, 2° –15° and >15° are plotted in yellow, red and blue, respectively. Bands of fine grains delineated by high angle boundaries (>15°) corresponding to the laser scan path in the sectioned plane are seen in both cases, though these bands are considerably wider (occupying a larger volume fraction of the material) for the samples printed under high oxygen level conditions. Also in the low oxygen cases there are several regions

IOP Conf. Series: Materials Science and Engineering **1310** (2024) 012007 doi:10.1088/1757-899X/1310/1/012007

with only a low area-per-volume density of very low angle misorientations (0.6° to 2°; plotted as yellow lines in figures 2c and 2f), whereas for the high oxygen level case very few such regions are found. The EBSD observations indicate a much more disrupted microstructure in terms of both high and low angle boundaries, though still heterogenous on the scale of the hatch spacing.

The large difference in microstructure between the high and low oxygen level conditions could in theory have its origin either in the influence of oxide particles on grain coarsening, or via some change in the solidification conditions. For the former it should be expected that increasing the oxygen level from 150 ppm to 1500 ppm should lead to a continuous change in the microstructure, whereas a change in solidification conditions should be expected to lead to a sharper transition in microstructure with increasing oxygen level. Figure 3 shows the variation of grain size, measured both in the build plane and in the cross-section plane, with increasing oxygen level. Here the grain size is given as the equal circle diameter, measured based on a grain reconstruction method using a definition of 15°, where the graph shows both the average value and the spread in values. The results very clearly show two microstructural regimes, with a sharp transition between chamber oxygen levels of 450 ppm and 550 ppm, suggesting that the difference may arise due to a change in solidification mechanism. It can be noted also here that microstructures resulting from 3D metal printing can only be imperfectly quantified by a single grain size parameter, on account of both the complex 3D grain shapes and the fact that the grain size distribution is typically not unimodal. Nevertheless, the data in figure 3 clearly show a transition in the microstructure above a certain oxygen chamber content. It should be noted also that Ar has a higher density than oxygen, and the bleed value is at the located above the print chamber. A further possible explanation for the observations is therefore that a minimum opening of the pressure release value is required before full mixing of the gasses in the chamber takes place. This possibility is currently being investigated by measurements of the oxygen content in all printed samples using glow discharge optical emission spectroscopy (GD-OES) to see if this also shows a sharp transition or smooth variation.

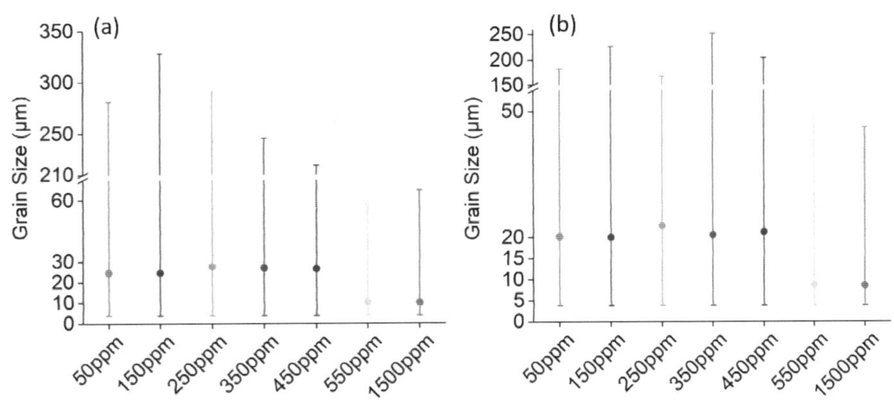

Figure 3. Graphs showing the variation of grain size (equal circle diameter using a 15° criterion) as a function of chamber oxygen content from measurements in (a) the plane with normal parallel to the build direction, and (b) in a cross-section plane containing the build direction (error bars show max. and min. values)

Figure 4a,b shows TEM micrographs showing the fine scale oxide particles in samples printed under chamber oxygen levels of 150 ppm and 1500 ppm, respectively. At the higher oxygen level the oxides have a finer average size (34 nm vs. 8 nm) and a higher number density (1.2×10^{21} m^{-3} vs. 1.3×10^{22} m^{-3}). It can also be noted here that the oxygen sensor is located in the chamber, but

IOP Conf. Series: Materials Science and Engineering **1310** (2024) 012007 doi:10.1088/1757-899X/1310/1/012007

above the build plate, so the actual oxygen levels at the sample may differ to those measured on the sensor. The GD-OES measurements currently underway will provide complementary information on the oxygen pick up in bulk volumes for each sample. Note that for these TEM micrographs observations were made in an under-focused to show the oxide particles more clearly. Nevertheless, the images also show that the dislocation cell structure in both samples is of a similar length scale (in agreement with the SEM observations in figure 1). Many of the oxide particles are located at the dislocation cell boundaries, indicating some degree of interaction between these features. The similar length scale of the cell structure for samples printed under low and high oxygen conditions suggests however that the oxide particles do not have an influence on the formation of the solidification cell structure.

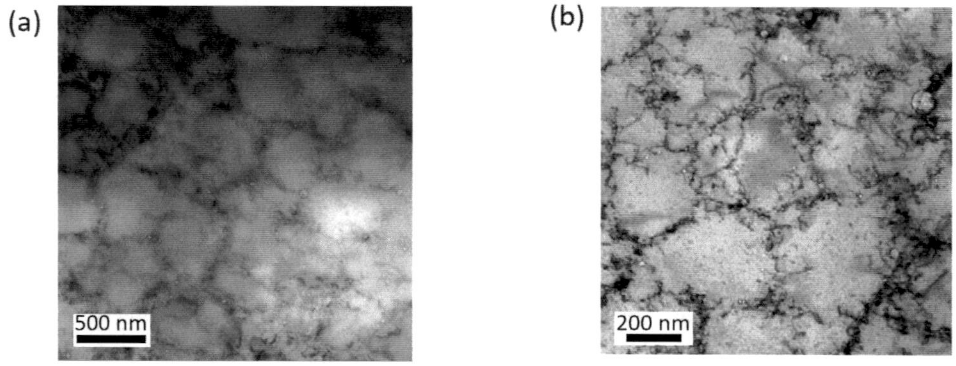

Figure 4. TEM micrographs showing nanoscale oxide particles and the dislocation cell structure in samples printed under chamber oxygen levels of (a) 150 ppm and (b) 1500 ppm.

3.2 Tensile properties

Tensile flow curves for samples printed under different oxygen chamber levels, from 50 ppm to 1500 ppm, are plotted in figure 5. For each oxygen level tensile tests were carried out with the tensile axis taken parallel to the build direction (||BD) and with the tensile axis perpendicular to the build direction (⊥BD). In common with previous observations on metal samples prepared by L-PBF for a range of metal systems, the samples show a clear anisotropy in mechanical properties in terms of both the yield strength and stress and work hardening rate, with testing perpendicular to the build direction showing both higher yield strength and higher work-hardening rates for most of the flow curve. In terms of the influence of oxygen print level on mechanical properties the data very clearly fall into two groups - with samples printed under oxygen levels of 550 ppm and above showing strength values of approximately 100 MPa greater at all strain levels to samples printed under oxygen levels of 450 ppm and below. Although the samples printed at the higher oxygen levels still show obvious mechanical anisortopy, the magnitude of this anisotropy is lower than in the samples printed at lower oxygen contents and with a coarse scale microstructure.

Within each of these two groups, however, the flow curves are almost independent of the oxygen print level. This suggests, combined with the microstructural observations, that the tensile properties, at least in terms of strength, are primarily correlated with the scale of the microstructure (which also shows a transition in for an oxygen print level between 450 ppm and 550 ppm). This is somewhat unexpected, as in previous studies on L-PBF 316L an additive model, including a contribution from precipitates/dispersoids, has been found to provide a good description of the mechanical strength [14]. It can be noted, however, that these earlier studies only cover a limited range of grain sizes.

IOP Conf. Series: Materials Science and Engineering **1310** (2024) 012007 doi:10.1088/1757-899X/1310/1/012007

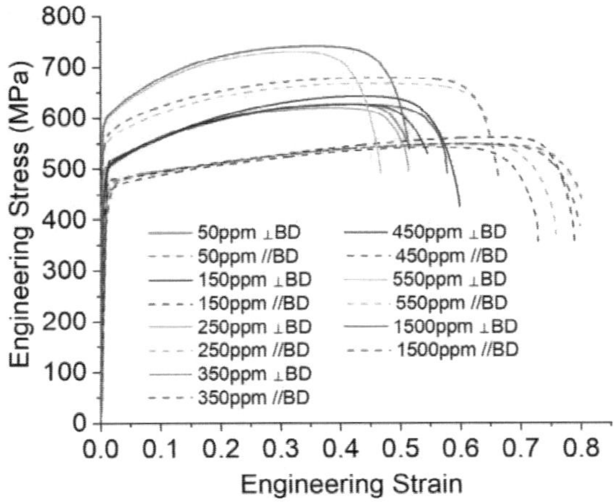

Figure 5. Tensile flow curves for samples printed under a range of chamber oxygen levels from 50 ppm to 1500 ppm, tested with the tensile direction either parallel to the build direction (⊥BD) or perpendicular to the build direction (||BD). Samples printed at oxygen levels of 550 ppm and 1500 ppm show higher strength levels and smaller mechanical anisotropy than samples printed at lower oxygen levels.

3.3 Influence of oxygen print content on thermal stability

The influence of oxygen level on thermal stability of the microstructure in the 3D-printed samples, specifically the influence on the onset and progress of recrystallization, has been explored by isothermal annealing of samples at 1000 °C for durations of 1 to 6 hours. The results are shown in figure 6, showing the microstructural evoluation during annealing for the cases of 150 ppm and 1500 ppm oxygen. Recrystallization results in the removal of volumes containing low angle misorientations (shown in the EBSD maps as yellow lines) so that the area density of such misorientations can be used as a proxy to track the recrystallized fraction during isothermal annealing [13]. As such it can be seen that that recrystalliztion proceeds more quickly in the sample printed under an oxygen content of 150 ppm than in the sample printed under an oxygen content of 1500 ppm. Interestingly, however, the microstructure after full recrystallization is similar in both cases, indicating that the main effect of the higher oxide particle density is on the kinetics of the recrystallization process.

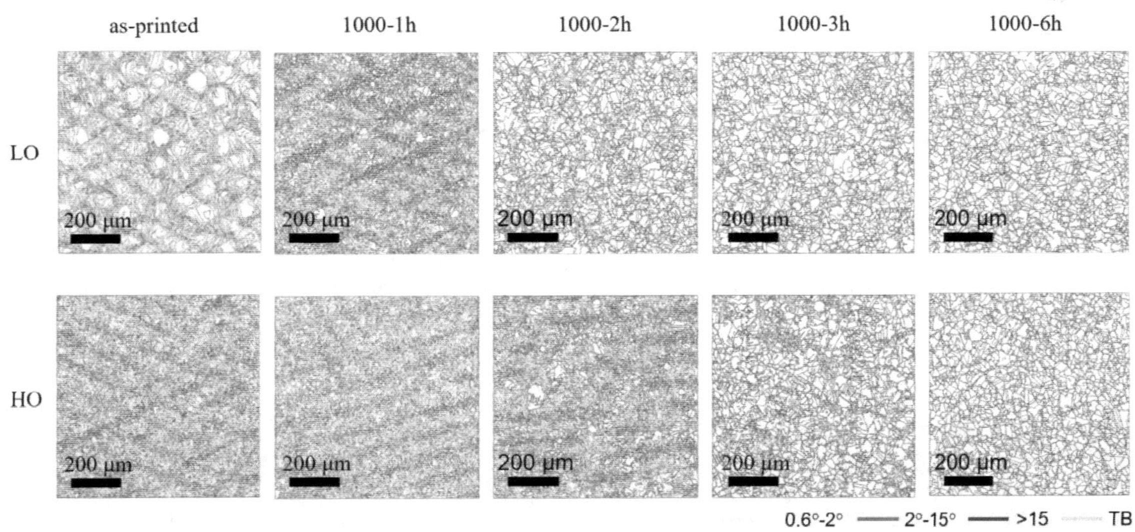

Figure 6. EBSD misorientation maps tracking the evolution of microstructure during isothermal annealing at 1000 °C of samples printed at chamber oxygen levels of 150 ppm (LO = "low oxygen") and 1500 ppm (HO = "high oxygen").

4. Summary

The influence of oxygen chamber content on the microstructure, and resulting mechanical properties, of 316L produced by L-PBF has been investigated. Greater levels of oxygen content result in an increase in the number density of oxide particles in the as-printed samples. All samples show a typical cellular structure within each soldified grain, where the scale of this cell structure is independent of oxygen content. A relatively sharp transition in grain size is observed for oxygen chamber contents of above approx. 450 ppm, suggesting that oxygen levels above this value trigger a change in the solidification pattern, rather than influencing the grain size via a Zener pinning effect. Samples printed with a finer grain size printed at high oxygen chamber content show both greater strength and lower mechanical anisitropy than samples printed at lower oxygen levels. The tensile strength is, however, not directly related to the oxide content, suggesting that the cell structure and grain size are the controlling factors in determining the meachanical properties of theses samples.

Acknowledgements

This research was funded by the National Natural Science Foundation of China (project number 51971117).

References

[1] Herzog D, Seyda V, Wycisk E and Emmelmann C 2016 *Acta Mater.* **117** 371–92
[2] Bajaj P, Hariharan A, Kini A, Kürnsteiner P, Raabe D and Jägle E A 2020 *Mater. Sci. Eng. A* **772** 138633
[3] Wang Y M et al. 2018 *Nat. Mater.* 2018 **17** 63–71
[4] Liu L, Ding Q, Zhong Y, Zou J, Wu J, Chiu, Y-L, Li J, Zhang Z, Yu Q and Shen Z 2018 *Materials Today* **21** 354–61
[5] Yang D, Yin Y, Kan X, Zhao Y, Zhao Z and Sun J 2021 *Mater. Res. Express* **8** 096510
[6] Motaman S A H and Kibaroglu D 2024 *J. Mater. Sci. Technol.* **181** 240–56
[7] Kok Y, Tan X P, Wang P, Nai M L S, Loh N H, Liu E and Tor S B 2018 *Mater. Design* **139** 565–86
[8] Kong D, Ni X, Dong C, Zhang L, Man C, Cheng X and Li X 2019 *Mater. Lett.* **235** 1–5
[9] Chandra S, Wang C, Tor S B, Ramamurty U and Tan X 2024 *Nat. Commun.* **15** 3094
[10] Zhang J, Liu Y, Bayat M, Tan Q, Yin Y, Fan Z, Liu S, Hattel J H, Dargusch M and Zhang M-X 2021 *Scr. Mater.* **191** 155–60
[11] Li W, Meng L, Wang S, Zhang H, Niu X and Lu H 2023 *Mater. Sci. Eng. A* **884** 145557
[12] Zhai W, Zhou W and Nai S M L 2022 *Mater. Sci. Eng. A* **840** 142912
[13] Wang W-Y, Godfrey A and Liu W 2023 *Metals* **13** 1062
[14] Saeidi K, Kvetková L, Lofaj F and Shen Z 2015 *RSC Advances* **5** 20747–50
[15] Deng P, Karadge M, Rebak R B, Gupta V K, Prorok B C and Lou X 2020 *Addit. Manuf.* **35** 101334
[16] Chia H Y, Wang L and Yan W 2023 *Acta Mater.* **249** 18824

Optical Microscopy on Machined Surfaces in Ti-6Al-4V Reveals Pore Statistics that Correlate with Fatigue

J Choi[1*], K Zhou[1], E Adcock[1], J Pauza[1], A Ngo[2], T Swierzewski[3], J Lewandowski[2], and A D Rollett[1]

[1] Materials Science & Engineering, Carnegie Mellon University, Pittsburgh, PA, 15213, USA
[2] Materials Science & Engineering, Case Western Reserve University, Cleveland, OH, 44106, USA
[3]Mechanical Engineering, Carnegie Mellon University, Pittsburgh, PA, 15213, USA

*E-mail: jechoi@andrew.cmu.edu

Abstract. Optical microscopy on the machined surfaces of 4-point bend fatigue samples 3D printed in Ti-6Al-4V is used to characterize pores larger than ≈15 microns. A large enough number of distinct scan speeds and fatigue tests that span the process window provides a statistically useful dataset. The variation in stress at which runouts (to 10 million cycles) are observed is anti-correlated with the pore density and the same inverse relationship exists for the maximum pore size observed within the area exposed to maximum tensile stress in each test. Total fatigue life is also anti-correlated with maximum defect size albeit with considerable scatter.

1. Introduction

Additive manufacturing (AM) of metals is rapidly growing due to increased design freedom, part reliability, and complex structure manufacturing and reduced waste, production time, and cost, compared to traditional manufacturing processes. Among AM processes, laser powder bed fusion (L-PBF) is currently the dominant technology. It uses high-power and focused laser beam to selectively melt a spread layer of powder, so it is also known as Selective Laser Melting (SLM), and it builds parts layer-by-layer using a computer-aided-design (CAD) model. It provides the most versatility and a wide range of applications, as it has the capacity to fabricate geometrically complex parts from a CAD that are applicable to customized patient-specific biomedical applications[1] and topology optimization in aerospace applications[2]. As for the material, one of the most common alloys manufactured by L-PBF is Ti-6Al-4V (Ti64), an alpha-beta titanium alloy known for its high strength and low density, high corrosion resistance, heat treatability, and excellent biocompatibility. Therefore, it is of highest interest in medical, aviation, and power/energy industry[3,4] and widely used in implants and prostheses[5]. For these reasons, we explore the defect content of L-PBF printed Ti64 as found on the machined surfaces of test bars.

Process-induced defects in metal AM are the main initiation point for failure under cycling loading, assuming that surface roughness removal via machining is a routine procedure in industry for safety critical components. Defects in metal AM are reported to be the cause of the poor performance and data scatter[6-11]. They reduce fatigue performance compared to the defect-free condition[12-14] and cause an order of magnitude[9] or even greater dispersion in the stress-life (S-N) data and largely overwhelm the microstructural effects in the AM as-printed parts[15]. One of the main challenges to the widespread use of metal AM to produce safety critical structural components has been widely recognized as the issue of material properties and repeatability. As a result, industry takes a relatively conservative stance in the acceptance of L-PBF materials and printed components, which slows the widespread adoption of this advanced manufacturing technology for innovations. Several studies have been conducted to overcome the engineering uncertainty or understand the underlying failure mechanism for L-PBF parts. Various types of defects are typically noted as being present in L-PBF metals, namely, lack-of-fusion (LoF) porosity, keyhole (KH) porosity, balling, residual porosity from powder[16], and gas porosity.

Product qualities for L-PBF metal AM builds are governed by various processing parameters, where the most commonly considered are beam power (P), scan velocity (V), hatch spacing (H), layer thickness (L), laser spot size, and scan strategy. Optimizing these parameters is crucial for qualities such as density, minimization of defects, enhancement of build rate, and acceptable material properties. Specifically, factors such as beam power (P) and scan velocity (V) can minimize defects of the printed material[17-19] since altering local melting and fusion process of the powder leads to variable porous structures as presented in figure 1[20]. We have successfully studied power-velocity (P-V) mapping with respect to LoF, KH, and Process Window porosities at power-speed combinations within the balling boundary. 4-point bend (4PB) fatigue tests were performed on a large number of samples (≈1,000), where four samples were typically fabricated for each P-V combination. In almost all cases, the fatigue crack initiated from a pore exposed on the surface subject to tensile stress.

Both LoF and KH porosity are commonly found in L-PBF that are particularly deleterious to mechanical behavior and define an effective "process window" between them, meaning printed to almost full density, > 99.9 % volumetric density, although large defects are still possible. LoF porosity occurs when the melt pools do not overlap enough[21] leaving some regions un-melted from insufficient energy density, so their pores are typically elongated with sharp corners containing un-melted powder particles. Also, LoF pores are generated around spatter particles

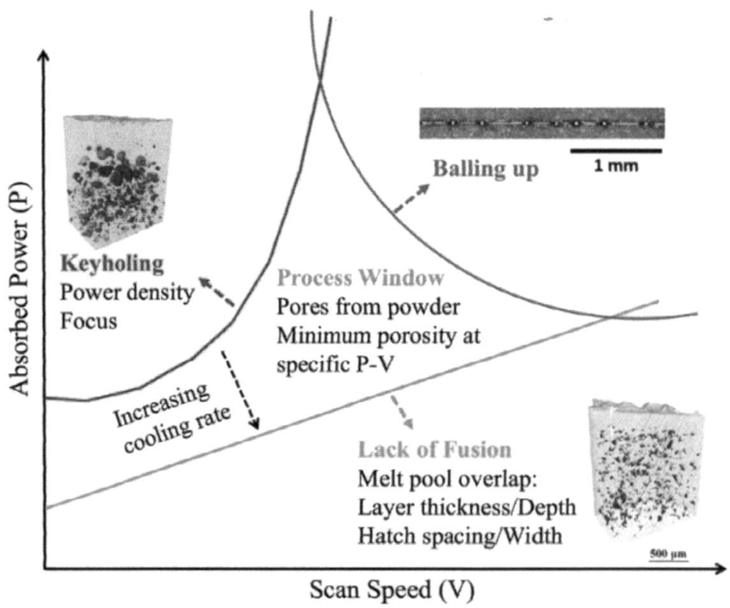

Figure 1. Schematic process window encircled by keyhole porosity (top left), lack of fusion porosity (bottom right) and bead-up (top right) in power-velocity (P-V) space. Figure from Narra et al.[20].

IOP Conf. Series: Materials Science and Engineering **1310** (2024) 012008 doi:10.1088/1757-899X/1310/1/012008

that have landed on the layer surface especially when a large spatter is not fully melted[22]. KH pores are formed when there is a deep and narrow vapor cavity created by strong recoil pressure from rapid evaporation of metal, which pushes the surrounding melt liquid downward[23.24]. At high power density (power divided spot area) and low scan speed, this leads to instabilities which leave pinched-off spherical pores trapped in the re-solidified metal. The process window, which is demarcated by KH, LoF, and bead-up porosity boundaries, where both spherical and irregular shaped pores were present[25].

Our objective is to establish a robust correlation between the porosity of the LPBF-printed materials and their mechanical properties, which then enables evaluation of part quality based on the print parameters used. Computed tomography (CT) is effective in revealing porosity throughout a sample, but its resolution limit depends on sample thickness and the characteristics of the X-ray source. Certain LoF pores may be hard to detect because they are too thin to give enough contrast despite being larger in extent than the nominal detection limit. Serial sectioning has been shown to be more capable of finding "sliver" defects[26] than CT, i.e., an optical microscope, can have higher resolution. Particularly in the case where fatigue cracks originate from defects exposed at the surface which is almost invariably the case for the 4PB fatigue tests analyzed here, then we present evidence that optical microscopy is a reliable guide to porosity levels.

Statistical-approaches have been used effectively for correlating fatigue life with the initial distribution of porosity obtained. Extreme Value Statistics (EVS) methods estimate the largest expected defect in a larger volume from the upper tail of the defect size distribution for a limited number of samples. Murakami developed an EVS method to predict the maximum inclusion size in a large volume of steel by adopting block maxima sampling and interpolating the data with Gumbel distribution[27]. Shi et al. later reported a variation of EVS based on the Generalized Pareto Distribution (GPD) to carry out this prediction[28]. In this paper, we use our own statistical approach that is based on a histogram of pore counts binned by equivalent spherical diameter (ESD) to predict the likely maximum pore size in 4PB fatigue tests and show its correlation to fatigue life.

Classically, Murakami in the 1980s demonstrated defects close to the surface are more prone to initiate crack growth than defects further from the surface[29]. In accordance with this approach, we focus on surface porosity. However, he defines the area of those surface defects from which a crack emanates to be onto a plane perpendicular to the maximum principal stress[30]. We instead analyzed surface pore areas on all four planes of each bar where (only) one face was subject to tensile stress: this provides a pore size distribution that is expected to approach that for the bulk material. Reddy et al.[31] used a combination of CT and conventional cross-sectional microscopy to correlate bulk porosity to the same set of fatigue results. They noted that the machined bars exhibit minimal spatial variation in porosity, thanks to machining off approximately 500 microns. In addition, we made a separate analysis of only those surface pores within the area that experienced the maximum tensile stress, which is about 10 × 5 mm, i.e., much smaller than the full extent of the surface but corresponds to the region from which the crack actually originates. This is quite different from standard approaches where the porosity in the entire sample is considered which then necessitates sampling of the larger distribution to obtain a maximum expected defect size.

In summary, to address the above-mentioned challenges, we investigated the effects of surface pore characteristics of LPBF-printed Ti64 parts with varying the P-V space parameters after surface roughness removal by machining on (total) fatigue life using 4PB fatigue tests. To present the surface pore characteristics, the statistics of the surface pore density, size (equivalent spherical diameter, ESD) and shape are systematically analyzed. We emphasize that this paper investigates porosity on the surfaces of the machined samples as opposed to, say, CT of bulk material or metallographic characterization of cross-sections.

2. Materials and Methods

2.1 Power-velocity (P-V) mapping

P-V map is visual snapshot of the LPBF processing regime, plotted at constant hatch spacing (H), layer thickness (L), and laser spot size (figure 2). The LoF region was calculated using the Rosenthal-based melt pool equations and the LoF offset criterion by assuming semi-elliptical melt pool geometry and counting its validity only at lower energy density values[32,33]. The calculations for the KH threshold were based on the normalized enthalpy values[34], along with information from the experimental study by Cunningham et al.[16].

2.2 Sample printing and postprocessing

The gas atomized Ti64 extra-low interstitial (ELI) powder for fabrication had a particle size from 24 μm (D10) to 61 μm (D90) with nominal composition of the powder listed in Table 1 manufactured by ATI Powder Metals.

Table 1: Chemical composition of Ti64 ELI powder in wt.%

Al	V	Fe (max)	O (max)	C (mac)	N (max)	H (max)	Ti
5.50 – 6.50	3.50 – 4.50	0.25	0.08	0.05	0.01	0.01	Balance

The fatigue test samples with 6 mm × 6 mm × 73 mm were fabricated on an EOS M290 L-PBF machine equipped with a ytterbium fiber laser capable of 400 W maximum power and a nominal laser spot diameter of 100 μm at Carnegie Mellon University by holding a constant hatch spacing of H = 140 μm, layer thickness of L = 30 μm, preheat of 180 °C, and a 67° scan rotation between layers. 73 mm is the length parallel to the build direction. These conditions are maintained while varying speed from 800 mm/sec to 1950 mm/sec in increments of 50 mm/sec at constant power of 370 W (figure 2), meaning total 24 P-V combinations, *aka* "CMU" groups that have moats. It also shows the process window is the widest where the CMU group #7 (370 W and 1100 mm/sec) lies on the KH boundary and the CMU group #22 (370 W and 1850 mm/sec) lies on the LoF boundary. All the samples were then stress relieved at 593 °C for 2 hours ± 0.25 and furnace cooled in air to eliminate a large fraction of the residual stress as per AMS 2801 standard and recommendation of ULI project industry partners[20,35]. They are removed from build plate by wire electrical discharge machining.

Each P-V combination or group has 4 samples to ensure repeatability. The *moat* represents a 1 mm thin wall which was added 5 mm from the sample edge to minimize local temperature build up by having additional time at the turnaround. Overheating has the effect of narrowing the process window by shifting the operating window lower towards the LoF region[36]. Therefore, as the moat guards against overheating, it also guards against KH porosity. Note that no evidence for gross keyhole porosity was found in any of the samples examined here.

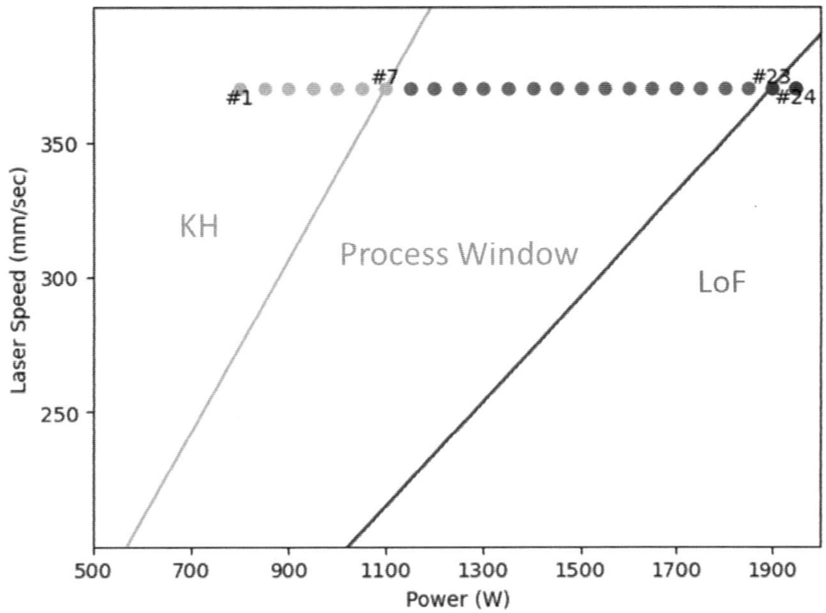

Figure 2. Power-velocity (P-V) space into KH porosity, process window, and LoF porosity regimes for CMU groups that were fabricated with moats on an EOS M290 L-PBF machine. Porosity boundaries are denoted by two lines, which are colored orange and blue. P-V combinations are presented as solid circles with their group numbers assigned.

2.3 Sample machining

To remove surface roughness, all samples were machined to the identical surface finishing as per ASTM E466-15 (2021) standard, where fatigue bars had small removals until 125 μm left, then 50 μm per pass for flats with a final 25 μm with longitudinal polishing to $R_a < 0.2$ μm. Overall, 0.5 mm was removed from the as-built 6 mm × 6 mm × 73 mm rectangular bars to eliminate any surface porosity induced by the contouring, where the machining alignment was such that all grinding marks were parallel to the length of the bar to minimize the tendency to initiate cracks from those marks.

2.4 Sample characterization

Optical microscope (Zeiss Axio Imager.Z2 Vario) was used to characterize the metal surface via brightfield reflection imaging. All images were taken at the same magnification (20X). We characterized all four surface areas of a square bar, meaning these are machined plane surfaces that contain both the stress axis and the build direction (colored grey in figure 3). The segmented areas are then converted into ESD using the equation[37] where particle size characteristics are described in the SIO 9276-6:2008[38], along with aspect ratio.

$$ESD = \sqrt{\frac{4 \cdot A}{\pi}} \qquad \text{Equation 1}$$

$$aspect\ ratio, or \frac{B}{L} = \frac{x_{Fe\ min}}{x_{Fe\ max}} \qquad \text{Equation 2}$$

Table 2: List of processing parameters used to fabricate Ti64 fatigue test samples.

Group Number	Power, P (W)	Velocity, V (mm/sec)	Expected Regime	LOF Index	N1 = ESD at 1 pore/255.5 mm^2 for all four surfaces	Max. ESD (μm) at F3, the fatigue surface	Max. ESD (μm) at 4PB fatigue test areas (red boxes) inside F3
#1	370	800	KH	0.38	54	80.97	29.88 and 60.02
#2	370	850	KH	0.40	-	-	-
#3	370	900	KH	0.42	22.8	36.37	20.49 and 28.88
#4	370	950	KH	0.44	-	-	-
#5	370	1000	KH	0.47	22.5	30.14	24.61 and 30.14
#6	370	1050	KH	0.49	-	-	-
#7	370	1100	KH	0.52	-	-	-
#8	370	1150	PW	0.54	-	-	-
#9	370	1200	PW	0.57	17	27.37	6.52 and N/A
#10	370	1250	PW	0.59	-	-	-
#11	370	1300	PW	0.62	-	-	-
#12	370	1350	PW	0.65	18.7	25.57	15.63 and 13.24
#13	370	1400	PW	0.68	-	-	-
#14	370	1450	PW	0.71	-	-	-
#15	370	1500	PW	0.74	30	33.16	11.05 and N/A
#16	370	1550	PW	0.78	-	-	-
#17	370	1600	PW	0.81	22.5	34.88	21.87 and 34.88
#18	370	1650	PW	0.85	-	-	-
#19	370	1700	PW	0.88	-	-	-
#20	370	1750	PW	0.92	30	33.8	15.97 and 16.94
#21	370	1800	PW	0.96	25	37.67	19.15 and 37.38
#22	370	1850	PW	1.01	27.5	32.84	17.40 and 16.78
#23	370	1900	LoF	1.05	-	-	-
#24	370	1950	LoF	1.10	41	47.35	31.69 and 17.10

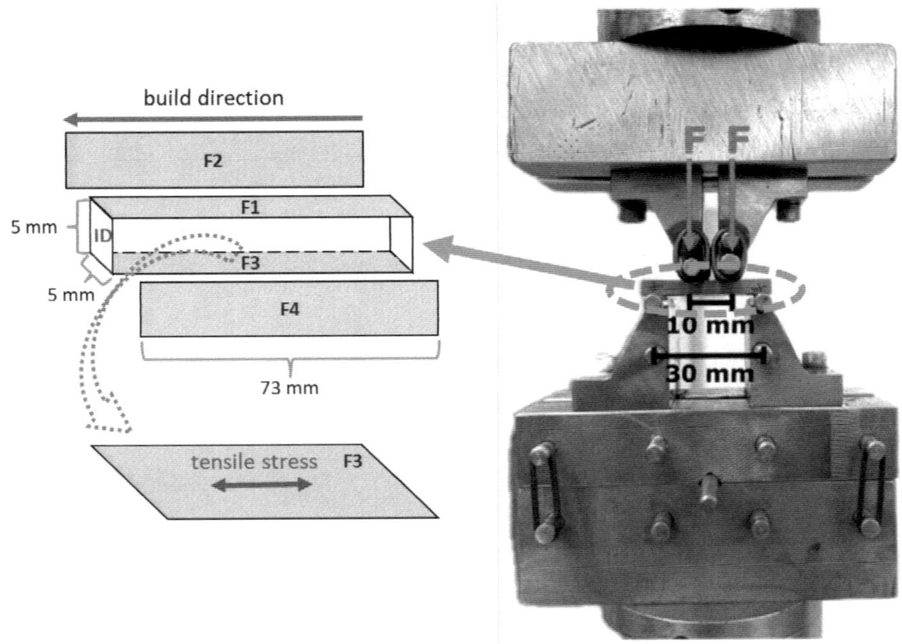

Figure 3. Four surface areas of a bar are characterized via brightfield reflection imaging by using Zeiss optical microscope. These four machined plane surfaces contain the stress axis, which is parallel to the build direction. The bars were consistently aligned with F3 facing down such that this face experienced tensile stress, based upon figure reproduced from Reddy et al.[31]. F4 faces forward and F2 faces backward while F1 faces up.

2.5 Stress-controlled Fatigue Testing

4PB fatigue tests were performed to study effect of porosity content on fatigue performance, especially of a process window, using MT servo-hydraulic machines at Case Western Reserve University, where R = 0.1, i.e., wholly tensile stress, and frequency was fixed at 20 Hz to maintain contact between the fixture and the sample. Before the tests, bars were chamfered at 45° at each edge to reduce the stress concentration from sharp corners. Two tests were conducted per bar as illustrated in figure 4, in which each test exerts a uniform maximum tensile stress on an area of 5 mm × 10 mm during the loading cycle.

In our 4PB fatigue tests, the tensile stress (σ) in MPa at the fatigue surface (F3) in figure 3 is derived from beam theory equation[39]:

$$\sigma_{tensile} = \frac{PL}{bd^2} \qquad \text{Equation 3}$$

Figure 4. 4PB fatigue test specimen geometry and testing setup. Each sample was long enough that two tests were run on each of two sections (the red boxes that are denoted as test #1 and test #2) over F3: each section of 5 mm × 10 mm experiences a uniform maximum tensile stress.

IOP Conf. Series: Materials Science and Engineering **1310** (2024) 012008 doi:10.1088/1757-899X/1310/1/012008

where P is the applied load by the rollers (N), L is the outer span length (mm), b is the width of the sample (mm), and d is the sample depth (mm). They are L = 30 mm, b = 5 mm, and d = 5 mm,

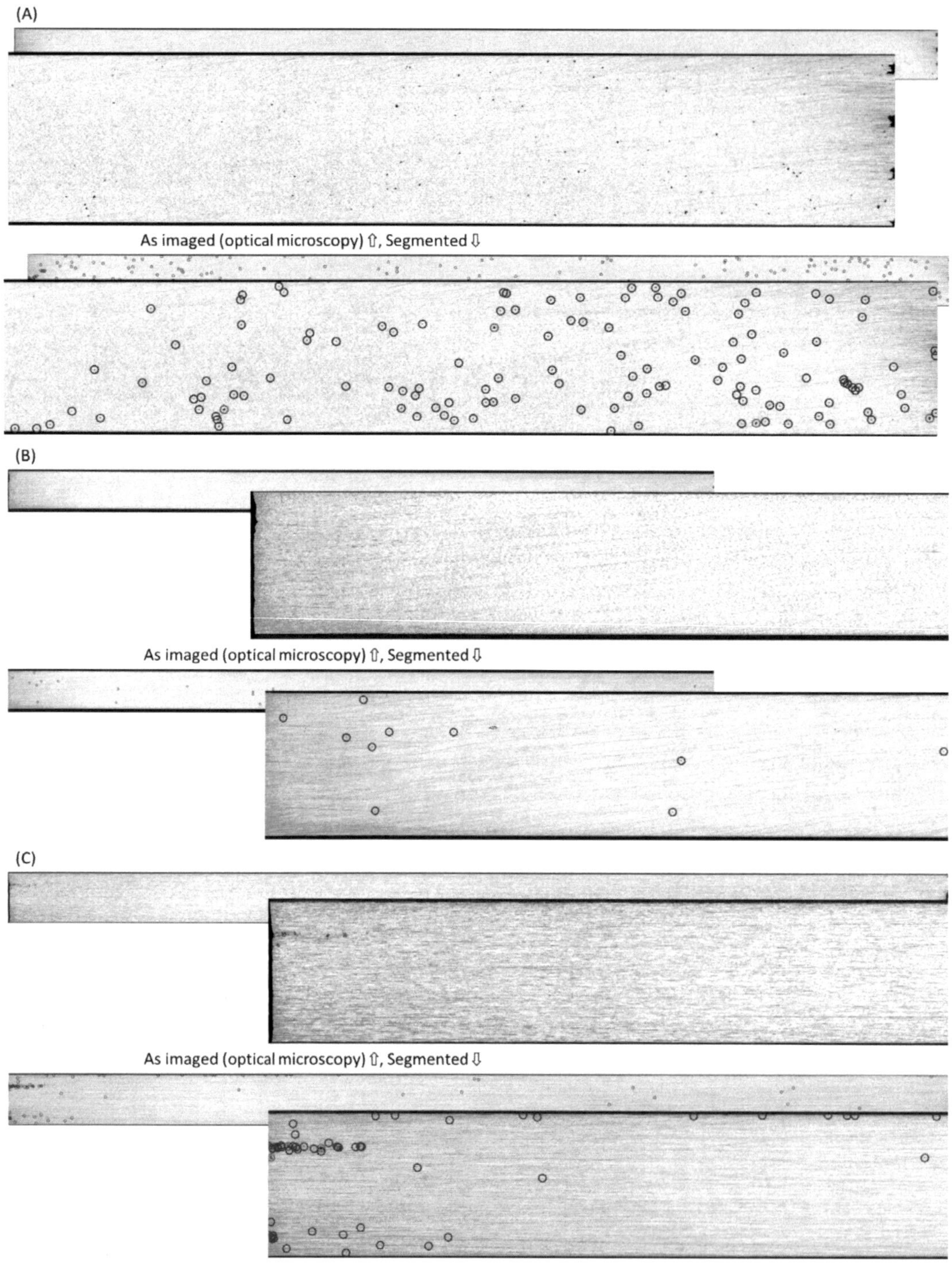

Figure 5. Optical microscopy and segmented images of (A) KH (CMU group #1, 800 mm/sec, F4), (B) process window (CMU group #9, 1,200 mm/sec, F2), and (C) LoF (CMU group #24, 1,950 mm/sec, F2).

as shown in figure 4.

One outcome of the 4PB fatigue testing was a determination of the peak stress values without failure runout to 10^7 cycles for each P-V combination. The peak stress values were chosen based on the prior work[20] and, if no runout occurred at the given stress level, subsequent tests were run with lower stress until the samples reached the runout target 10^7 cycles: where the test was stopped at 10^7 cycles, it was classified as a "runout". The corresponding peak stress values became the fatigue strength. Fractographic analysis using a scanning electron microscope (Thermo Scientific Apreo 2 at 30 kV and 0.1 nA) was performed for those samples that prematurely failed before reaching 10^7 cycles. Overall, each P-V combination has 4 samples that and each sample has 2 tests, so 8 tests total at each P-V combination to ensure test repeatability.

2.6 Segmentation

Each original Zeiss image (czi format) that presents one face is of order 50 GB, so all the images are compressed to BMP format to avoid losing quality. The Otsu method, referred to as standard, is used to segment the pores from the machining marks, and blurring is implemented so unmelted particles inside a pore are included in as the whole pore. Parameters are set manually to confine detection to pores only among the densely spaced machining scratches that would be detected by any automatic thresholding. Scikit's *regionprops* package is performed to get pore information from computer vision/image processing-based pore analysis. Pore analyses are performed over 1) all the four surfaces, 2) only at the F3, and 3) 4PB fatigue test area (see figure 4) over F3. Verification of segmentation was performed via Zeiss' Zen Blue 3.5 software. See figure S1 for detail on samples and how many cases.

2.7 Number density analysis and Extraction of Largest Expected Pore (N1)

The number density distribution over all four surfaces is plotted with pore counts as a function of pore ESD to correlate the expected maximum pore ESD within the 4PB fatigue test area, N1, to fatigue strength. It is plotted by normalizing the counts by the inspected area to obtain the number of pores per unit area as a function of pore ESD and split into equally sized bins (figure 6). We then find the pore ESD values at an area number density of 1 pore/50 mm^2, which are termed *N1*, where 50 mm^2 is a 4PB fatigue test area. Therefore, *N1* is a statistically based estimate of the likely maximum pore ESD within the 4PB fatigue test area for each P-V combination (or, later, for each individual scan).

2.8 Annotation of plots

It is noted that sample names are annotated in addition to laser speed values for some of the plots, such as figures 6, 7B, 8A, and 11B. Figures 8A and 11B specifically have 4PB fatigue test #1 or 2 annotated after laser speed values and sample names are given.

3. Results

We plot pore density over all four surfaces based on the values at the bin centers. The variation in sizes (figure 6A) is substantial. The count values are comparable because the area surveyed is nearly constant across samples. The pore number density shows a gentler decrease when transitioning from LoF to Process Window (righthand side) than for KH to Process Window (lefthand side). As mentioned above, CMU groups that are expected to be in the process window region have the lowest pore density along with size distribution.

As laser speed decreases (at constant power) from within the process window (moving right to left), interaction time and energy density increase, increasing front wall angle and vapor

IOP Conf. Series: Materials Science and Engineering **1310** (2024) 012008 doi:10.1088/1757-899X/1310/1/012008

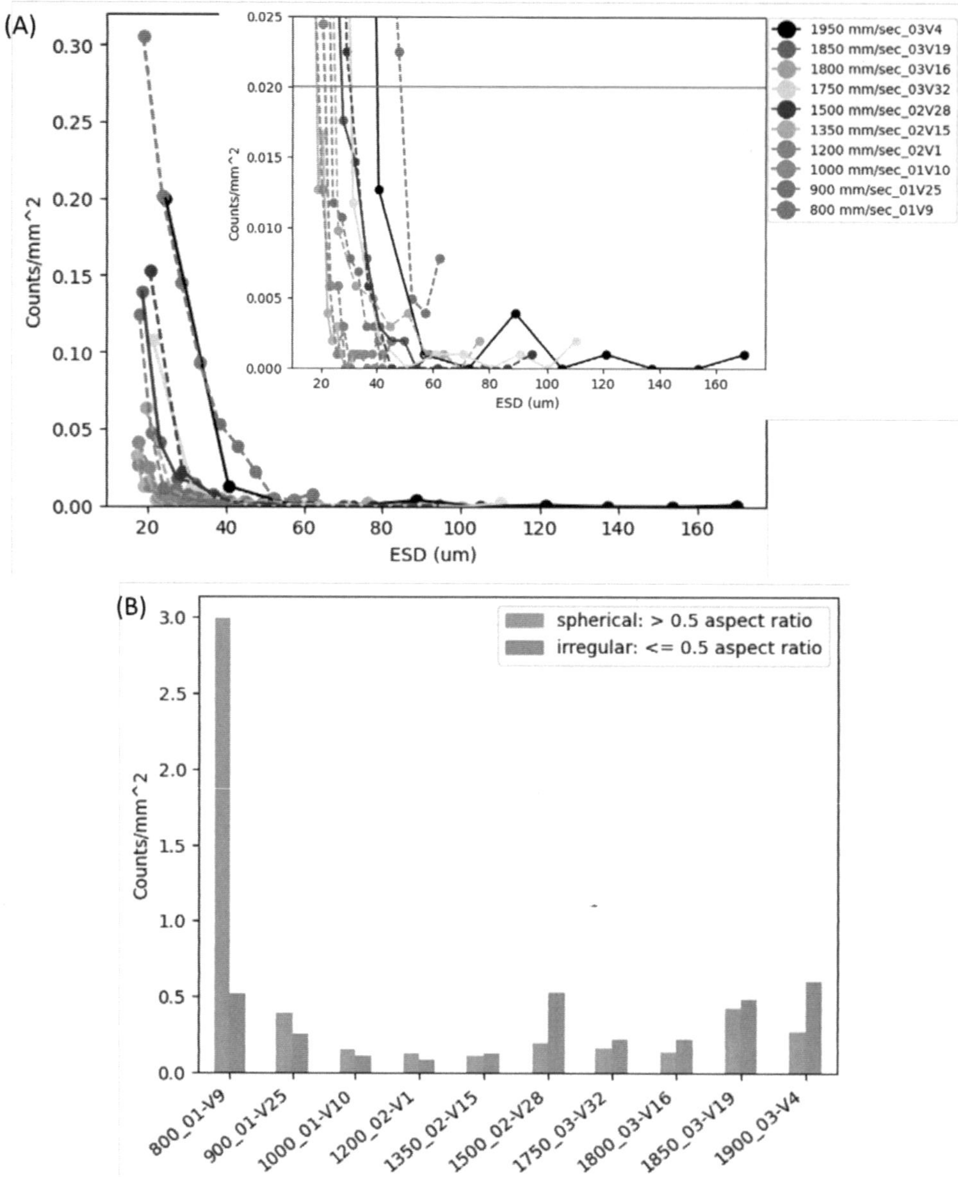

Figure 6. Pore characteristics as a function of laser speed for CMU groups. Plots of (A) pore density and (B) pore density divided according to aspect ratio (also as a function of laser speed). For (A) pore density plot, a horizontal line is drawn at 1 count/50 mm^2 to extrapolate N1 values, statistically estimated as the largest expected pore ESD within the 4PB fatigue test area, obtained by drawing a vertical line through the point of intersection. Pores of all shapes occur under essentially all conditions. For (B) aspect ratio plot, pores are classified into two types: (a) spherical (> 0.5 aspect ratio) and irregular (≤ 0.5 aspect ratio).

depression depth, meaning keyhole instabilities that generate pores increase. Therefore, as shown in figure 6A, CMU groups #6 and 7 that are on/near the boundary between KH and process window have comparable pore density and size distribution to process window (CMU groups #8 – 13, 1,150 – 1,400 mm/sec) but much lower pore density and size distribution than KH (CMU groups #1 – 5, 800 – 1,000 mm/sec). As laser speed increases (at constant power) from within the process window, LoF porosity (CMU groups #23 – 24, 1,900 – 1,950 mm/sec) occurs since melt pool size decreases reducing overlap between melt pools, leaving more un-melted areas; in this range the Tang et al. criterion[21] for LoF porosity is satisfied, *i.e.*, LoF index > 1.

Although CMU groups #14 – 20 (1,450 – 1,750 mm/sec) are nominally inside the process window below the LoF boundary and have LoF index less than 1, they have a higher pore density and size than the other process window groups (CMU groups #8 – 13, 1,150 – 1,400 mm/sec) and fluctuate noticeably compared to the small rate of increase.

The aspect ratios of pores are plotted as a function of laser speed in figure 6B. We classified pores into two types: (a) spherical (> 0.5 aspect ratio) and irregular (≤ 0.5 aspect ratio). The pores become more irregular and elongated with sharp corners with increasing laser speed starting from the middle of the process window (CMU group #14, 1,450 mm/sec) to LoF regions, leading to the aspect ratio (Eq. 2) to be ≤ 0.5. Many spherical shapes, where their aspect ratios being > 0.5, are present in the KH samples. The samples in between LoF and KH (CMU groups # 8 – 13, 1,150 – 1,400 mm/sec, show similar number densities of spherical and irregular pores.

In summary, our pore analyses align with the expected trends as a function of P-V combinations, except that CMU group #14 – 20 (1,450 – 1,750 mm/sec) exhibit fluctuations with higher pore density and size distribution although both were expected to be the lowest because they were anticipated to fall within the process window (figure 2). The LoF boundary shift is indicated based on our experimental results, different from the theoretically calculated boundary shown in figure 2.

3.1 Largest Expected Pore Size, N1

Figure 7A plots N1 values, which were estimated as the maximum pore ESD values within the 4PB fatigue test areas by drawing a horizonal line at an area density of 1 counts/50 mm^2 and drawing intersecting vertical lines on each curve in figure 6A. Pore densities for KH (CMU group #1, 800 mm/sec) are much greater than LoF (CMU group #24, 1,950 mm/sec) almost 4 times more (figures 6A and B) and distributed compactly up to their maximum pore ESD values, shifting up its N1 value, whereas LoF has a lower N1 value. Therefore, figure 7A reflects a "bathtub" curve with the KH groups' N1 values rapidly increasing whereas process window group's N1 values towards LoF increase more slowly with scan speed, that the magnitude in the LoF side doesn't reach such large sizes as KH sizes.

Figure 7B presents box plots of all four faces' pore ESD values of a bar as a function of laser speed. Both the box plots (figure 7B) and pore density plot's right tails (figure 6A) show that maximum pore ESD values in the LoF range are larger than those of KH. However, because LoF pore densities are sparse over a large pore ESD values, larger standard deviations are found for higher speeds near the LoF boundary, starting above ≈1,700 mm/sec (figure 7B), compared to KH.

To study the effect of porosity content on fatigue performance, three types of pore ESD values are presented first as in figure 8A as a function of laser speed: 1) N1 values and maximum pore ESD values on 2) F3 only and on 3) all four 4PB fatigue test areas. Along with N1 values (extracted from each size distribution), maximum pore ESD values on F3 show the U-shapes with the right side being shorter than the left side (figure 8A).

Colors represent 5 maximum stress values used for 4PB fatigue tests, whereas shapes represent 11 scan speeds. Tests at 667 MPa (yellow filled shapes from figure 8B) exhibit runouts

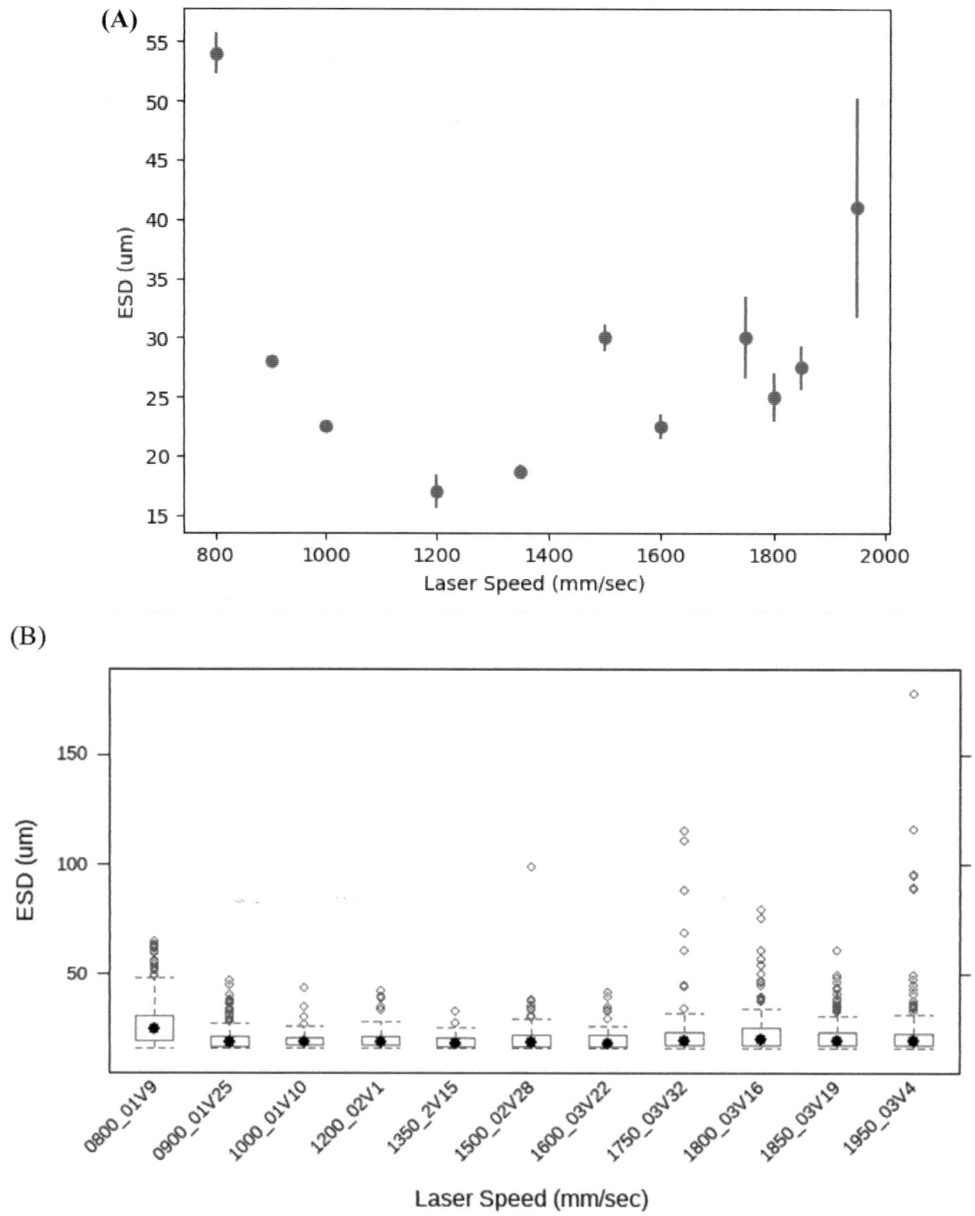

Figure 7. Pore characteristics as a function of laser speed for CMU groups. (A) N1 values with standard deviation bars and (B) box plots as a function of laser speed. For (B) box plots, all ESD values are plotted for each laser speed, like how the Figure 6 (A) pore density plot is, but with respect to pore sizes, instead of counts. Note the trend of decreasing then increasing ESD, i.e., a "bathtub curve."

between 900 mm/sec KH and 1,500 mm/sec process window. Tests at 1,200 MPa (gray filled shapes from figure 8B) have low median N_f values which generate difficulty in understanding pore ESD effects on median N_f before 1,000 mm/sec KH and after 1,600 mm/sec process window. The

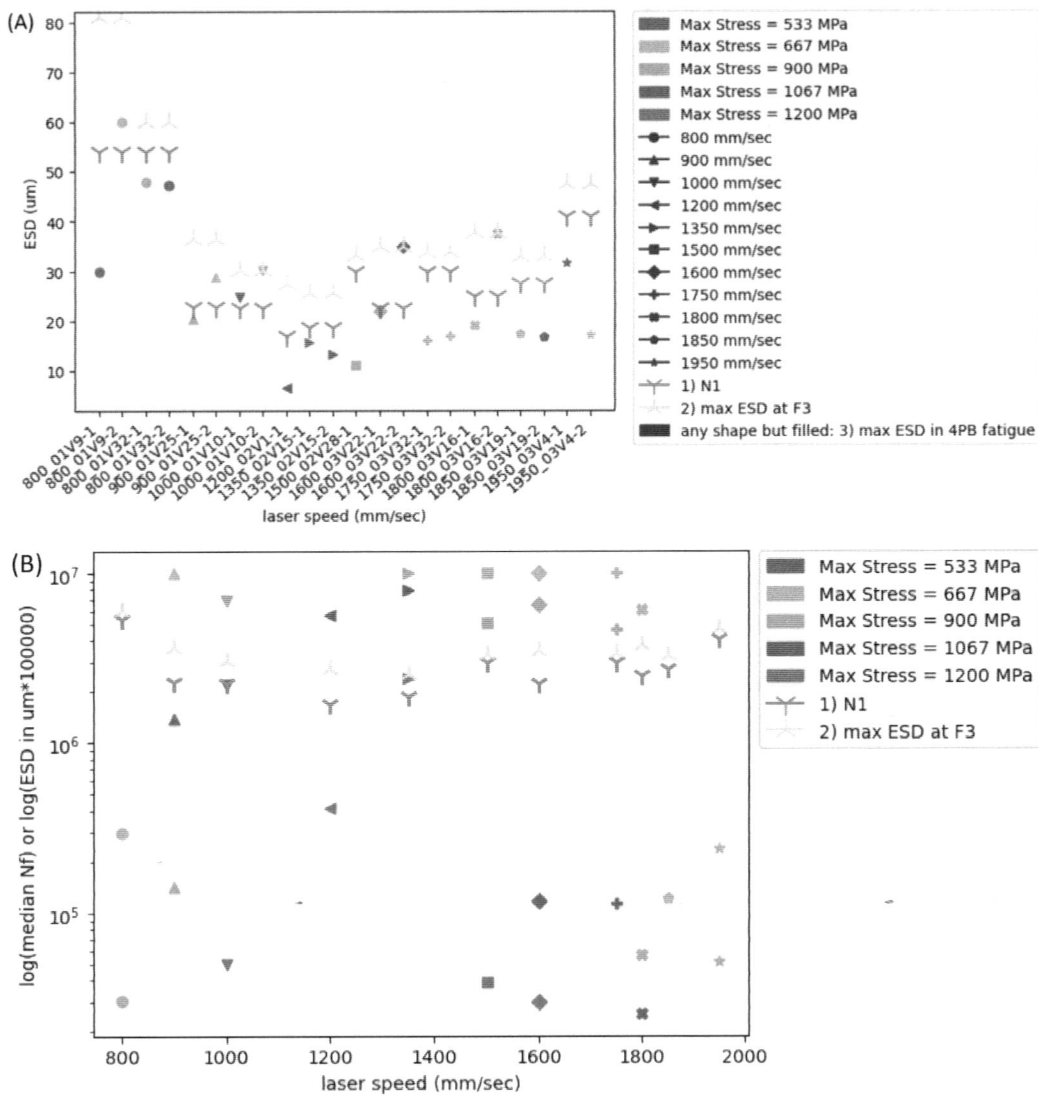

Figure 8. Correlation between pore characteristics and high-cycle fatigue (HCF) results as a function of laser speed for CMU groups. (A) ESD and number of cycles to failure at (B) 667 MPa stress, 900 MPa, 1,067 MPa, and 1,200 MPa as a function of laser speed. For ESD values, there are 3 types: 1) N1 values and maximum ESD values at 2) F3 and 3) 4PB fatigue test area.

same applies to tests at 1,067 MPa (blue filled shapes from figure 8B) considering that 1,800/sec process window has a low median N_f value. Tests at 900 MPa (green filled shapes from figure 8B) have median N_f values that are not as low, except 800 mm/sec KH, the slowest scan speed in the KH range. Therefore, we chose 900 MPa to be the appropriate max stress to analyze how median N_f changes as a function of laser speed, where inverse of the "bathtub" curve with shorter right side has been detected with both N1 and maximum pore ESD values.

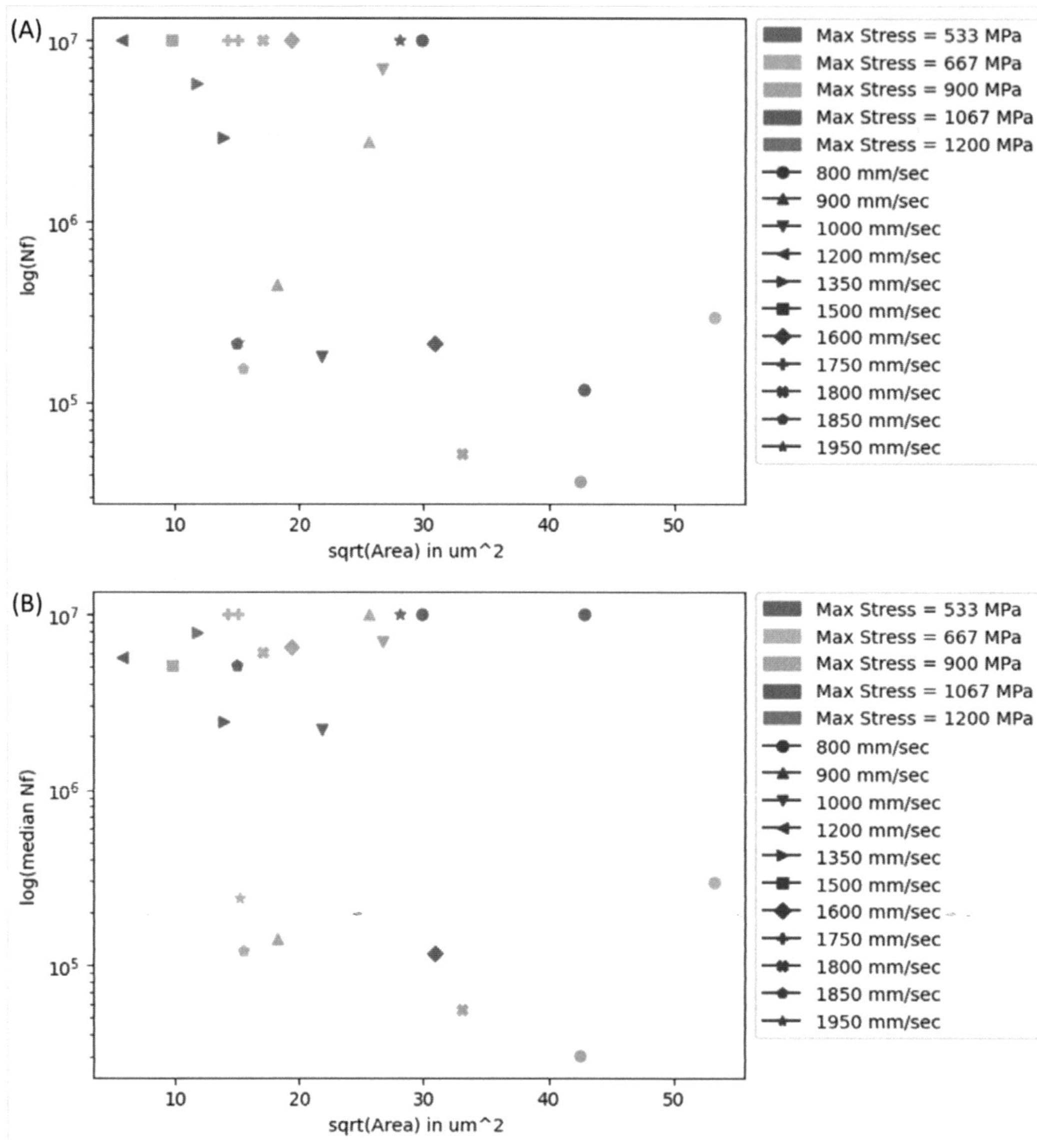

Figure 9. HCF results as a function of $\sqrt{\text{pore area}}$ for CMU groups. (A) Individual and (B) median number of cycles to failure as a function of the largest $\sqrt{\text{pore area}}$ inside fatigue testing region specifically.

Thus, these N1 values and maximum pore ESD values at F3 show similar trends that match with how median N_f values change with respect to laser speed, that even maximum pore ESD values at the F3 show potential to correlate their relationship with median N_f values without scanning four faces.

Also, at maximum stresses of 900, 1,067, and 1,200 MPa, the median N_f values vary strongly with a peak that is most obviously centered on 1,350 mm/sec.

To study the relationship between porosity content and fatigue performance in more detail, figure 9A presents the maximum pore ESD values for the individual 4PB fatigue test areas, each

IOP Conf. Series: Materials Science and Engineering **1310** (2024) 012008 doi:10.1088/1757-899X/1310/1/012008

matched with the corresponding fatigue life. Except for the highest and lowest (max) stresses, a correlation is evident for each maximum stress value for its $log(N_f)$ vs. $\sqrt{pore\ area}$, *i.e.*, there is an approximate log-linear relationship. This (anti-)correlation is best for the intermediate stresses of 900 & 1,067 MPa and is better than plotting the maximum pore ESD values at the 4PB fatigue test areas, figure 8A. If instead, we match the median life with each 4PB fatigue test region's largest pore area (instead of individual fatigue results) we obtain figure 9B. Because of scatter in N_f values that can be affected by pore areas, if we plot median Nf instead of individual maximum pore ESD values, a slightly weaker correlation appears.

In general, there is a decreasing trend of N_f with increasing maximum size of pore within the 4PB fatigue test area as expected. However, there is substantial scatter despite the approximately linear relationships. Also, at lower stresses, runouts are observed despite large pores, so 900 MPa would be the appropriate max stress to study N_f vs. $\sqrt{pore\ area}$.

4. Discussion

4.1 Boundaries of process window and the KH and LoF boundaries

At the outset of the work, we expected a minimum pore density inside the boundaries of KH at low speed and LoF at high speed, which were calculated to be around 1,100 and 1,850 mm/sec (dashed lines in figure 10), discussed in the method section for figure 2. It is important to note that the KH boundary was estimated from prior work[20] which was an empirical estimate that is located at higher scan speeds than the boundary determined from direct observations with high-speed x-ray visualization[24].

It is significant that the LoF boundary found from our *N1* values and standard deviation bars was also different. Starting at 1,500 mm/sec, N1 values change from the lowest pore density to higher one and sparsely distributed larger pores appear, where their aspect ratios indicate their shapes are irregular and elongated with sharp corners, supported by figures 6 and 7. Therefore, our experiments demonstrate LoF boundary to be around 1,500 mm/sec, resulting in a narrower process window than its theoretical calculation being 1,850 mm/sec.

This shifted LoF boundary value matches with our process window boundaries based on our experimental runout peak stresses (MPa) instead of median N_f values as a function of laser speed, leading to step plot (figure 10). The sharp decline in runout stress towards the LoF boundary begins at a much lower scanning velocity, 1,450 mm/sec, further away from the LoF boundary calculated value from the P-V map (figure 2), 1,850 mm/sec. Moreover, when laser speed decreases from CMU group #6 (1,050 mm/sec), the runout stress declines sharply from 1,067 MPa runout peak stress that CMU group #6 – 14 (1,050 – 1,450 mm/sec) have (bisque shaded region in figure 10). This decline at 1,050 mm/sec is in good agreement with the KH boundary defined by figures 2, 6, and 7, meaning calculated value from the P-V map and observed value based on N1 plot with standard deviation bars. Also, it is noted that therefore 900 MPa is the only max stress that does not contain any runouts.

Along with our process window boundaries, N1 values are compared with our experimental runout peak stress. Between KH and process window regimes, the fatigue strength was found to increase with decreasing N1 value, whereas between process window and LoF, despite larger deviations, the fatigue strength was found to decrease with increasing N1 value. Therefore, a non-linear inverse relationship exists between fatigue results, in either N_f (figure 8B) or stress (figure 10), and N1 values.

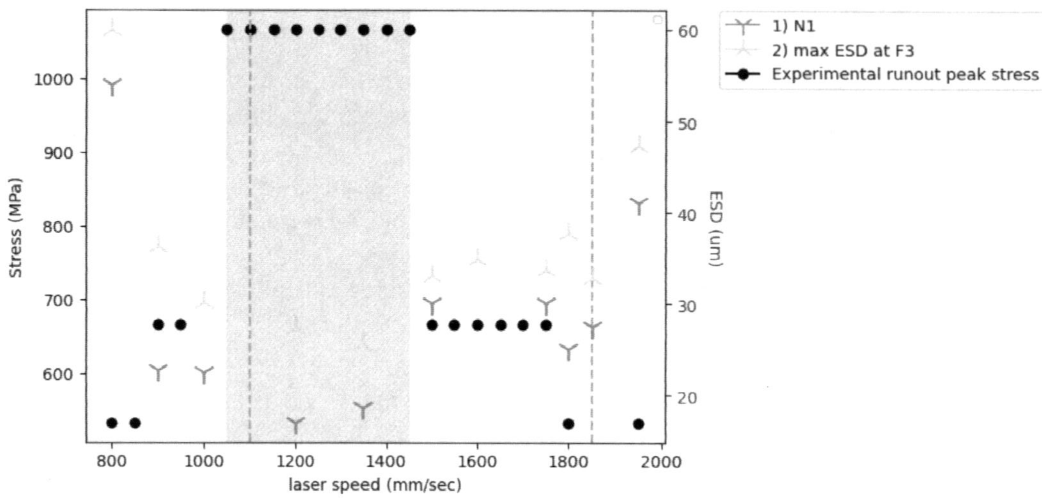

Figure 10. Experimentally determined fatigue runout stress and pore characteristics as a function of scanning velocity. The bisque shaded region between V = 1,050 – 1,450 mm/sec denotes the fatigue-based process window. Dashed lines between V = 1,100 and 1,850 mm/sec demarcate defect-based process window boundaries from figure 2. The secondary y-axis represents the N1 values measured from number density plots.

The ESD values observed on both the KH and LoF sides exhibit a sharp decline from 1067 to 533 MPa, over a range of 150 mm/sec range on the KH side and over 300 mm/sec on the LoF side. This aligns with observation that decreasing the laser speed closer to the KH boundary results in a more rapid decrease in fatigue strength than increasing the laser speed closer to the LoF boundary. However, because the runout stress values are step functions based on the limited number of experimental runouts rather than actual lifetimes, the plots may be overly optimistic about fatigue life. Thus, the correlation between pore characteristics and fatigue results as a function of laser speed is described more accurately in figure 8B than in figure 10.

Our experimental LoF boundary shift results based on multiple factors. The main factor is from laser line for printing not being uniform as laser speed increases, so it would create un-melted regions due to the melt pools occasionally not overlapping enough. The criterion of Tang et al.[21] therefore needs refinement because of the variability of the melt pool shape. The variability of melt pool size has been documented by, e.g., Scime & Beuth[41] and very recently its effect on porosity formation has been noted by Subraveti et al.[40].

4.2 Comparison of fatigue data to surface scanning and CT of pores on the four long surfaces of the 4PB bars

Not only do N1 values show strong correlation with fatigue results (figure 8), but they also lie within the boundaries of the crack-initiation pore ESD from fractographic analysis using SEM as a function of laser speed (figure 11A). This demonstrates that scanning four faces of a sample bar and its N1 value derived from their number density can be effective enough to predict the range of fatigue crack-initiation pore sizes, as well as a given area's likelihood pore size ranges.

Figure 11A shows that the maximum pore ESD values found on F3, i.e., the face subject to the maximum tensile stress, lie above the initiating defect sizes. This can be explained by remembering that the area associated with F3 is approximately 73 x 5 mm, *i.e.*, the full length of

IOP Conf. Series: Materials Science and Engineering **1310** (2024) 012008 doi:10.1088/1757-899X/1310/1/012008

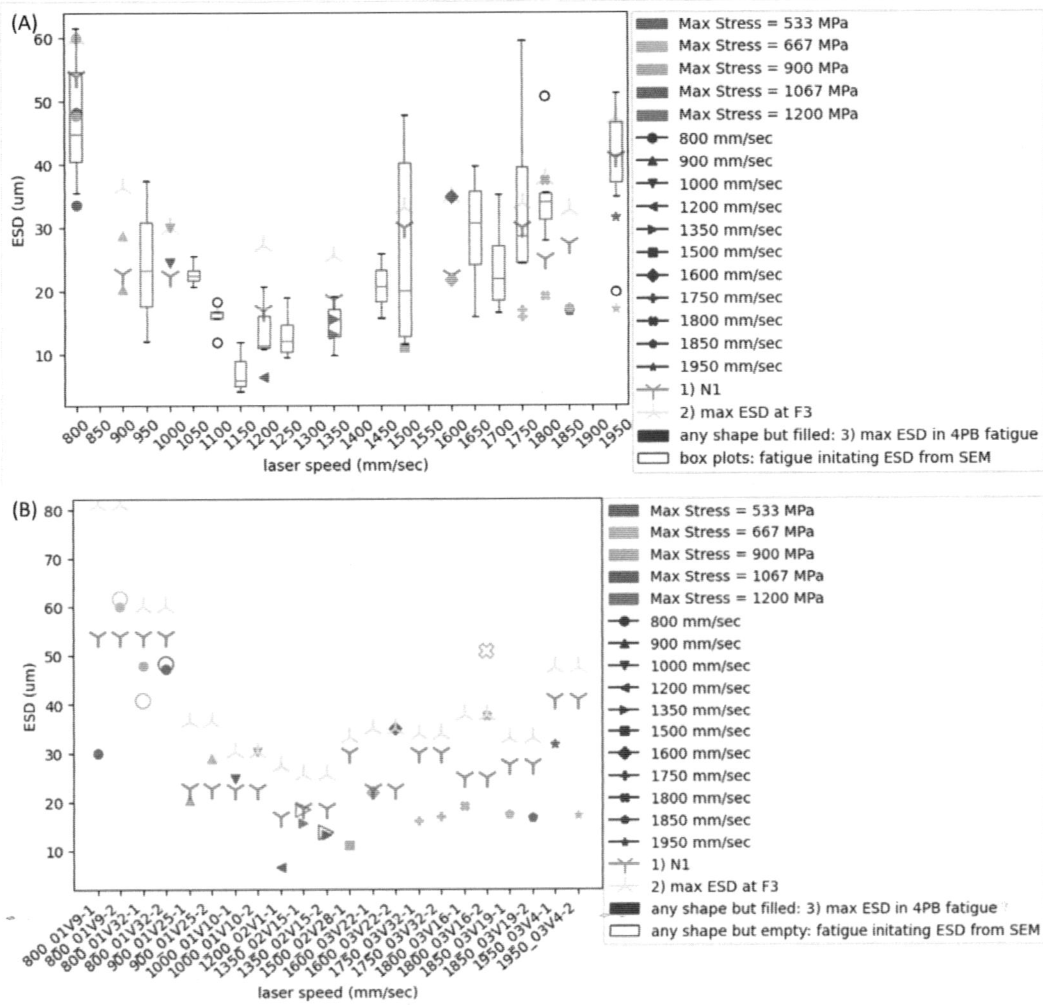

Figure 11. Pore characteristics as a function of laser speed for CMU groups compared to our fatigue initiating pores determined from fractographical analysis using SEM. The difference between (A) and (B) presentation is that (A) is based on fatigue initiating ESD range from the SEM at varied laser speed, whereas (B) is established on individual fatigue initiating ESD from the SEM at varied laser speed according to sample names and 4PB fatigue test numbers.

the bar, whereas the two sub-sections that experience the highest stress occupy only 2 x 10 x 5 mm which is ¼ of the full area. Therefore, it is not surprising that the larger sampling area finds larger maximum values than the smaller stressed area associated with the actual fatigue tests. Nevertheless, both N1 and F3 both show similar U-shaped or "bathtub" curves with a broad minimum in the process window and a steeper increase on the keyhole side than the LoF. The net result is that direct observations on only the stressed face (F3) should be sufficient to define the expected fatigue performance.

The fatigue initiating pores determined from fractographic analysis using SEM (figure 11A) has the inverse relationship to their 900 MPa max stress plot (figure 8B), where a "bathtub" curve like ours was found with smaller values on the right side (LoF) than the steep rise on the left side

(KH). Its interquartile ranges follow the same trend. This pattern is detected for all three measures, viz., N1, the maximum ESD on F3, and the maximum ESD in 4PB fatigue (figure 11), but with CT's N1 values their trend the "bathtub" curve with right side being longer than the left side[31].

4.3 Surface scanning of pores on the four machined surfaces of the 4PB bars

The maximum pore ESD values in the scanned areas of the 4PB fatigue samples are in good agreement with our fatigue initiating pore ESD results measured (figure 11B) and we also found that most of the fatigue initiating pores were at or near the surface. This aligns with analysis by Murakami et al., where they stated that the fatigue crack is expected to initiate at the largest defect on the surface[42,30] meaning less than a diameter away from the surface. In general, printed

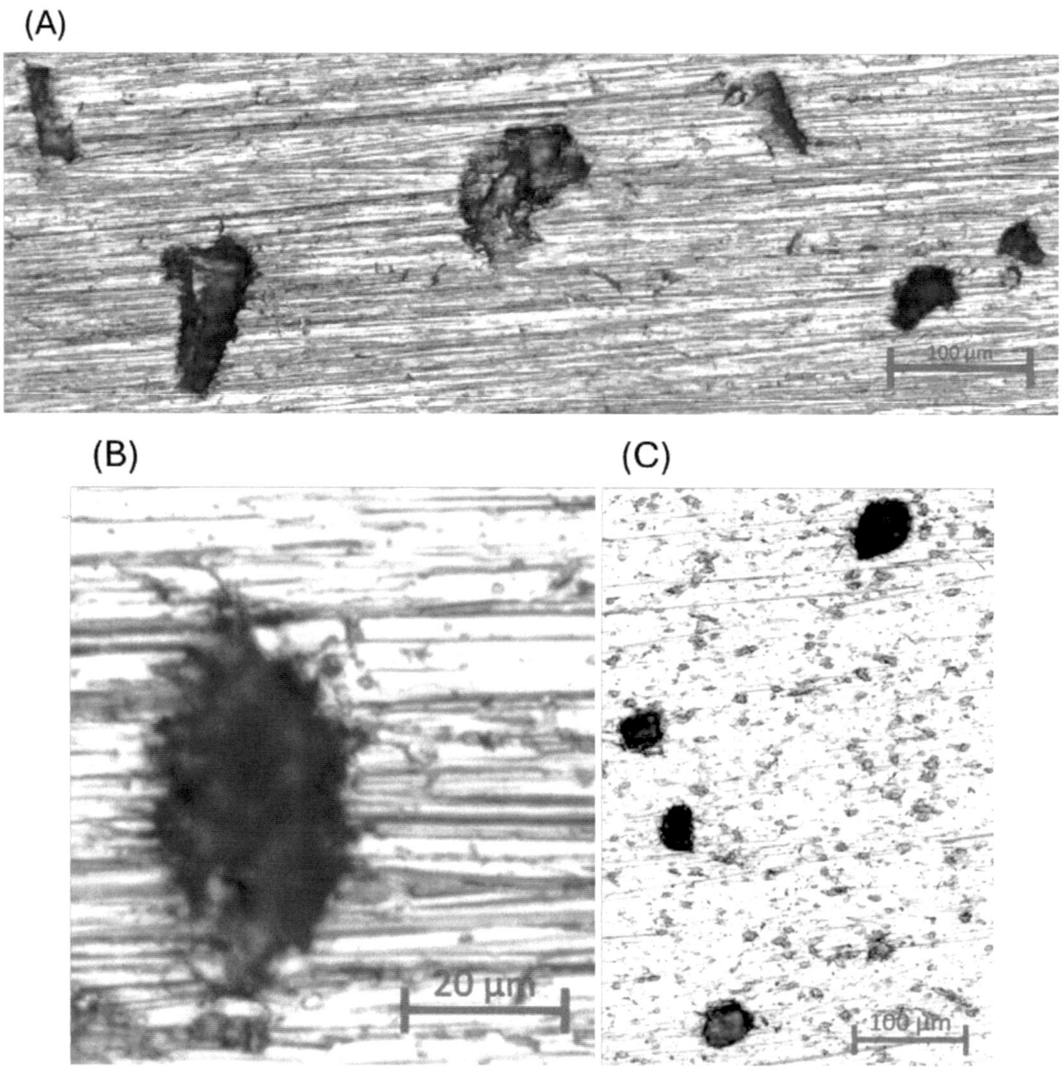

Figure 12. Optical microscopy of representative large pores for samples (A) within the LoF regime (CMU group #21, 1,800 mm/sec), (B) within the process window (CMU group #12, 1,350 mm/sec), and (C) in the keyhole region (CMU group #1, 800 mm/sec). The tensile axis and build direction are parallel to the machining marks.

IOP Conf. Series: Materials Science and Engineering **1310** (2024) 012008 doi:10.1088/1757-899X/1310/1/012008

samples have high number densities of defects such that there is almost always a near-surface defect determines the fatigue life.

One of the LoF samples, CMU group #21, 1,800 mm/sec, exhibits pores that are irregular and elongated with sharp corners, (figure 12A) but their number densities are low (figure 6B). However, even for samples with low pore number densities in process window (figure 6B) and those in the keyhole range have more circular pores (figures 12B and C), i.e., almost the same dimension parallel and perpendicular to the tensile stress. Clearly both size and pore shape affect the stress concentration, i.e., their potency as crack initiators, so the alignment of the pores with respect to the tensile axis in the sample at the LoF boundary is a suitable topic for further investigation (figure 11B).

To bring together all the observations from a processing-structure-properties perspective, we re-state the basic result that larger pores are strongly associated with shorter fatigue life, *i.e.*, poorer performance. Stress (& strain) concentration around a void is directly correlated with size, which takes us back to basic considerations of fracture toughness and the Griffith experiments. As analyzed in detail by Tammas-Williams *et al.*[8] in their work on AM Ti64, cracks tend to initiate in regions of high stress intensity and their figure 10(a) shows an excellent (linear) fit between $\sqrt{A_n}$ and N_f where A_n is the area of the initiating defect; they limited the data to samples tested at a particular alternating stress and cases of initiation from surface defects, which is directly relevant to our observations on surface porosity. Note the underlying assumption that total fatigue life is only partly controlled by crack growth and N_f often has a large fraction ascribed to initiation in which no crack is visible. The link between initiating flaw size and fatigue life has been extensively investigated by Murakami *et al.*[14,29,30] and is further borne out by observations in fatigue testing on AM samples as in Tammas-Williams *et al.*[8], Kantzos *et al.*[43], and Tang *et al.* [44]. In other words, the results reported here demonstrate that observations of surface porosity as exposed by machining have predictive value for fatigue life. This points to a broader applicability in that inspection of machined surfaces in locations that are known to be subject to high (tensile) stress may have value as a non-destructive inspection technique. Of course, AM parts are intended to be net shape so the technological value of this work is limited to those cases where machining of fatigue-sensitive locations is cost-effective.

5. Conclusions

We studied 24 P-V combinations across three main regimes in L-PBF of Ti64 at constant power but varied laser speed with respect to pore populations (number density and size distributions) and 4PB fatigue performance (S-N properties and runout stress) across the range of scan speed, namely KH, process window, and LoF.

Machined surfaces expose porosity at least in the case of low-stress grinding. This porosity can be imaged with conventional optical microscopy. Furthermore, despite the machining marks, pores can be resolved to 15 μm. This avoids the additional work involved in polishing numerous small cross-sections of a bar, which, for low pore densities under optimal printing conditions yield low pore counts. For example, Reddy et al.[31] used at least nine micrographs of cross-sectioned images at 500 μm intervals along the build direction and still found that CT provided much better statistics.

Pore distributions obtained from surface segmentations show good agreement with the data on fatigue initiating pores determined from fractographical analysis using SEM. They also show agreement with measurements using CT, but CT's N1 values are larger than the fatigue initiating pore sizes in most cases with KH.

From the pore size distributions, we extracted a maximum pore size consistent with 50 mm^2 area of high tensile stress in 4PB fatigue test, which varied systematically with scanned speed at constant power. The variation was U-shaped indicating location of P-V space of minimum size for constant number density.

The variation in pore size as noted above anticorrelates with fatigue performance as indicated for example by the variation in runout stress.

We conclude that, validated by fatigue life minimal defect densities are observed in a narrow window around 1350 mm/sec which suggests that these conditions will generate the most acceptable product.

Acknowledgments

This work was supported by the National Aeronautics and Space Administration (NASA) University Leadership Initiative (ULI) program under grant 80NSSC19M0123. The authors acknowledge the use of the Materials Characterization Facility at Carnegie Mellon University supported by grant MCF-677785.

References

[1] Liu S, Shin YC. Additive manufacturing of Ti6Al4V alloy: a review. Mater Des 2019; 164: 107552.
[2] Frazier WE. Metal additive manufacturing: a review. J Mater Eng Perform 2014; 23 (6): 1917-28.
[3] Wohlers TT, et al. Wohlers report 2018: 3D printing and additive manufacturing state of the industry annual worldwide progress report. Fort Collins: Wohlers Associates; 2018.
[4] Kannan GB, Rajendran DK. A review on status of research in metal additive manufacturing. In: Advances in 3D Printing and Additive Manufacturing Technologies. Singapore: Springer; 2016: 95–100.
[5] Alipour S, Nour S, Attari SM, Mohajeri M, Kianersi S, Taromian F, et al. A review on in vitro/in vivo response of additively manufactured Ti-6Al-4V alloy. J Mater Chem B 2022; 10 (46): 9479–534.
[6] Leuders S, Thöne M, Riemer A, Niendorf T, Tröster T, Richard HA, et al. On the mechanical behaviour of titanium alloy TiAl6V4 manufactured by selective laser melting: fatigue resistance and crack growth performance. Int J Fatigue 2013; 48: 300–7.
[7] Edwards P, O'Conner A, Ramulu M. Electron beam additive manufacturing of titanium components: properties and performance. J Manuf Sci Eng 2013; 135 (6): 061016.
[8] Tammas-Williams S, Withers PJ, Todd I, Prangnell PB. The influence of porosity on fatigue crack initiation in additively manufactured titanium components. Sci Rep 2017; 7 (1): 1–13.
[9] Wycisk E, Solbach A, Siddique S, Herzog D, Walther F, Emmelmann C. Effects of defects in laser additive manufactured Ti-6Al-4V on fatigue properties. Phys Procedia 2014; 56: 371–8.
[10] Wan H, Wang Q, Jia C, Zhang Z. Multi-scale damage mechanics method for fatigue life prediction of additive manufacture structures of Ti-6Al-4V. Mater Sci Eng: A 2016; 669: 269–78.
[11] Kasperovich G, Hausmann J. Improvement of fatigue resistance and ductility of TiAl6V4 processed by selective laser melting. J Mater Process Technol 2015; 220: 202–14.
[12] Greitemeier D, Palm F, Syassen F, Melz T. Fatigue performance of additive manufactured TiAl6V4 using electron and laser beam melting. Int J Fatigue 2017; 94: 211–7.
[13] Edwards P, Ramulu M. Fatigue performance evaluation of selective laser melted Ti-6Al-4V. Mater. Sci. Eng. A 2014; 598: 327–37.
[14] Masuo H, Tanaka Y, Morokoshi S, Yagura H, Uchida T, Yamamoto Y, Murakami Y. Influence of defects, surface roughness and HIP on the fatigue strength of Ti-6Al-4V manufactured by additive manufacturing. Int. J. Fatigue 2018; 117: 163–79.

[15] Becker TH, Kumar P, Ramamurty U. Fracture and fatigue in additively manufactured metals. Acta Mater 2021; 219: 117240.
[16] Cunningham R, Narra SP, Montgomery C, Beuth J, Rollett AD. Synchrotron-based X-ray microtomography characterization of the effect of processing variables on porosity formation in laser power-bed additive manufacturing of Ti-6Al-4V. JOM 2017; 69: 479–84.
[17] Qiu C, Panwisawas C, Ward M, Basoalto HC, Brooks JW, Attallah MM. On the role of melt flow into the surface structure and porosity development during selective laser melting. Acta Mater 2015; 96: 72-9.
[18] Panwisawas C, Qiu CL, Sovani Y, Brooks JW, Attallah MM, Basoalto HC. On the role of thermal fluid dynamics into the evolution of porosity during selective laser melting. Scr Mater 2015; 105: 14-7.
[19] Gong H, Rafi K, Gu H, Starr T, Stucker B. Analysis of defect generation in Ti6Al4V parts made using powder bed fusion additive manufacturing processes. Addit Manuf 2014; 1–4: 87-98.
[20] Narra SP, Rollett AD, Ngo A, Scannapieco D, Shahabi M, Reddy T, et al. Process qualification of laser powder bed fusion based on processing-defect structure-fatigue properties in Ti6Al-4V. J Mater Process Technol 2023; 311: 117775.
[21] Tang M, Pistorius PC, Beuth JL. Prediction of lack-of-fusion porosity for powder bed fusion. Addit Manuf 2017; 14: 39–48.
[22] Guo Q, Zhao C, Escano LI, Young Z, Xiong L, Fezzaa K, et al. Transient dynamics of powder spattering in laser powder bed fusion additive manufacturing process revealed by in-situ high-speed high-energy x-ray imaging. Acta Mater 2018; 151: 169–80.
[23] Zhao C, Parab ND, Li X, Fezzaa K, Tan W, Rollett AD, Sun T. Critical instability at moving keyhole tip generates porosity in laser melting. Science 2020; 370: 1080–6.
[24] Cunningham R, Zhao C, Parab N, Kantzos C, Pauza J, K. Fezzaa, T. Sun, A. Rollett, Keyhole threshold and morphology in laser melting revealed by ultrahigh-speed x-ray imaging. Science 2019; 363: 849–52.
[25] Gordon JV, Narra SP, Cunningham RW, Liu H, Chen H, Suter RM, et al. Defect structure process maps for laser powder bed fusion additive manufacturing. Addit Manuf 2020; 36: 101552.
[26] Sundar V, Snow Z, Keist J, Jones G, Reed R, Reutzel E. Flaw Identification in Additively Manufactured Parts Using X-ray Computed Tomography and Destructive Serial Sectioning. J Mater Eng Performance 2021; 30(7): 4958–64.
[27] Murakami Y. Inclusion rating by statistics of extreme values and its application to fatigue strength prediction and quality control of materials. J Res-Natl Instit Standards Technol 1994; 99: 345-51.
[28] Shi G, Atkinson H, Sellars C, Anderson C. Application of the generalized Pareto distribution to the estimation of the size of the maximum inclusion in clean steels. Acta Mater 1999; 47(5): 1455–68.
[29] Murakami Y, Endo M. Quantitative evaluation of fatigue strength of metals containing various small defects or cracks. Eng Fract Mech 1983; 17: 1–15.
[30] Murakami Y, Endo M. Effects of defects, inclusions and inhomogeneities on fatigue strength. Int J Fatigue 1994; 16: 163–82.
[31] Tharun Reddy, Austin Ngo, Justin P. Miner, Christian Gobert, Jack L. Beuth, Anthony D. Rollett, John J. Lewandowski, Sneha P. Narra, Fatigue-based process window for laser beam powder bed fusion additive manufacturing. Int J Fatigue 2024;108428.
[32] Rosenthal D. Mathematical theory of heat distribution during welding and cutting. Weld J 1941; 20: 220–34.
[33] Tang M, Pistorius PC, Beuth J. Geometric model to predict porosity of part produced in powder bed system. Mater Sci Technol Proc 2015; 129–136.
[34] King WE, Barth HD, Castillo VM, Gallegos GF, Gibbs JW, Hahn DE, et al. Observation of keyhole-mode laser melting in laser powder-bed fusion additive manufacturing. J Mater Process Technol 2014; 214: 2915–25.
[35] SAE International Material Specification. Heat Treatment of Titanium Alloy Parts. SAE Standard AMS2801C 2022.
[36] Suter RM, Liu H., Rollett AD. Towards optimal processing of additive manufactured metals for high strain rate properties. Paper presented at: SSSAP Chicago; 2017 April 12–13; Chicago, IL.

[37] Mitterlehner M, Danninger H, Gierl-Mayer C, Gschiel H, Martinez C, Tomisser M, et al. Comparative Evaluation of Characterization Methods for Powders Used in Additive Manufacturing, J Mater Eng Performance 2021; 30: 7019-34.
[38] International Organization for Standardization (ISO), 2008, Representation of Results of Particle Size Analysis—Part 6: Descriptive and Quantitative Representation of Particle Shape and Morphology, ISO 9276-6:2008, Geneva, Switzerland.
[39] ASTM International, 2020, Standard Test Method for Flexural Properties of Unreinforced and Reinforced Plastics and Electrical Insulating Materials by Four-Point Bending, ASTM D6272-17, West Conshohocken, PA, United States.
[40] Subraveti V, Richter B, Yeratapally SR, Oskay C. Three-Dimensional Prediction of Lack-of-Fusion Porosity Volume Fraction and Morphology for Powder Bed Fusion Additively Manufactured Ti–6Al–4V. Integr Mater Manuf Innov 2024;13: 511-525.
[41] Scime L, Beuth J. Melt pool geometry and morphology variability for the Inconel 718 alloy in a laser powder bed fusion additive manufacturing process. Addit Manuf 2019; 29: 100830.
[42] Murakami Y, Usuki H. Quantitative evaluation of effects of non-metallic inclusions on fatigue-strength of high strength steels. II: Fatigue limit evaluation based on statistics for extreme values of inclusion size. Int J Fatigue 1989; 11: 299–307.
[43] Kantzos, C., Pauza, J., Cunningham, R., Narra, S. P., Beuth, J., & Rollett, A. An Investigation of Process Parameter Modifications on Additively Manufactured Inconel 718 Parts. J Mater Eng Performance 2019, 28: 620-626.
[44] Tang M, Pistorius PC Fatigue life prediction for AlSi10Mg components produced by selective laser melting. Int J Fatigue 2019; 125: 479–490.

The Metallurgy of Additive Manufacturing: Potentials and Challenges towards Industrialisation

P Mayr[1], S Rauh[1,2], G Matheson[1,3], S Rotzsche[1], S Hartmann[1,4], and E Kabliman[1]

[1]Chair of Materials Engineering of Additive Manufacturing, Department of Materials Engineering, TUM School of Engineering and Design, Technical University of Munich, Freisinger Landstrasse 52, Garching b. Munich, 85748, Bavaria, Germany
[2]Fraunhofer UMSICHT, Fraunhofer Institute for Environmental, Safety, and Energy Technology, An der Maxhütte 1, Sulzbach-Rosenberg, 92237, Bavaria, Germany
[3]Oerlikon AM Europe GmbH, Freisinger Landstrasse 52, Garching b. Munich, 85748, Bavaria, Germany
[4]Siemens AG, Erlangen, Bavaria, Germany

E-mail: evgeniya.kabliman@tum.de

Abstract. The present paper discusses the potential and challenges of processing metallic materials using additive manufacturing. Particular focus is given to laser powder bed fusion (PBF-LB/M) and the use of traditional alloy powders such as Al alloys and Ni-based superalloys, as well as novel materials such as metal-matrix composites. The research includes the improvement of the processability of these alloys using PBF-LB/M and optimizing material properties such as strength, creep resistance, and thermal conductivity of printed parts for various applications. Another important aspect presented within this manuscript is the digital representation of advanced manufacturing systems to improve manufacturability and enable advanced quality control. Herein, the development of a digital twin through in-situ process monitoring for the direct energy deposition process of laser metal deposition is presented. In the last part, the future of materials development for additive manufacturing is discussed, focusing on applying material computational techniques. All demonstrated examples result from the successful cooperation between the Chair of Materials Engineering of Additive Manufacturing, TUM, and its industrial and research partners.

1. Introduction

Additive manufacturing (AM) of metallic components has rapidly developed over the last decades and attracted the interest of scientists and industry simultaneously. Although, this technology has been known for more than 100 years, when the first metal welding techniques were applied to generate 3-dimensional objects, the boom over the last 15 years has resulted in substantial technological advancements in various areas of industry and also our daily lives [1, 2, 3]. Further advantages such as cost reduction, customization, increased part complexity, supply chain optimization, promoting sustainability and circular economy have been discussed widely in the literature [4, 5]. The boom of additive manufacturing, especially with metals, can be seen in the steadily increasing number of published scientific papers. Also, the number of scientific conferences in this area has skyrocketed. From an industry standpoint, many companies have already included AM in their production portfolio. Also, founders see additive manufacturing as an enabler technology for their start-ups or directly found companies to further develop and promote this new production technology.

While the fantastic opportunities provided through AM have attracted a lot of interest, industrialization is still behind the anticipated growth rate. The reasons for this can be found in various aspects. First, it should not be forgotten that our existing conventional advanced manufacturing technologies, like casting, forging, welding, etc. have developed and been improved over more than a century. Also, AM is a throughout digital technology, and new developments in other sectors like IT, sensing and control, design, robotics, AI, etc. are implemented daily. This, on the one hand, leads to a steady improvement in AM with uncountable possibilities, but on the other hand also gets AM in conflict with traditional manufacturing philosophies, especially quality control. There are still several critical topics that have to be solved to advance AM further. The development of regulatory and legal frameworks, quality assurance and certification, scalability and production efficiency, process control, and standardization, post-processing and finishing, and supply chain integration are just a few but major areas that have to be addressed. Only if the whole value chain of AM is mastered, the casualness and professionalism, known from other advanced manufacturing technologies, will allow the further implementation of AM.

Within this manuscript, research along the process chain of AM, starting with materials development for advanced metallic alloys and improved manufacturability, is presented. The critical role of post-processing and heat treating is shown for a Ni-based alloy used in critical high-temperature applications. How digitalization and simulation can support advancements along the process chain is presented for computational alloy design for AM and the establishment of a digital twin for the directed energy deposition process using a laser beam (DED-LB/M).

2. Laser powder bed fusion of copper-tungsten composites for use in power electronics

Copper-tungsten (Cu/W) composites combine the properties of two metals. The excellent thermal and electrical conductivity of copper (Cu) with the high density, high melting point, and a low coefficient of thermal expansion of tungsten (W). Due to the mutual solid insolubility of Cu and W [6], it is a metal matrix composite (MMC) which is also called a pseudo alloy [7]. These composite materials are used as electrodes for arc welding [8, 9] and for electrical discharge machining (EDM) [10], as high-performance contacts [11] and heat sinks [12, 13, 14]. Commonly Cu/W composites are manufactured by liquid-phase sintering of Cu/W powders compacts or infiltration of porous W skeleton with liquid Cu. However, those processes limit the component geometry and result in costly post-machining [15]. In contrast to this, the powder bed fusion-laser beam (PBF-LB) process can produce components with highly complex geometries. Additively manufactured MMCs can combine multiple properties in a material with a nearly limitless geometry design, tailored to the specific application. So far, MMCs have not been regularly used in industry in the PBF-LB process.

There are several publications on the processing of Cu/W composites by the PBF-LB process. Gu et al. [16] processed a W-20wt%Cu sub-micrometer powder in means of a 2000-watt laser and achieved a relative density of up to 95%. By using a 100-watt laser, Li et al. [17] fabricated building parts with a chemical composition of W–10wt%Cu. Hereby, they measured a relative density of up to 66%. A solution to increasing the densification of Cu/W composite is to use nickel (Ni). However, this leads to a reduction in electrical and thermal conductivity. Such W-Ni-Cu alloys were processed by the PBF-LB process in a few publications [18, 19, 20]. Yan et al. [21] processed several compositions of Cu/W composites with Cu contents of 15-40wt%. They worked with a 200-watt laser system and achieved a relative density up to 98%. They also analyzed thermal conductivity, thermal expansion, roughness, and hardness. Furthermore, there is investigation of the absorptivity of Cu/W powders. The densification of the Cu/W MMC highly depends on absorptivity of the radiation. Qin et al. [22] determined an absorptivity of 56% for W-20wt%Cu and 51% for W-40wt%Cu at a wavelength of 1070 nm. By using a 500-watt laser, they produced Cu/W composites with a relative density of up to 99.6%. In contrast, Zheng et al. [23] processed Cu/W with low W content of approximately 0.5-2wt%. The Cu/W powder was manufactured in a solution with ammonium tungstate and subsequently reduced through a calcination process. For the Cu-0.5wt%W powder, they measured an absorptivity of 53.1% at a wavelength of 1070 nm. Using a 500-watt laser, they generated test specimens with a relative density of up to 99.5%.

In this study, the processability of copper-tungsten (Cu/W) composites for the PBF-LB process was investigated. A chemical composition of Cu-35wt%W was used. Therefore, sub-micrometer W particles were processed with micrometer Cu particles. The long-term goal is the production of additively manufactured MMC heat sinks for high-performance electronics with a reduced coefficient of thermal expansion to minimize thermal stress in the multi-layer compound. The elements of power device, substrate and heat sink of the multi-layer compound should be able to endure as many heat cycles as possible before failure. Moreover, the MMC should retain high thermal conductivity. Therefore, the process parameters were adjusted to minimize porosity in the test specimen.

2.1. Process setup and characterization methods

MMC powder was processed through a milling process (Pulverisette 5/4, Fritsch). Cu powder with a mean particle diameter (d_{50}) of 25.5 µm and W powders with a d_{50} of 0.7 µm were mixed. 65 wt% Cu and 35 wt% W were weighed in the glove box and modified in a planetary ball mill. To manufacture the test specimens a PBF-LB system (ProX DMP 200, 3D Systems) was used. A maximum laser power of 262 watts, hatch distance of 60 µm, and layer thickness of 30 µm was adjusted. The scan velocity was varied from 500 to 900 mm/s. Argon with 700 ppm O_2 was used as process gas and the specimens were produced on steel plates. Furthermore, a piston reducer (3D Systems) was installed to process reduced amounts of MMC powder. Cuboid samples were manufactured to optimize the process parameters. The scan velocity was varied for the optimization of relative density. Cylindrical specimens were fabricated to analyze thermal conductivity. The relative density was determined optically using light microscopy (Eclipse LV 150, Nikon). The microstructure was analyzed by a scanning electron microscope (JSM-6490, JEOL) and a digital microscope (VHX-2000, Keyence). Thermal diffusivity was measured using laser flash analysis (LFA 457, Netzsch). In addition, the coefficient of thermal expansion was determined using dilatometry (L75 Platinum, Linseis).

2.2. Manufacturing of PBF-LB test specimens and porosity measurements

In figure 1(a), the relative density is plotted as a function of the scan velocity. Incomplete powder layer coverage occurred during the PBF-LB process, leading to the creation of irregular pores due to the lack of material. These types of pores must be differentiated from the lack of fusion porosity, as they are not caused by incomplete melting due to insufficient volume energy density.

The incorrect powder layering could not be eliminated during the experiments. One reason is the poor flowability of the Cu/W composite powder, which hinders a uniform powder layer. According to DIN EN ISO 4490:2018 (Hall flowmeter), the powder is non-flowable. Another negative effect on the powder layer distribution is the application of a piston reducer. Hereby the powder bed is not limited to the sides, in contrast to the standard building plate. The powder can be pushed away from the build plate on both sides during the layering process. In general, powder layering is the dominant factor in the formation of porosity. However, a range from 600 to 900 mm/s as scan velocity can be identified as moderate to manufacture test specimens with high relative density. The experiments also showed that test specimens fabricated in position 1

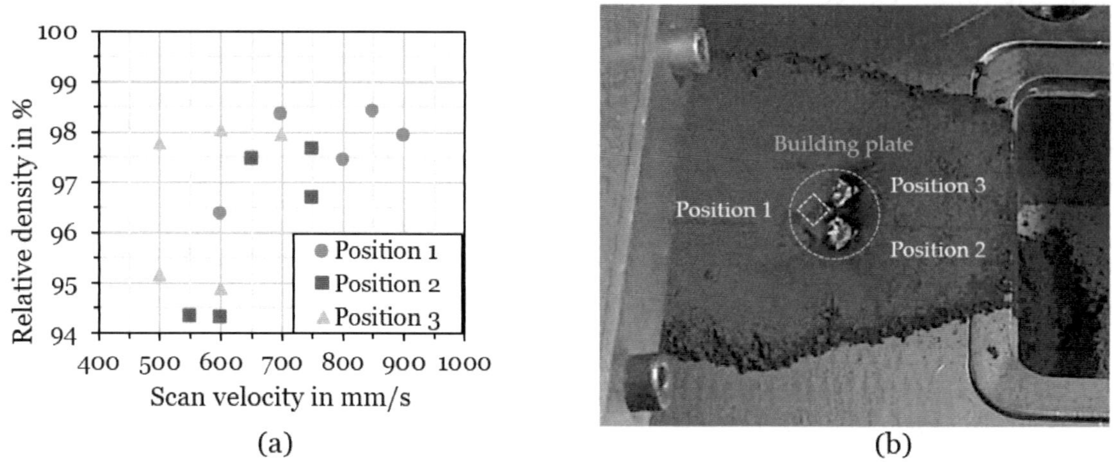

Figure 1. (a) Relative density as a function of the scan velocity; (b) Incomplete powder layering during the PBF-LB process.

had higher relative densities. This position was less affected by incorrect powder layering than positions 2 and 3, as shown in figure 1(b).

2.3. Microstructure

Figure 2(a) shows the microstructure of a Cu-35wt%W composite, analyzed by a scanning electron microscope (SEM). The light grey spots are W and the dark grey area is Cu. Some W particles are melted into phases, surrounded by a Cu matrix. The W phases are separated from the Cu matrix. The distribution of the W phases is heterogeneous. Most of the pores are located between the W phases. The reason could be the high surface tension of Cu. Also, figure 2(b) shows the microstructure of a Cu-35wt%W composite. The grey spots are the W phases and the brown area is the Cu matrix. The black spots are pores.

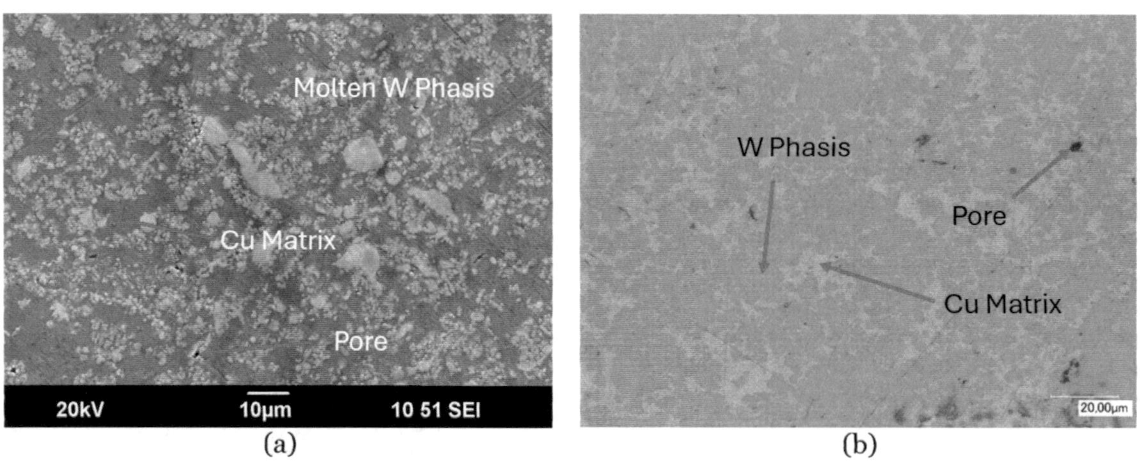

Figure 2. (a) SEM image of Cu-35wt%W composite. (b) Digital microscope image of Cu-35wt%W composite.

2.4. Coefficient of thermal expansion and thermal conductivity

In figure 3(a), the coefficient of thermal expansion (CTE) is plotted as a function of the temperature. The cuboid test specimens, which were measured by dilatometry, are shown in figure 3(b). An average CTE of $14.0 \cdot 10^{-6} K^{-1}$ (at 30°C to 300°C) was achieved with Cu/W composite. The Cu reference specimen had a CTE of $18.1 \cdot 10^{-6} K^{-1}$ (at 30°C to 300°C). In comparison to the Cu reference, the Cu/W composite has a reduced CTE of 22.7%.

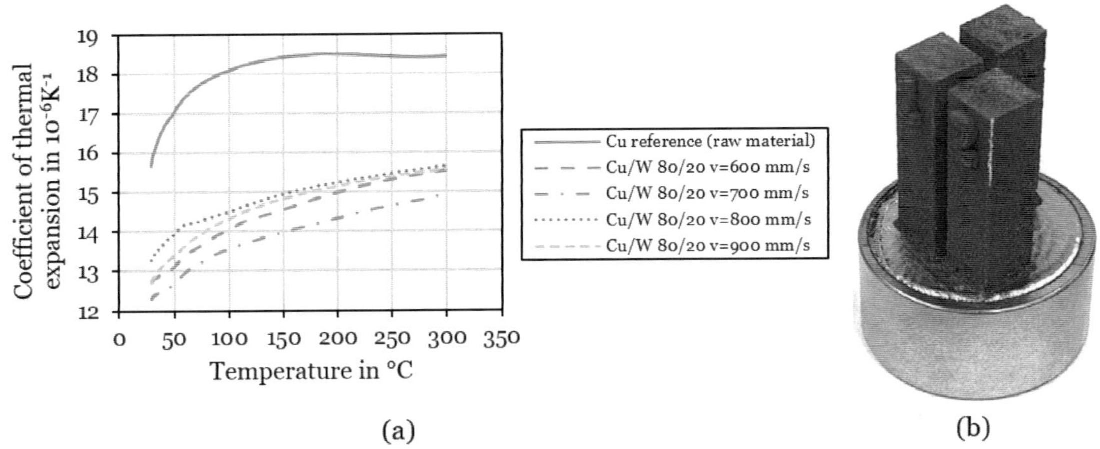

Figure 3. Coefficient of thermal expansion as a function of the temperature (a) measured for the PBF-LB manufactured cuboid test specimens (b).

The thermal conductivity as a function of temperature is shown in figure 4(a). It was measured for cylindrical specimens produced using different scan velocities. The additively manufactured cylindrical test specimen is shown in figure 4(b). As a result, the thermal conductivity of Cu/W composite up to 238.7 W/(m·K) and 396.0 W/(m·K) for the Cu reference was measured. Compared to the Cu reference, a thermal conductivity of 60.3% was achieved.

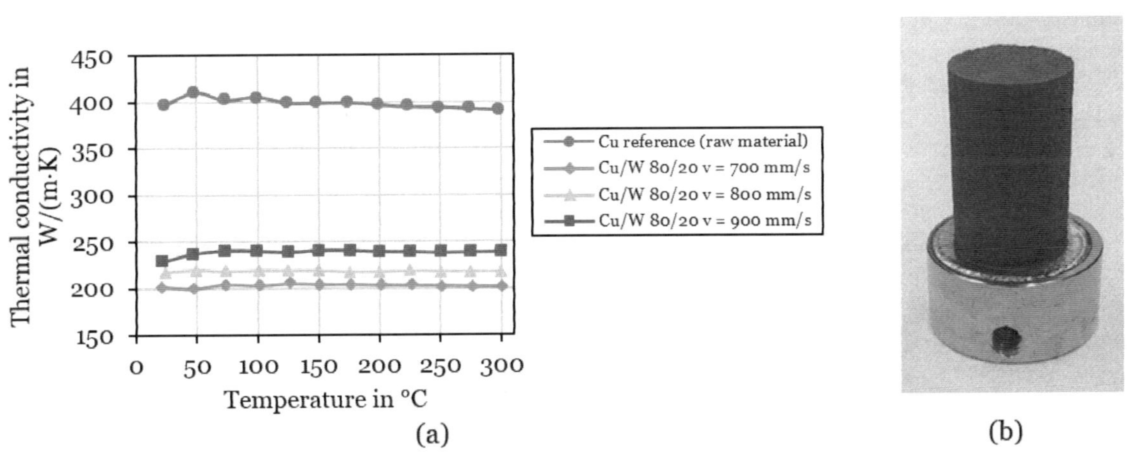

Figure 4. Thermal conductivity as a function of the temperature (a) measured for the PBF-LB manufactured cylindrical test specimens (b).

2.5. Summary and outlook

In summary, a relative density in the range of 96% to 98% was achieved. Furthermore, a coefficient of thermal expansion of $14.0 \cdot 10^{-6} K^{-1}$ (at 30°C to 300°C) was determined. Compared to pure Cu, the thermal expansion coefficient was reduced by 22.7%. Moreover, a thermal conductivity of up to 238.7 W/(m·K) was determined. In comparison to pure copper, 60.3% of the thermal conductivity was retained. Possible measures to increase thermal conductivity by reducing porosity will be implemented. In further experiments, the piston reducer will be removed and samples will be manufactured on the conventional build plate. To increase the flowability of the MMC powder, coarser W powder can be used. After optimizing the process parameters to reduce porosity, demonstrators will be built to show the limits of geometry design.

3. Design of high-strength Al alloys for PBF-LB processing

The rapid solidification rate and complex thermal cycling inherent to the PBF-LB process lead to notably different microstructures in metal alloys compared to cast or wrought material with the same composition.[2] Particular characteristics of the rapidly cooled material are the presence of non-equilibrium phases, fine microstructural features, and a supersaturated matrix [24, 25]. These unique solidification conditions provide major challenges related to cracking but also opportunities related to improved heat treatments when adapting high-strength Al alloys to the PBF-LB process.

3.1. Process induced cracking and routes to printability

In traditional high-strength aluminum alloys such as the age hardenable Al-Cu (2xxx series), Al-Mg-Si (6xxx series), and Al-Zn-Mg (7xxx series) systems, directional solidification in PBF-LB leads to columnar grained microstructures susceptible to intergranular hot tearing [26, 27, 28] This results in parts that are industrially useless due to the high crack density and poor mechanical properties. Strategies to reduce the formation of columnar grains and eliminate cracking are based either on the modification of the processing parameters to reduce thermal gradients, or modification of the chemical composition to refine the grain structure. Heating the building platform to 500°C has been shown to eliminate cracking in the AA6061 alloy [29] while both slow scan speeds [30] and laser defocusing [31] have been used to successfully fabricate crack-free 2xxx series material. However, these processing methods present major limitations for industrial adoption due to low productivity and powder degradation concerns. Modification of the chemical composition to induce a fine equiaxed grain structure, particularly via the introduction of grain refining elements or particles such as Sc, Ti, Zr, TiB2, is a promising solution to render these known alloy systems printable [25]. For example, figure 5 shows the powder feedstock and PBF-LB processed microstructure of AA7075 with and without the addition of 1.2 weight percent titanium (Ti) particles smaller than 25 micrometers via powder blending. A crack-free equiaxed grain structure formed in the printed material following the addition despite incomplete Ti melting seen in figure 5(f). The Ti addition takes the alloy out of standard industrial specification ranges for AA7075 chemical composition and therefore can be considered a new design for PBF-LB Al alloy. Blending powders is a useful way to screen new compositions, but the added process step, more complex powder recycling, and the presence of inclusions lead the industry to prefer pre-alloyed powders.

3.2. Heat treatments of high strength PBF-LB Al alloys

The typical heat treatment for age-hardenable wrought and cast high-strength Al alloys is a two-stage sequence notated as T6 when peak mechanical properties are achieved [32]. The first step is a solution heat treatment meant to dissolve alloying elements into a solid solution, followed by water quenching to trap solutes in the Al matrix. An aging step at a lower temperature is then performed to allow nanometric secondary phase precipitation from solid solution resulting

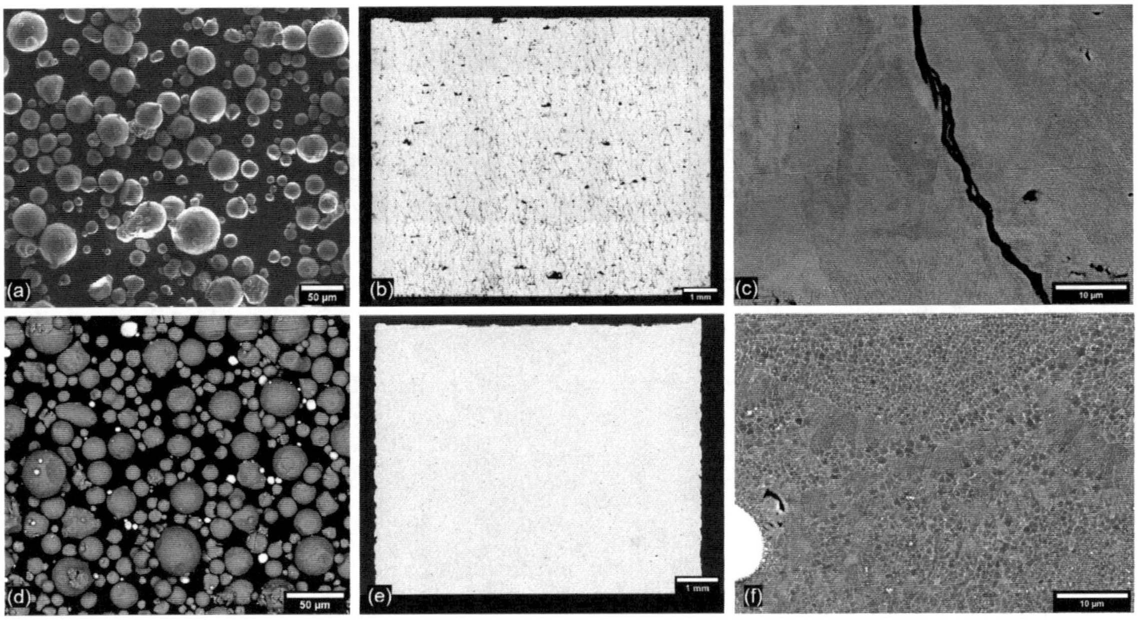

Figure 5. AA7075 powder and bulk micrographs (a,b,c) with no addition and (d,e,f) with 1.2 wt% titanium particle addition.

in hardening. These T6 heat treatments were developed for traditionally fabricated Al material with lower solidification rates than occur in PBF-LB, and where a solution and quenching operation is necessary to obtain a saturated solid solution. The as-built microstructures of PBF-LB Al alloys already have a supersaturated matrix and therefore require novel heat treatments to reach optimized mechanical properties [33]. Moreover, a solution heat treatment step can coarsen fine grains, and result in coarser precipitates which harm the mechanical properties compared to the as-built microstructural state [25]. Direct aging - a single-step heat treatment, notated as T4, to relieve residual stress and trigger precipitation is generally found to be a better choice for PBF-LB Al alloys [34, 35, 36]. Moreover, skipping the water quenching step in the standard T6 reduces risk for industrial users. Since PBF-LB is a near-net shape manufacturing technique that often takes advantage of complex features and thin walls, water quenching can cause unacceptable geometric distortion in critical components.

In the case of the AA7075 alloy with blended Ti additions shown in figure 5, uniaxial tensile testing was performed on samples in the as-built, T4 and T6 heat treatment cases shown in figure 6. The single-step T4 heat treatment resulted in increased mechanical properties (413MPa yield strength, 6.7 % elongation) compared to the T6 heat treatment (373 MPa yield strength, 5.8 % elongation), These results illustrate that direct aging treatments can provide superior mechanical properties than standard two-step heat treatments on grain refined AA7075. Heat treatment optimization needs to be conducted for each alloy based on its as-built microstructure. Such optimizations should be aided by kinetic modeling of precipitate phases as described in section 6 of this paper.

3.3. Summary and outlook

In summary, adapting crack-susceptible high-strength Al alloys to PBF-LB via the addition of grain refiners is an effective solution to solve the cracking challenge - demonstrated here with the AA7075 alloy rendered printable via the blending of 1.2 wt% Ti particles into the feedstock

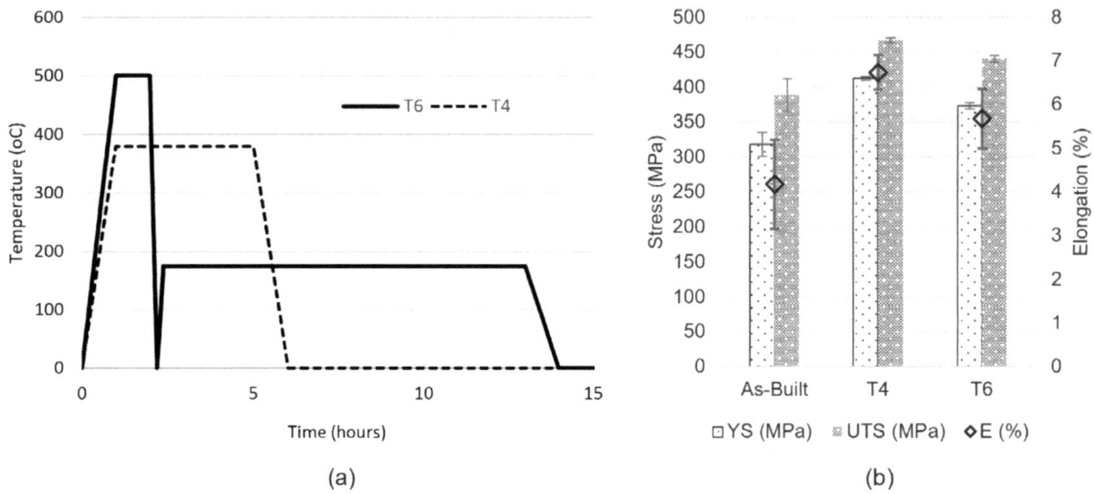

Figure 6. Holding temperatures and duration of T4 and T6 heat treatments applied to AA7075 + 1.2 wt% Ti particles (a) associated yield strength, ultimate tensile strength and elongation at break in the as-built and heat treated conditions (b).

powder. The particularities of the AM microstructure mean that traditional heat treatments are not well suited for these alloys and new heat treatments must be developed to take advantage of the processing technique.

4. Optimization of heat treatment for creep-resistant Ni-based superalloys

Due to the wide range of possible applications in the field of AM, many research questions regarding the process and materials remain unanswered. This section investigates the specific challenges of developing AM alloys for high-temperature applications. In particular, we focus on IN738LC alloy and its modification to improve the processability when using PBF-LB/M.

4.1. Challenges of AM in high-temperature applications

The main challenge of using AM in high-temperature applications lies in the following aspects. It begins with a question on the transferability of existing alloys typically used in conventional manufacturing such as casting or welding for AM including their processability and the quality of the final components. In particular, the resulting microstructure of AM components is typically characterized by elongated grains with a small width-to-length ratio that has a detrimental effect on high-temperature properties such as creep. It can be explained by the fact that grain boundaries exhibit higher diffusion rates of alloying elements resulting in grain boundary sliding at elevated temperatures. Thus, minimizing the fraction of grain boundaries in AM components is essential to enhance resistance against creep. It can be achieved by creating a coarse, equiaxed grain structure to mitigate these effects. In conventional casting processes, grain size can be controlled within the range of several hundred μm to mm by adjusting the casting temperature and preheating the mold. Additionally, resistance against creep can be enhanced through post-heat treatment by adjusting the distribution of the precipitation particles, their size, and morphology. IN738LC alloy, in particular, possesses its high-temperature resistance through a bimodal distribution of primary (cubic) and secondary (spherical) γ' precipitates of $Ni_3(Al,Ti)$ achieved by applying a standard heat treatment consisting of solution annealing at 1120°C for 2 hours (AC) followed by aging at 850°C for 24 hours.

Further insights into the application of conventional alloys in AM can be reached by analyzing their weldability, given the similarity of the respective processes. Alloys used in high-temperature industrial applications, such as CM247LC, and IN738LC, are generally considered difficult to weld [37]. This difficulty is reflected in the AM process by the increased formation of hot cracks and pores during the build process, which can be mitigated, for instance, by altering the chemical composition of the alloy. Unlike casting, where post-process grain coarsening through heat treatment is unnecessary, the resulting grain size from the PBF-LB/M process is in the low micrometer range with a high aspect ratio, coupled with strong grain texturing along the $< 001 >$-direction. Consequently, additional heat treatment is required for AM components to achieve grain coarsening and equiaxed grain orientation, ensuring isotropic mechanical properties. However, applying the aforementioned standard heat treatment to AM components does not lead to grain growth in IN738LC due to the restriction of grain boundary migration by finely precipitated metal-carbon (MC) carbides. This phenomenon is attributed to the Zener pinning effect, which is enhanced by the fine, dispersed precipitation of MC carbides.

4.2. Promoting grain growth in IN738LC alloy

To promote grain growth in additively manufactured IN738LC components, heat treatment at higher temperatures (around 1200°C) and dwell times of several hours is currently applied (see table 1). Upon heating the sample above 1200°C, MC carbides partially dissolve or grow depending on their initial size after the build process [38]. Consequently, the number of carbides, and thus the Zener pinning force, is minimized, facilitating grain growth [38, 39]. Another approach is the transformation of the metastable MC carbides to $M_{23}C_6$ carbides by applying a two-step heat treatment of 950°C for 24 hours followed by 1230°C for 10 hours. This results in a grain enlargement of approximately 400% [39, 40]. Herein, MC carbides are transformed into $M_{23}C_6$ carbides in the first heat treatment step through carbon diffusion along the grain boundaries. In the second step, when the temperature is raised to 1230°C, the $M_{23}C_6$ carbides are dissolved, and coarse MC carbides with a lower density are formed. Thus, the Zener pinning force is reduced allowing for further grain growth [39]. Finally, Peachey et al. [41] demonstrated that also directional recrystallization on AM samples can produce coarse grains with lengths of 650 µm and widths of 150 µm. In this process, rod material is drawn from a cooled zone through a hot zone. However, this method presents a disadvantage compared to furnace heat treatment as it severely restricts component geometry, thus offsetting the advantages of AM.

Table 1. Various strategies of post heat treatment applied to IN738LC after PBF-LB/M.

Heat treatment	Average grain size
1195°C, 6 hours	<60 μm [38]
1245°C, 6 hours	86 μm [38]
1230°C, 10 hours	$\sim$ 40 μm [39]
950°C, 24 hours + 1230°C, 10 hours	$\sim$ 100 μm [39]

4.3. Modification of IN738LC alloy and its heat treatment

The processability of conventional alloys can be enhanced through the modification of their chemical compositions. In this paper, we investigate the IN738LC alloy whose original chemical composition [42] was modified by the addition of 4 wt.% Fe to minimize susceptibility to hot cracking (see table 2) during PBF-LB. The resulting alloying alloy composition demonstrated a hot cracking index of 1.03 and could be manufactured with less than 1 $\frac{Crack}{mm}$ depending on the process parameters on an unheated substrate plate. The samples were additionally subjected to hot isostatic pressing (HIP) to close residual pores and microcracks as well as to reduce the risk of crack propagation when applying a heat treatment (950°C, 24 hours + 1230°C, 10 hours).

Table 2. Nominal chemical composition of modified IN738LC in wt.% [43]

Cr	Al + Ti	Co	Fe	W	Mo	Ta	C	Ni
16	7	9	4	3	2	2	< 0.1	Bal.

Figure 7 illustrates the optical micrographs of as-built and HIPed samples before and after the heat treatment. Using the line intercept method according to ASTM-E112, grain sizes of 25 μm and 165 μm were determined for the as-built and as-heat-treated samples, respectively. The grain size after the HIP treatment was determined to be 29 μm, and subsequent heat treatment resulted in a grain size of 98 μm. As can be seen, grain growth was observed for both as-built and HIPed samples. However, the percentage of grain growth was significantly lower in samples that have undergone HIP treatment. This is attributed to the fact that the preceding HIP treatment reduces the driving force for grain growth, as it removes the stored internal energy stemming from the building process. To tailor the optimal precipitate distribution after the grain growth, the standard heat treatment (1120°C, 2 hours (AC) + 850°C, 24 hours) for IN738LC should be applied after the 2-step heat treatment (950°C, 24 hours + 1230°C, 10 hours) required for grain growth.

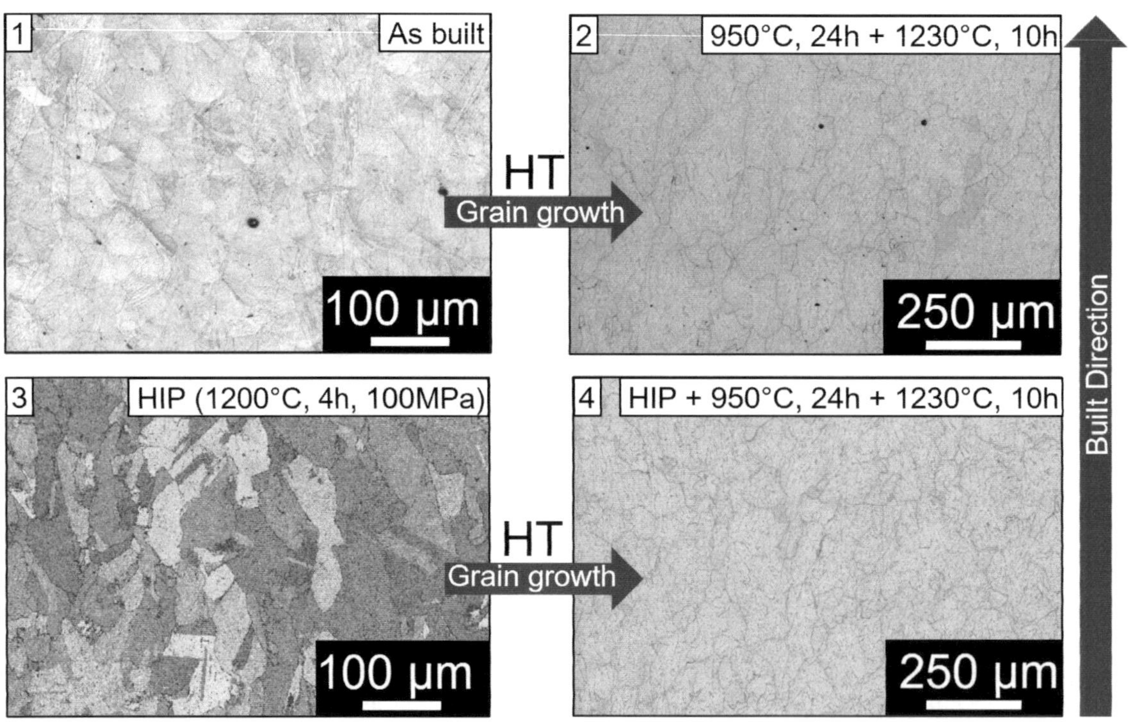

Figure 7. Optical micrographs of (1) as-built sample, (2) as-built + heat-treated sample, (3) HIPed sample, (4) HIPed + heat-treated sample.

4.4. Summary and outlook

The current study demonstrated that high-temperature alloys fabricated by AM, particularly PBF-LB/M, must undergo a post-heat treatment to optimize creep properties by promoting grain growth. The typical temperatures for IN738LC exceed 1200°C. It is associated with the dissolution or transformation of MC carbides and facilitates effective grain growth in PBF-LB/M fabricated samples, leading to increased grain boundary mobility. In particular, the 2-step heat treatment at 950°C for 24 hours + 1230°C for 10 hours presents a viable solution, although with a total duration of 34 hours, posing a significant time factor for industrial applications. Further research questions remain to be investigated, such as detailed investigations of the influence of HIP treatments on grain growth and the addition of 4 wt.% Fe on long-term creep properties.

5. Towards digital twins through in-situ process monitoring

As mentioned above, the extreme physical conditions during metal AM lead to unexpected process results. For DED-L, these anomalies can result in porosities, distortions, and lack of fusion amongst other defects in the final part [44]. A better understanding of the process-structure-property relationship is generally viewed as the most promising possibility to address these persisting issues for DED-L [45].

5.1. Digital Twins

Digital twins (DTs) can help to improve this understanding by using a virtual representation of the respective physical entity [46]. Such virtual representations should combine physics-based simulations to model the behavior of the real object while updating its state based on data derived from the real entity [47]. Applied to manufacturing processes, DT can lead to increased transparency about the respective processes while improving its resilience and flexibility via data-based process adjustments [48]. In the case of metal AM, Gaikwad et al. [49] pointed out that the most significant process improvements can be expected when physics-driven predictions and in-situ anomaly detection from process data are combined within one DT framework.

5.2. Digital Twins in DED-L

Multiple research efforts have already presented approaches towards such a DT framework for DED-L. The work of DebRoy et al. [50] opened the field for DTs in AM. They also proposed a potential use case for DED-L where they showcased how a DT can be employed to reduce the amount of trial and error tests inherent with this technology at the moment. Gaikwad et al. [49] utilized machine learning to fuse physical simulation data with in-situ monitoring data. In their conclusion, the authors pointed out that the prediction accuracy of a DT is enhanced by combining simulation and monitoring data. Regarding the simulation data, Mukherjee and DebRoy [51] highlighted the necessity of a precise mechanistic simulation model and outlined the parameters that such a model should be capable of predicting. With regards to the monitoring data, Gunasegaram et al. [52] deemed a proper validation of the developed DT crucial for obtaining reliable process predictions. In this regard, the works of Hartmann et al. [53] and Chen et al. [54] demonstrated the applicability of combining data streams from multiple sensors with the Tool Center Point (TCP) of the DED-L nozzle. Based on this applicability, another work by Hartmann et al. [55] proposed a more sophisticated monitoring framework based on an edge-cloud computing methodology to systematically create, store, and analyse digital representations of DED-L manufacturing processes. With such a DT of the actual manufacturing process, anomalies can be spotted in situ and location-based correlations between part defects and process faults can be realized. Other examples of applying DTs in DED-L include the work of Zou et al. [56] who measured strain and temperature profiles by embedding fiber optic sensors in an Inconel deposition. The monitoring data was compared to a DT model based on FEA yielding errors of less than 10%. Another research work was published by Hermann et al. [57], where the authors showcased the first DT framework combining data-driven and physical simulation models to predict the geometry of single tracks. Karkaria et al. [58] also used a data-driven model based on a recurrent neural network to predict temperatures during DED-L deposition in real time. The predictions were then used to adapt the printing process based on Bayesian optimization, but the approach was not tested on a physical DED-L system. Feng et al. [59] proposed a methodology for creating a DT for DED-L. The authors dissect the creation process into different steps starting with the virtual assembly of the equipment, the motion settings, collision inspection, and PLC debugging. However, their work is limited to only providing a concept without testing it with real experiments. The same limitation is valid for the research work conducted by Figueroa et al. [60] who presented a DT concept that was based on the ISO-standard 23247. Their concept combined a KUKA robot with a Meltio deposition head and used IoT tools and protocols to collect and visualize the process data in real-time. While presenting a sophisticated data architecture, their efforts also did not include an experimental use case collecting real data from a real DED-L system.

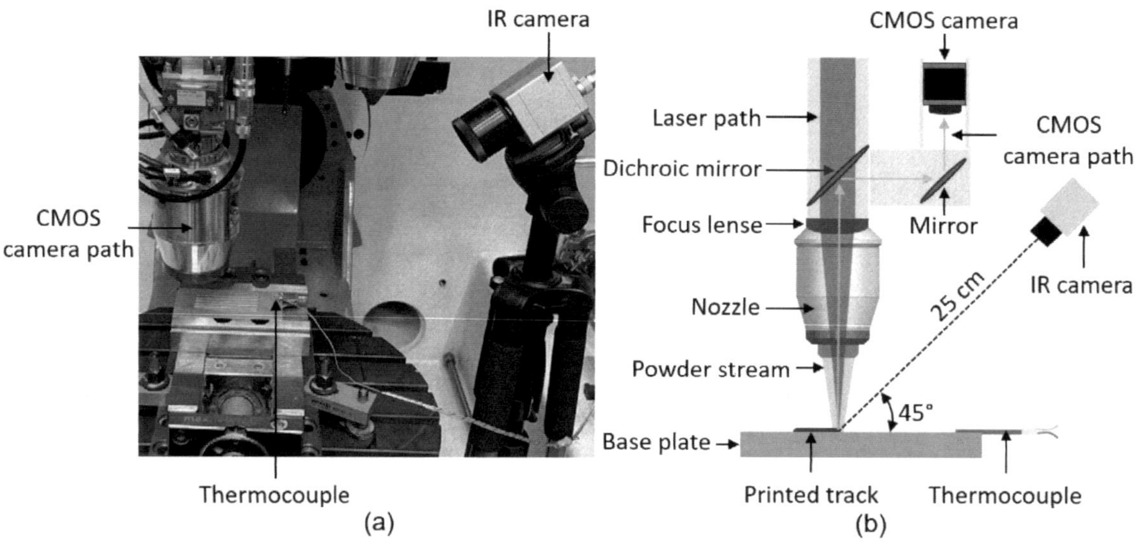

Figure 8. Monitoring setup: (a) employed sensor systems inside of the building chamber, and (b) schematic front view of the monitoring setup (reproduced from [53]).

5.3. Digital Twin validation through in-situ process monitoring

Upon reviewing the literature in the field of DTs, it becomes evident that proper validation with in-situ data coming from physical systems can be very beneficial for the creation of accurate DTs. For DED-L, however, there is a notable absence of such validations in the published research works. To address this gap, Hartmann et al. [53] presented a DT approach, where they compared DT data coming from a multiscale simulation model with in-situ monitoring data from an industrial-grade DED-L system. The monitoring setup included a CMOS camera capable of capturing grayscale images of the melt pool through the optical channel, as well as an off-axis IR camera measuring the temperature of the melt pool. In addition, a thermocouple that constantly measures the temperature of a fixed position was mounted onto the baseplate during the experiments. Inconel 625 powder was used as feedstock material and was melted onto a stainless steel 304 base plate with 100x50x10 mm^3 in dimensions. The experimental setup can be viewed in figure 8.

The sensors were integrated into a BeAM Modulo 400 that was equipped with a SINUMERIK ONE controller and an Industrial Edge IPC to collect the data from the machine and sensors. All data streams were fused so that each position during the build job held information about the respective laser power, feed rate, and sensor values measured. An example following this geo mapping approach can be seen in figure 9. The melt pool width, obtained by the grayscale camera, is depicted for an experiment where five tracks with different process parameters were deposited. The monitoring data is used to validate the DT by comparing its simulation results of the melt pool temperature and width with the IR camera and CMOS camera respectively. Moreover, the capability of the DT to calculate the heat dissipation throughout the build plate is tested with the thermocouple for one fixed position during the experiments. As the DT is also capable of calculating the final deposition's width, height, and cooling rate, metallographic inspections were carried out for the first two experiments to assess those parameters.

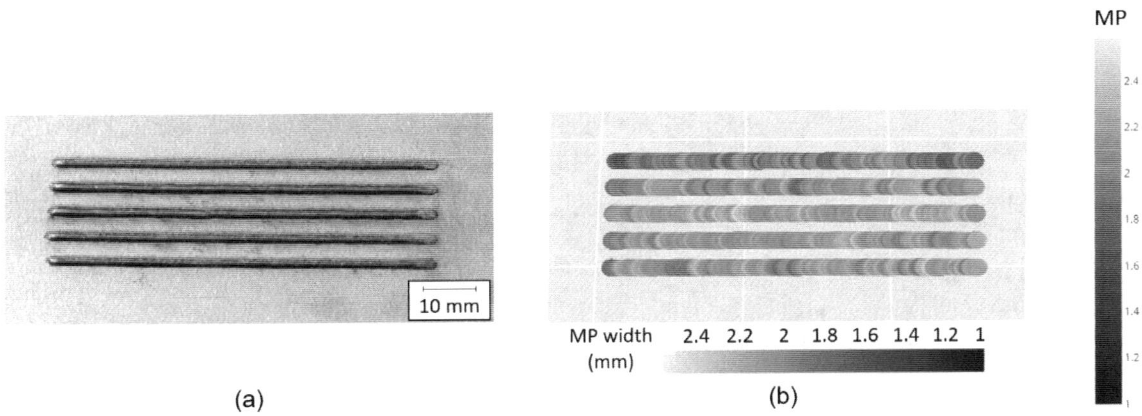

Figure 9. Comparison between the (a) real process part and (b) the geo mapping of the sensor data (reproduced from [53]).

In total, four experiments were defined to test the DT with in-situ monitoring and metallographic data. At first, a five-track sample with varying process parameters was printed twice with different void movement velocities to test the DT's capability for varying process parameter sets and heat dissipation dynamics. For these first two experiments, metallographic data was also obtained via multiple cross-sections to measure the width and height of the clads' cross-sections taken at multiple positions as well as the secondary dendrite arm spacing that correlates with the cooling rate after the deposition process. For the in-situ monitoring data, the DT's average errors were 4.62% for the melt pool temperature, 4.04% for the melt pool width, and 3.75% for the heat dissipation measured with the fixed thermocouple. For the metallographic inspections, the average errors amounted to 4.29% for the clad width and 6.98% for the clad height.

As the DT is also capable of calculating the secondary dendrite arm spacing (SDAS) during DED-L, the evaluation of its accuracy is done via analysis of the microstructures of the produced parts. The SDAS provides valuable information about the cooling rate of the solidification during DED-L. The cooling rate is an important factor influencing the microstructural and mechanical properties of the DED-L-produced parts. In the DT framework, the SDAS is employed to quantify the cooling rate and can be measured in the final parts. Figure 10 depicts the comparison between the simulated and actual SDAS representation.

A total of 16 individual measurements of the SDAS were carried out with a Leica DCM-3D microscope after a Krailing 2 agent was applied to each polished clad. The spacing between the

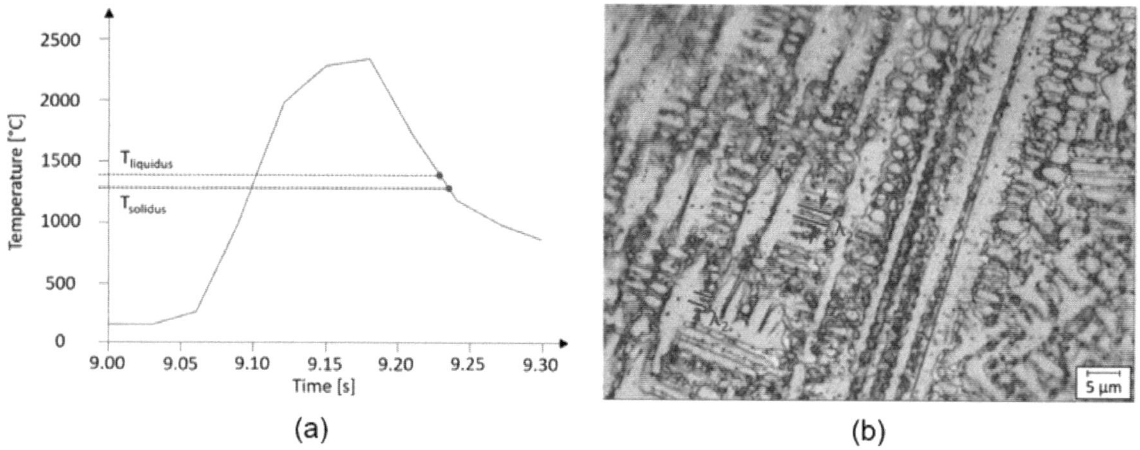

Figure 10. (a) Simulated CR for the SDAS calculation of clad 2.3 and (b) SDAS representation in the metallography of clad 2.3. (reproduced from [53]).

secondary dendrite arms of each sample is compared to the predictions of the DT. The maximum error of 0.36 µm can be deemed acceptable while the average error of 0.018 µm indicates highly precise predictions of the DT.

The validity of the DT for single-clad depositions was hereby ensured. To assess whether the DT is also capable of simulating error-prone build strategies in DED-L, a U-shape and a single-track wall consisting of five layers were chosen as the third and fourth experiments. Here, the in-situ monitoring data was assessing the DT accuracy for the melt pool temperature, dimension, and overall heat dissipation. The average errors amounted to 7%, 4.5%, and 6.19%, respectively.

5.4. Summary and outlook

The systematic collection of in-situ monitoring data enabled a scientific assessment of the predictive capabilities of the proposed DT for DED-L. Moreover, the fusion of sensor and machine data facilitated this assessment by directly comparing spatiotemporal positions within the DT simulation with their corresponding counterparts during the production process. As a result, the DT can be utilized to predict the essential parameters before the manufacturing process. By that, extensive trial-and-error cycles caused by a switch in process parameter sets or manufacturing geometries can be minimized. The capabilities of the DT will be further assessed for more complex geometries and different materials.

6. Computational alloy design for additive manufacturing

The future of materials design involves the intensive application of computational techniques and material models that allow predicting the material properties based on theoretical backgrounds from many disciplines like chemistry, physics, materials science, and engineering. The ability to simulate the manufacturing process and predict the material's properties using virtual (computational) tools can drastically reduce the use of resources such as raw materials and speed up the development process avoiding the time-consuming experimental trials. The requirement is, however, the application of time-efficient computational techniques.

6.1. CALPHAD and its application to AM

Among a large variety of computational materials science methods, which are used to describe the material's behavior, at various time scales, the calculation of phase diagrams known as CALPHAD [61] has found a wide application in many areas of metal manufacturing, including AM [62]. Using CALPHAD, it is possible to theoretically predict the phases and their fractions that can precipitate in multi-component multi-phase materials such as metallic alloys [63]. It is based on the minimization of the Gibbs free energy which is written in the special form accounting for molar Gibbs energies of individual phases as well as configuration effects and physical contributions. Since its development, CALPHAD has been implemented in many open-source and commercially available software such as OpenCalphad, PanDat, Thermo-Calc, and MatCalc to name a few.

When being applied to additive manufacturing, several specifics of the process should be considered. First of all, rapid solidification where cooling happens under non-equilibrium conditions. Thus, the equilibrium state conditions commonly assumed in CALPHAD do not apply here. Alternatively, the Scheil-Gulliver equation can be used which assumes infinite diffusion in the liquid state and restricted diffusion in the solid state. The next challenge is that during AM the already deposited layers are repeatedly heated by the addition of the upper layers. It leads first to the remelting of the underlying layers, and then to the heat treatment of the already solidified layers. This results in the cooling having a non-linear form. Finally, to achieve the target material property values (e.g. strength) the post-heat treatment of the as-built state is often required. This leads to a very complex thermal treatment of the material that is schematically illustrated in figure 11(a). While the solidification step can be approximated using the Scheil-Gulliver approach, special kinetic models are required to calculate the phase fractions in the solid state. These steps should be combined in such a way that the outcome of the Scheil-Gulliver calculation is an input for the kinetic modeling in the solid state.

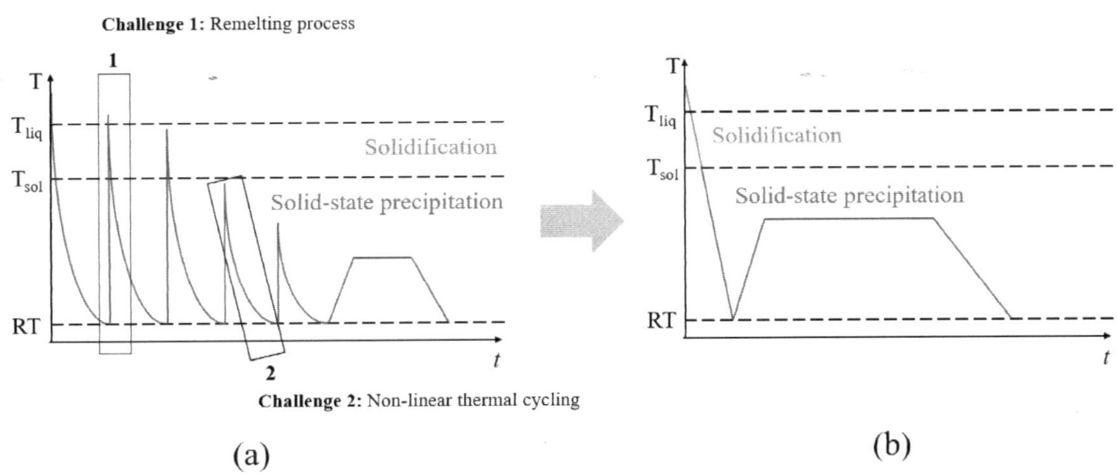

Figure 11. Schematic representation of thermal treatment during additive manufacturing (a) and its common simplification during the CALPHAD modeling (b).

The knowledge of the time-temperature history for each deposition layer or specific sample area is crucial for the correct prediction of phase precipitation in additively manufactured metals. Obtaining this data is, however, not trivial. Therefore, the intrinsic cyclic heat treatment is often approximated by a single-step rapid cooling requiring only a single Scheil-Gulliver calculation (see figure 11(b)).

6.2. CALPHAD-based high-throughput screening

The other strength of the computational materials modeling, e.g. CALPHAD, is the ability to be combined with the high-throughput screening to test many variations of chemical compositions and process parameters. In particular, in AM, several uncertainties can influence the material quality in the end. For instance, the uncertainty in the feedstock material can result in different quality of the final part depending on the supplier [64]. The other challenge is to find the optimal heat treatment of the as-built samples to reach the target material properties (e.g. strength values). While the experimental investigation is usually limited in the number of trials and associated costs, the computational screening is limited mainly by the available computational power. Otherwise, hundreds to millions of combinations can be tested. To do such high-throughput screening, special tools are required. At the Chair of Materials Engineering of Additive Manufacturing, we are developing a tool called CAROUSEL - an open-source framework for high-throughput microstructure simulations [65]. The tool uses MatCalc [66] as the CALPHAD implementation and offers such features as open-source implementation, scripting tools, graphical user interface, task distribution system, data management, and data visualization. Using CAROUSEL, we can perform multiple CALPHAD-based calculations of a different type: equilibrium, non-equilibrium (Scheil), and kinetic simulations as well as combine different steps to realize the through-process modeling.

Figure 12 demonstrates the results of the screening performed for 705 combinations from chemical composition values and heat treatment conditions (temperature, duration, time) for a post-heat treatment of Al-Mg-Si-Ti-Fe alloy after PBF-LB. For simplicity, we restricted the selected phases to the liquid, fcc (Al), Mg_2Si_B, and Al_3Ti_L which were of the most interest for the present study. During the post-heat treatment, it is expected to generate the nano-scale particles that contribute to the strength, herein the secondary Al_3Ti_L particles. However, no precipitation was observed in the calculation confirmed by the experimental measurements that could detect only the coarsening of the secondary Mg_2Si_B particles. This effect was confirmed by computational screening as demonstrated in figure 12. While increasing the holding time at 500 °C, the amount of the secondary Mg_2Si_B increased due to the increase of their mean radius. At the same time, while decreasing the heat treatment temperature from 550 °C to 350 °C, the amount of these particles decreased. Herein, each point on the graph is the result of a single combined calculation using the Scheil-Gulliver equation for solidification and kinetic models for the precipitation in the solid state.

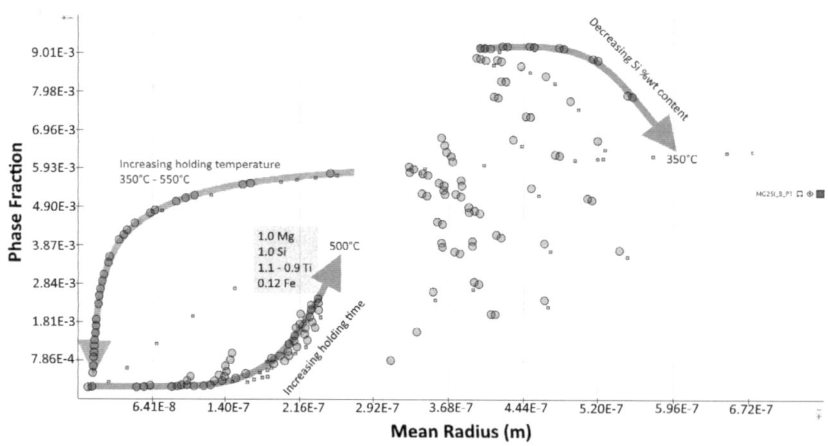

Figure 12. Calculated phase fractions vs. mean radius (reproduced from [65]).

6.3. Summary and outlook
CALPHAD and CALPHAD-based models offer an effective methodology for fast theoretical prediction of the microscopic properties of the materials during AM. Combined with the high-throughput approach, this method opens the opportunity for a rapid design of novel materials and optimization of the related process parameters. However, one should be aware of the shortcomings in both the methodology and application of the CALPHAD to AM. In particular, the available thermo-dynamic databases should be critically evaluated since they were historically validated for manufacturing processes rather than AM. This requires some prior knowledge of the investigated system since simulation with all phases from the database can lead to numerical issues and an inability to find the global minimum. Finally, the process specifics should be taken into account such non-equilibrium conditions.

7. Summary and conclusions
This manuscript presented the results of several successful research activities at the Chair of Materials Engineering of Additive Manufacturing of TUM in cooperation with industry and research partners. The main focus of the chair research strategy is the development of structure-process-property relationships for advanced metallic materials and various additive manufacturing processes. For example, the successful processing of CuW metal matrix composites for power electronics applications using laser powder bed fusion (PBF-LB/M) was demonstrated. It was possible to produce relatively dense parts (98%) with a reduced coefficient of thermal expansion and retaining a good thermal conductivity of 60% compared to pure copper. The improved absorbance of the laser beam's energy for the CuW-alloy system allows additive manufacturing of almost dense CuW parts. The CuW MMC, in combination with the advantages of additive processing, will allow many new applications in power electronics.

Additive manufacturing using PBF-LB/M offers many benefits. However, the intrinsic ultra-fast thermal cyclic heat treatment can result in high porosity and cracking of susceptible alloy systems, such as high-strength aluminum alloys. Within this manuscript, the potential of modifying conventional Al-alloys, including grain-refining elements such as 1.2 wt% titanium, was shown. Crack-free parts with good mechanical properties based on an ultra-fine microstructure have been printed. It was shown that classical heat treatments to improve mechanical properties, such as a T6 heat treatment consisting of solution annealing and aging, can be detrimental to parts produced using PBF-LB/M. The solution annealing leads to significant grain coarsening and inferior mechanical properties. It can be concluded that for additive manufacturing, not only does the chemical composition of alloys have to be adjusted, but special heat treatment cycles also have to be developed as was demonstrated also for the Ni-based superalloys. Additive manufacturing of these materials for high-temperature applications using PBF-LB/M has attracted much interest. However, the application of traditional Ni-based alloys used for forged parts creates problems in additive manufacturing. Aside from a strong susceptibility to hot cracking during printing, PBF-LB/M results in large columnar grains, which result in strong anisotropy of mechanical properties and reduced creep resistance. Here, it was shown that for the alloy IN738LC, special heat treatment procedures must be developed to promote grain coarsening and improve creep properties.

Furthermore, additive manufacturing is a fully digital technology that goes from the design step to the final quality assurance process. This offers the possibility of creating a digital twin fusing machine data, sensor data, and simulation results to create a predictive tool for quality control. This manuscript presents the approach to creating a digital twin of the laser beam melting process. The first results for single-bead deposits are promising and show the potential of establishing a digital twin on process and quality control.

Overall, the huge potential of additive manufacturing of challenging alloy systems for high-tech applications was shown. The specifics of various additive manufacturing processes require at least modifications to alloy systems and heat treatment routines developed for conventional manufacturing methods. In the opinion of the authors, the full potential of additive manufacturing can only be exploited if tailored alloys are developed. Unique microstructural features in additively manufactured parts due to the complex time-temperature history also require novel heat treatments. Herein, computational and high-throughput approaches offer significant advantages over conventional strategies in alloy development and heat treatment design.

Acknowledgments

The study "Laser powder bed fusion of copper-tungsten composites for use in power electronics" was funded by the German Federal Ministry of Economics and Climate Protection (BMWK) as part of the AiF-IGF program as part of a transnational CORNET project (No. 250 EN) based on a resolution by the German Bundestag. Simon Rauh thanks his team at Fraunhofer UMSICHT Lioba Fischer, Shashank Deepak Prabhu, and Gerhard Wolf for their support.

TUM and Oerlikon AM Europe GmbH would like to thank the TUM-Oerlikon Advanced Manufacturing Institute for the support of the research. Oerlikon AM Europe GmbH would like to thank for the support by the Bayerisches Staatministerium für Wirtschaft, Landesentwicklung und Energie (grant number NW-1901-0012) as part of this study.

References

[1] Derekar K S 2018 *Materials Science and Technology* **34** 895–916
[2] DebRoy T, Wei H, Zuback J, Mukherjee T, Elmer J, Milewski J, Beese A, Wilson-Heid A, De A and Zhang W 2018 *Progress in Materials Science* **92** 122–224
[3] Huang S H, Liu P, Mokasdar A and Hou L 2013 *The International Journal of Advanced Manufacturing Technology* **67** 1191–1203
[4] Yakout M, Elbestawi M A and Veldhuis S C 2018 *Solid State Phenomena* **278** 1–14
[5] Leino M, Pekkarinen J and Soukka R 2016 *Physics Procedia* **83** 752–760
[6] Subramanian P R and Laughlin D E 1991 Cu-W (Copper-Tungsten) *Phasen Diagrams of Binary Tungsten Alloys* ed Nagender Naidu S V and Rama Rao P (Calcutta: Indian Institute of Metals) pp 76–79 ISBN 978-8185307107
[7] Wang Y, Zhuo L and Yin E 2021 *International Journal of Refractory Metals & Hard Materials* **100** 105648
[8] Duan J, Guo X, Huang T, Song K, Feng J, Wang X, Zhong J, Duan K and Zhang Y 2023 *Materials Today Communications* **34** 105173
[9] Shao Y, Liu Y, Wu D, Song Z, Guo F and Zou Y 2023 *International Journal of Refractory Metals & Hard Materials* **115** 106315
[10] Wang C C and Lin Y C 2009 *International Journal of Refractory Metals & Hard Materials* **27** 872–882
[11] Liu T, Han Y, Jia D, Pang Z, Fu Y, Song Z and Ding Y 2024 *Applied Sciences* **14** 2731
[12] Rape A, Chanthapan S, Singh J and Kulkarni A 2011 *Journal of Material Science* **46** 94–100
[13] Tejadoa E, Müller A, You J and Pastora J 2018 *Materials Science & Engineering A* **712** 738–746
[14] v Müller A, Böswirth B, Cerri V, Greuner H, Neu R, Siefken U, Visca E and You J H 2020 *Physica Scripta* **2020** 014015
[15] Dong L L, Ahangarkani M, Chen W G and Zhang Y S 2018 *International Journal of Refractory Metals & Hard Materials* **75** 30–42
[16] Gu D and Shen Y 2009 *Journal of Alloys and Compounds* **473** 107–115
[17] Li R, Shi Y, Liu J, Xie Z and Wang Z 2010 *The International Journal of Advanced Manufacturing Technology* **48** 597–605
[18] Zhang D, Cai Q, Liu J, He J and Li R 2013 *The International Journal of Advanced Manufacturing Technology* **67** 2233–2242
[19] Yan A, Wang Z, Yang T Wang Y and Ma Z 2017 *The International Journal of Advanced Manufacturing Technology* **90** 657–666
[20] Wang M, Li R, Yuan T Chen C, Zhang M, Weng Q and Yuan Y 2018 *International Journal of Refractory Metals & Hard Materials* **70** 9–18
[21] Yan A, Wang Z, Yang T, Wang Y and Ma Z 2016 *Materials & Design* **109** 79–87
[22] Qin Y, Wu B, Wang G, Song P, Förster D J, Huang M and Yang S 2023 *Optics & Laser Technology* **162** 109243

[23] Zheng R, Li S F, Misra R D K, Kondoh K and Yang Y F 2023 *Journal of Materials Processing Technology* **322** 118169
[24] Li G, Brodu E, Soete J, Wei H, Liu T, Yang T, Liao W and Vanmeensel K 2021 *Additive Manufacturing* **47** 102210
[25] Rometsch P A, Zhu Y and Xinhua Wu and A H 2022 *Materials & Design* **219** 110779
[26] Martin J H, Yahata B D, Hundley J M, Mayer J A, Schaedler T A and Pollock T M 2017 *Nature* **549** 365–369
[27] Mishra R S and Thapliyal S 2021 *Materials & Design* **204** 109640
[28] Riener K, Nagler Aand Letofsky-Papst I and Leichtfried G 2022 *Alloys* **132** 277–287
[29] Uddin S Z, Murr L E, Terrazas C A, Morton P, Roberson D A and Wicker R B 2018 *Additive Manufacturing* **22** 405–415
[30] Guercio G D, McCartney D, Aboulkhair N, Robertson S, Maclachlan R, Tuck C and Simonelli M 2022 *Additive Manufacturing* **54** 102776
[31] Nie X, Chen Z, Qi Y, Zhang H, Zhang C, Xiao Z and Zhu H 2022 *Virtual and Physical Prototyping* **15** 325–339
[32] Kumar N S, Dhruthi, Pramod G, Samrat P and Sadashiva M 2022 *Materials Today: Proceedings* **58** 71–79
[33] Pereira J C, Gil E, Solaberrieta L, San Sebastián M, Bilbao Y and Rodríguez P P 2020 *Materials Science and Engineering: A* **778** 139124
[34] Mehta B, Frisk K and Nyborg L 2022 *Journal of Alloys and Compounds* **920** 165870
[35] Opprecht M, Roux G, Garandet J P and Flament C 2022 *Journal of Materials Engineering and Performance* **32** 1840–1855
[36] Pauzon C, Buttard M, Després A, Charlot F, Fivel M, Chehab B, Blandin J J and Martin G 2023 *Acta Materialia* **258** 119199
[37] Attallah M M, Jennings R, Wang X and Carter L N 2016 *MRS bulletin* **41** 758–764
[38] Risse J 2019 *Additive manufacturing of nickel-base superalloy IN738LC by laser powder bed fusion* PhD thesis Rheinisch-Westfälische Technische Hochschule Aachen Aachen URL `https://publications.rwth-aachen.de/record/764478`
[39] Song H, Lam M, Chen Y, Wu S, Hodgson P, Wu X, Zhu Y and Huang A 2022 *Journal of Materials Science & Technology* **112** 301–314
[40] Messé O, Muñoz-Moreno R, Illston T, Baker S and Stone H 2018 *Additive Manufacturing* **22** 394–404
[41] Peachey D D, Carter C P, Garcia-Jimenez A, Mukundan A, Leonard D N, Charpagne M A and Cordero Z C 2022 *Additive Manufacturing* **60** 103198
[42] Vogekpoth A, Sharma S and Schleifenbaum J H 2018 Development of Adapted Ni-738 Powder for Additive Manufacturing *Euro PM2018 proceedings: 14-18 October 2018, Bilbao Exhibition Centre (BEC), Bilbao, Spain / Euro PM2018 Congress & Exhibition; European Powder Metallurgy Association* (Shrewsbury, United Kingdom: European Powder Metallurgy Association (EMPA)) URL `https://publications.rwth-aachen.de/record/751611`
[43] MetcoAdd™ Metal Powder Portfolio for Additive Manufacturing last viewed 2024-06-14 URL `https://www.oerlikon.com/ecoma/files/AMPowderPortfolioflyer.pdf`
[44] Svetlizky D, Das M, Zheng B, Vyatskikh A L, Bose S, Bandyopadhyay A, Schoenung J M, Lavernia E J and Eliaz N 2021 *Materials Today* **49** 271–295
[45] Ahn D G 2021 *International Journal of Precision Engineering and Manufacturing-Green Technology* **8** 703–742
[46] Mykoniatis K and Harris G A 2021 *Journal of Intelligent Manufacturing* **32** 1899–1911
[47] Tao F, Cheng J, Qi Q, Zhang M, Zhang H and Sui F 2018 *The International Journal of Advanced Manufacturing Technology* **94** 3563–3576
[48] Lu Y, Liu C, Wang K I K, Huang H and Xu X 2020 *Robotics and Computer-Integrated Manufacturing* **61** 101837
[49] Gaikwad A, Yavari R, Montazeri M, Cole K, Bian L and Rao P 2020 *IISE Transactions* **52** 1204–1217
[50] DebRoy T, Zhang W, Turner J and Babu S S 2017 *Scripta Materialia* **135** 119–124
[51] Mukherjee T and DebRoy T 2019 *Applied Materials Today* **14** 59–65
[52] Gunasegaram D R, Murphy A B, Barnard A, DebRoy T, Matthews M J, Ladani L and Gu D 2021 *Additive Manufacturing* **46** 102089
[53] Hartmann S, Murua O, Arrizubieta J I, Lamikiz A and Mayr P 2024 *Simulation Modelling Practice and Theory* **132** 102881
[54] Chen L, Bi G, Yao X, Tan C, Su J, Ng N P H, Chew Y, Liu K and Moon S K 2023 *Robotics and Computer-Integrated Manufacturing* **84** 102581
[55] Hartmann S, Vykhtar B, Möbs N, Kelbassa I and Mayr P 2024 *Processes* **12** 1180
[56] Zou R, Liang X, Chen Q, Wang M, Zaghloul M A S, Lan H, Buric M P, Ohodnicki P R, Chorpening B, To A C and Chen K P 2020 *Journal of Lightwave Technology* **38** 6402–6411

[57] Hermann F, Chen B, Ghasemi G, Stegmaier V, Ackermann T, Reimann P, Vogt S, Graf T and Weyrich M 2022 *Procedia CIRP* **107** 83–88
[58] Karkaria V, Goeckner A, Zha R, Chen J, Zhang J, Zhu Q, Cao J, Gao R X and Chen W 2024 *Journal of Manufacturing Systems* ISSN 0278-6125
[59] Feng A, Chen C, Wu C, Wei Y and Wang Y 2022 *Metals* **12** 169
[60] Figueroa B S, Araújo L and Alvares A 2024 Development of a digital twin for a laser metal deposition (lmd) additive manufacturing cell *Advances in Automation and Robotics Research* (*Lecture Notes in Networks and Systems* vol 940) ed Cardona M N, Baca J, Garcia C, Carrera I G and Martinez C (Cham: Springer Nature Switzerland) pp 68–76 ISBN 978-3-031-54762-1
[61] Kaufman L and Bernstein H 1970 *Computer Calculation of Phase Diagrams* (Academic Press NY) ISBN 0-12-402050-X
[62] Olson G and Liu Z 2023 *Calphad* **82** 102590
[63] Kattner U R 2016 *Tecnol. Metal. Mater. Min.* **13** 3–15
[64] Wang X and Xiong W 2020 *npj Computational Materials* **6** 188
[65] Carrion Ständer S, Barschkett N and Kabliman E 2023 *Integrating Materials and Manufacturing Innovation* **12** 289–300
[66] Matcalc: Matcalc—solid state and kinetics precipitation last viewed 2024-06-14 URL `https://www.matcalc.at/`

Assessing and controlling microstructure heterogeneity in fusion-based additive manufacturing

M Seita

Department of Engineering, University of Cambridge, Cambridge, United Kingdom

E-mail: ms2932@cam.ac.uk

Abstract. One of the defining features of fusion-based additive manufacturing (AM) is the localized melting of metal by a high-energy source, which fuses the material together point by point and layer by layer into a 3D object. The rapid solidification velocity, directional thermal gradients, and site-specific thermal build-ups produced by this process yield parts with complex and heterogeneous microstructure. This heterogeneity is a double-edged sword. On the one hand, it leads to large property scatter and casts uncertainty over parts performance, hindering the adoption of additive technologies by the industry. On the other hand, it may impart exceptional mechanical properties and new functionalities, which are not found in conventionally produced materials. In this paper, we present two ongoing research endeavours aimed at mitigating the detrimental effects of microstructure heterogeneity in AM, and at capitalizing on the opportunities it offers in the design of novel metal alloys, respectively. The first consists of developing a high-throughput characterization technique to enable large-scale microstructure analysis of AM builds. The second consists of a new strategy to control the material's microstructure site-specifically during laser powder bed fusion.

1. Introduction

Metal additive manufacturing (AM), or 3D printing, has garnered significant attention, even among the general public, due to its revolutionary potential for producing objects with complex and previously unattainable geometries. Notable examples demonstrating these capabilities include the MX3D bridge in Amsterdam [7] and the structural components on the Perseverance Rover [8]. Despite its prominence, engineering applications relying on AM-produced parts remain limited, primarily confined to sectors with substantial capital investment.

One of the main barriers to the widespread adoption of AM is the very feature that makes it unique: the assembly of material "brick by brick." This hierarchical process spans several orders of magnitude in size and is prone to multi-scale variability, as not all "bricks" are equal. This material heterogeneity translates into parts with inconsistent mechanical performance, posing significant challenges for part certification and quality control, particularly in safety-critical applications like aerospace.

Figure 1 illustrates this issue in two different metal alloys produced using different AM technologies. For instance, the block of Inconel 718 produced via directed energy deposition (DED) exhibits a ~10% difference in hardness along the build direction (BD) and a ~20% difference in strength along the transverse direction [4]. Similarly, an array of stainless-steel tensile specimens produced by laser powder bed fusion (LPBF) shows a ~10% difference in strain to failure and a ~25% difference in strength [6].

These issues necessitate the use of specialized practices, such as material-specific heat treatments, aimed at homogenizing the microstructure of parts produced by fusion-based AM technology to improve consistency. These post-processing steps account for a large portion of the production cost, making AM prohibitive for many engineering applications. Addressing the issue of microstructure heterogeneity requires innovations on multiple fronts.

First, there is the practical challenge of assessing this heterogeneity—both within individual parts and across multiple batches—in a time- and cost-effective manner. This is difficult due to the inherently low measurement throughput of conventional electron and X-ray diffraction-based methods, which are typically used for microstructure analysis. Another challenge is understanding the origin of this microstructure heterogeneity during AM, which is a complex function of the processes involved in material formation. This knowledge could be used to control the material's microstructure and produce parts with more consistent and reliable behavior. This paper addresses both aspects by presenting methods for analyzing and controlling the microstructure of metal alloys produced by fusion-based AM.

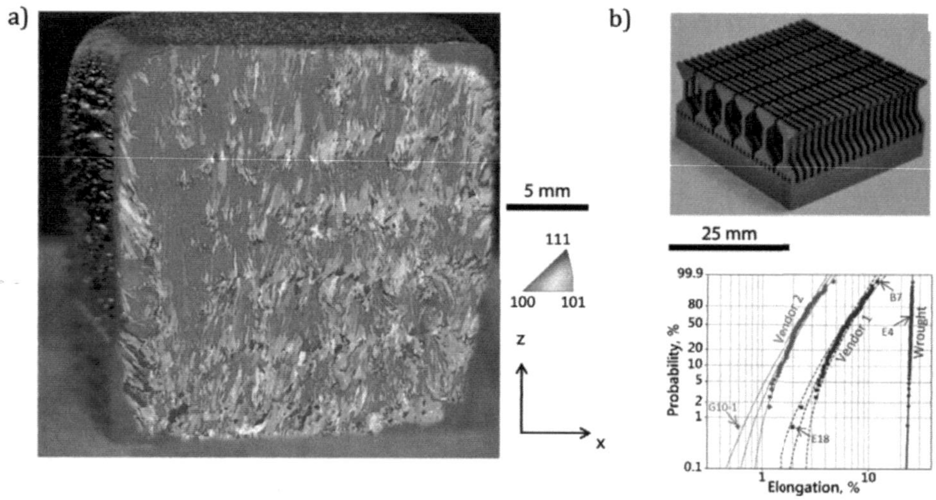

Figure 1. Examples of microstructure heterogeneity in fusion-based AM, and its effect on mechanical properties. a) block of Inconel 718 made by directed energy deposition, which is characterized by electron backscatter diffraction (EBSD). Colours indicate crystal orientation along the z-axis and evince large scale variability in grain size, shape and orientation. Adapted from [4]. b) Example of microstructure heterogeneity across nominally identical specimens produced in the same batch by laser powder bed fusion. Adapted from [6].

2. Advanced optical microscopy for high-throughput microstructure analysis

Traditional microstructure analysis relies on diffraction-based electron and X-ray microscopy techniques. One of the most powerful and widely used among these is electron backscatter diffraction (EBSD) [9], which produces grain orientation maps like the one shown in figure 1.

However, using these techniques to assess microstructure heterogeneity in AM is both time-consuming and resource-intensive. For example, the measurements in figure 1 took approximately 12 hours to complete, at an average cost of $200 per hour[1].

To address these limitations, we developed a novel method called directional reflectance microscopy (DRM). DRM provides the same information as EBSD but at a fraction of the time and cost, and can cover much larger areas [2]. As such, DRM is a promising technique for analyzing the large-scale microstructure heterogeneity found in metal AM [10]. Figure 2 compares EBSD and DRM measurements of an Inconel 718 sample similar to the one in figure 1 [1].

2.1 The working principle of DRM

Directional Reflectance Microscopy (DRM) is entirely based on optical microscopy. Figure 3 shows

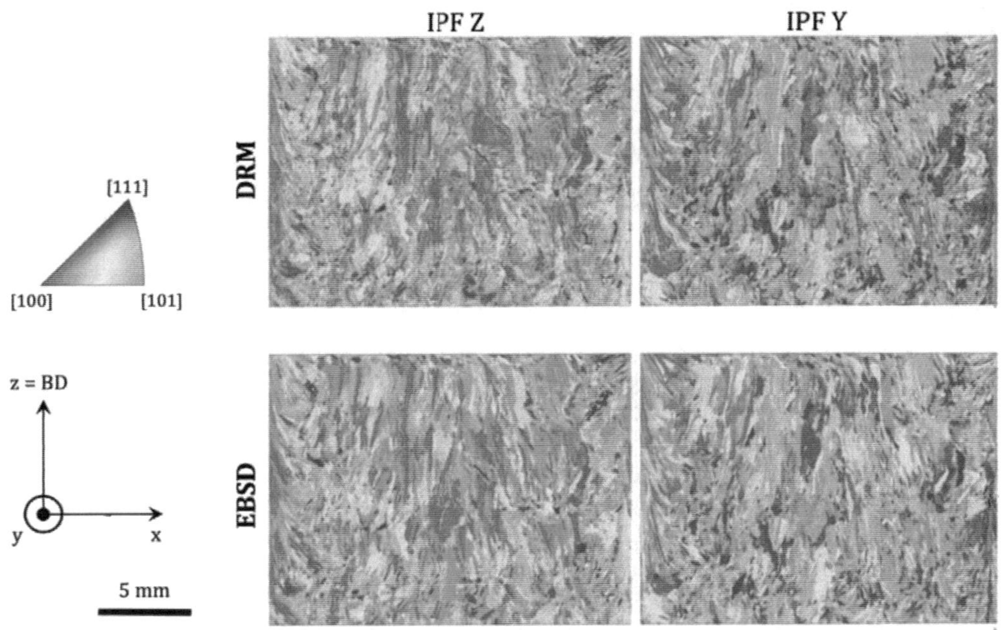

Figure 2. Comparison between DRM and EBSD orientation maps of an Inconel 718 sample produced by DED. The inverse pole figure (IPF) maps show orientation along the z- and y-axis, respectively. The strong similarities between these two sets of measurements demonstrate that DRM can be used to assess the complete orientation information from polycrystalline surfaces. Adapted from [1].

a composite image of a DRM setup, which includes a stereo microscope and a motorized stage that drives a light source around a chemically etched crystalline sample to vary the illumination direction. Due to the local surface topography introduced by the chemical etching process, each grain in the sample reflects light with varying intensity as the light source moves [11]. During

[1] Estimate based on average user facility fee.

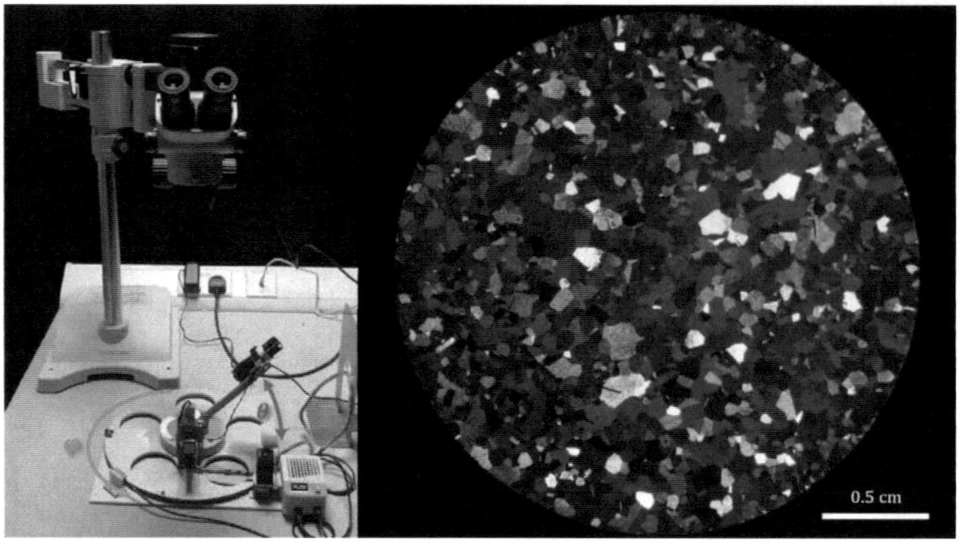

Figure 3. (left) a DRM set up consisting of an optical microscope and a motorized stage to control the illumination angle. (Right) optical micrograph of a chemically etched polycrystalline nickel sample.

DRM, this "directional reflectance" is quantified and compiled into directional reflectance profiles (DRPs) for each grain on the sample surface.

Figure 4 presents several example DRPs from a pure nickel sample alongside the corresponding etch-induced surface topography. DRPs depict the reflection intensity of each point (or grain) on the surface as a function of the light source azimuth (φ) and elevation angle (θ). Notably, DRPs are hollow because, at $\theta > 70°$, the light source obscures the optical microscope camera, making it impossible to capture a micrograph of the etched sample. Peaks in DRPs identify a set of φ and θ angles at which the corresponding grain reflects light into the microscope camera, appearing bright. Conversely, "cooler" regions in DRPs indicate the angles along which the grain appears dark as it reflects light away from the microscope.

DRPs can be considered the optical equivalent of diffraction patterns in EBSD, as they encode crystallographic orientation information. The etch-induced topography on the surface of different grains follows the underlying atomic lattice orientation. By analyzing DRPs, it is possible to derive the geometry of the local surface topography and, knowing its crystallography, determine the orientation of the crystal (figure 4). This process is known as orientation indexing of DRPs. Various methods can be employed to index DRPs, depending on the complexity of the optical signal [12].

2.2 Indexing DRPs in Inconel 718

A suitable chemical etchant for DRM of Inconel 718 is Kalling's 2 reagent. Etching with this reagent creates a surface topography dominated by second-phase intermetallic precipitates (Nb_3Ni). These precipitates form due to higher solute concentration at the boundaries of dendrites that grow during solidification, as shown in figure 5 [4]. These precipitates are responsible for directional reflectance in this alloy. Because they are aligned along the {111}

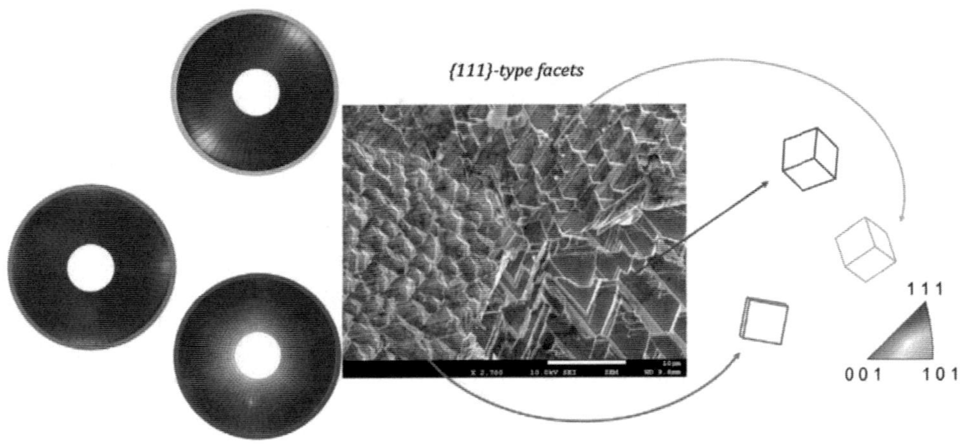

Figure 4. Relationships between directional reflectance, surface topography, and crystal orientation in three example grains from the polycrystalline nickel sample shown in Figure 3. Adapted from [2].

planes of the γ matrix, their reflections can be interpreted as originating from these crystallographic planes, enabling orientation mapping from DRP analysis.

Despite this crystallographic relationship, DRP indexing in this system is challenging. Compared to the DRPs shown in figure 4, the optical signal generated by etched Inconel grains appears "noisy" and is difficult to interpret accurately using traditional image analysis methods. Therefore, DRP indexing of Inconel 718 is best achieved via a machine learning approach [1].

To this end, we conduct both DRM and EBSD measurements on a set of samples produced using different processing conditions (via DED technology) to account for sample variability. We then register the DRM and EBSD datasets taken from each of these samples to relate individual DRPs to the corresponding crystal orientation measured by EBSD. This paired dataset is used to train a neural network—called EulerNet—which we then employ to index DRPs from DED Inconel 718 specimens not included in the initial training set. The orientation maps shown in figure 2 were produced using this approach.

2.3 DRM accuracy

As DRM relies solely on optical microscopy, it doesn't offer the same spatial and angular resolution as EBSD, which boasts a resolution of ~30 nm and <0.1°, respectively. To assess DRM's accuracy, we compute the misorientation angle between the EBSD-measured grain orientations and those predicted by EulerNet using the measurements from figure 2. The histogram in figure 5 illustrates the distribution of these misorientation angles. While the error distribution peaks around 5°, it extends to larger angles (>50°).

The peak reflects the true angular resolution of DRM. Notably, when DRM is applied to pure metals, resulting in clearer DRPs, this figure decreases to approximately 3° [2]. The origin of this discrepancy in performance is unlcear at the moment. Although we have not yet assessed the effect of etching time on the indexing performance of DRM on Inconel 718, previous work on pure aluminum and nickel suggests this as not being a plausible reason [12]. Instead, we believe that the lower performance could stem from the limited size of the training set used for building EulerNet. A larger training set reduces the likelihood of encountering orientations not covered

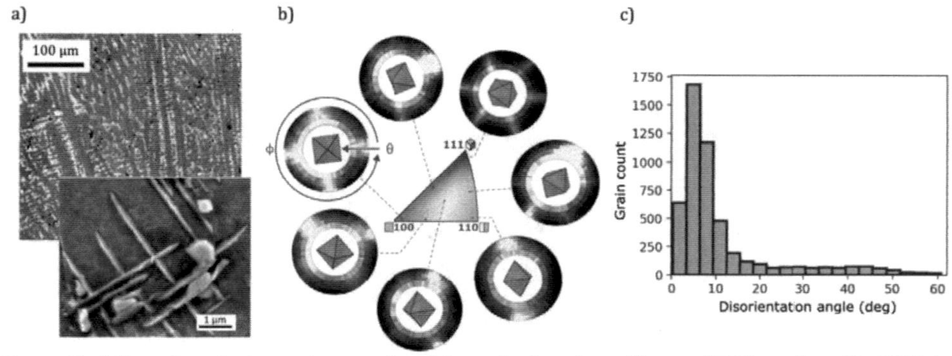

Figure 5. a) Scanning electron micrographs of the etched surface of Inconel 718 produced by DED. In bright contrast are the Ni_3Nb platelets, which precipitate during the aging treatment at locations with higher concentration of Nb (in-between dendrites). b) a sequence of DRPs from grains with different orientation. Unit cells consisting of only {111} planes are visible at the centre of DRPs to highlight the link between crystal orientation and directional reflectance signal. c) DRM measurement error analysis on Inconel 718. Adapted from [1].

during training. Another possible strategy to improve DRM performance on multi-phase alloys would be to capture multiple DRM measurements on the same surface prepared using different etching techniques. The acquired datasets could carry complementery information (e.g., from different phases), which could be used to improve accuracy when combined together. The larger errors in figure 5(c), by contrast, may arise from minor misalignments between the DRM and EBSD datasets. During dataset registration for calculating misorientation angles and error assessment, small grains might be inaccurately paired, introducing uncertainty in DRP indexing. Additionally, regions at grain boundaries are consistently misindexed, as the local DRP captures directional reflectance from both adjacent grains, yielding ambiguous results.

Another aspect to consider is spatial resolution. The pixel size in DRM datasets typically ranges from 5 μm to 12 μm, depending on optical magnification and camera resolution. This is significantly lower than the sub-micron spatial resolution provided by EBSD. For fundamental materials science studies in academia, DRM spatial resolution may be sufficient for grain reconstruction [13] and even intragranular strain mapping [14]. Moreover, even in industrial applications where data acquisition speed and a large field of view are of high priority, compromises in spatial and angular resolution may be acceptable. Finally, it is noteworthy that initial screening of parts by DRM could always be refined using follow-up EBSD on selected regions of interest. In this sense, DRM and EBSD complement each other in microstructure analysis.

2.4 Practical applications of DRM

DRM stands out as a high-throughput characterization technique capable of effectively assessing microstructure heterogeneity in parts produced by AM. Regardless of sample size, a DRM dataset can be acquired in approximately 12 minutes at a fixed field of view. However, training an EulerNet for different materials or when employing various AM technologies is a meticulous process. It entails capturing EBSD and DRM datasets from multiple samples to enhance the robustness of orientation predictions. EulerNet training may take up to 1 hour on a standard laptop computer. Nonetheless, once the training is complete, the same EulerNet can swiftly process any new sample produced, generating orientation maps in just about 12 seconds.

Thanks to its cost-effectiveness and swift processing, DRM proves to be a valuable asset for part certification in industry [15] and for delving into processing-microstructure relationships in academic research.

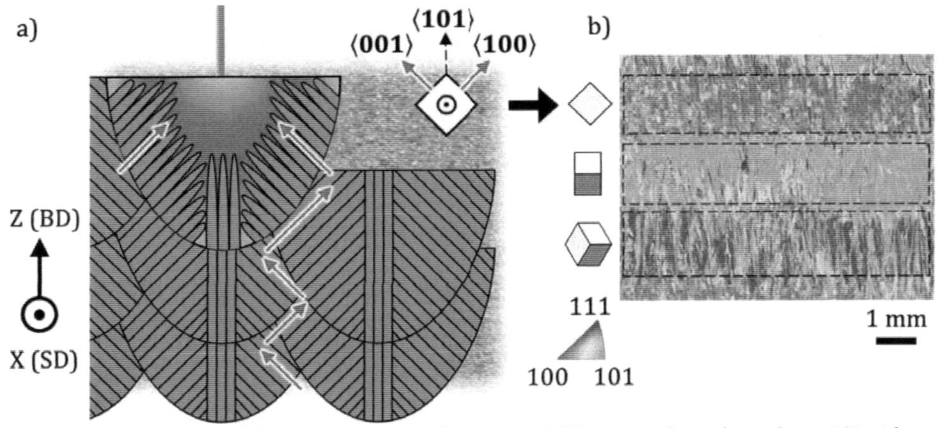

Figure 6. a) schematics of the LEGO process, showing solidification of a melt pool at ±45° with respect to the build direction and zig-zag epitaxial growth when the laser scan angle is kept constant. This process results into the formation of a (quasi-) single crystal with <101> texture along the BD. b) Stack of (quasi-) single crystal with different orientation produced by varying the laser scan angle every 50 layers. The EBSD analysis shows that the regions made keeping laser scan angle constant are not uniform in colour, suggesting that the crystallographic orientation is not exact (hence these blocks are referred to as quasi-single crystal). Adapted from [3].

3. Crystallographic texture control in laser powder bed fusion of stainless steel

Understanding the mechanisms governing microstructure formation during fusion-based AM, such as laser powder bed fusion (LPBF), unlocks the potential to finely adjust processing parameters. This adjustment allows for the design and production of parts with precisely controlled microstructure and reliable behavior. Within this framework, microstructure control serves dual purposes: creating homogeneous microstructures akin to those achieved by conventional manufacturing processes or deliberately engineering heterogeneous microstructures. The latter approach is particularly unique to AM, as its "brick by brick" material assembly process facilitates the modification of the part's constituent "bricks" as needed. In the broadest sense, a "brick" could correspond to a region with consistent phase and crystal orientation content—such as a single crystal grain. Here, we present a specialized LPBF strategy known as layerwise engineering of grain orientation (LEGO) [3], which enables the production of (quasi-) single crystal grains with controlled orientation and morphology.

3.1 LEGO's working principle

During LPBF, a recoater arm traverses a platform, spreading a thin layer of powders. Subsequently, a high-power laser melts the powders by scanning back and forth over areas that need to be consolidated, following a geometry defined in a CAD file. This process involves adjusting parameters such as laser power, velocity, and scan angle to achieve a fully dense part. Zooming in on a single melting event—namely the melt pool generated when the laser strikes and melts the powders—we can observe the solidification of the material through the nucleation and growth of dendrites or cells (figure 6(a)). This solidification occurs along steep thermal gradients, which are typically perpendicular to the melt pool boundaries. As a result, the solidification structures forming will grow inward from these boundaries toward the top-central region of the melt pool.

IOP Conf. Series: Materials Science and Engineering **1310** (2024) 012010 doi:10.1088/1757-899X/1310/1/012010

Within a single layer, adjacent melt pools need to be overlapped by a certain, adjustable amount to prevent leaving unmelted powders in the build. Once a layer is complete, the build platform is lowered by a few tens of microns, the laser scans over the new areas to be consolidated, and the entire sequence is repeated layer by layer.

The crystallographic texture control strategy presented here relies on two fundamental principles:

1. Fast-growing cells/dendrites have a fixed crystallographic orientation.
2. Epitaxial growth of existing crystals is more energetically favorable than the nucleation of new ones.

The first principle stems from the concept that the growth rate of cells/dendrites is anisotropic and fastest along specific crystallographic directions, such as <100> for cubic materials like austenitic stainless steel [16]. Consequently, the majority of the melt pool volume in these materials consists of fast-growing, <100>-textured solidification structures, with varying angles depending on their location along the melt pool boundary, following the thermal gradient maximum.

The second principle implies that when melt pools overlap, the solidifying cells/dendrites inherit the crystallographic orientation of the pre-existing structures in the solid material, which may grow epitaxially via side-branching [17]. Leveraging these two principles allows us to

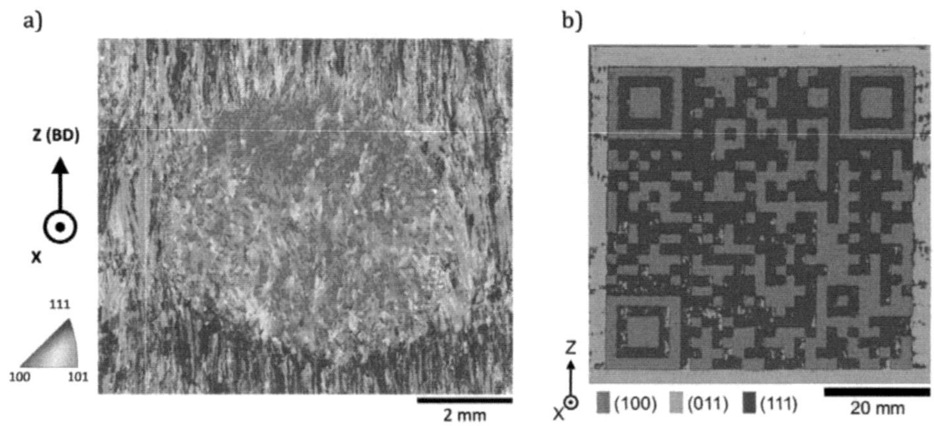

Figure 7. a) EBSD orientation map along the x-axis of a LEGO microstructure with cylindrical grain and curved boundaries. Adapted from [3]. b) DRM orientation map along the x-axis of a large block of steel in which a QR code was encoded utilising texture as the fundamental unit of information. Adapted from [5].

produce (quasi-) single crystal grains by LPBF. By adjusting the laser power and scanning velocity, we can change the shape of the melt pool and thus the angle of the <100>-textured solidification structures. We utilize a combination of parameters that results in a significant portion of these structures being oriented at ±45° with respect to the build direction (BD). Maintaining the same laser scan angle as we process subsequent layers allows these structures to grow epitaxially in a zig-zag fashion throughout the volume (figure 6(a)). By restricting epitaxial growth to occur only along these two directions, we effectively select one single crystal orientation characterized by two <100> axes at ±45° with respect to the build direction.

3.2 Heterogeneous LEGO microstructures

According to the theory outlined in section 3.1, altering the laser scan angle during the build process would disrupt the continuous zig-zag growth and epitaxy. In such a case, new cells/dendrites would need to nucleate, and if the laser scan angle were then kept constant again, they would form a new (quasi-) single crystal. This concept is exemplified in figure 6(b), which displays builds featuring different blocks of (quasi-) single crystals produced using the LEGO strategy. Essentially, this technique enables the fabrication of (quasi-) polycrystals from the bottom-up with controlled orientation, size and morphology. The LEGO microstructure shown in figure 7(a), for instance, contains a cylindrical grains with non-flat boudnaries.

The exploration of these novel microstructures and their properties is still in its infancy and will likely require extensive research efforts due to the vast design space enabled by LEGO. Nevertheless, the potential to construct polycrystals "brick by brick" with highly controlled mechanical or functional properties is exceedingly promising. This advancement not only reduces uncertainty regarding the lifespan and performance of parts manufactured by AM, thereby promoting the adoption of these manufacturing processes in industry, but also unlocks entirely new microstructure designs unattainable via conventional manufacturing methods. These innovations may lead to the development of new engineering applications that have yet to be conceived. An illustrative example of this is depicted in figure 7(b), where the LEGO process is utilized to encode a QR code into a block of steel [5].

4. Conclusions

Microstructure heterogeneity in metal AM is a double-edged sword. While it presents significant challenges, it also offers opportunities for innovation. Techniques like DRM offer promising avenues for understanding the origins of microstructure heterogeneity and enhancing quality control in part production. Meanwhile, processing strategies like LEGO can harness this heterogeneity to develop advanced materials with enhanced properties or novel functionalities.

References

[1] M. Wittwer and M. Seita, "A machine learning approach to map crystal orientation by optical microscopy," *npj Computational Materials,* vol. 8, no. 1, p. 8, 2022, doi: 10.1038/s41524-021-00688-1.

[2] B. Gaskey, L. Hendl, X. Wang, and M. Seita, "Optical Characterization of grain orientation in crystalline materials," *Acta Mater.,* vol. 194, pp. 558-564, 2020/08/01/ 2020, doi: https://doi.org/10.1016/j.actamat.2020.05.027.

[3] K. A. Sofinowski, S. Raman, X. Wang, B. Gaskey, and M. Seita, "Layer-wise engineering of grain orientation (LEGO) in laser powder bed fusion of stainless steel 316L," *Additive Manufacturing,* vol. 38, p. 101809, 2021/02/01/ 2021, doi: https://doi.org/10.1016/j.addma.2020.101809.

[4] Y. Y. Chen *et al.*, "Multiscale microstructural heterogeneity and mechanical property scatter in Inconel 718 produced by directed energy deposition," *J Alloy Compd,* p. 161426, 2021/08/02/ 2021, doi: https://doi.org/10.1016/j.jallcom.2021.161426.

[5] K. Sofinowski, M. Wittwer, and M. Seita, "Encoding data into metal alloys using laser powder bed fusion," *Additive Manufacturing,* vol. 52, p. 102683, 2022/04/01/ 2022, doi: https://doi.org/10.1016/j.addma.2022.102683.

[6] B. C. Salzbrenner *et al.*, "High-throughput stochastic tensile performance of additively manufactured stainless steel," *Journal of Materials Processing Technology,* vol. 241, pp. 1-12, 3// 2017, doi: http://doi.org/10.1016/j.jmatprotec.2016.10.023.

[7] MX3D Bridge last viewed on 07/06/2024 available online: https://mx3d.com/industries/mx3d-bridge/

[8] PIXL MARS ROVER - PRECISION ADDITIVE MANUFACTURING FOR MISSION-CRITICAL PERFORMANCE last viewed on 07/06/2024 available online: https://www.carpenteradditive.com/news-events/pixl-mars-rover

[9] F. J. Humphreys, "Quantitative metallography by electron backscattered diffraction," (in English), *Journal of Microscopy-Oxford,* vol. 195, pp. 170-185, Sep 1999. [Online]. Available: <Go to ISI>://WOS:000082895200002.

[10] M. Seita, M. Wittwer, and X. Wang, "Optical Metallography of Fusion-Based Additively Manufactured Metals," in *Reference Module in Materials Science and Materials Engineering*: Elsevier, 2020.

[11] M. Seita, M. M. Nimerfroh, and M. J. Demkowicz, "Acquisition of partial grain orientation information using optical microscopy," *Acta Mater.,* vol. 123, pp. 70-81, 1/15/ 2017, doi: http://dx.doi.org/10.1016/j.actamat.2016.10.021.

[12] C. Zhu and M. Seita, "A physics-based model for crystal orientation dictionary indexing by directional reflectance microscopy," *Acta Mater.,* vol. 231, p. 117863, 2022/06/01/ 2022, doi: https://doi.org/10.1016/j.actamat.2022.117863.

[13] M. Wittwer, B. Gaskey, and M. Seita, "An automated and unbiased grain segmentation method based on directional reflectance microscopy," *Mater Charact,* vol. 174, p. 110978, 2021/04/01/ 2021, doi: https://doi.org/10.1016/j.matchar.2021.110978.

[14] X. Wang, S. Gao, E. Jain, B. Gaskey, and M. Seita, "Measuring crystal orientation from etched surfaces via directional reflectance microscopy," *J. Mater. Sci.,* 2020/05/05 2020, doi: 10.1007/s10853-020-04734-z.

[15] A. T. Clare, M. Seita, A. Speidel, P. Collins, and M. Clark, "Driving next generation manufacturing through advanced metals characterisation capability," *Scripta Mater,* vol. 246, p. 116009, 2024/06/01/ 2024, doi: https://doi.org/10.1016/j.scriptamat.2024.116009.

[16] W. A. Tiller, "Preferred growth direction of metals," *Jom-Us,* vol. 9, no. 7, pp. 847-855, 1957/07/01 1957, doi: 10.1007/BF03397928.

[17] M.-S. Pham, B. Dovgyy, P. A. Hooper, C. M. Gourlay, and A. Piglione, "The role of side-branching in microstructure development in laser powder-bed fusion," *Nat. Commun.,* vol. 11, no. 1, p. 749, 2020/02/06 2020, doi: 10.1038/s41467-020-14453-3.

Building microstructures by welding millions of little bits of metal together: measurement approaches, model validation, and post-build processing

L E Levine[1*], E J Schwalbach[2] and F Zhang[1]

[1] Materials Measurement Laboratory, National Institute of Standards and Technology, Gaithersburg, MD, USA
[2] Materials and Manufacturing Directorate, Air Force Research Laboratory, Dayton, OH, USA

*E-mail: Lyle.Levine@nist.gov

Abstract. Over the past decade, immense, world-wide research and development efforts have attempted to bring additive manufacturing (AM) of metals into broad industrial use. Although major advances in our understanding and control over metal AM have accrued, its practical application has been underwhelming. The slow pace of commercial adoption can be traced to numerous factors, including poor build reproducibility, sensitivity to heterogeneous local processing conditions during the build, complications in developing and validating suitable processing-structure-property-performance (PSPP) simulation capabilities (physics-based models, surrogate models, and machine learning models), the small number of alloys suitable for metal AM, and the need for developing new alloy-specific post-processing protocols. All these factors negatively impact the purely business decision of what manufacturing approach should be used to produce a given component. Computational materials engineering approaches could play a major role in accelerating the adoption of metal AM, but rigorous model validation will be necessary to make this a reality. Here, discussion will focus on development and deployment of measurement approaches for model validation, and methodologies for accelerating development of post-build heat treatment through thermo-kinetic modelling and in situ measurements.

1. Introduction

Metals additive manufacturing (AM) is a broad set of manufacturing methods that additively build three-dimensional (3D) metal parts from a digital design. Most metal AM approaches use an energy source (e.g., laser, electron beam, electrical arc) to locally melt a feedstock material (e.g., powder, wire), which then resolidifies. Thus, the parts and the material are produced at the same time. Because the local processing conditions depend upon variables such as the scan path and the local part geometry, the microstructures and the resulting material properties are typically heterogeneous within the part. This variation in material properties greatly complicates part qualification, which typically relies on extensive testing of standard coupons. Such tests are of little value if the microstructures and material properties of the test coupons do not match the material within the component.

Incorporation of computational materials (CM) engineering methods into the qualification and certification (Q&C) processes for AM has been put forward as an approach for dramatically decreasing the time and cost of AM Q&C. Recently, a multi-year collaborative effort in the United States including representatives from industry, regulatory agencies, national laboratories, and universities completed work on a comprehensive document describing how CM engineering approaches can be incorporated into the formal AM Q&C process for aviation [1,2]. A key requirement for incorporating CM engineering methods into Q&C is rigorous model validation.

Model validation measurements are distinct from more traditional exploratory experiments; instead of being conducted to provide insight into underlying physical processes, validation measurements are specifically designed to test key aspects of predictive models. Critical requirements include:

- quantitative comparison metrics,
- measurement uncertainties,
- adequate information to fully define the model boundary conditions.

When validating multi-scale and multi-physics simulation capabilities, such as those typically required for AM, additional requirements include:

- measurements for validating the inputs and outputs of the coupled codes being utilized,
- multiple coordinated validation measurements covering the full multi-scale and multi-physics range (here referred to as hierarchical measurements).

Here, examples of validation measurements for AM-built nickel alloy 625 test specimens are taken from the National Institute of Standards and Technology (NIST) led Additive Manufacturing Benchmark Test Series (AM Bench), including 3D microstructure characterization of as-built material using automated serial sectioning and in situ characterization of precipitate and phase evolution during heat treatment using synchrotron-based ultra-small angle X-ray scattering (USAXS), small-angle X-ray scattering (SAXS), and wide-angle X-ray scattering (WAXS). The USAXS/SAXS/WAXS measurements are used to validate post-build heat treatments and can also be used to extract activation energies for the corresponding microstructural changes.

2. Alloy 625 builds

The builds and measurements described here are part of broad sets of hierarchical measurements designed to provide coupled validation datasets encompassing the full range of processing-structure-properties (PSP). The validation measurements were conducted as part of AM Bench 2018 [3] and AM Bench 2022 [4], within the AMB2018-01 and AMB2022-05 sets of benchmarks. For detailed information on these and related benchmarks, see the AM Bench website at <www.nist.gov/ambench>, the overview papers [3,4], and the measurement papers referenced therein. Links to all the published digital datasets may be found at <www.nist.gov/ambench/direct-am-bench-data-links-and-referencing-guidance>.

The alloy 625 samples were built using laser powder bed fusion (PBF-LB). Full details of the scan strategy, process parameters, and extensive in situ measurements conducted during the build may be found in [5]. Figure 1 shows the bridge geometry of the test specimens.

3. Serial sectioning measurements

As part of AM Bench 2018, extensive scanning electron microscopy measurements were conducted using secondary electron imaging (SE) and backscattered electron imaging (BSE)

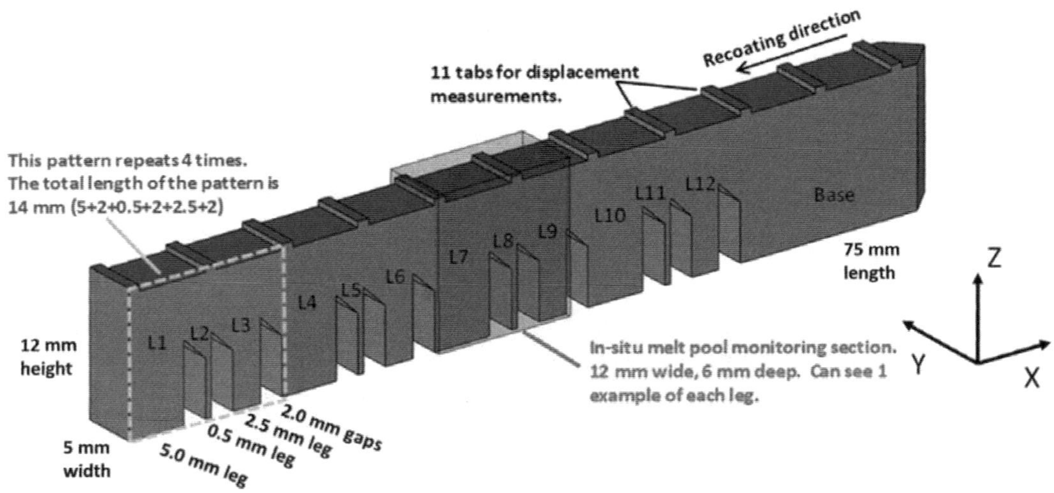

Figure 1. Geometry of alloy 625 PBF-LB samples built for these benchmark measurements.

techniques, energy dispersion spectroscopy (EDS), and electron backscatter diffraction (EBSD) on prepared specimen cross sections [6]. These measurements were used to characterize the melt pools, grain structure, and sub-grain structure from legs of different thickness both perpendicular and parallel to the build direction.

Continued interaction with the modeling community identified additional microstructure measurements that were needed, including larger-area SEM measurements that characterize the microstructural changes from the build plate to above the legs into the bridge region. The data from these measurements may be found online in reference [7]. Another set of measurements that was frequently requested was three-dimensional (3D) characterization of the grain shapes. Synchrotron X-ray methods were considered to obtain these data, but the high dislocation density of the as-built parts would smear out the diffraction patterns making such measurements untenable. Automated serial sectioning with SEM-based characterization could provide the required data and a collaboration between NIST, the U.S. Air Force Research Laboratory (AFRL), and Lawrence Livermore National Laboratory (LLNL) was assembled to design and execute the necessary experiments. Similar serial sectioning measurements have been reported in the literature [8-10].

Figure 2 shows the geometry of the serial sectioned region of the sample. Each serial sectioning step proceeds by first removing approximately 2 μm of material, including a final polish with 40 nm colloidal silica in an automated polishing system. The specimen is then transferred to an optical microscope to collect bright-field images that are combined into a full cross section. Next, the specimen is transferred to an SEM for BSE and EBSD imaging. The individual datasets are post-processed to produce a 3D reconstruction of the region of interest (ROI) using a range of software tools. Detailed descriptions of the measurements and processing steps used to achieve the final 3D reconstruction are available online [11] under benchmark AMB2022-03 and in reference [12]. All the raw and processed measurement data and 3D reconstructions have been published online [13].

Figure 3(a) and 3(b) show two views of the 3D reconstruction using an inverse pole figure

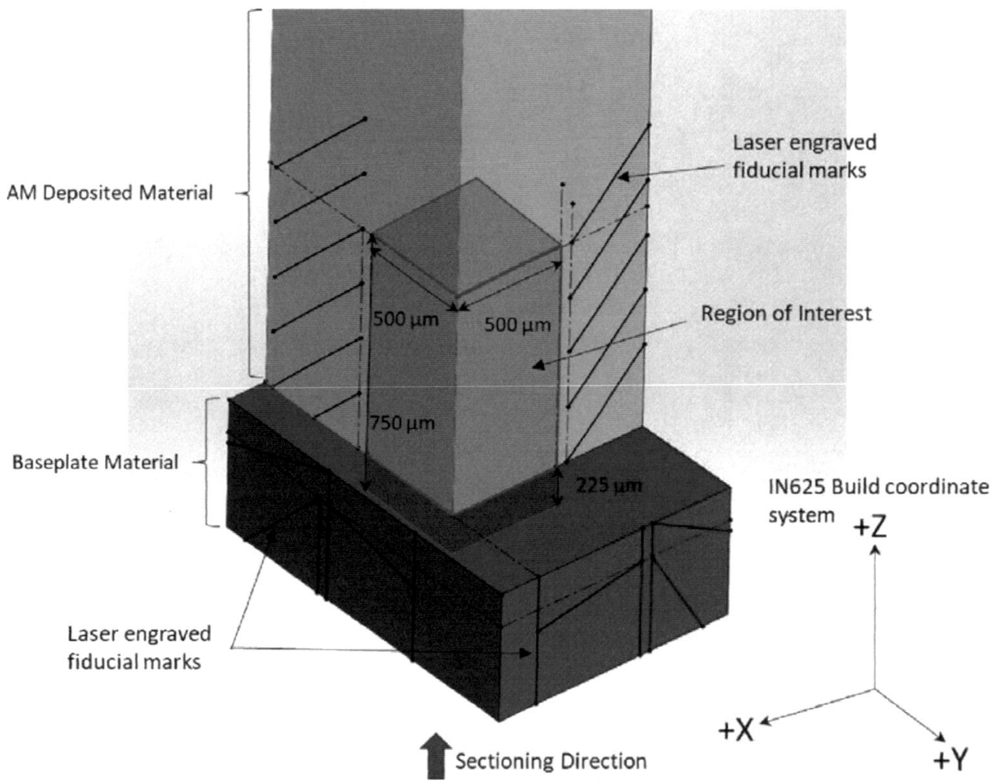

Figure 2. Location on Leg 12 where sample is extracted for serial sectioning measurements.

(IPF) colormap relative to the Z direction (IPFZ). In all, 510 sections were measured and the volume spans 750 μm × 750 μm × 918 μm in X, Y, Z, respectively. The spatial resolution along these directions is 1 μm × 1 μm × 1.8 μm. Figure 3(c) is a cross sectional view perpendicular to the X direction that shows the transition from the small-grain baseplate microstructure to large-grain as-built material that is elongated in the Z (build) direction. Figure 3(d) shows the largest grains where grain boundaries are defined by misorientations between adjacent pixels of ≥ 2°. Using this definition, these large grains all start near the transition to the baseplate and extend beyond the measurement field of view in the Z direction. Another key observation is that the local orientation of a single grain can change gradually with position, especially along the build direction. These grain rotations can be quite large, extending over tens of degrees. To the best of our knowledge, no microstructure evolution or property prediction codes account for these rotations. The gradual accumulation of geometrically necessary dislocations during the build process is one possible mechanism.

Figure 3. Views of the 3D reconstruction using an IPF colormap relative to the Z (build) direction. Panels a) and b) show images of the ROI from two orientations, panel c) shows a cross section that includes the transition from the baseplate into the as-built region, and panel d) shows a view of the largest grains in the measured region.

4. USAXS/SAXS/WAXS measurements

Microstructure measurements such as those described above characterize the local grain shapes, sizes, and crystallographic orientation of a given sample. However, if information is required concerning the evolution of the microstructure as a function of both time and temperature, conducting such measurements for the requisite large number of sample conditions can be prohibitively expensive and time consuming. This is particularly true when statistically reliable information is needed on aspects such as phase fractions and precipitate densities. The development of new heat treatments for AM-built alloys is a good example where in situ measurements using synchrotron X-ray approaches can provide significant reductions in the development time.

Heat treatments for metal alloys have been developed over many years and assume starting microstructures typical for long-used manufacturing approaches such as casting, forging, and conventional welding. However, the high cooling rates for AM processes produce microstructures that are considerably different from those produced by these more traditional approaches; new heat treatments are required to produce material with the required properties and reliability. A good example is alloy 625 produced using PBF-LB.

A typical heat treatment for wrought alloy 625 to relieve residual stress is 870 °C for 1 h [14]. However, using this heat treatment for alloy 625 components built using PBF-LB produces large amounts of the deleterious δ-phase, approximately 2 % volume fraction [15]. This δ-phase precipitation is facilitated by the existence of strong compositional gradients in the solidification microstructure produced by the high cooling rates [16], typically up to 1 x 10^6 °C/s. To reduce the amount of δ-phase in the stress relieved material, Lass et al. [17] recommended decreasing the annealing temperature to 800 °C and demonstrated that 4 h at this temperature produced less δ-phase than 1 h at 870 °C.

To provide δ-phase benchmark data for this 800 °C stress relief heat treatment, USAXS/SAXS/WAXS measurements were conducted on material from the AM-built bridge specimens described above using the beamline 9-ID-C USAXS facility at the Advanced Photon Source (APS) [18]. To explore the possible effect of build thickness, the as-built samples were prepared from central regions of both 0.5 mm thick and 5.0 mm thick legs (see figure 1). The USAXS measurements were conducted using a Bonse-Hart-type optics geometry [19] with a beam size of 0.8 mm x 0.8 mm and a sample thickness of ≈ 40 μm. Thus, the measurements probed a statistically significant volume of ≈ 2.6 x 10^7 μm^3. The combined accessible range of scattering vector, q, from USAXS and SAXS was from 1×10^{-4} $Å^{-1}$ to 1.5 $Å^{-1}$. Here, $q=4\pi \sin(\theta)/\lambda$, where θ is one half of the scattering angle and λ is the X-ray wavelength. The WAXS measurements were conducted using a single-photon sensitive area detector giving a q range of 1.4 $Å^{-1}$ to 6.8 $Å^{-1}$ that overlaps the q range of the combined USAXS/SAXS measurements. The WAXS measurements used a beam size of 0.2 mm x 0.8 mm. A full set of USAXS/SAXS/WAXS measurements took ≈ 5 min and each in situ annealing experiment lasted at least 2 h. Detailed descriptions of the measurements and analysis may be found in reference [20].

Figure 4(a) shows WAXS data acquired from an AM alloy 625 sample during a 2 h anneal at 800 °C. The sample was produced from one of the 5.0 mm thick legs. The inset panel shows the steady growth of peaks associated with the δ-phase. At the beginning of the anneal, there are no measurable δ-phase peaks, indicating that this phase was not present in the as-built condition. Measurable amounts of δ-phase appear after just five minutes of annealing. Figure 4(b) shows the corresponding USAXS data acquired during the same experiment. The changing shapes of the scattering curves are consistent with growing disk-shaped precipitates that are also consistent

IOP Conf. Series: Materials Science and Engineering **1310** (2024) 012011 doi:10.1088/1757-899X/1310/1/012011

with electron microscopy measurements. The average thickness and diameter of the precipitates can be extracted by fitting scattering models to the data.

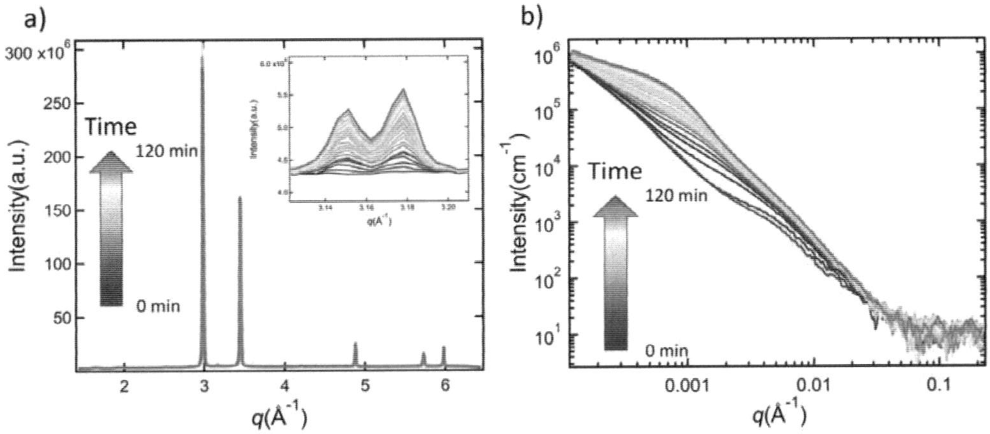

Figure 4. a) Time resolved WAXS data acquired during anneal of AM-built alloy 625 at 800 °C. b) USAXS data acquired during the same experiment shown in a). Modified from reference [19].

Figure 5 shows various aspects of the evolving precipitates as determined from the WAXS and USAXS data shown in figure 4. Figure 5(a) plots the integrated intensity of the δ-phase {211} reflection as a function of annealing time. This intensity is approximately proportional to the volume of δ-phase within the probed sample volume and it increases steadily throughout the experiment. Figures 5(b) and 5(c) show the extracted average precipitate thickness and diameter, respectively, as a function of annealing time. The precipitate thickness initially grows rapidly and then slows, reaching a maximum thickness of ≈ 62 nm. After the first 5 min, the average

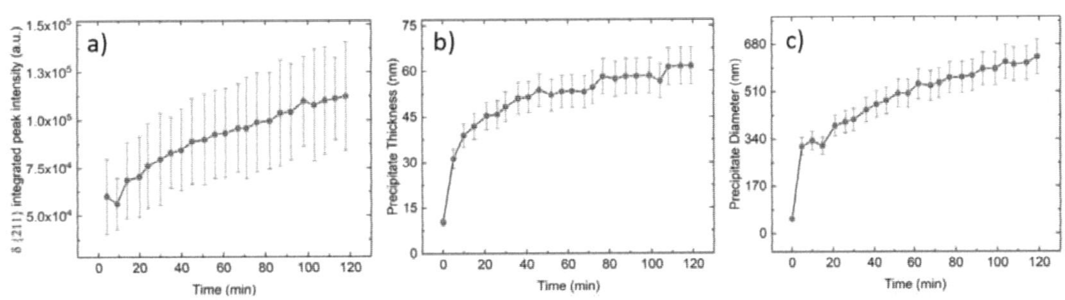

Figure 5. Aspects of the δ-phase precipitates as determined from the WAXS and USAXS data shown in Figure 4. These include, a) the integrated intensity of the {211} reflection, b) the precipitate thickness, and c) the precipitate diameter. The measurement uncertainties are described in the text. Modified from reference [19].

precipitate diameter increases at a slowly decreasing rate up to ≈ 631 nm after 2 h. The measurement uncertainties in figure 5 are one standard deviation and were determined from fits to the data. For the WAXS data, the uncertainties come from peak fitting. For the USAXS data, the analysis works by randomly selecting input data within the estimated uncertainty of the intensity,

IOP Conf. Series: Materials Science and Engineering **1310** (2024) 012011 doi:10.1088/1757-899X/1310/1/012011

considering q to be absolute calibrated. The analysis is then run multiple times (typically 10 to 25 iterations). From these runs, the average and standard deviation are calculated.

5. Conclusions

As described in the Introduction, the serial sectioning and USAXS/SAXS/WAXS results presented here are just a small part of a broad set of continuing measurements designed to guide and validate computational simulations of AM processes. For AM Bench 2018 and 2022, these include measurements for both metals and polymers and cover the full range of processing-structure-properties-performance (PSPP). Some of the microstructure-focused measurements described here are also part of formal challenge problems for the AM community. For example, the USAXS/SAXS/WAXS measurements are part of a broader phase evolution challenge (CHAL-AMB2018-01-PFRS), where participants were asked to predict the evolution of phases and phase fractions, including major precipitates, as a function of time for residual stress anneals of alloy 625 and 15-5 stainless steel, from transverse specimens cut from thick and thin legs.

Looking ahead, the next round of AM Bench is scheduled for late summer of 2025, and extensive sets of benchmark measurements are underway. Because many of these measurements are tied to formal AM Bench challenge problems for the modeling community, detailed information cannot be provided prior to our formal challenge problem release currently scheduled for the end of September, 2024. Speaking in general terms, there are three main new areas that AM Bench is working to expand into:

- New material class: AM Bench 2018 and 2022 only supported metals and polymers AM. In addition to these material classes, one of the upcoming sets of 2025 benchmarks is currently targetting ceramics AM.
- New build methods: For metals, AM Bench 2018 and 2022 concentrated on PBF-LB and this work will continue for AM Bench 2025. However, new efforts with new partners are underway to support larger builds produced using wire feedstock materials.
- Greater emphasis on industrial challenges: Although AM Bench has always been focused on industry's needs, most of the earlier benchmarks had a strong fundamental component. While continuing to support measurements of the underlying physics of AM, AM Bench will build upon this foundation to directly address issues important to industry, including broader measures of mechanical behavior and the role of composition and build defects.

In addition to being the world's largest provider of rigorous measurement data for AM simulations worldwide, AM Bench is committed to building bridges between communities. AM Bench challenge problems are providing focus and direction to simulation developments across the world and allow the strengths and weaknesses of differing approaches to be evaluated. These efforts are serving to bring the AM measurement and simulation communities together. AM Bench has also worked to build bridges between the United States and other parts of the world. Numerous research groups from Europe and Asia have submitted simulation results to AM Bench challenge problems, and the AM Bench conference series has recently reached a partnership agreement with the European AM conference series, Sim-AM, where both conferences would host joint symposia. AM Bench would also be open to collaborating more closely with the European community on developing and executing future benchmark measurements in Europe. Lastly, AM

Bench is strongly motivated to provide value to industries who are working to increase the role of AM in manufacturing. Many of the AM Bench measurements and challenge problems are developed in response to industrial feedback, and AM Bench 2022 hosted an embedded workshop on AM Q&C for the aviation industry.

If you have ideas for useful AM benchmark measurements or if you want to become involved more closely as an AM Bench partner, please contact the AM Bench organizers.

References

[1] Levine L, Lane B, Glaessgen E, Gorelick M 2024 *JOM* **76** 1897–1904 https://doi.org/10.1007/s11837-024-06388-7

[2] Glaessgen E, Gorelick M, Levine L, Battaile C, Kane M, Phan N, Plotkowski A, Schwalbach E, Lamm D, Raghavan N, Ashmore N, Barto R, Fodran E, Heinimann M, Kantzos P, Peralta-Duran A, Popelar C, Toivonen P, Venkatesh V, Vogel B, Whitis D, Millwater H, Rollett A, Holm E, Mahadevan S, Oskay C, Palmer T, Wagner G. In preparation.

[3] Levine L E, Lane B M, Heigel J C, Migler K B, Stoudt M R, Phan T Q, Ricker R E, Strantza M, Hill M R, Zhang F, Seppala J E, Garboczi E J, Bain E D, Cole D P, Allen A J, Fox J, Campbell C E 2020 *Integr. Mater. Manuf. Innov.* **9** 1–15 https://doi.org/10.1007/s40192-019-00164-1

[4] Levine L E, Lane B M, Becker C, Belak J, Carson R, Deisenroth D, Glaessgen E, Gnaupel-Herold T, Gorelik M, Greene G, Habib S, Higgins C, Hill M, Hrabe N, Killgore J, Kim J W, Lemson G, Migler K, Rowenhorst D, Moylan S, Pagan D, Phan T, Praniewicz M, Simonds B, Stoudt M R, Schwalbach E, Seppala J, Weaver J, Yeung H, Zhang F. *Integr. Mater. Manuf. Innov.* submitted

[5] Heigel J C, Lane B M, Levine L E 2020 *Integr. Mater. Manuf. Innov.* **9** 31–53 https://doi.org/10.1007/s40192-020-00170-8

[6] Stoudt M R, Williams M E, Levine L E, Creuziger A, Young S A, Heigel J C, Lane B M, Phan T Q 2020 *Integr. Mater. Manuf. Innov.* **9** 54-69 https://doi.org/10.1007/s40192-020-00172-6

[7] Levine L, Rowenhorst D, Maich A, Creuziger A, Carson R, Raghavan N 2023 AM Bench 2022 Microstructure Measurement Follow-On for AM Bench 2018 IN625 3D Builds, National Institute of Standards and Technology, https://doi.org/10.18434/mds2-2712 (Accessed 2024-04-19)

[8] Chapman M G, Shah M N, Donegan S P, Scott J M, Shade P A, Menasche D, Uchic M D 2021 *Integr. Mater. Manuf. Innov.* **10** 129–141 https://doi.org/10.1007/s40192-021-00212-9

[9] Nguyen L T, Rowenhorst D J 2021 *JOM* **73** 3272–3284 https://doi.org/10.1007/s11837-021-04865-x

[10] Polonsky A T, Lang C A, Kvilekval K G, Latypov M I, Echlin M P, Manjunath B S, Pollock T M 2019 *Integr. Mater. Manuf. Innov.* **8** 37–51 https://doi.org/10.1007/s40192-019-00126-7

[11] <www.nist.gov/ambench/am-bench-2022-measurements-and-challenge-problems>

[12] Schwalbach E J, Chapman M G, Shah M N, Uchic M D, Levine L E, Hrabe N, Kafka O L, Moser N H, Carson R, Belak J *Integr. Mater. Manuf. Innov.* In preparation

[13] Schwalbach E J, Chapman M G, Shah M N, Uchic M D, Levine L E, Hrabe N, Kafka O, Moser N, Carson R, Belak J 2023 AM Bench 2022 IN625 3D Microstructure Reconstructions, National Institute of Standards and Technology, https://doi.org/10.18434/mds2-2765 (Accessed 2024-04-19)

[14] Chandler H (ed) 1996 *Heat treater's guide: practices and procedures for nonferrous alloys* (Materials Park, OH: ASM International) p 657

[15] Stoudt M R, Lass E A, Ng D S, Williams M E, Zhang F, Campbell C E, Lindwall G, Levine L E 2018 *Metall. Mater. Trans. A* **49** 3028–3037 https://doi.org/10.1007/s11661-018-4643-y

[16] Keller T, Lindwall G, Ghosh S, Ma L, Lane B M, Zhang F, Kattner U R, Lass E A, Heigel J C, Idell Y, Williams M E, Allen A J, Guyer J E, Levine L E 2017 *Acta Mater.* **139** 244-253 https://doi.org/10.1016/j.actamat.2017.05.003

[17] Lass E A, Stoudt M R, Williams M E, Katz M B, Levine L E, Phan T Q, Gnaeupel-Herold T H, Ng D S 2017 *Metall. Mater. Trans. A* **48** 5547–5558 https://doi.org/10.1007/s11661-017-4304-6

[18] Ilavsky J, Zhang F, Andrews R N, Kuzmenko I, Jemian P R, Levine L E, Allen A J 2018 *J. Appl. Crystallogr.* **51** 867–882 https://doi.org/10.1107/S160057671800643X

[19] Ilavsky J, Jemian P R, Allen A J, Zhang F, Levine L E, Long G G 2009 *J. Appl. Crystallogr.* **42** 469–479 https://doi.org/10.1107/S0021889809008802

[20] Zhang F, Levine L E, Allen A J, Young S W, Williams M E, Stoudt M R, Moon K W, Heigel J C, Ilavsky J 2019 *Integr. Mater. Manuf. Innov.* **8** 362-377 https://doi.org/10.1007/s40192-019-00148-1

IOP Conf. Series: Materials Science and Engineering **1310** (2024) 012012 doi:10.1088/1757-899X/1310/1/012012

Dislocation structure evolution during metal additive manufacturing

M V Upadhyay[1*], S Gaudez[1] and W Pantleon[2]

[1] Laboratoire de Mécanique des Solides (LMS), École Polytechnique, Institut Polytechnique de Paris, CNRS UMR 7649, Palaiseau 91128, France
[2] Department of Civil and Mechanical Engineering, Technical University of Denmark (DTU), Kongens Lyngby, Denmark

*E-mail: manas.upadhyay@polytechnique.edu

Abstract. Dislocation structures are abundantly present in any additively manufactured alloy and they play a primary role in determining the mechanical response of an alloy. Until recently, it was understood that these structures form due to rapid solidification during AM. However, there was no consensus on whether they evolve due to the subsequent solid-state thermal cycling that occurs with further addition of layers. In order to design alloy microstructures with desired mechanical responses, it is crucial to first answer this outstanding question. This question was answered in a recent work [1] involving a novel experiment employing high resolution reciprocal space mapping, a synchrotron-based X-ray diffraction technique, *in situ* during AM of an austenitic stainless steel. The study revealed that dislocation structures formed during rapid solidification undergo significant evolution during subsequent solid-state thermal cycling, in particular during addition of the first few (up to 5) layers above the layer of interest. A summary of the findings of this study are presented in this work. A possible pathway (involving experiment and modelling synergy) to better understanding dislocation structure formation during AM is presented.

1. Introduction

Additive manufacturing (AM), or 3D printing, of alloys is a revolutionary technology involving reading a 3D part geometry from a computer-aided design file and building it in a layer-by-layer manner by locally melting a feedstock (e.g., powder or wire) using a moving heat-source (e.g., laser or electron beam). AM has the unique ability to simultaneously build complex part geometries and create appropriate material microstructure without requiring additional tooling. In comparison to conventional processing techniques, AM can save significant amounts of energy and cost during manufacturing. It is becoming a top-choice advanced manufacturing technique for specialized applications such as: (i) utilizing new alloy systems, (ii) building lighter parts with novel geometries, (iii) replacing existing parts with better performing ones, (iv) repairing damaged or broken parts, among others.

During AM, just after melting of the feedstock, the material undergoes melt-pool dynamics and rapidly solidifies within a few milliseconds. For the rest of the building time, the material undergoes heating-cooling cycles in the solid-state, i.e. solid-state thermal cycling (SSTC), also known as intrinsic heat treatment, at varying temperature amplitudes and heating/cooling rates. These non-equilibrium processes result in the formation of hierarchical microstructures exhibiting physical and chemical heterogeneity at multiple length scales [2–9] that are much more varied than those in their conventionally manufactured counterparts.

Designing microstructures that exhibit desired mechanical responses is a tremendous challenge. It requires altering the AM parameters to obtain appropriate microstructures. Achieving such control needs a thorough understanding of the role of the different non-equilibrium processes occurring during AM on microstructure genesis and evolution.

Among the different possible microstructural features, dislocations are abundantly present in any as-built alloy. They mainly organize in intragranular cell structures with a high dislocation density in cell walls and a low density in cell interiors. These dislocation cells play a primary role in determining type III residual stresses and enhancing the yield strength of the material, which is often accompanied by a lower ductility [5–7]. However, until recently, the role of the different non-equilibrium processes during AM on the formation of these cells was an open question.

There were two main hypotheses on their origin in as-built alloys. Both hypotheses were based on post-built electron microscopy studies. One hypothesis was that these cells form in inter-dendritic regions during rapid solidification and do not significantly evolve, if at all, during SSTC. This hypothesis was based on observations of dislocation cells coinciding with microsegregations in inter-dendritic regions [6,10]; the latter are a signature of directional solidification in AM alloys. The opposing hypothesis is that the thermo-mechanical forces occurring during SSTC result in significant changes in the dislocation structures formed during rapid solidification. This hypothesis is motivated from other post-build electron microscopy studies [11–15] presenting observations that dislocation cells do not always coincide with microsegregations. Since both hypotheses are based on studies performed on as-built samples, it is impossible to use the findings to univocally separate the role of SSTC from that of solidification on dislocation structure formation and evolution during AM. A clear insight into the effects of the different processes can only be gained by tracking the evolution of the dislocation structure during AM.

To that end, Gaudez et al. [1] recently conducted a novel experiment involving high-resolution reciprocal space mapping (HRRSM), a synchrotron-based X-ray diffraction technique, while building samples from an austenitic stainless steel. HRRSM involves using a narrow beam (slightly larger than the grain size) of high-energy monochromatic X-rays to record the diffraction pattern of a single reflection from a single grain embedded in the bulk material, and generating a 3D reciprocal space map of the crystal lattice of that grain [16–20]. From these maps, the intragranular distributions of lattice strain and orientation can be extracted. Tracking the evolution of the latter, Gaudez et al. [1] provided an unambiguous answer on the importance of SSTC during 3D printing for the evolution of the dislocation structures obtained after solidification. In what follows, we briefly recall the main result of this experiment. Then, we discuss the questions that have arisen from this study and the possible routes that can be taken to address them.

2. Dislocation structure evolution due to SSTC during AM revealed via high resolution reciprocal space mapping

2.1 Materials and Methods

A miniature laser direct energy deposition (mini-DED) device (figure 1a), developed and developed in [1], was used for the HRRSM experiments. Inert gas atomized 316L austenitic stainless steel powder with composition (in wt%) Fe–17.34Cr–12.55Ni–2.34Mo–1.4Mn–0.49Si–

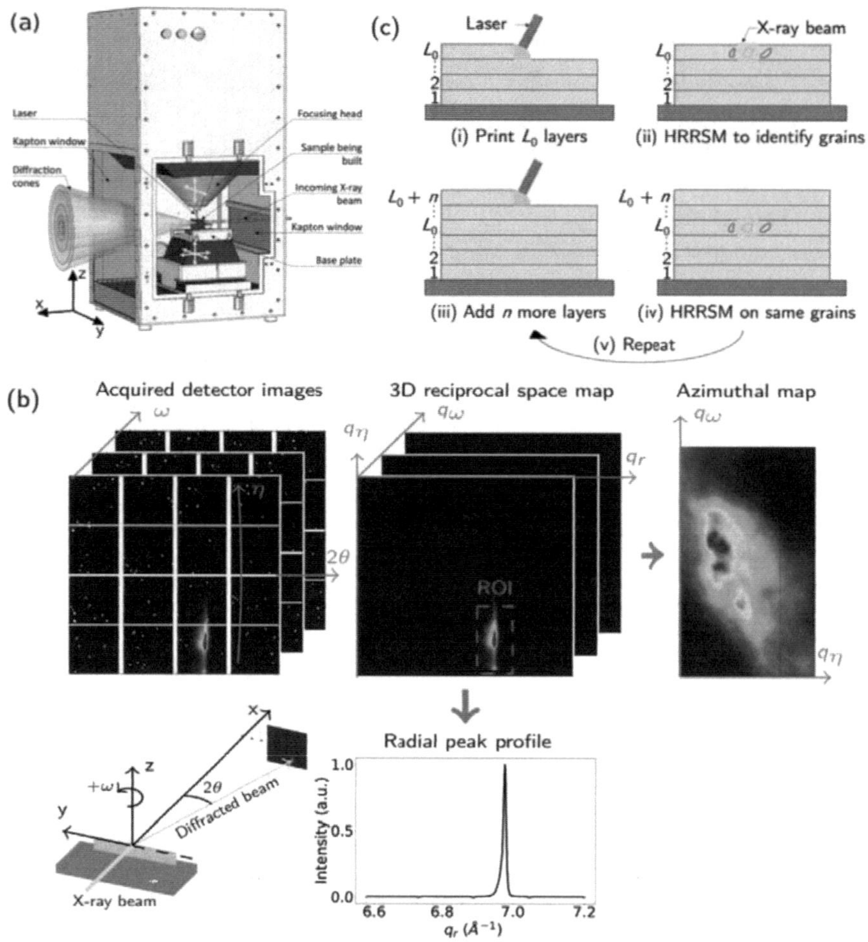

Figure 1. Setup and experimental procedure: (a) Mini-LMD device in a typical operating condition building a sample with an X-ray beam interacting with the sample. (b) HRRSM procedure and data analysis for a grain: a sample is continuously rocked about the z axis. 3D (q_r, q_η, q_ω) reciprocal space map of a 400 diffraction peak from a grain with one of its <100> directions closely aligned with the printing direction is obtained from the acquired raw data. Projections along the radial direction (q_r) and onto the azimuthal plane (q_η, q_ω) are derived, and only the latter is studied in this work, which reveals the intragranular orientation distribution. (c) A depiction of the experimental procedure. The figure has been adapted from [15].

0.08N–0.04Cu–0.03O and trace amounts of P (0.01 wt%), Si (less than 0.01 wt%), and C (less than 0.01 wt%) was used. Thin-wall type samples were built on 10 mm thick 316L plates using the following DED strategy and parameters: Unidirectional printing with single line per layer, 100 W laser power, 600 mm/min deposition speed, 7 g/min powder flow rate, and 0.25 mm upward movement of focusing head between each layer.

HRRSM experiments during DED were performed at the P21.2 beamline of the PETRA III synchrotron of Deutches Elektronen-Synchrotron (DESY, Germany). Diffraction patterns of 400 (γ-austenite) diffraction peaks in the horizontal diffraction plane were acquired using a 2D detector Eiger CdTe 4M (Dectris) positioned at 6.48 m (along the beam direction x) downstream of the sample and at ~1.75 m from the incident beam direction (along the printing direction, $-y$ direction as shown in the bottom of figure 1b. All investigated grains had one of their <100> directions close to the printing direction. Monochromatic X-ray beam of energy 52.5 keV was used to perform HRRSM. The beam size was 300 × 300 μm^2 and its bandwidth was 0.009 keV (narrow) with negligible divergence. The 2D detector records diffraction patterns along the radial direction corresponding to the diffraction angle 2θ (related to the interplanar spacing d_{hkl} through Bragg's law), and the azimuthal direction corresponding to angle η around the incident beam ($+x$ axis). The specimen is rocked around the $+z$ axis by ±2° to obtain information along the other azimuthal direction corresponding to the rocking angle ω (figure 1b) with images acquired in intervals of 0.02°. This setup provided an angular resolution of $\Delta 2\theta$ = 0.0026°, $\Delta\eta$ = 0.0025°, and $\Delta\omega$ = 0.02°. More details on the setup, data acquisition and treatment to obtain 3D reciprocal space maps from each studied grain are given in [1].

The following experimental procedure was used (as sketched in figure 1c): (i) Print 20 layers (L_0) using the aforementioned strategy and let the sample cool down to room temperature. (ii) Identify grains of interest in the topmost layer (L_0) close to the centre of the sample and perform HRRSM for each grain; the 3D reciprocal space maps of these grains act as reference for the as-solidified state. (iii) Add n layers above L_0 using the same printing strategy. (iv) Re-perform HRRSM on each of the previously identified grains. (v) Repeat the last two steps until the end of fabrication. One of the two samples was built in intervals of n = 5 layers to a total of L_0 + 60 layers i.e., 80 layers. This sample is denoted as denoted as S5L. The other sample was built in intervals of n = 2 layers to a total of L_0 + 20 i.e., 40 layers. In what follows, only the results from the S5L sample are presented and discussed. They are representative of the observations reported in [1].

2.2 Results

Three grains viz., S5L-G1, S5L-G2 and S5L-G3 were identified in the 20th (L_0) layer for S5L. However, only results of two of them (S5L-G1 and G3) are presented and discussed here. All grains were smaller in size than the incoming X-ray beam; they were in the range [38, 110] μm, thus in each case the entire grain was illuminated. Projections along the radial direction (q_r) and in the azimuthal plane (q_η, q_ω) were derived from each 3D reciprocal space map (figure 1b) and their evolution was followed as a function of added layers in [1].

The 2D diffraction pattern of S5L-G1 was followed live on the Eiger detector operating at 100 Hz during the addition of the first five layers above the layer of interest. The printing causes the diffraction peak of S5L-G1 to move and spread in 2θ and η and also causes a change in its overall intensity, however, the peak never vanishes, which indicates that the grain has either melted partially or not at all. A quantitative analysis of the *operando* measurements will be undertaken in a future work.

From the radial line profile of a grain, one can derive statistical measures such as average peak position and integral widths for a particular grain. The average peak position provides information on the grain averaged lattice strain. The integral widths contain convolved information on instrumental broadening, grain size, internal (dislocation) structure size, and heterogeneous strain distribution. While deconvolution of the contribution of instrumental

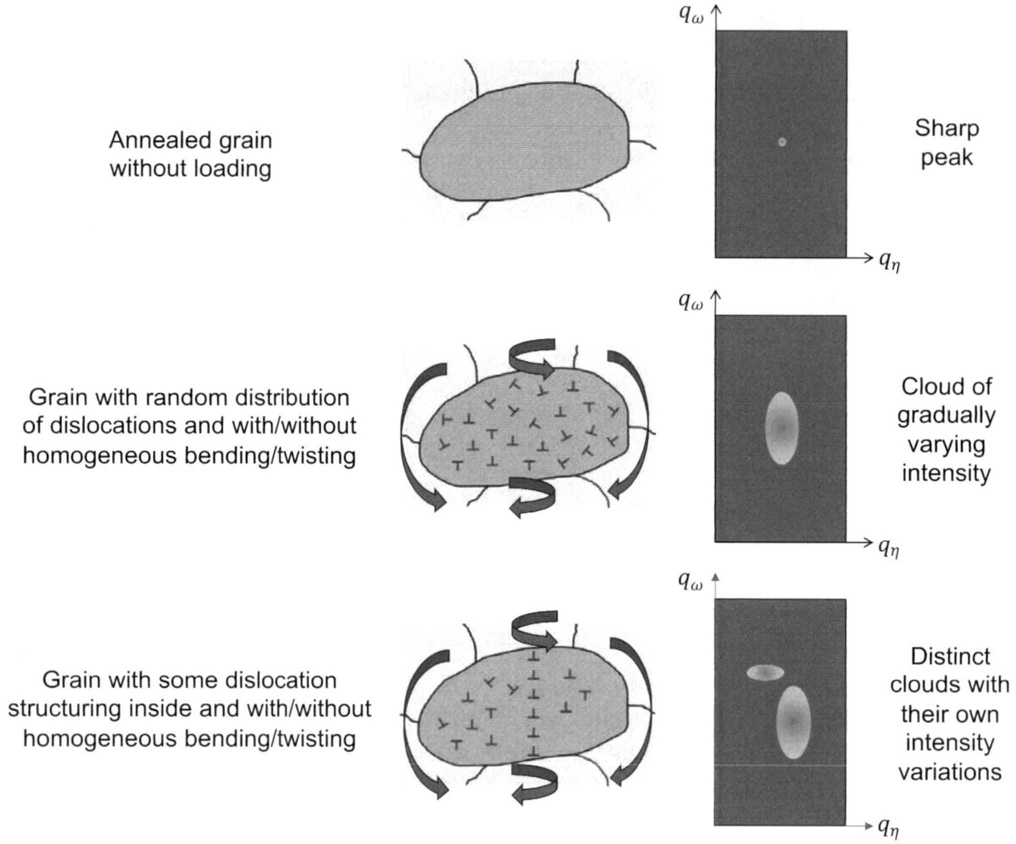

Figure 2. Illustration to demonstrate some possible correlations between an investigated grain and its corresponding azimuthal map.

broadening is possible, the path to deconvolution of size and heterogeneous strain contributions would require additional assumptions that may be difficult to justify.

Instead, azimuthal maps (q_η, q_ω) containing information on intragranular orientation distribution can be studied to univocally prove that dislocation evolution indeed occurs due to SSTC during further addition of layers. Each azimuthal map may contain clouds of non-zero intensity that depend on multiple factors. Taking the example of a single-phase material, several scenarios illustrated in figure 2 must be considered: (i) A dislocation-free, annealed grain subjected to no bending or twisting would result in a single sharp peak in the azimuthal map. (ii) A dislocation-free, annealed grain subjected to homogeneous bending and twisting due to type II stresses [21] from its neighbourhood would result in an extended cloud with gradually varying intensity as the intensity is determined by the affected volumes. Similarly, local bending and twisting by randomly distributed dislocations or thick dislocation walls would result in further intensity variation within the cloud. (iii) A grain containing localized finite orientation differences induced by dislocation boundaries will result in distinct intragranular orientation domains (ODs) [17,20]. In the azimuthal maps, such ODs appear as distinguishable regions of higher intensity

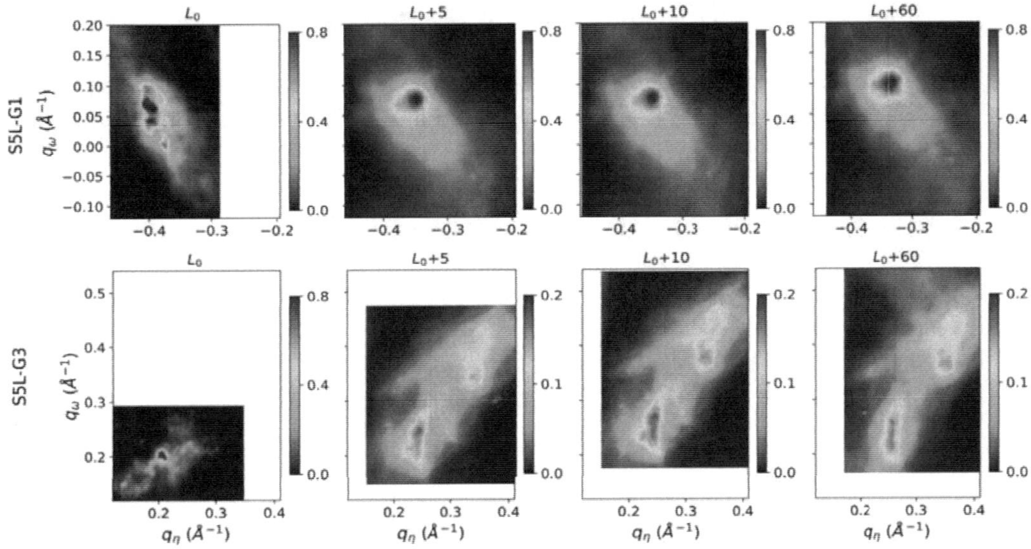

Figure 3. Azimuthal maps (q_η, q_ω) of the 400 diffraction peaks of S5L-G1 and S5L-G3: when L_0 is the topmost layer, after the addition of 5, 10, and 60 layers above L_0. Within each azimuthal map, the intensity has been normalized as presented in [15]. Vertical lines with low intensity are a consequence of interpolation over inactive areas on the detector. The figure has been adapted from [15].

around local intensity maxima separated from other ODs by regions of lower intensity. (iv) The presence of many individual dislocation cells (subgrains) together with thick dislocation walls would manifest themselves in a cloud of smoothly varying intensity superimposed by individual sharp peaks representing individual subgrains. Such individual peaks may only be observed, if their disorientation is large enough and not too many of them contribute to the diffraction peak. Despite the obtained high resolution in reciprocal space, individual dislocation cells could not be observed in the investigated grains due to the small mutual disorientation between subgrains.

Type II stresses and dislocation structures inside a grain are expected to evolve under the action of thermomechanical driving forces, e.g. due to SSTC during AM. Changes in the intragranular structure are expected to result in the evolution of the azimuthal maps, which can be interpreted as follows: (i) A rigid displacement of the entire intensity distribution in q_η and/or q_ω implies the rotation of the entire grain in η and/or ω either due to a rigid body rotation of the sample or due to a rotation of the grain with respect to its neighbourhood due to a change in type II stresses. (ii) A broadening of the intensity distribution reflects an increase in the bending or twisting of the grain (locally or uniformly). (iii) A change in the total intensity corresponds to an increase or decrease in the diffracting volume. (iv) Any other change should result from dislocation structure evolution.

Azimuthal maps of S5L-G1 and S5L-G3 acquired at room temperature at L_0, L_0 + 5, L_0 + 10 and L_0 + 60 are shown in Figure 3. Several high intensity regions with distinct maxima can be identified in the as-solidified state (L_0). Their presence is indicative of the existence of individual ODs that differ slightly in their orientations due to the presence of dislocation boundaries. After adding 5 layers (L_0 + 5), the azimuthal maps of all grains have considerably evolved. For S5L-G1, a single OD with a large spread and one maximum replaces the distinct ODs with multiple local maxima.

Meanwhile, S5L-G3 shows the opposite trend with the single OD splitting into distinct ODs with their own local maxima and orientation spread. These changes make it evident that dislocation structures have significantly evolved due to SSTC; this result is particularly striking since other intragranular heterogeneities such as microsegregations in 316L are rather stable during SSTC as shown by Voisin et al. [22]. A closer inspection of the azimuthal maps reveals that the first set of added layers results in a strong increase in the azimuthal spread for all grains but subsequent addition of layers has only a minor effect. Collecting all these results makes it evident that dislocation structures significantly evolve during the initial stages of SSTC, specifically during the addition of up to 5 layers above the layer of interest (under the studied manufacturing conditions), but they do not significantly evolve with further layer addition.

Gaudez et al. [1] obtained the same results for the grains studied in the other printed wall S2L. A subsequent thorough investigation of the ODs and their lattice strains and integral widths made it clear that dislocation structure evolution was the main contributor to the change in integral widths of the radial profiles.

3. A roadmap to better understanding dislocation structure formation during AM

3.1 Possible experimental approaches to understand the role of solidification and the first cooling down to room temperature on dislocation structure formation

The HRRSM experiments have proven that dislocations structures generated during solidification and the first cooling down to room temperature evolve significantly due to subsequent SSTC during AM. The next step would be to understand whether they form during solidification and rearrange during the subsequent cooling down to room temperature or whether they first form during the cooling down.

Studying the mechanism of dislocation structure formation due to solidification or the initial cooldown and of their evolution due to subsequent SSTC is a daunting task. An attempt of such an endeavour with HRRSM of a single diffraction peak seems unrealistic as the accessible azimuthal positions (η, ω) are limited by the position of the detector. Answering the question of dislocation structure formation requires using time-resolved equipment to track the appearance of entire diffraction patterns with complete Debye-Scherrer rings during the deposition of a layer. The use of a narrow X-ray beam (size comparable to grain size) at high energy and high flux remains important because this allows penetrating AM samples and generating spotty Debye-Scherrer rings from which diffraction peaks with sufficient intensity belonging to individual grains can be discerned. The success of such an experiment depends on the type of detector used. Some existing 2D detectors (e.g., Eiger CdTe 4M used for the HRRSM experiment or a modified Eiger detector used in [23] but limited to energies bellow 25 keV.) provide sufficiently high acquisition rates (1 kHz for the HRRSM experiment and 40 kHz used in [24]) to resolve the emergence of solid phases during the liquid-to-solid transformation amid DED-type AM, where maximum heating/cooling rates of the order of 10^3 – 10^4 K/s are typically encountered [25]. However, detectors with high frame rates often lack the number of pixels required to resolve sufficiently well individual diffraction spots when put closer to the sample for acquiring entire Debye Scherrer rings; the reciprocal space mapping with high resolution is achieved by using such detectors far from the sample limiting the usable coverage in reciprocal space. Supposing that an ideal combination of X-ray source and detector were to be available, one could track the evolution of the diffraction peaks from their first appearance after the liquid-to-solid phase transition until room temperature, e.g. as done by Dass et al. [26]. Based on the finding in section 2.2 that most changes to the integral width of the radial peak profile are due to dislocation structure evolution and density change, one could deduce the contribution of solidification on dislocation structure formation.

A significant challenge to such a study arises from the solidification path of the material investigated. Depending on the composition and cooling rates, solid-state phase transformations may occur, particularly in the case of stainless steels of the 300 series. The solidification path (including the solid-state cooling down to room temperature) of stainless steels depends on the chromium equivalent to nickel equivalent ratio (Cr_{eq}/Ni_{eq}) and the cooling rates. For the 316L composition studied in section 2, following Schaeffler [27] and Delong [28], the Cr_{eq}/Ni_{eq} is 1.45 and 1.23, respectively. With these ratios and the typical initial cooling rates encountered during DED, either the austenite (A) or primary austenite and secondary ferrite (AF) solidification mode is expected to occur [29]. According to the Schaeffler diagram, a small amount of ferrite could be expected at room temperature. Meanwhile, the Delong diagram predicts a fully austenitic microstructure. Due to the important role of N in stabilizing austenite, the Delong diagram is preferred here and this is corroborated by the fact that 99.9% of our 316L is austenite [1]. In steels with Cr_{eq}/Ni_{eq} above 1.5, however, several different transformation modes are possible [29]: primary ferrite (F), primary ferrite with secondary austenite (FA), primary ferrite with massive transformation into austenite (F/MA) or primary ferrite with Widmanstätten austenite. Such solidification paths are not suitable for an envisaged study because the transformation of ferrite into austenite can contribute to appearance/disappearance and changes in diffraction peaks.

Compositions that result in either A or F solidification modes are preferable to isolate the contribution of solidification and the first cooling down to room temperature on dislocation structure evolution; the 316L composition studied in [1] remains a suitable candidate. Ideally, using mono-element metals is desirable because they do not result in microsegregations structures and solid-state phase transformations; however, oxide or nitride formation during printing may be unavoidable.

3.2 Probing mechanisms of dislocation structure evolution

Dislocations are lattice defects whose interaction range spans from a few nanometres to several tens of micrometres. A comprehensive understanding of their behaviour requires studying their evolution and interactions at the micrometre scale with nanometre resolution in 3D. Furthermore, since AM is a rapid process, time-resolved equipment is necessary to study these mechanisms. To the best of our knowledge, synchrotron X-ray based approaches are unable to fulfil the stringent spatio-temporal resolution requirements to monitor the dislocation behaviour. To that end, the recent development [30] of X-ray microscopy at X-ray free electron laser (XFEL) sources seems promising.

Due to the high photon flux at XFELs, a temporal resolution down to 100 femtoseconds can be achieved. Combining dark-field X-ray microscopy (DFXM), which allows mapping individual dislocations in samples with several hundreds of micrometres of resolution [31], with X-rays at XFEL sources holds the promise to achieve the spatio-temporal resolution necessary to track rapid phenomena. However, accomplishing operando DFXM with femtosecond temporal resolution at XFELs during AM still remains an impossible task due to the non-periodic changes happening throughout the process. Obtaining such high temporal resolutions typically involves a pump-probe scheme, i.e., the measurement is repeated multiple times by steadily increasing the time interval between the initiation of the phenomenon under investigation and the X-ray pulse. Although several order of magnitudes larger, temporal resolution down to the tenth of microseconds during AM could be considered at the European XFEL. In addition, it has to be mentioned that maximum X-ray energy at XFELs is currently limited to 25 keV restricting sample thickness and material investigated during metal AM. Finally, DFXM using any particular diffraction peak requires selecting a rather narrow (2θ, η, ω) interval for imaging in the diffraction plane. While in this manner dislocation dynamics might be followed during SSTC, monitoring the formation of dislocation structures during the liquid-to-solid transformation and rapid cooling is hindered in the same manner as for HRRSM.

Beyond the challenge of realizing an *operando* experiment, another challenge is treatment of the amount of data generated. Printing a layer requires several hundreds of milliseconds to seconds, whereas a temporal resolution of hundreds of femtoseconds to a few picoseconds is necessary to study dislocation dynamics and interactions.

3.3 Simulating dislocation structure formation and synergy with experiments

An approach to gain insight into the mechanisms of dislocation structure formation is to develop and exploit experiment, modelling and simulation synergistically. Simulating dislocation-based mechanisms during AM requires models that can resolve nano-/micro-scale defect evolution and interactions but with a temporal resolution of microseconds or higher. It is important to note that there is no ideal modelling and simulation approach that can be entirely relied upon to gain this understanding and the domains of applicability of each approach must be understood. For example, molecular dynamics simulations are versatile enough to study nano-scale phenomena, but they require tens of femtoseconds of resolution in time, which makes them impractical for AM applications. A possibility is to use discrete dislocation dynamics approaches, which are significantly faster, but they rely on asymptotic solutions to stress and displacement fields and predetermined rules of dislocation interactions that are based on isothermal studies and may not be applicable under complex thermomechanical loadings occurring during AM.

A solution to this problem is to use the Phase Field Crystal (PFC) approach [32,33]. PFC is a special class of phase field approaches that introduces an order parameter describing the temporally coarse-grained atomic density of a crystal with its peaks arranged in crystalline order, and a free-energy functional (Lagrangian) that can be obtained from classical density functional theory (DFT) [34,35]. Capturing the crystal structure in this manner has many advantages e.g., dislocation glide on correct slip planes is a prediction in PFC unlike discrete dislocation dynamics models. Over the last 20 years, PFC has emerged as a useful tool to study mechanisms where the atomic and continuum scales are strongly coupled such as dislocation dynamics [36], grain growth during solidification [34], grain boundary motion [35], etc. Furthermore, PFC can also be spatially coarse-grained [37], which facilitates studying defect structure formation/interaction up to several micrometres covering a range of phenomena during AM. However, while PFC captures the crystal structure well, it does not appropriately treat conservation laws (particularly, momenta) and elasticity because the phase field evolution is entirely dissipative as shown in some recent studies [38,39]. This is particularly problematic when one wants to appropriately capture thermomechanically driven mechanisms e.g. dislocation dynamics, phase transformations, etc.

PFC must be augmented to capture conservation of momenta and to evolve dislocations under imposed thermomechanical boundary conditions. As a first step, Acharya and Viñals [38] proposed to remedy this problem by strongly coupling the isothermal field dislocation mechanics (FDM) model and PFC. FDM is a continuum theory [40–46] in which thermal lattice vibrations are not considered. The theory focuses rather on the longer time scale of dislocation motion and chooses as a primary field the polar dislocation density tensor and its associated conservation law of topological charge [44,47–49]. The theory also includes nonlinear, anisotropic elastic fields, and therefore it can account for topological defects in atomic configurations of nonlinear elastic media. The fundamental equations of the theory are directly derived from classical balances of mass and momentum, an elastic incompatibility equation linking the finite elastic distortion to the dislocation density tensor, and a conservation law for Burgers vector in the body. An energy dissipation inequality based on the Second Law of thermodynamics leads to the equations for the various fields [46,50]. Recently, the theory was extended to account for the role of thermal strain incompatibilities on the conservation of the Burgers vector and for temperature evolution due to dislocation activity through the first law of thermodynamics [51,52] and the dislocation density transport equation. The latter is a first-order hyperbolic partial differential equation, which was stably solved using the discontinuous Galerkin finite element approach [53]. The FDM theory requires some minimal, but essential constitutive input: The mobility law of single defects, and its constitutive nonlinear (and non-monotonous) elastic behaviour (i.e., nonlinear elasticity and appropriate generalized stacking fault energies [54–56]). Nevertheless, the theory does not require the slip systems to be known *a priori* in order to define an energy density. Importantly, the dissipation inequality yields the proper nonlinear generalization of the Peach-Koehler force field acting on a dislocation segment.

Upadhyay and Viñals [39] have recently numerically implemented the statics part of the coupling between PFC and FDM proposed by Acharya and Viñals [38] to demonstrate the need for a full dynamic coupling. The next steps involve the full dynamic implementation with the expectation that PFC can predict dislocation velocities correctly. Then, based on the generalization of the isothermal FDM to the fully-coupled thermomechanical FDM (TFDM) model [51,52], the next step will be to extend the aforementioned coupling to a PFC-TFDM framework, which will form the basis to study dislocation dynamics under thermomechanical loadings e.g., those occurring during AM.

Once these developments are completed, the simulation predictions need to be compared with experiments to establish domains of applicability. For example, one possibility could be to seed the initial microstructure of these simulations in such a way that the virtual diffraction profiles in the azimuthal plane match those from the initial HRRSM measurements from section 2.2. The simulation predictions could then be compared with the profiles after adding a few layers.

Another possibility of comparison with experiments involving *in situ* laser scanning inside a scanning electron microscope was recently proposed by Mohanan et al. [57]. They pursued a novel experiment-driven modelling and simulation approach to study the formation of intergranular stresses and plastic strains during laser scanning of additively manufactured 316L stainless steel. Laser scanning experiments were performed using a unique coupling between a continuous-wave laser and a scanning electron microscope, called CW Laser-SEM [58]. These experiments formed the basis for the development of a thermo-elasto-viscoplastic polycrystal model endowed with the minimum necessary physics of the problem in order to facilitate simulating sufficiently large domains to obtain reasonable stress magnitudes. The case of a single laser line scan performed in vacuum is analysed in detail. Polar dislocation density magnitudes and distributions predicted from the simulation are compared with those measured experimentally. Results reveal that for 93% of the grains in the lasered zone, a statistical measure of the predicted (Nye's) dislocation density tensor lied within a factor of 2 (higher or lower) of the experimentally determined one [57]. This result sets the domain of applicability of the polycrystal model as well as a benchmark for future experiment-simulation comparisons of this kind. The aforementioned coupling between the PFC and FDM model to study dislocation structure evolution could also benefit from comparing the dislocation density with the one obtained from CW Laser-SEM experiments.

4. Concluding remarks

The results of a recent experiment [1] involving high resolution reciprocal space mapping (HRRSM), a synchrotron X-ray diffraction technique, to probe whether dislocation structures formed during the deposition of a layer during AM significantly evolve due to SSTC caused by addition of more layers above the layer of interest, are presented. Based on the evidence collected from the 3D reciprocal space maps, in particular from the azimuthal maps, it is apparent that the dislocation structures that manifest just after solidification evolve significantly during the initial stages of SSTC, i.e. after adding a few layers (up to 5) on top of the layer investigated. In other words, the dislocation structures obtained at the end of an AM process are not the same as those obtained just after the first solidification and rapid cooling down to room temperature.

Backed by this experiment, a possible pathway to a better understanding of dislocation structure formation during AM is discussed. It demands using experiment and modelling in a synergistic manner for the following purposes: (i) to decipher the role of solidification and the first cooling down to room temperature on the formation of the initial dislocation structure during AM, and (ii) to gain insight into the dislocation evolution mechanisms during subsequent SSTC.

Acknowledgements

SG and MVU are grateful to the European Research Council (ERC) for their support through the European Union's Horizon 2020 research and innovation programme for project GAMMA (Grant agreement No. 946959) awarded to MVU. The HRRSM experiments were carried out at the P21.2 beamline of the PETRA III synchrotron. Beamtime had been allocated for proposal no. I-20210525 EC. WP acknowledges support from the EuroTech Visting Researcher Program and thanks the Danish Agency for Science, Technology, and Innovation for funding the instrument centre DanScatt.

References

[1] Gaudez S, Abdesselam K A, Gharbi H, Hegedüs Z, Lienert U, Pantleon W and Upadhyay M V 2023 High-resolution reciprocal space mapping reveals dislocation structure evolution during 3D printing *Addit. Manuf.* **71** 103602

[2] Amato K N, Gaytan S M, Murr L E, Martinez E, Shindo P W, Hernandez J, Collins S and Medina F 2012 Microstructures and mechanical behavior of Inconel 718 fabricated by selective laser melting *Acta Mater.* **60** 2229–39

[3] Gorsse S, Hutchinson C, Gouné M and Banerjee R 2017 Additive manufacturing of metals: a brief review of the characteristic microstructures and properties of steels, Ti-6Al-4V and high-entropy alloys *Sci. Technol. Adv. Mater.* **18** 584–610

[4] Martin J H, Yahata B D, Hundley J M, Mayer J A, Schaedler T A and Pollock T M 2017 3D printing of high-strength aluminium alloys *Nature* **549** 365–9

[5] Gray G T, Livescu V, Rigg P A, Trujillo C P, Cady C M, Chen S R, Carpenter J S, Lienert T J and Fensin S J 2017 Structure/property (constitutive and spallation response) of additively manufactured 316L stainless steel *Acta Mater.* **138** 140–9

[6] Wang Y M, Voisin T, McKeown J T, Ye J, Calta N P, Li Z, Zeng Z, Zhang Y, Chen W, Roehling T T, Ott R T, Santala M K, Depond P J, Matthews M J, Hamza A V and Zhu T 2018 Additively manufactured hierarchical stainless steels with high strength and ductility *Nat. Mater.* **17** 63–71

[7] DebRoy T, Wei H L, Zuback J S, Mukherjee T, Elmer J W, Milewski J O, Beese A M, Wilson-Heid A, De A and Zhang W 2018 Additive manufacturing of metallic components – Process, structure and properties *Prog. Mater. Sci.* **92** 112–224

[8] Hosseini E and Popovich V A 2019 A review of mechanical properties of additively manufactured Inconel 718 *Addit. Manuf.* **30** 100877

[9] Liu S and Shin Y C 2019 Additive manufacturing of Ti6Al4V alloy: A review *Mater. Des.* **164** 107552

[10] Barkia B, Aubry P, Haghi-Ashtiani P, Auger T, Gosmain L, Schuster F and Maskrot H 2020 On the origin of the high tensile strength and ductility of additively manufactured 316L stainless steel: Multiscale investigation *J. Mater. Sci. Technol.* **41** 209–18

[11] Smith T R, Sugar J D, San Marchi C and Schoenung J M 2019 Strengthening mechanisms in directed energy deposited austenitic stainless steel *Acta Mater.* **164** 728–40

[12] Birnbaum A J, Steuben J C, Barrick E J, Iliopoulos A P and Michopoulos J G 2019 Intrinsic strain aging, Σ3 boundaries, and origins of cellular substructure in additively manufactured 316L *Addit. Manuf.* **29** 100784

[13] Bertsch K M, Meric de Bellefon G, Kuehl B and Thoma D J 2020 Origin of dislocation structures in an additively manufactured austenitic stainless steel 316L *Acta Mater.* **199** 19–33

[14] Yamanaka K, Kuroda A, Ito M, Mori M, Bian H, Shobu T, Sato S and Chiba A 2021 Quantifying the dislocation structures of additively manufactured Ti–6Al–4V alloys using X-ray diffraction line profile analysis *Addit. Manuf.* **37** 101678

[15] Prasad K, Obana M, Ito A and Torizuka S 2021 Synchrotron diffraction characterization of dislocation density in additively manufactured IN 718 superalloy *Mater. Charact.* **179** 111379

[16] Jakobsen B, Poulsen H F, Lienert U, Almer J, Shastri S D, Sørensen H O, Gundlach C and Pantleon W 2006 Formation and Subdivision of Deformation Structures During Plastic Deformation *Science* **312** 889–92

[17] Jakobsen B, Poulsen H F, Lienert U, Huang X and Pantleon W 2007 Investigation of the deformation structure in an aluminium magnesium alloy by high angular resolution three-dimensional X-ray diffraction *Scr. Mater.* **56** 769–72

[18] Wejdemann C, Lienert U and Pantleon W 2010 Reversal of asymmetry of X-ray peak profiles from individual grains during a strain path change *Scr. Mater.* **62** 794–7

[19] Pantleon W, Wejdemann C, Jakobsen B, Lienert U and Poulsen H F 2010 Advances in characterization of deformation structures by high resolution reciprocal space mapping *Proceedings Risø International Symposium on Materials Science* vol 31 pp 79–100

[20] Diederichs A M, Thiel F, Lienert U and Pantleon W 2018 In-situ investigations of structural changes during cyclic loading of aluminium by high resolution reciprocal space mapping *Int. J. Fatigue* **117** 206–16

[21] Macherauch E, Wohlfahrt H and Wolfstieg U 1973 Zur zweckmäßigen Definition von Eigenspannungen *HTM J. Heat Treat. Mater.* **28** 201–11

[22] Voisin T, Forien J-B, Perron A, Aubry S, Bertin N, Samanta A, Baker A and Wang Y M 2021 New insights on cellular structures strengthening mechanisms and thermal stability of an austenitic stainless steel fabricated by laser powder-bed-fusion *Acta Mater.* **203** 116476

[23] Tinti G, Bergamaschi A, Cartier S, Dinapoli R, Greiffenberg D, Johnson I, Jungmann-Smith J H, Mezza D, Mozzanica A, Schmitt B and Shi X 2015 Performance of the EIGER single photon counting detector *J. Instrum.* **10** C03011

[24] König H-H, Pettersson N H, Durga A, Van Petegem S, Grolimund D, Chuang A C, Guo Q, Chen L, Oikonomou C, Zhang F and Lindwall G 2023 Solidification modes during additive manufacturing of steel revealed by high-speed X-ray diffraction *Acta Mater.* **246** 118713

[25] Upadhyay M V, Ben Haj Slama M, Gaudez S, Mohanan N, Yedra L, Hallais S, Héripré E and Tanguy A 2021 Non-oxide precipitates in additively manufactured austenitic stainless steel *Sci. Rep.* **11** 10393

[26] Dass A, Tian C, Pagan D C and Moridi A 2023 Dendritic deformation modes in additive manufacturing revealed by operando x-ray diffraction *Commun. Mater.* **4** 1–10

[27] Schaeffler A L 1949 Constitution diagram for stainless steel weld metal *Met. Prog.* **56** 680-680B

[28] Delong W T 1974 Ferrite in Austenitic Stainless Steel Weld Metal *Weld. J. Weld. Res. Suppl.* **53** 273–86

[29] Lippold J C Solidification behavior and cracking susceptibility of pulsed-laser welds in austenitic stainless steels *Weld. J. Weld. Res. Suppl.* **73** 129–39

[30] Dresselhaus-Marais L E, Kozioziemski B, Holstad T S, Ræder T M, Seaberg M, Nam D, Kim S, Breckling S, Choi S, Chollet M, Cook P K, Folsom E, Galtier E, Gonzalez A, Gorkhover T, Guillet S, Haldrup K, Howard M, Katagiri K, Kim S, Kim S, Kim S, Kim H, Knudsen E B, Kuschel S, Lee H J, Lin C, McWilliams R S, Nagler B, Nielsen M M, Ozaki N, Pal D, Pablo Pedro R, Saunders A M, Schoofs F, Sekine T, Simons H, van Driel T, Wang B, Yang W, Yildirim C, Poulsen H F and Eggert J H 2023 Simultaneous bright- and dark-field X-ray microscopy at X-ray free electron lasers *Sci. Rep.* **13** 17573

[31] Jakobsen A C, Simons H, Ludwig W, Yildirim C, Leemreize H, Porz L, Detlefs C and Poulsen H F 2019 Mapping of individual dislocations with dark-field X-ray microscopy *J. Appl. Crystallogr.* **52** 122–32

[32] Elder K R, Katakowski M, Haataja M and Grant M 2002 Modeling Elasticity in Crystal Growth *Phys. Rev. Lett.* **88** 245701

[33] Elder K R and Grant M 2004 Modeling elastic and plastic deformations in nonequilibrium processing using phase field crystals *Phys. Rev. E* **70** 051605

[34] Elder K R, Provatas N, Berry J, Stefanovic P and Grant M 2007 Phase-field crystal modeling and classical density functional theory of freezing *Phys. Rev. B* **75** 064107

[35] Wu K-A and Voorhees P W 2012 Phase field crystal simulations of nanocrystalline grain growth in two dimensions *Acta Mater.* **60** 407–19

[36] Skaugen A, Angheluta L and Viñals J 2018 Dislocation dynamics and crystal plasticity in the phase-field crystal model *Phys. Rev. B* **97** 054113

[37] Salvalaglio M and Elder K R 2022 Coarse-grained modeling of crystals by the amplitude expansion of the phase-field crystal model: an overview *Model. Simul. Mater. Sci. Eng.* **30** 053001

[38] Acharya A and Viñals J 2020 Field dislocation mechanics and phase field crystal models *Phys. Rev. B* **102** 064109

[39] Upadhyay M V and Viñals J 2024 Coupling Phase Field Crystal and Field Dislocation Mechanics for a consistent description of dislocation structure and elasticity

[40] Acharya A 2001 A model of crystal plasticity based on the theory of continuously distributed dislocations *J. Mech. Phys. Solids* **49** 761–84

[41] Acharya A 2003 Driving forces and boundary conditions in continuum dislocation mechanics *Proc. R. Soc. Lond. Ser. Math. Phys. Eng. Sci.* **459** 1343–63

[42] Acharya A 2004 Constitutive analysis of finite deformation field dislocation mechanics *J. Mech. Phys. Solids* **52** 301–16

[43] Acharya A and Roy A 2006 Size Effects and Idealized Dislocation Microstructure at Small Scales: Predictions of a Phenomenological Model of Mesoscopic Field Dislocation Mechanics: Part I *J. Phys. Mech. Solids* **54** 1687–710

[44] Acharya A 2011 Microcanonical Entropy and Mesoscale Dislocation Mechanics and Plasticity *J. Elast.* **104** 23–44

[45] Zhang X, Acharya A, Walkington N J and Bielak J 2015 A single theory for some quasi-static, supersonic, atomic, and tectonic scale applications of dislocations *J. Mech. Phys. Solids* **84** 145–95

[46] Arora R, Zhang X and Acharya A 2020 Finite element approximation of finite deformation dislocation mechanics *Comput. Methods Appl. Mech. Eng.* **367** 113076

[47] Rickman J M and Viñals J 1997 Modelling of dislocation structures in materials *Philos. Mag. A* **75** 1251–62

[48] Kosevich A M 1979 Crystal dislocations and the theory of elasticity *Dislocations in Solids, Nabarro, F. R. N. (editor)* vol 1 (Amsterdam: North Holland Publishing Company) pp 33–141

[49] Perreault B, Viñals J and Rickman J M 2016 Impact of lattice rotation on dislocation motion *Phys. Rev. B* **93** 014107

[50] Arora R and Acharya A 2020 A unification of finite deformation *J*2 Von-Mises plasticity and quantitative dislocation mechanics *J. Mech. Phys. Solids* **143** 104050

[51] Upadhyay M V 2020 On the thermo-mechanical theory of field dislocations in transient heterogeneous temperature fields *J. Mech. Phys. Solids* **145** 104150

[52] Lima-Chaves G D and Upadhyay M V 2024 Finite element implementation of the thermal field dislocation mechanics model: Study of temperature evolution due to dislocation activity *Comput. Methods Appl. Mech. Eng.* **421** 116763

[53] Upadhyay M V and Bleyer J 2022 Dislocation transport using a time-explicit Runge–Kutta discontinuous Galerkin finite element approach *Model. Simul. Mater. Sci. Eng.* **30** 034002

[54] Vítek V 1968 Intrinsic stacking faults in body-centred cubic crystals *Philos. Mag.* **18** 773–86

[55] Vitek V 1998 Atomic structure of dislocations in intermetallics with close packed structures: a comparative study *Intermetallics* **6** 579–85

[56] Vitek V and Paidar V 2008 Chapter 87 - Non-planar Dislocation Cores: A Ubiquitous Phenomenon Affecting Mechanical Properties of Crystalline Materials *Dislocations in Solids* A Tribute to F.R.N. Nabarro vol 14, ed J P Hirth (Elsevier) pp 439–514

[57] Mohanan N, Macías J G S, Bleyer J, Helfer T and Upadhyay M V 2024 Intergranular stress and plastic strain formation during laser scanning of additively manufactured stainless steel: An experimentally-driven thermomechanical simulation study *Materialia* 102082

[58] Santos Macías J G, Chen K, Tanguy A, Vallet M, Cornet L, Michel V and Upadhyay M V 2024 Post-Process Lasering Improves Strength-Ductility Tradeoff and Fatigue Limit of Additively Manufactured Stainless Steels

Control of crystallographic textures by metal additive manufacturing – A review

Takayoshi Nakano[1,2*]

[1] Division of Materials and Manufacturing Science, Graduate School of Engineering, Osaka University, Suita, Japan
[2] Anisotropic Design and Additive Manufacturing Research Center, Osaka University, Suita, Japan

*E-mail: nakano@mat.eng.osaka-u.ac.jp

Abstract. Metal additive manufacturing (AM), a unique manufacturing method that stacks powder materials layer-by-layer to fabricate products with complex shapes and high precision, allows for a wide range of control over metallurgical microstructures. Metal AM defines solidification units with specific solidification directions and fast cooling, thereby enabling the control of the resulting metallurgical microstructure. However, the studies investigated texture control and utilizing textured microstructure are lacking in the literature. This review describes some of the results of our research on the control of crystallographic textures by laser powder bed fusion (LPBF), focusing on (1) the effect of powder properties on densification and crystallographic texture formation, (2) the effect of the melt pool shape and crystallographic characteristics of the starting material on the formation and orientation of single crystals, and (3) the successful application of alloy design to the preparation of highly functional single-crystalline-like textured biological high-entropy alloys considering specific solidification fields under LPBF.

1. Overview

The basic concept of additive manufacturing (AM) is the fabrication of three-dimensional parts in a two-dimensional plane using a layer-by-layer approach [1]. In 2013, over 30 years after the concept was first introduced, former US President Barack Obama suggested that "3D printing has the potential to revolutionize the way we make almost everything." Indeed, the number of academic publications on AM has increased annually (figure 1, red columns), conferences and research groups on AM have been established in many countries, and several exhibitions and international conferences have been held. In recent years, the market for AM has rapidly increased worldwide.

Metal AM is expected to be applicable to a wide range of industries, including aerospace, automotive manufacturing, energy, chemicals, medical/healthcare, and consumer products. When combined with topology optimization technology, AM may be able to overcome several issues often encountered in conventional subtractive manufacturing, such as the manufacture of complex shapes, flexible customization of structures, and significant reductions in weight, while maintaining the required strength under specific mechanical environments. The significance of and expectations for the social implementation of AM are important from the viewpoint of realizing a carbon-neutral society.

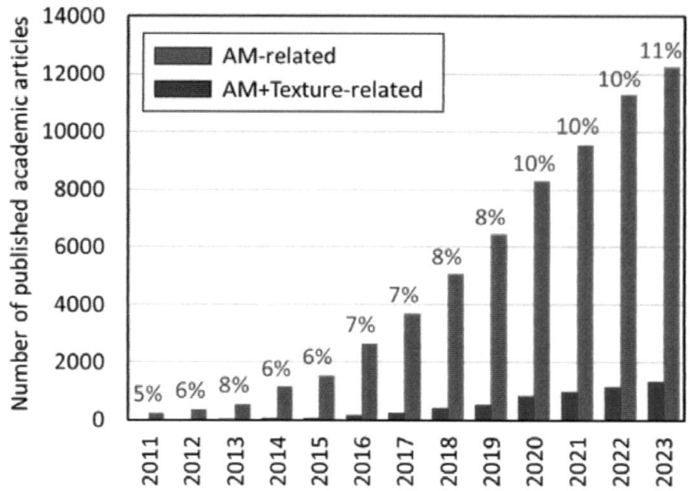

Figure 1. Trends of publications on additive manufacturing (AM) in general (red) and those focusing on texture-related AM research (blue). We searched for publications containing the phrase "additive manufacturing" in their title, abstract, and keywords for the former and those containing "additive manufacturing" AND texture OR orientation for the latter in Scopus on April 13, 2024. The numbers at the top the red columns indicate the percentage of texture-related publications relative to the number of AM publications for the given year.

AM, especially powder bed fusion (PBF), which uses fine powder as the starting material, was originally developed to fabricate complex three-dimensional shapes with high precision. In recent years, however, the importance of the characteristics of the metallic material, rather than its shape, has gradually been recognized. For example, alloy design, which involves the design of alloys that support various conditions inherent to AM, such as ultrarapid melting and solidification, layer-by-layer manufacturing, and the occurrence of thermal and residual stresses, is expected to become a major trend in future AM research [2].

Our research group has focused on material microstructures exhibiting anisotropic characteristics, including cubic or fiber texture formations, as a material property of metals, and has been conducting research aimed at the artificial control of such microstructures [3, 4]. In 2017, we successfully formed a single-crystalline-like highly oriented microstructure and controlled its orientation using different laser scanning strategies [3, 4]. Since then, we have developed a new method for controlling the orientation of crystals in various metals and alloys [5]. We previously investigated crystallographic textures and the control of mechanical properties based on crystal orientation in various metal and alloy systems [5, 6]. As shown in figure 1, the proportion of publications on crystalline microstructures in relation to AM has steadily increased in recent years, reaching 11% in 2023. Thus, the control of crystallographic textures may be considered one of the most important and popular research topics in AM today., of which there are still challenges to control the texture of metal materials with different crystal structures.

This review presents some of the results of our research on metal AM, focusing on the crystallographic textures produced by laser PBF (LPBF) as a powder-based process. Specifically, this article discusses: (1) the effect of powder properties on densification, which is essential for the formation of crystallographic textures, (2) the effect of the melt pool shape and crystallographic characteristics of the starting material on the formation and orientation of single crystals, and (3) the successful application of alloy design for the formation of single crystals in biological high-entropy alloys (BioHEAs) by LPBF.

2. Importance of powder properties to composite quality

Material microstructures with anisotropic properties, that is, exhibiting crystallographic textures, are formed via epitaxial growth between adjacent scan tracks and subsequent melt pool layers. Therefore, obtaining a dense compact is essential to control crystal growth. Similar to that in other powder metallurgical methods [7,8], the properties of the starting powder material also influence the process performance and quality of products fabricated using the PBF method [9]. In particular, for difficult-to-print materials, the properties of the powder have a pronounced influence on product quality.

Figure 2 shows SEM images of tungsten powders prepared by (a) crushing, (b) gas atomization, and (c) plasma atomization. The particle size distributions of these particles were analyzed by laser diffraction, and the results are shown in figure 2(d). Tungsten solidifies rapidly owing to its extremely high melting point (3422 °C), high thermal conductivity, and high melt viscosity [10, 11], rendering PBF processing challenging. The crushed powder particles are polygonal in shape and agglomerate to form secondary particles, whereas the atomized powder particles are generally spherical. Notably, the plasma-atomized powder particles are nearly spherical in shape, with hardly any satellite particles [12].

According to the results of analyses of the particle size distribution, powder bed density, and powder flowability characteristics of the tungsten powders (table 1), plasma-atomized powders have a high powder bed density and excellent flowability (Hauser ratio = tapped density/bulk density close to 1, with a small avalanche angle, angle of repose, and surface fractal). In other words, plasma-atomized powders form highly dense and uniform powder beds [12].

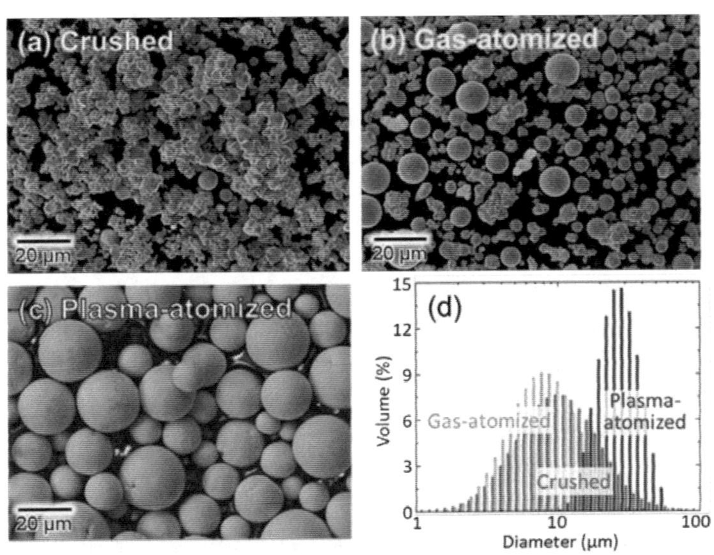

Figure 2. Morphology of (a) crushed, (b) gas-atomized, and (c) plasma-atomized tungsten powders. (d) Particle size distributions of the tungsten powders. Adapted with modifications from reference [12], published under the CC-BY-4.0 license.

Table 1. Size distribution, density, and flowability of crushed, gas-atomized, and plasma-atomized tungsten powders. Adapted with modifications from reference [12], published under the CC-BY-4.0 license.

Property	Crushed	Gas-atomized	Plasma-atomized
D10 (μm)	4.5	4.1	17.7
D50 (μm)	10.7	8.3	27.4
D90 (μm)	24.6	16.5	41.8
Bulk density (g/mL)	3.61	7.50	13.64
Tapped density (g/mL)	4.84	9.09	15.00
Hauser ratio	1.34	1.21	1.10
Avalanche angle (deg.)	59.3	47.9	32.4
Angle of repose (deg.)	41.3	40.4	25.5
Surface fractal	4.25	2.99	2.21

Figure 3(a) shows the relative densities of the products fabricated using the three types of tungsten powders depending on the laser energy density. SEM images of the cross-section of the products prepared at a laser energy density of 375 J/mm^3 (laser power = 360 W, scanning speed = 600 mm/s, hatch spacing = 0.08 mm, layer thickness = 0.02 mm) and their Vickers hardness are shown in figure 3(b–d). The densities of the products depend on the properties of the starting powders, with plasma-atomized powders consistently exhibiting the highest densities. Macroscopic defects, such as large lack of fusion regions, attributed to a low powder bed density are observed in the crushed powders, resulting in low hardness in the final product. By contrast, macrodefects are almost nonexistent in the atomized powders but cracks parallel to the building direction (BD) are formed, preventing densification [12]. As shown in figure 4, cracks parallel to the BD are formed at the center of the melt pool, where <100>//BD-oriented grains extending in the BD are found, resulting straight high-angle grain boundaries which are stated to be sensitive for crack initiation and propagation due to residual stress during fast solidification [12]. These cracks are located at high-angle grain boundaries on both sides of the <100>//BD grains. The formation of a crystallographic lamellar microstructure in which <100>//BD grains extend in the BD has been reported for SUS316L stainless steel [13, 15] and nickel-based alloys [16,1 7]. This phenomenon has been shown to occur when the radius of curvature at the bottom of the melt pool is large and the downward heat flow is stabilized [16]. The formation of such <100>//BD-oriented lamellar layers and their associated lamellar boundaries contributes to the strengthening of alloys with excellent ductility [13, 16]. However, in materials such as tungsten, which demonstrates significant brittleness at its grain boundaries, the formation of <100>//BD lamellae must be prevented because lamellar boundaries serve as crack initiation and propagation points. As mentioned above, the density and flowability of the starting powders directly influence the quality of the final product, including its densification and mechanical properties. At present, the development of powders that can achieve improved product quality is being vigorously pursued [18].

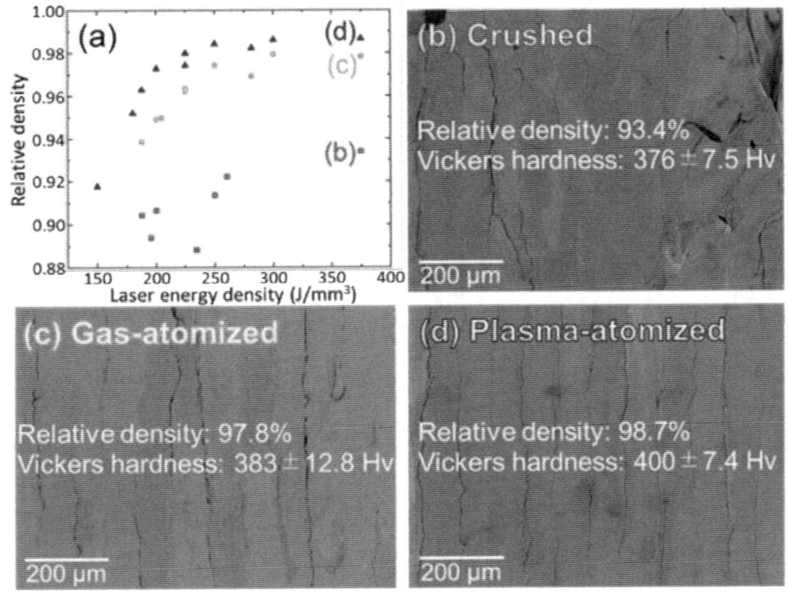

Figure 3. Dependence of product quality on powder properties. (a) Relative density of as-prepared tungsten as a function of the laser energy density. (b–d) SEM images and Vickers hardness of as-prepared products obtained at a laser energy density of 375 J/mm^3. Adapted with modifications from reference [12], published under the CC-BY-4.0 license.

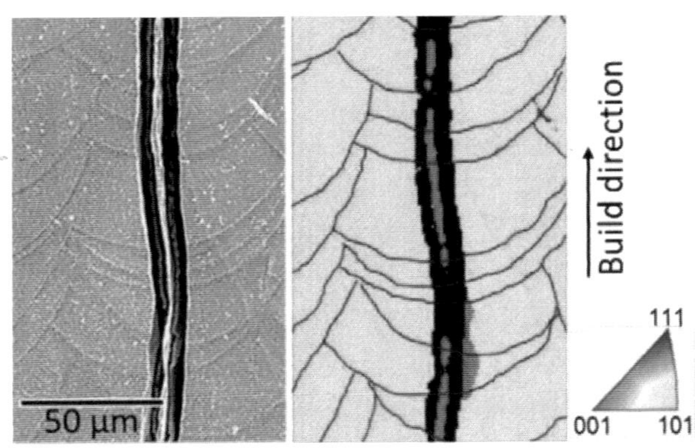

Figure 4. SEM image and corresponding inverse pole figure map of the morphology of the melt pools and crystallographic lamellae observed on the y-z plane in a product fabricated using atomized tungsten powders. <100>//build direction-oriented lamellae occur at the centre of the bottom of the melt pool, with cracks on both sides (high-angle grain boundaries).

3. Control of crystallographic texture formation by PBF method

The crystallographic texture (distribution and accumulation of crystalline orientation) of a material is an important metallurgical factor that controls both its mechanical properties such as strength, ductility, Young's modulus, and wear resistance, as well as its functional properties such as magnetism and corrosion resistance. Therefore, high functionality can be achieved in

IOP Conf. Series: Materials Science and Engineering **1310** (2024) 012013 doi:10.1088/1757-899X/1310/1/012013

anisotropic structures by the suitable control of their crystallographic orientation [19]. Crystallographic texture can be mainly altered between two strategies: formation of randomly oriented polycrystals to obtain isotropic properties and formation of either single-crystalline-like or fiber textures (columnar crystals) to obtain anisotropic properties. These structures must be freely controlled to enable the preparation of high-performance materials. Fortunately, AM has recently been shown to be a promising method for realizing such control.

The development of the crystallographic texture of a material is governed by the superposition of the crystallographic features of the material itself, such as the preferred direction of crystal growth and its multiplicity and symmetry, and the solidification behavior associated with the process conditions, such as solidification in a melt pool (temperature gradient of the solid–liquid interface, movement speed of the solid–liquid interface), melt pool shape, and heat flow direction. The following sections discuss the development of crystallographic texture in LPBF-processed materials depending on the solidification conditions, melt-pool properties, and crystallographic features.

3.1 Single- and polycrystal formation depending on the solidification conditions

In general, when forming single crystals, the temperature gradient G at the solid–liquid interface is increased and the solid–liquid interface velocity (crystal growth rate) R is decreased to suppress compositional undercooling and control the solidification conditions outside the columnar-equiaxed transition (CET) line. Recent digital twin engineering simulations have predicted that the solidification conditions in LPBF are located near the CET line [20, 21]. In other words, single or polycrystals can be produced by precisely controlling the solidification conditions according to the laser properties, such as the laser power P and scanning speed v, during fabrication.

Figure 5 shows the formation of single-crystalline-like and polycrystalline textures in β-type titanium alloys considering the CET drawn with black line indicating change in grain morphology associated with solidification rate (G/R) and the relationship between the laser conditions and solidification conditions [22]. Condition A has a higher energy density, higher power, and lower velocity than Condition B. The G–R distribution under Condition A, in which a single-crystalline-like structure is produced, is on the lower R side than that under Condition B, in which a polycrystal-like structure is formed. This result indicates that Condition A is a solidification condition under which smooth interfacial growth is relatively easy to achieve. Analyses of the effects of P and v on G and R using temperature-field simulations and statistical methods have shown that higher P and v decrease and increase R, respectively, but have no significant effect on G [22].

The position of the CET line depends on material-specific properties, such as the solidus–liquidus temperature difference and nucleation density of the solid phase. However, the laser conditions and tendency of various metallic materials to form single-crystalline or polycrystalline textures coincide [14-17]. When the laser conditions are adjusted to the range in which a single-crystalline-like texture can be formed, the crystallographic orientation can be controlled, depending on the melt pool properties and material type.

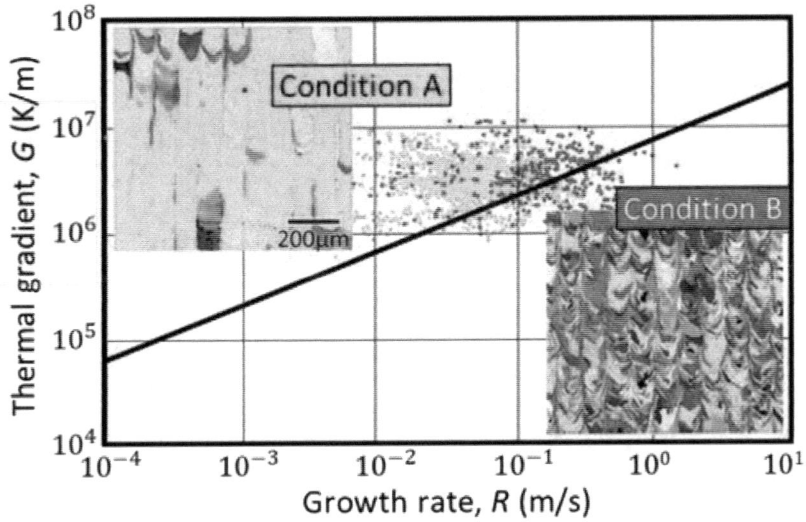

Figure 5. Single-crystalline-like and polycrystalline texture formation depending on the solidification conditions, which are further determined by the laser properties during laser powder bed fusion. Condition A: 360 W and 1200 mm/s; Condition B: 180 W and 1400 mm/s. Adapted with modifications from reference [22], published under the CC-BY-4.0 license.

3.2 Crystallographic texture formation in cubic structures

The formation of microstructures depends largely on the crystal structure of the starting material. In cubic metals with BCC or FCC structures, the direction of preferential crystal growth during solidification is generally parallel to the <100> plane; therefore, given the rapid cooling employed in PBF, especially LPBF, columnar cellular structures and/or secondary dendrite arrays extending toward the <100> plane are formed in the melt pool. Preferential orientation primarily adopted by the cell growth which is dendrite without secondary arm formation due to fast cooling rate during LPBF. Therefore, the texture formation strongly affiliated to cell orientation.

During solidification in the melt pool, columnar cells aligned to the <100> plane tend to grow parallel to the heat flow direction and perpendicular to the tangential direction of the melt pool [23, 24] if the effect of epitaxial growth is negligible. However, in practice, epitaxial growth occurs from the underlying and adjacent solidification zone [25], and the combination of preferential growth direction, melt pool shape based on the heat flow direction, and crystal orientation at the solidification zone determines the final crystallographic texture and preferred orientation of the obtained material [26]. In other words, the artificial adjustment of these geometrical relationships could enable the control of single-crystalline-like or fiber textures and even their crystal orientations, such as the <100>, <110>, and <111> planes, parallel to the BD. The scan strategy, which refers to the combination of laser scanning directions per layer [3, 4, 27], may be a representative LPBF parameter that can be controlled for this purpose.

Figure 6 shows the scan-strategy dependence of the crystallographic textures of LPBF fabrications of β-titanium alloys and tungsten in terms of inverse pole figure maps and {100} pole diagrams [4, 28]. The preferential grain orientation changes depending on the scan strategy during the LPBF fabrication of β-titanium alloys; specifically, a crystallographic texture with a {110}z<100>x orientation is formed under scan strategy X, and a crystallographic texture with a

IOP Conf. Series: Materials Science and Engineering **1310** (2024) 012013 doi:10.1088/1757-899X/1310/1/012013

{100}z<100>x orientation is formed under scan strategy XY. By contrast, the crystallographic texture of tungsten, which forms a cubic structure, similar to β-titanium alloys, is independent of the scan strategy, forming a {100}z<110>x orientation in the scan strategies X and XY. Moreover, scan strategy X indicated tilted grain morphology towards the scan track direction for tungsten, which can be identified in IPF map and pole figure represented in figure 6.

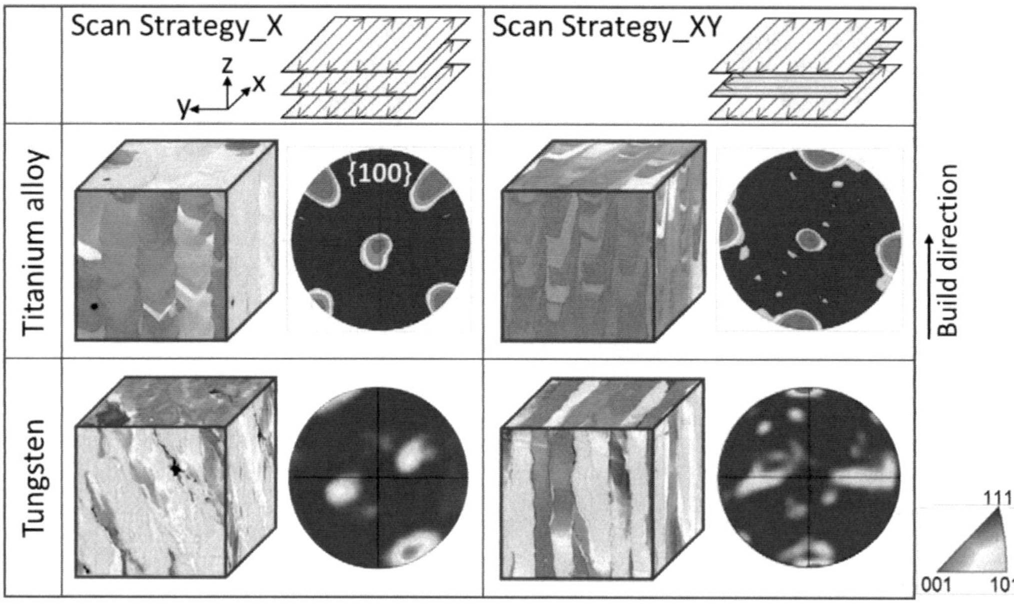

Figure 6. Variation of the formation of crystallographic texture during laser powder bed fusion depending on the starting material (β-titanium alloy or tungsten) and scan strategy (X or XY). Adapted with modifications from references [4] and [28], published under the CC-BY-4.0 license.

The difference in crystallographic texture orientation between the two materials can be attributed to differences in their melt pool shape: the melt pool formed by β-titanium alloys has a long tail, whereas the melt pool formed by tungsten hardly shows this tail and has a hemispherical shape [28]. This difference in melt-pool shape is due to the large difference in the melting point and thermal conductivity between the two materials.

Figure 7(a, b) shows optical microscopy images of the y-z section of the sample surface produced under scan strategies X and XY after chemical etching. A certain melt pool shape and cell structure can be observed owing to the segregation of a small amount of the constituent elements [29]. The direction of cell elongation differs between the left and right sides of the melt pool, each having an angle of approximately ±45° relative to the BD, and the left and right solid–liquid interfaces meet at an angle of approximately 90° at the center of the melt pool (the meeting interface is indicated by the white arrow in the figure). Thus, a grain boundary with a low angle misorientation exists approximately at the center of the melt pool. The crystalline orientation formed during LPBF self-adjusts to reduce the orientation difference at the center of the melt pool, indicating a cell growth interface from the side branches of the melt pool. Therefore, the {110}z<100>x orientation becomes stable, the crystal orientation is transferred to the adjacent melt pool by side-branching or to the next melt pool layer (as indicated by the • in figure 7(a, b)), solidification occurs at the melt pool boundary, and the growth direction of the cell changes by 90° and maintains its crystal orientation [25].

Although epitaxial growth under scan strategy XY is possible, it results in a different crystal orientation. The laser scanning direction in scan strategy XY alternates from the x direction (where <100>//x and <011>//y are the stable orientations in the melt pool) to the y direction (where <100>//y and <011>//x are the stable orientations) such that the <100> plane in the solidification zone and the <011> plane in the front layer compete in texture formation depending on the laser scanning direction. Because the growth rate of the <100> plane is greater than that of the <011> plane [30], the <100> plane is first oriented in the two scanning directions (x and y directions), followed by the z direction, stabilizing the {001}z<100>x crystallographic texture. The crystal orientation selection of the adjacent melt pool and subsequent layer is driven by the reduction in the crystal orientation difference at the solid–liquid interface in the transverse section of the melt pool.

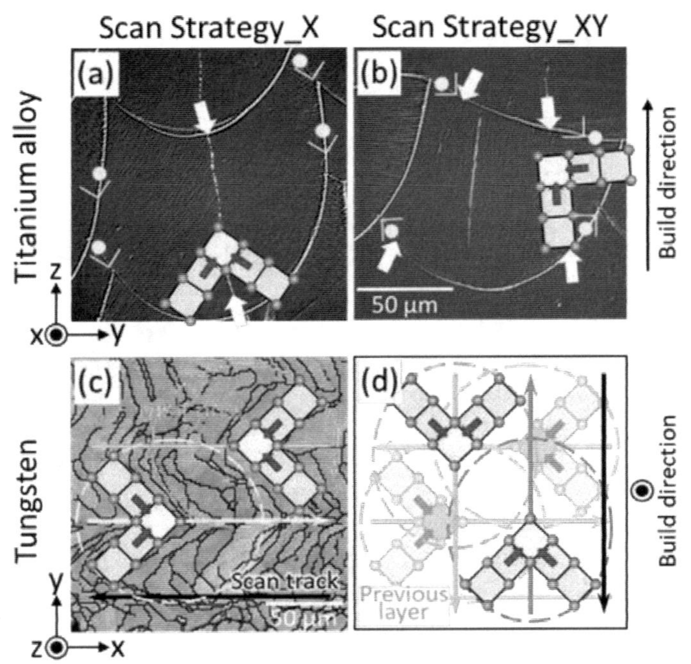

Figure 7. Melt pool shape, solidification plane, and <100> growth direction on this plane in titanium alloy and tungsten. Adapted with modifications from references [4] and [28], published under the CC-BY-4.0 license.

Solidification is known to occur on the x-y plane in hemispherical melt pools of tungsten [28]. In pure metals, no segregation occurs; therefore, the direct observation of the cell and melt-pool shape is not possible. However, the direction of solidification can be estimated from the grain geometry and temperature-field simulations. Scan strategy X results in <100> growth in the x-y plane at an angle of ±45° with respect to the laser scanning direction owing to the direction of G, forming a crystal orientation of {100}z<110>x, as shown in figure 7(c). In this case, the crystal growth direction in the subsequent layer corresponds to that in the previously melted layer in scan strategy XY, as shown in figure 7(d), and no change in the direction of the crystal orientation occurs between the two scan strategies[2] [28].

The melt pool shape varies with the physical properties of metallic materials; however, the melt pool shape and resulting crystallographic texture orientation also vary depending on the laser scanning conditions and energy distribution of the heat source [31-33].

3.3 Crystallographic orientation formation in hexagonal crystal structures

Materials with a hexagonal crystal structure show a crystallographic texture that completely differs from that of materials with a cubic crystal structure; figure 8 shows the scan-strategy dependence of the crystallographic texture of intermetallic $NbSi_2$, which has a hexagonal C40 structure [34]. This hexagonal intermetallic compound shows preferential growth in the <0001> direction. The multiplicity of the <100> plane in the cubic crystal system is three if the positive and negative directions are not distinguished; by contrast, the multiplicity of the <0001> plane in the hexagonal crystal system is one and shows low symmetry. In other words, the crystalline orientation around the <0001> direction is ambiguous and arbitrary, and a single-crystalline-like microstructure is not achieved, although fiber texture formation with a <0001> orientation parallel to the BD occurs. Because the crystal orientation around the <0001> orientation is not determined, a similar crystallographic texture develops independently of scan strategy [34], as shown in figure 8, while scan strategy X for $NbSi_2$, exhibited similar tilted grain, similar to the case of tungsten, towards scan track direction.

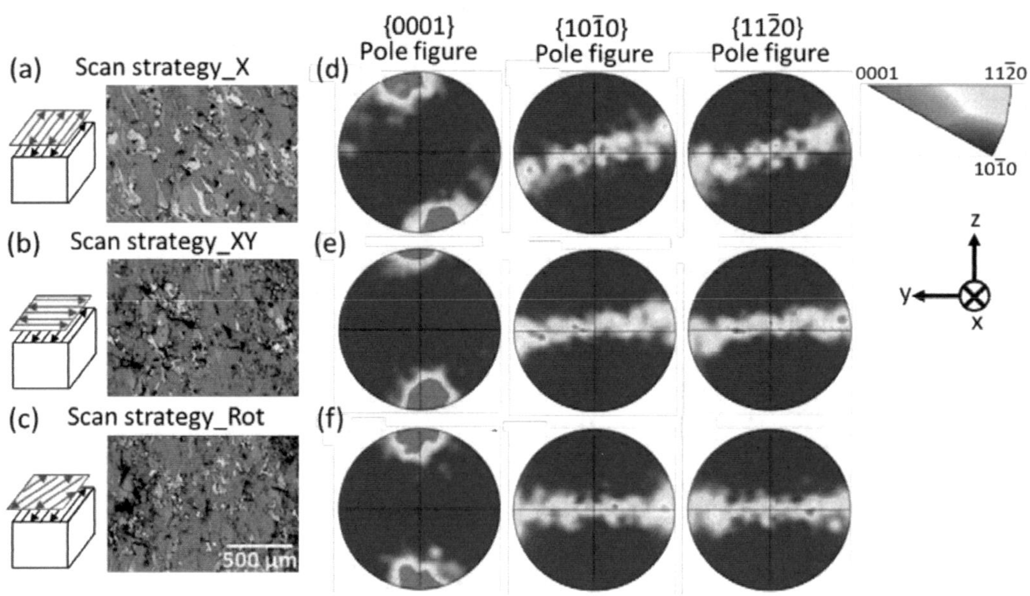

Figure 8. Dependence of crystallographic texture formation in hexagonal $NbSi_2$ on the laser bed powder fusion scan strategy. Adapted with modifications from reference [34], published under the CC-BY-4.0 license.

However, even in the case of hexagonal crystal systems, single-crystalline-like textures can be successfully formed by applying a special scan strategy that takes symmetry into account [35]. The formation behaviors of the crystallographic textures of cubic and hexagonal crystal systems are summarized in figure 9.

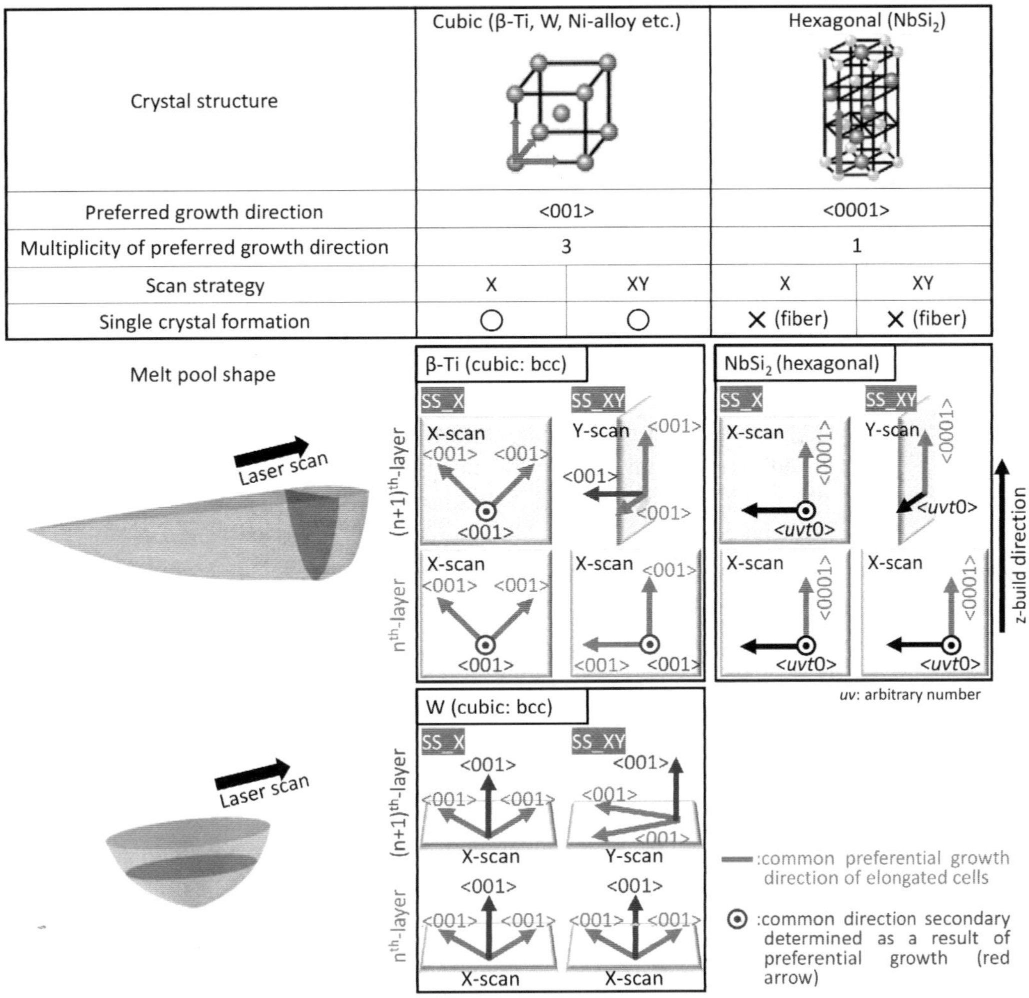

Figure 9. Relationships among crystal structure, preferential growth direction, and texture formation under scan strategies X and XY during laser powder bed fusion. Adapted with modifications from reference [34], published under the CC-BY-4.0 license.

4. Alloy design for the preparation of single-crystalline-like biological high-entropy alloys by laser powder bed fusion

AM realizes unique microstructures and functionalities in existing alloys that cannot otherwise be obtained via conventional methods by employing a unique thermal history, such as ultrarapid cooling and reheating in track-by-track and layer-by-layer melt pools [16, 36, 37]. Future research and development efforts will focus on creating novel functionalities by designing alloys that are tailored to the characteristics of AM to realize their maximum potential[2].

Our research group developed a quintic $(TiZr)_{30.43}(NbTaMo)_{13.04}$ BioHEA with a BCC crystal structure as a new bone biomaterial [38, 39]. By utilizing ultrarapid LPBF cooling to suppress segregation, which cannot be avoided in casting (figure 10(b)), and maximize solid-solution strengthening, we obtained an ultrahigh-strength metal material with good biocompatibility [40].

However, a low Young's modulus, which is essential for biomaterials used in bones, must be achieved to avoid stress shielding.

To achieve a low Young's modulus, we attempted to simultaneously (1) reduce the number of total electrons (VEC) per atom in the valence band and (2) achieve Young's modulus anisotropy (low Young's modulus at <100>) via single-crystalline-like texture formation [41]. Single-crystalline-like texture formation was not observed in the aforementioned quintic alloy (figure 10(d)) because segregation could not be completely suppressed even under ultrarapid cooling during LPBF, resulting in dendrite formation (figure 10(c)). Moreover, high-melting-point dendrites remained in the melt at the bottom of the melt pool, forming heterogeneous nucleation sites and generating equiaxed grains (figure 10(d)), which are detrimental to the formation of a single-crystalline-like texture [40]. To avoid the formation of equiaxed grains, we designed the composition of $(TiZrHf)_{28.33}(NbTa)_{6.74}Mo_{1.55}$ to satisfy condition (1) while reducing the temperature range (ΔT) in the solid–liquid coexistence region and the deviation of the equilibrium distribution coefficient from 1 (figure 10(e)), thus suppressing the segregation tendency of the alloy [41]. ΔT was reduced from approximately 400 to 50 K and VEC was reduced from 4.52 to 4.17, both of which are significantly reductions. As a result, as shown in figure 10(h), no equiaxed grains were formed at the melt boundary owing to segregation, and a {100}z<100>x single-crystalline-like texture similar to that of the β-type titanium alloy shown in figure 6 was successfully obtained under scan strategy XY. The single-crystalline-like textured BioHEA fabricated by LPBF exhibited a yield strength greater than 1350 MPa, which is approximately 1.4 times greater than that of the cast material, a low Young's modulus of less than 90 GPa, and biocompatibility equivalent to that of pure Ti [41]. Thus, a useful metal biomaterial with multiple functionalities was obtained. These results provide a good example of the preparation of highly functional materials by utilizing the specific characteristics of LPBF.

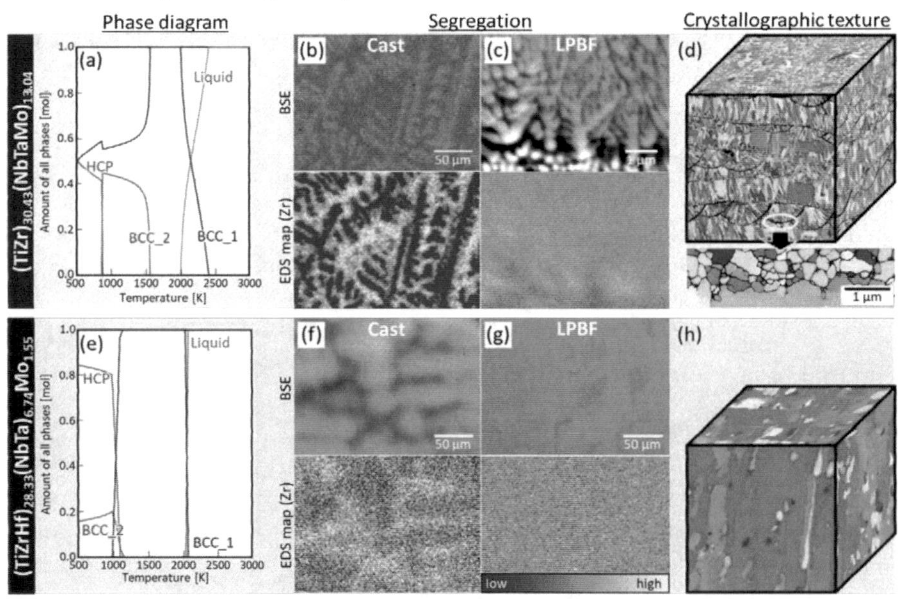

Figure 10. Alloy design for the creation of biological high-entropy alloys (BioHEAs) with a single-crystalline-like texture. (a–d) Alloy with segregation and failed single-crystalline texture formation. (e–h) BioHEA with a single crystalline-like texture formed by suppressing segregation via alloy design. Adapted with modifications from references [40] and [41], published under the CC-BY-4.0 license.

5. Summary and future recommendations

This article reviews recent findings on the control of material microstructures exhibiting anisotropic properties by metal AM, particularly LPBF. Various crystallographic texture formation mechanisms have been reported for a wide range of materials, but the artificial and flexible control of such microstructures has not yet been fully established. In particular, the control of crystallographic textures in materials that undergo solid–solid transformation, including Ti alloys [42] and TiAl [43] containing an α-phase such as Ti-6Al-4V and martensitic stainless steels [44], requires the control of not only the crystal orientation during solidification but also the variant selection during phase transformation. The latter is a challenging issue because it requires limiting the selectivity of the variants during phase transformation, in addition to controlling the crystal orientation during solidification. Many materials and products made from materials in which the crystal orientation is directly related to their functionality, including metals, ceramics, and polymers, have been reported. Control of microstructure with anisotropic properties is expected to become a major guideline for future research and development efforts in AM. Moreover, design aspects from the viewpoint of material control, in addition to shape control, are expected become increasingly important.

Acknowledgements

This work was supported by a Grant-in-Aid for Scientific Research (JP23H00235) from the Japan Society for the Promotion of Science (JSPS) and by CREST-Nanomechanics: Elucidation of macroscale mechanical properties based on understanding the nanoscale dynamics of innovative mechanical materials (Grant Number: JPMJCR2194) from the Japan Science and Technology Agency (JST).

References

[1] Kodama H 1981 *Rev. Sci. Instrum.* **52** 1770
[2] Clare A T, Mishra R S, Merklein M, Tan H, Todd I, Chechik L, Li J, Bambach M 2022 *J. Mater. Proc. Technol.* **299** 117358
[3] Sun S-H, Hagihara K, Nakano T 2017 *Mater. Des.* **140** 307
[4] Ishimoto T, Hagihara K, Hisamoto K, Sun S-H, Nakano T 2017 *Scr. Mater.* **132** 34
[5] Ishimoto T, Nakano T 2022 *J. Jpn. Inst. Light Metals* **72** 327
[6] Nakano T *et al* 2024 Materia **63** 36
[7] Nakayama H, Yamagami K, Kyogoku H, Komatsu S 1997 *J. Jpn. Soc. Powder Powder Metall.* **44** 427
[8] Shinagawa K 2019 *J. Jpn. Soc. Powder Powder Metall.* **66** 387
[9] Vock S, Klöden B, Kirchner A, Weißgärber T, Kieback B 2019 *Prog. Addit. Manuf.* **4** 383
[10] Zhou X, Liu X, Zhang D, Shen Z, Liu W 2015 *J. Mater. Proc. Technol.* **222** 33
[11] Li K-L, Chen J-H, Zhao C-C, Shen Z-J, Liu W 2021 *Tungsten* **3** 218
[12] Gokcekaya O, Ishimoto T, Todo T, Wang P, Takayoshi N 2021 *Addit. Manuf. Lett.* **1** 100016
[13] Sun S-H, Ishimoto T, Hagihara K, Tsutsumi Y, Hanawa T, Nakano T 2019 *Scr. Mater.* **159** 89
[14] Ishimoto T, Wu S, Ito Y, Sun S-H, Amano H, Nakano T 2020 *ISIJ Int.* **60** 1758
[15] Tsutsumi Y, Ishimoto T, Oishi T, Manaka T, Chen P, Ashida M, Doi K, Katayama H, Hanawa T, Nakano T 2021 *Addit. Manuf.* **45** 102066
[16] Gokcekaya O, Ishimoto T, Hibino S, Yasutomi J, Narushima T, Nakano T 2021 *Acta Mater.* **212** 116876
[17] Hibino S, Todo T, Ishimoto T, Gokcekaya O, Koizumi Y, Igashira K, Nakano T 2021 *Crystals* **11** 1064
[18] Sugitani Y 2021 *J. Jpn. Thermal Spray Soc.* **58** 154
[19] Hagihara K, Nakano T 2022 *JOM* **74** 1760
[20] Miyata Y, Okugawa M, Koizumi Y, Nakano T 2021 *Crystals* **11** 856
[21] Koizumi Y, Okugawa M 2022 *ISIJ Int.* **62** 2183
[22] Ishimoto T, Suganuma R, Nakano T 2023 *Mater. Lett.* **349** 134835

[23] Ekubaru Y, Gokcekaya O, Ishimoto T, Sato K, Manabe K, Wang P, Nakano T 2022 *Mater. Design* **221** 110976
[24] Ishimoto T *et al* 2021 *Scr. Mater.* **194** 113658
[25] Pham M-S, Dovgyy B, Hooper P A, Gourlay C M, Piglione A 2020 *Nat. Commun.* **11** 749
[26] Ishimoto T, Hagihara K, Hisamoto K, Nakano T 2021 *Addit. Manuf.* **43** 102004
[27] Marattukalam J, Karlsson D, Pacheco V, Beran P, Wiklund U, Jansson U, Hjörvarsson B, Sahlberg M 2020 *Mater. Design* **193** 108852
[28] Todo T, Ishimoto T, Gokcekaya O, Oh J, Nakano T 2022 *Scr. Mater.* **206** 114252
[29] Okugawa M, Saito K, Yoshima H, Sawaizumi K, Nomoto S, Watanabe M, Nakano T, Koizumi Y 2024 *Addit. Manuf.* **84** 104079
[30] Messler RW 2008 *Principles of Welding* (New York: Wiley)
[31] Liu Y, Nose K, Okugawa M, Koizumi Y, Nakano T 2023 *Mater. Trans.* **64** 1135
[32] Sun S-H, Hagihara K, Ishimoto T, Suganuma R, Xue Y-F, Nakano T 2021 *Addit. Manuf.* **47** 102329
[33] Bi J, Wu L, Li S, Yang Z, Jia X, Starostenkov M D, Dong G 2023 *J. Mater. Res. Technol.* **26** 4606
[34] Hagihara K, Ishimoto T, Suzuki M, Ozasa R, Matsugaki A, Wang P, Nakano T 2021 *Scr. Mater.* **203** 114111
[35] Ozasa R, Kizawa Y, Hagihara K, Sato K, Tobata T, Tokunaga T, Nakano T 2024 Mater Des submitted
[36] Yang J, Yu H, Yin J, Gao M, Wang Z, Zeng X 2016 *Mater. Des.* **108** 308
[37] Todai M, Nakano T, Liu T, Yasuda H Y, Hagihara K, Cho K, Ueda M, Takayama M 2017 *Addit. Manuf.* **13** 61
[38] Todai M, Nagase T, Hori T, Matsugaki A, Sekita A, Nakano T 2017 *Scr. Mater.* **129** 65
[39] Iijima Y, Nagase T, Matsugaki A, Wang P, Ameyama K, Nakano T 2021 *Mater. Des.* **202** 109548
[40] Ishimoto T *et al* 2021 *Scr. Mater.* **194** 113658
[41] Gokcekaya O *et al.* 2023 *Mater. Res. Lett.* **11** 274
[42] Amano H, Ishimoto T, Suganuma R, Aiba K, Sun S-H, Ozasa R, Nakano T 2021 *Addit. Manuf.* **48** 102444
[43] Park S-H, Gokcekaya O, Ozasa R, Oh M-H, Kim Y-W, Kim H-S, Nakano T 2023 *Met. Mater. Int.* **30** 1227
[44] Kannan R, Nandwana P 2022 *Sci. Rep.* **12** 6396

Microstructure evolution in laser-based powder bed fusion of metals

Venkata K Nadimpalli* **and Tianbo Yu***

Department of Civil and Mechanical Engineering, Technical University of Denmark, Kgs. Lyngby, Denmark

*E-mail: vkna@dtu.dk; tiyu@dtu.dk

Abstract. Laser-based powder bed fusion (LPBF) of metals offers the unique possibility of creating the microstructure voxel-by-voxel. The minimum voxel size in each direction is dependent on material dosing accuracy coupled with laser processing parameters. The rapid solidification conditions during LPBF lead to material heterogeneity coupled with hierarchical and non-equilibrium microstructures. The current paper delves into two different pathways available currently to control microstructure in LPBF, namely: in-situ microstructure control through material distribution to form functionally graded components with complex interfaces; application of post-processing thermo-mechanical treatments to control the microstructure. Unlike traditional manufacturing methods, each voxel in LPBF can be further processed multiple times after the first fusion process. Such in-situ processing presents further opportunity for tailoring the microstructure of each voxel in 3D. A future perspective is thus offered on the opportunities to control and engineer LPBF microstructures in metals.

1. Introduction

Additive Manufacturing (AM) represents a significant advancement in metal manufacturing. The AM process has major advantages compared to conventional manufacturing, including fast net-shape production, resulting in the use of less raw materials, energy savings [1], and a larger design-freedom for manufacturing of complex and customized components. For example, it has been shown that the fuel efficiency of nozzles can be improved by 15% through an updated design, which can only be produced by AM [2]. Furthermore, AM allows a shorter time from design to production. Among the various AM techniques, Laser-based Powder Bed Fusion (LPBF) stands out due to its precision and ability to produce complex geometries coupled with a high density. LPBF employs a high-powered focused laser to fuse powdered metal layer by layer, allowing for the creation of intricate components that are often impossible to achieve with traditional manufacturing methods. In certain cases, it means that previously unachievable performance can be unlocked.

The research on LPBF has historically focused on using one material and finding an optimum process parameter window that produces good density in components. The range of processing parameters providing >99% density is often termed as an optimal process window for a given

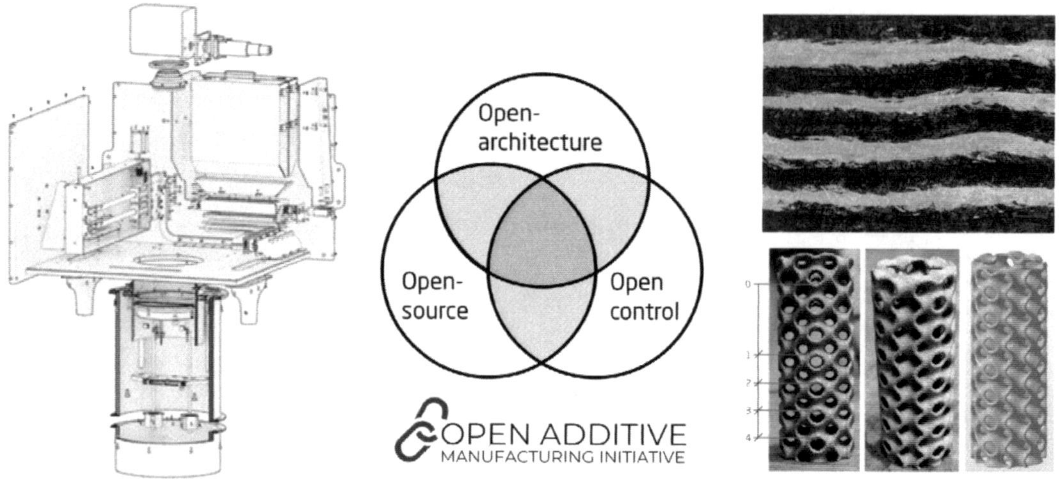

Figure 1. DTU Open AM system built and open-sourced in August 2024 as part of the Open Additive Manufacturing Initiative. On the right top, we see an example of a multi-material steel-steel composite and on the right bottom is a complex geometry which was within 100 μm geometric deviation on the outer surfaces in relation to the CAD model as measured using a 3D scanner. Adapted from [11].

material and is linked to the energy density which is a function of laser power, scan speed, hatch spacing and layer thickness. The energy density, or similar representations like normalized enthalpy [3] operate under the assumption that the bulk of a component consists of one material and therefore only needs to have one set of process parameters. In parallel, industrial and research efforts in improving AM components focused on the development of contour parameters, upskin and downskin parameters to improve geometric fidelity on the outer surface while retaining high density (>99%) in the bulk material [4]. Some research also investigates development of processing parameters for thin walled and lattice structures to enable higher precision [5–7].

While optimizing the bulk density and certain geometrical features of AM components have been the focus of LPBF machine tool development, the unique and hierarchical nature of LPBF microstructures caused by rapid solidification has been investigated across several materials and machine tools. The process-structure-property correlations for several common alloys have been widely reported, although the depth of microstructural information is sometimes lacking. The properties of AM components generally deviate from those expected for similar conventionally manufactured components [8]. The rapid solidification and cyclic heating taking place during AM create nano- and microstructural features that in some cases lead to improved mechanical properties, while other properties like corrosion (in austenitic steels) are poorer. Further, a deviation is observed in the measured properties [9,10] across different AM systems operating with the same process parameters as chosen by the user. One part of the deviation can be explained by changes in material composition based on powder supplier. Another part of the deviation is explained by specific physical machine conditions such as laser calibration, optical errors, cross-flow, recoating mechanism, inert gas and oxygen control etc. The final part of the

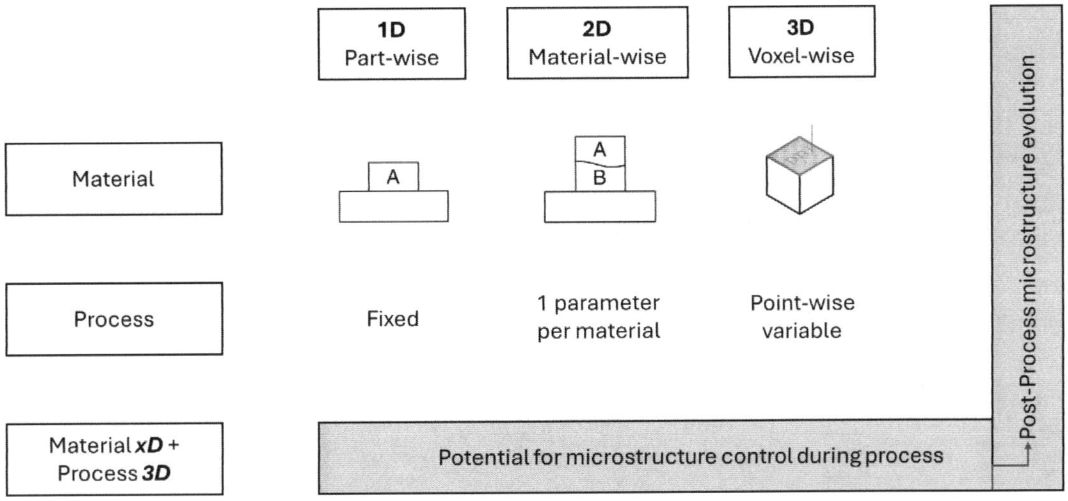

Figure 2. The material and process parameter space for controlling microstructure of AM. The material domain and process parameter domain can be either 1D (part-wise), 2D (material/layer-wise) or 3D (voxel-wise). 3D process parameters coupled with xD material can create heterogeneous and complex microstructures which can be tailored further during post-process treatments.

deviation comes from proprietary and closed-architecture software systems which mask the exact laser processing conditions for each track, which changes the local microstructure conditions.

The lack of access to open AM systems leads researchers to focus on optimizing the limited range of process parameter settings. LPBF process parameters can be varied point-by-point with a resolution limited by the electro-mechanical-optical systems. Typical lasers can be pulsed up to 100 kHz, while the galvo mirrors using a standard XY2-100 protocol can be given a command every 10 μs. At the Technical University of Denmark (DTU), we have worked to overcome the limitations imposed by closed systems by building our own open-architecture LPBF system. In figure 1, a schematic of the DTU open AM system is shown. All details regarding the system and the open-source files are available in [11].

Experimental open systems in research labs [12–14] including the DTU open AM system and a few commercially available systems (Aconity3D) offer the ability to utilize multiple materials during LPBF. Due to the nature of typical LPBF processing, multi-material processing creates mixed powder waste which needs to be separated subsequently. Multi-material processing is primarily limited by the design of LPBF machine tools and their dosing strategies which in turn determine the voxel size of material deposition. The fine-scale multi-material gradients produced by LPBF (compared to directed energy deposition) can unlock performance characteristics that are not possible elsewhere and hence are interesting for both research on the interfacial characteristics as well as their industrial applicability.

Material dosing combined with LPBF process parameters together form the microstructure of the as-built component. Both material and process parameters in theory can be controlled in 3 dimensions with varying voxel sizes. In figure 2, a schematic of material and process parameter changes across 3 dimensions are represented. Most LPBF systems are in theory capable of voxel-wise control of process parameters, but this has been hardly explored up to now. Moreover, the microstructure of as-built components can be further tailored through post-processing thermo-mechanical treatments, for example by hot isostatic pressing (HIP) and shot peening, as occurs in conventionally manufactured counterparts. However, due to the unique and hierarchical microstructures caused by the rapid solidification and cyclic heating during LPBF, those post-processing treatments must be finely tuned for microstructure and property optimization. In the current paper, we will first discuss the microstructural evolution during (in-situ) and after (post-processing) LPBF.

2. In-situ microstructure evolution

The in-situ microstructure evolution in LPBF is pivotal for attaining the desired properties in the final components. Both material deposition and processing parameters are highly interlinked to the microstructure. The rapid solidification and cyclic thermal conditions intrinsic to the LPBF process result in unique, hierarchical, and often non-equilibrium microstructures that are critical in determining the mechanical and physical properties of the built components. This section will examine three specific case studies illustrating the current methodologies employed in microstructural evolution.

2.1 Premixing case study (2D material, 1D process)

Powder mixing to create new alloys in LPBF involves blending different powders before the process to achieve a desired alloy composition. This method of in-situ alloying allows for the customization of material properties by varying the proportions of constituent powders. Further, the morphology and size distribution of the powders can be controlled to keep a reasonable flowability while improving packing density. The ability to fine-tune the powder mix enables the creation of novel alloys with tailored characteristics, such as improved strength, corrosion resistance, etc. This approach vastly expands the material palette available for LPBF, sometimes enabling the creation of non-equilibrium alloys. While there are two or more materials in this strategy, because they are always mixed together, they can be dosed together and treated as a single material. Hence, several commercial systems can handle these types of powders and this field of research has sparked a large interest from the research community. One prominent use case is the addition of Zr/Sc powders to high-strength aluminium alloys that are otherwise prone to solidification cracking [15]. Other studies investigated the creation and modification of high entropy alloys [16].

The current case study by the authors on powder premixing in LPBF focuses on the integration of chromium nitride (Cr_2N) with AISI 316L stainless steel powder to enhance corrosion resistance and mechanical properties [17]. By adding 2.5 wt% Cr_2N to the AISI 316L powder, the resultant mixture was subjected to LPBF, leading to the successful incorporation of both chromium and nitrogen into the fully austenitic microstructure of the stainless steel. This process significantly increased the nitrogen content from 0.09 wt% in the original powder to 0.31 wt% in the LPBF-fabricated component. The increased nitrogen content caused a slight expansion of the austenite lattice, which was observed as a shift in the X-ray diffraction peaks, seen in figure 3. The elongated austenite grains with an internal cellular substructure in the modified 316L indicated effective alloying and microstructural refinement.

The enhanced properties of the modified alloy were evident in its increased Vickers hardness, which rose by approximately 40 HV0.1 due to interstitial solid solution strengthening from the nitrogen. Additionally, the Cr_2N-modified 316L displayed superior corrosion resistance, demonstrated by its exceptional performance in a 5.6 wt% $FeCl_3$ solution after 10 days of immersion. The combination of improved hardness and corrosion resistance, while maintaining the manufacturability and characteristic cellular microstructure of LPBF 316L, showcases the

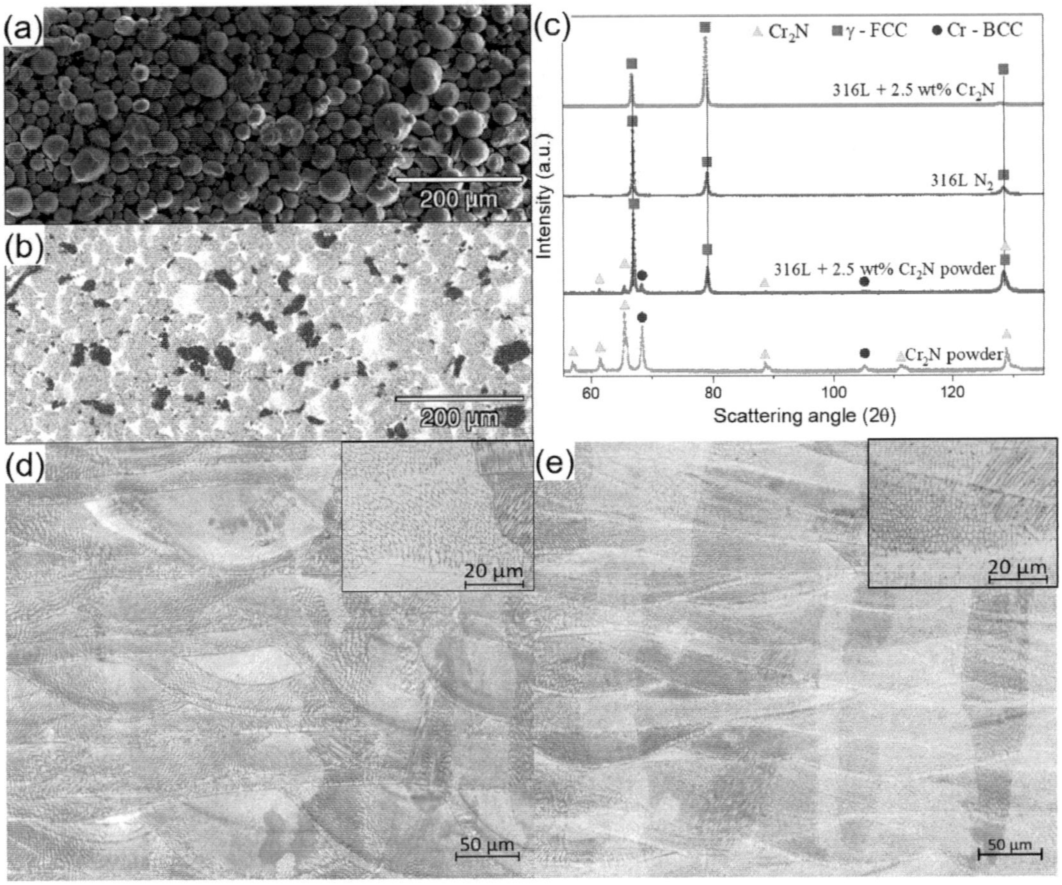

Figure 3. (a) SEM of 316L + 2.5 wt% Cr_2N mixed powder; (b) EDS map of area (a); (c) XRD of powders and printed specimen (316L N_2 represents the reference 316L sample printed in N_2 atmosphere); LOM micrographs of etched y-z cross sections from (d) L-PBF 316L specimens and (e) 316L + 2.5 wt% Cr_2N specimens. Reproduced from [17].

potential of powder premixing for alloy modification. This method allows for precise control over the composition and properties of LPBF-fabricated components, providing a versatile approach to enhancing the performance of stainless steel components in various applications. The method was also validated using a commercial system from Nikon SLM Solutions in an industrial setting and yielded a 15% increase in tensile strength with only a 7% reduction in elongation in as-built condition, coupled with superior corrosion performance.

2.2 In-situ mixing case study (2D material, 2D process)

In-situ mixing in additive manufacturing, particularly for functional metal components, harnesses the capability to blend different metal powders during the fabrication process. This method facilitates the creation of metal-metal composites with precisely tailored properties, optimizing them for specific applications. By incorporating multiple materials within a single build, functionally graded materials (FGMs) with gradual transitions in composition and microstructure can be engineered [18,19]. These gradients can enhance mechanical strength, thermal stability, while retaining good toughness. The fine control provided by LPBF systems enables complex

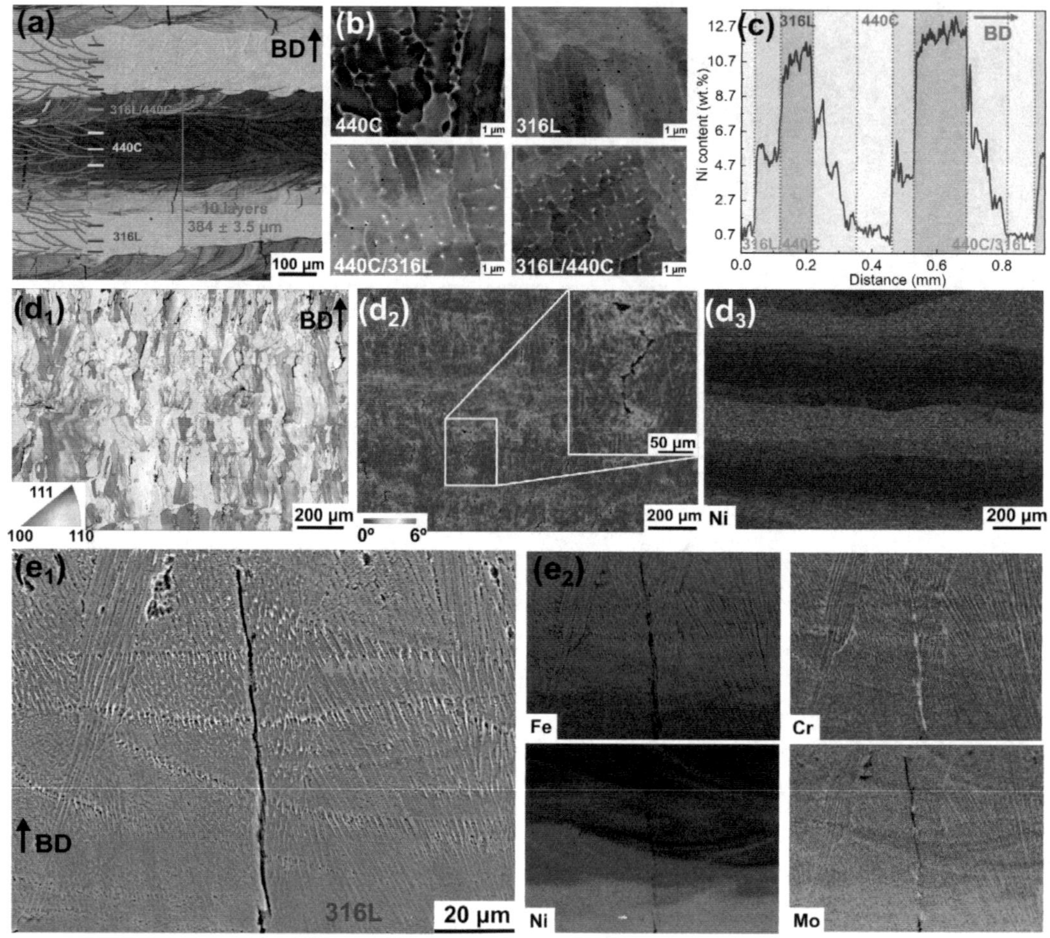

Figure 4. (a) An LOM image showing an overview of the layer structure sample (red lines indicate meltpools, purple dashes are 316L, white dashes are 440C, blue/orange dashes are interfaces); b) SEM images of different regions; (c) corresponding EDS line scan of Ni content; (d1-d3) A EBSD map showing the crystallographic orientation parallel to BD and KAM map with a threshold angle of 2.5° and also showing a region with a crack (e1) SEM image from interface zone with crack and (e2) corresponding EDS maps. Adapted from [21].

designs and precise material distribution, significantly expanding the potential for innovative multi-functional designs that were previously unattainable [20].

The current case study investigates the microstructural evolution of multilayered AISI 316L-440C steel composites manufactured using LPBF. The DTU open-architecture additive manufacturing system with a multiple powder delivery module was employed to fabricate these composites [21]. We found that more pronounced remelting occurs when 440C powder is deposited on the 316L band, leading to a thick transition region and a thin 440C band, see figure 4a. Cracks were primarily observed in the transition regions and were attributed to hot cracking mechanisms due to significant chemical segregation along the grain boundaries. The findings showed that the transition regions between 316L and 440C layers exhibited complex microstructural features and chemical segregation, particularly of chromium and molybdenum.

This segregation is linked to the initiation and propagation of cracks, mainly due to hot cracking susceptibility in these regions. figure 4 provides a detailed view of the resulting microstructure, revealing distinct bands of 316L and 440C, with transition regions exhibiting complex microstructural features due to the different thermal properties and solidification behaviours of the two materials. The 316L layers appeared brighter due to higher nickel content, while 440C layers, characterized by higher carbon and chromium contents which appears as darker regions in the LOM micrograph. It is interesting to note that the high-carbon steel (440C) itself doesn't exhibit cracking and prints in an austenitic phase, while the transition regions in the interface are prone to cracking. The study [21] provides insights into the processing-microstructure relationship in LPBF of multilayered steel composites and discusses challenges associated with defect formation and strategies for improving the quality of multi-material, multilayered composites.

2.3 In-situ mixing case study (3D material, 2D process)

Selective Powder Deposition (SPD) enables the simultaneous placement of different powders within a single layer, which is crucial for fabricating complex structures with distinct material phases in LPBF [22]. The SPD process begins with the strategic placement of different powders on the build plate, guided by a detailed CAD model that outlines the material layout. This is achieved by opening valves in the powder drums to release controlled amounts of each powder, creating a patterned powder bed. The precision of powder placement is typically achieving a resolution of approximately 0.5 mm, which allows for fine control over the material distribution with a voxel size of 0.5 mm*0.5 mm*layer thickness (figure 5).

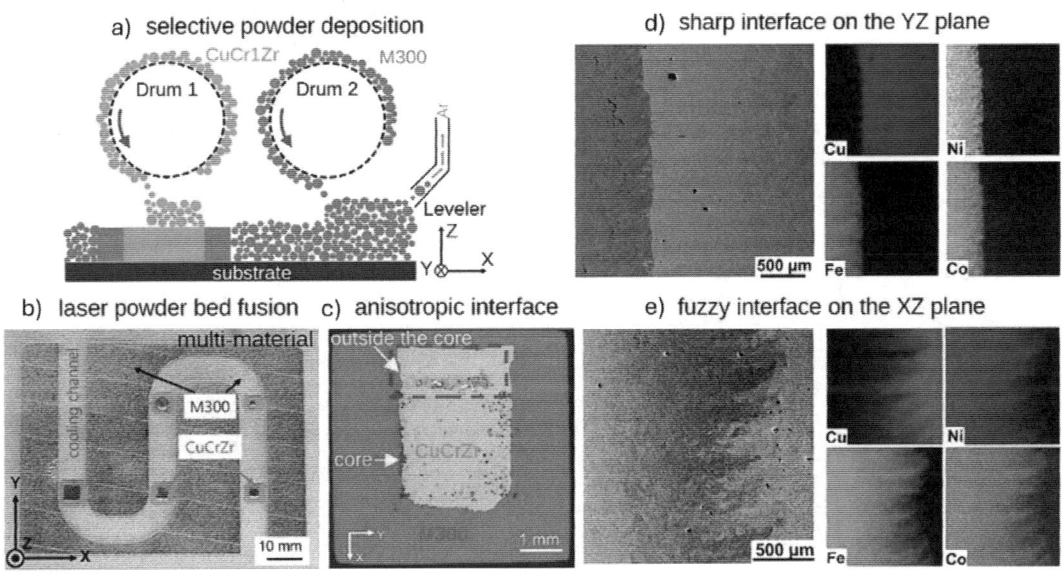

Figure 5. (a) Schematic of selective powder deposition (SPD), (b) manufacturing of mould tools with integrated cooling channels, (c) X-ray CT scan showing the defect distribution among the two materials (Cu, light blue and M300, dark blue) and the interface regions in X,Y directions, (d) Sharp interface along the YZ plane with an intermixed zone < 500 µm, (e) fuzzy interface along XZ plane with an intermixed zone larger than 1 mm, which can also be seen in EDS maps. Adapted from [22].

IOP Conf. Series: Materials Science and Engineering **1310** (2024) 012014 doi:10.1088/1757-899X/1310/1/012014

Currently, SPD is the state-of-the art system for multi-material dosing during LPBF. The current case study investigates the feasibility of manufacturing multi-material mould tools with vertical interfaces using SPD. The study focuses on integrating M300 tool steel and CuCrZr copper alloy, selected for their complementary properties—M300 for its high hardness and CuCrZr for its superior thermal conductivity to enable mould tools with conformal cooling. The components were built layer by layer with a laser scanning process tailored for each material (2D process parameters), see figure 2. Key process parameters, such as laser power, scanning speed, and hatch distance, were optimized separately for M300 and CuCrZr to maximize density. Microstructural analysis revealed that the interfacial regions have a significant material distribution anisotropy along the rolling direction of the recoater. This also leads to an anisotropy in porosity, primarily due to the distinct thermal conductivities and melting behaviours of M300 and CuCrZr. Thus, porosity was more prevalent along the recoating direction, with larger and less spherical pores forming caused by insufficient energy density during the melting of the CuCrZr regions, due to the different processing parameters for the two materials and the anisotropy in material dosing. In contrast, M300 regions displayed fewer defects and more consistent density.

The results indicated that while the overall metallurgical bonding was sound, the interfacial areas required further optimization to reduce porosity, micro-cracks and enhance the structural integrity of the multi-material components. Vickers microhardness testing after an ageing heat treatment showed significant differences between the M300 and CuCrZr regions, with the M300 areas achieving the targeted hardness of around 50 HRC after aging, while the CuCrZr regions had a much lower hardness due to over-aging effects. Despite the anisotropy at the interfaces, the components demonstrated good machinability, maintaining structural integrity during milling and other post-processing steps. The findings underscore the potential of PBF-LB for creating 3D tailored microstructures in multi-material components but highlight the need for further refinement in process parameters and scanning strategies to achieve defect-free interfaces. Given the current state-of-the-art in multi-material intralayer dosing, it is paramount to have voxel-wise process parameters that take into account the precise material distribution at each interface. Moreover, voxel-wise control can help mitigate common defects such as porosity and cracking. By fine-tuning the energy input and scanning strategy locally, the process can accommodate variations in thermal conductivity and melting points between different materials, ensuring uniform melting and solidification.

3. Post-processing treatment

The as-built components are often not ready for service. This is related to the characteristics of metal components produced by LPBF, including porosity, surface roughness, residual stress, microstructural anisotropy and heterogeneity [23,24]. Therefore, post-processing is often needed in order to improve their properties and performance [25,26]. For example, HIP may be applied to reduce the internal porosity of LPBF metal components. HIP is a special heat treatment that operates by applying high pressure and high temperature uniformly around a component to eliminate internal pores, thus increasing the density and improving the mechanical properties of the material. To reduce the surface roughness of LPBF components as required in certain applications, techniques like mechanical polishing, chemical finishing, electropolishing and laser polishing may be applied. Moreover, thermo-mechanical treatments, originally developed for conventionally manufactured metals, may be also employed in order to change the microstructure and residual stress of LPBF components.

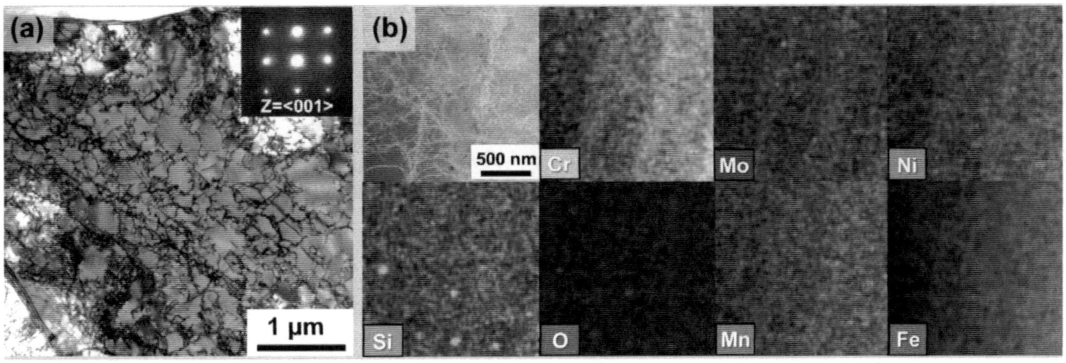

Figure 6. Microstructure of stainless steel 316L samples produced by LPBF after annealing at 800 °C for 0.5 h. (a) TEM bright-field micrograph showing the remaining dislocation network in a grain with one of its <001> axes aligned with the incident electron beam; (b) high-angle annular dark-field image and the corresponding energy dispersion X-ray spectroscopy maps showing the pinning effect of solute segregation on the dislocation network. Reproduced from [30].

3.1 Heat treatment

Heat treatments can be applied to practically all LPBF metallic materials, e.g. steels, aluminium alloys, titanium alloys and Ni-based superalloys. However, the choice of a particular heat treatment depends on the material to be treated and the microstructure to be tailored. It is worth noting that due to the unique microstructure produced by LPBF, the corresponding response to heat treatment may be different from that of conventionally manufactured materials of the same composition [27–29]. For example, the dislocation cell structure produced by LPBF shows slow recovery kinetics due to the pinning effect from the segregation network (figure 6) [30], and the pores also have a pinning effect on boundary migration during recrystallization [31].

One of the most studied materials used in LPBF is stainless steel 316L, which is an austenitic steel that cannot be heat treated for hardening. Therefore, besides HIP to reduce porosity, the two most relevant heat treatments for 316L are low temperature (<400 °C) stress relieving treatment and high temperature (>1000 °C) solution treatment. During the low temperature heat treatment, the residual stress is reduced while the microstructure including the dislocation cell structure is almost unchanged. During the high temperature heat treatment, the microstructure changes dramatically through homogenization and recrystallization. Homogenization leads to the dissolution of the segregation network originally formed during LPBF and results in a uniform distribution of the alloying elements. Meanwhile, the stored energy presented in the form of the characteristic dislocation cell structure in LPBF 316L serves as the driving force for recrystallization, which replaces the original grains of irregular shape with new grains of more equiaxed shape and with few interior dislocations, and thus relieving the original residual stress (figure 7) [30]. Moreover, this high temperature heat treatment results in dissolution of unwanted secondary phases (e.g. σ phase and carbides) that may form due to the intrinsic heating during manufacturing, as well as coarsening of nano-sized oxide inclusions. It should be noted that the high temperature heat treatment of LPBF 316L has to be followed by quenching as studies show that a prolonged soaking time at intermediate temperatures (e.g. 500–800 °C) leads to unwanted secondary phases [30]. It follows that after low temperature heat treatment, LPBF 316L still

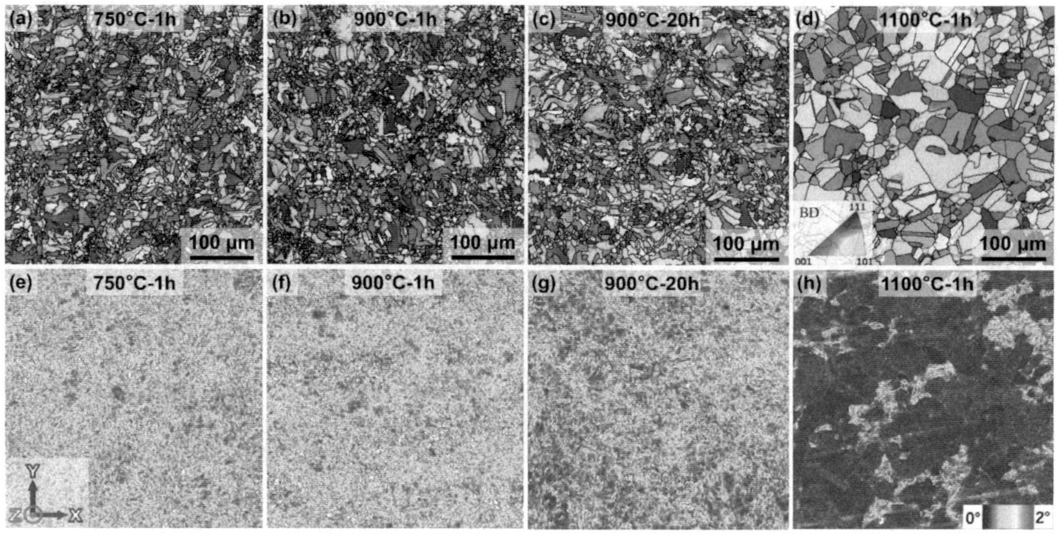

Figure 7. Microstructures of stainless steel 316L samples produced by LPBF after different heat treatments. (a–d) EBSD inverse pole figure colouring maps showing the grain structure; (e–h) the corresponding kernel average misorientation maps showing the orientation variation inside grains. Recrystallization occurred after annealing at 1100 °C for 1 h. Reproduced from [30].

possesses the original microstructure and a high yield strength, whereas after high temperature heat treatment, the microstructure becomes very similar to that of conventionally manufactured ones.

Many metallic materials manufactured by LPBF heavily rely on fine precipitates for their high strengths, for example precipitation hardening stainless steels, maraging steels, Inconel 718, AlSi10Mg, and Ti6Al4V. For these materials, heat treatments involving aging are necessary in order to achieve the required properties. The conventional heat treatment procedure starts with solution treatment at a high temperature, typically followed by quenching to room temperature to get metastable martensite or other phases supersaturated with solute elements, and finishes with aging at a relatively low temperature (e.g. 100–600 °C) to get fine precipitates. For components manufactured by LPBF, the standard heat treatment may be modified for property optimization and cost saving. For example, 18Ni-300 maraging steel has been manufactured by LPBF and its response to heat treatment has been documented in a number of studies. It was found that the rapid solidification inherent to the LPBF process results in supersaturated martensite, so the solution treatment before aging may be skipped. However, the maraging steel produced by LPBF often contains a small fraction of retained austenite, and aging without prior solution treatment introduces reverted austenite (figure 8) [32], so the final microstructure and properties are different from those of conventionally manufactured ones.

3.2 Mehcanical treatment

Surface deformation-based techniques, e.g. shot peening and laser shock peening, can alter the microstructure and residual stress at the surface layer significantly, thus having potential to improve the fatigue performance of LPBF components. The choice of treatment depends on the specific requirements of the application, including desired surface finish, mechanical properties,

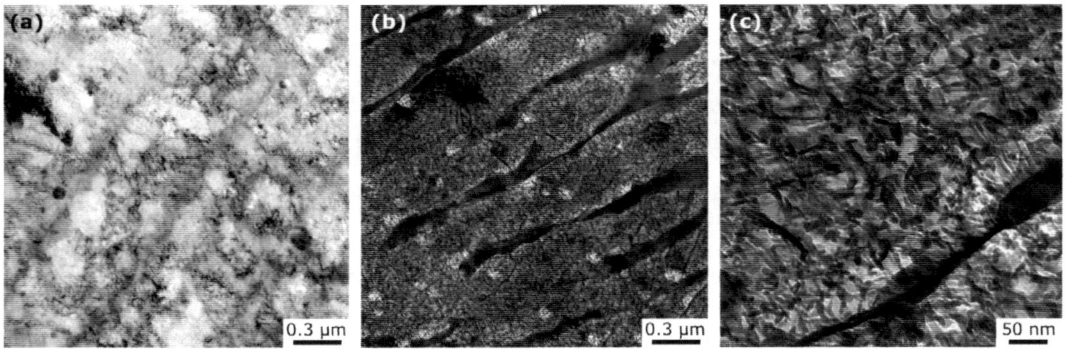

Figure 8. TEM bright-field micrographs showing the microstructure of 18Ni-300 maraging steel produced by LPBF. (a) High density of dislocations in the as-built condition; (b,c) high density of nano-sized precipitates and retained/reverted austenite after aging at 500 °C for 6 h. Reproduced from [32].

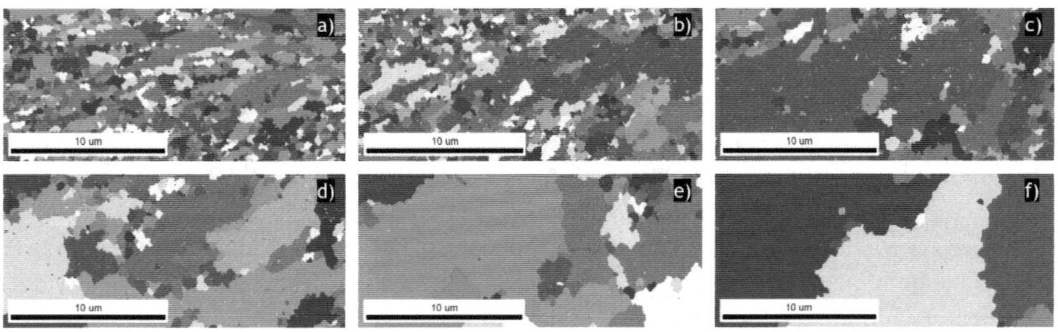

Figure 9. EBSD maps showing the grain structures in an AlSi10Mg sample produced by LPBF and subjected to multi-pass shot peening. Each panel represents a consecutive 10-μm-thick slice from the surface (a) to the interior (f). The grains are detected with a threshold misorientation angle of 5° and randomly coloured. Reproduced from [33].

and component geometry. Shot peening is a well-established surface treatment technique, which works by bombarding the surface of a component with small spherical media (shots) at high velocity. This process introduces a gradient microstructure and lateral compressive residual stresses at the surface layer, and thus improves the fatigue resistance, reduces the stress corrosion cracking, and enhances the overall surface hardness. For LPBF components, it can also improve the surface finish and close pores at the surface layer. In a recent study of multi-pass shot peening of LPBF AlSi10Mg alloy, it was found that shot peening lead to a fragmentation of the eutectic Si network and formation of a gradient microstructure in the surface layer, with the grain size decreased significantly at the top surface (figure 9) [33].

Due to the unique microstructure produced by LPBF, the deformation behaviour of as-built samples is significantly different from that of conventionally manufactured one. It was found in a recent study that deformation twinning occurs at lower strains and twins are thinner and have more variants in stainless steel 316L samples produced by LPBF, compared to that in conventionally manufactured ones [34]. It follows that more studies are needed in the application of surface deformation based post-processing techniques to LPBF metals in order to optimize the microstructure and properties.

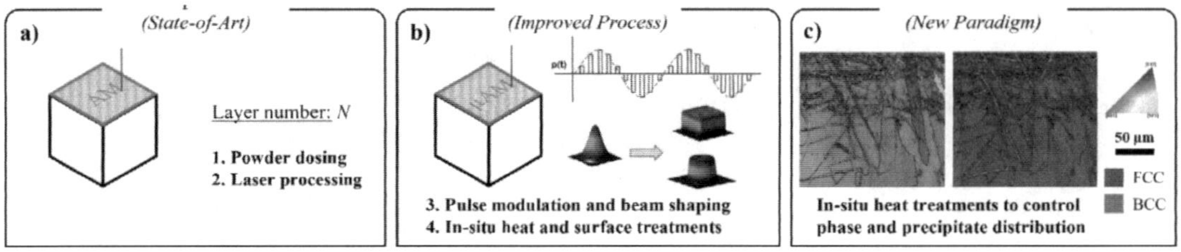

Figure 10. (a) Current state-of-the-art L-PBF process involving powder dosing and laser processing in each layer. (b) Improved process with the additional steps of in-situ pulse modulation and beam shaping during laser processing. (c) The new paradigm allowing for localized control of texture, phase and precipitate structure in AM materials. The example shows a high-carbon stainless steel (FCC in as-built condition) where in-situ heat treatments provoke a local martensitic transformation, which can be seen in three distinct layers in the top half of the orientation maps. The heat treatment involved defocusing of the laser beam and repeated scanning of the AM sample geometry at high scanning speeds.

4. Future perspective

The opportunity of microstructural engineering during the LPBF process is enormous (figure 10). The open-architecture LPBF systems, like the one we have in-house at DTU, enable us to tune freely all the process parameters, including laser scanning strategy, laser spot size, laser speed, hatch spacing, layer thickness, process atmosphere and material dosing in each voxel. The continued development of these systems also enables in-process treatments via in-situ beam shaping and closed-loop feedback control. We can control the laser power at very high speeds of up to 100 kHz, which can change the solidification from conduction to keyhole mode locally. Thereby, the melt pool shape and size (50–100 µm) can be changed every 10 µs, resulting in a spatial resolution (voxel) of around 50 µm. The control on this scale also has implications at the nano-scale, by changing the local dislocation density and cell boundary segregation. Moreover, it is also possible to perform in-situ heat and surface treatments using the LPBF laser source, to engineer the microstructure further before the next sample layer is manufactured. This offers a high flexibility in tailoring the local microstructure, precipitate distribution, texture and phase composition.

Post-processing treatments can further optimize the component density, internal microstructure, surface finish, as well as mechanical and corrosion properties. However, due to the significant difference in the microstructure between AM components and conventionally manufactured ones, a better understanding of the as-built microstructure is needed in order to select the optimal post-processing treatment for a specific metallic component fabricated by LPBF. A wide variety of post-processing treatments originally developed for conventionally manufactured metals may be optimized and employed for AM components, and new techniques may be also developed to suit the unique microstructure of AM components.

LPBF is an inherently complex and stochastic process, especially since many physical phenomena occur within very short timespans, e.g. evaporation, melting, solidification, ablation, laser-material interaction, multiple reflections, capillarity, and the Marangoni effect. The second challenge is LPBF's multi-scale nature, meaning that phenomena happening at the nano- to micrometre scales, have implications at the component-scale, i.e. the centimetre scale. Therefore, advanced numerical modelling that accounts for all these physical phenomena and scales, is necessary to pave the way for an in-depth understanding of LPBF. Currently, numerical simulations of LPBF rely on different coupling strategies, because they operate at different length-scales and have different fidelity levels. Therefore, a multi-scale modelling platform that establishes bi-directional coupling between the major modelling scales is needed. Such a platform also enables simulations of the impact of external stimuli (i.e. during post-processing treatments) on the evolution of the internal microstructure and local residual stress. Open-architecture LPBF systems enable the control of both material and process in 3D. However, it is in principle possible to apply local 3D process parameters on semi-open commercial systems. Therefore, it is paramount to develop a deep understanding of the relationships between the process and microstructure in each voxel, which can then be used to tune the microstructure in 3D.

Acknowledgements

This work was supported by the Villum Fonden [MicroAM—grant number VIL54495].

References

[1] Frazier W E 2014 Metal additive manufacturing: A review *J. Mater. Eng. Perform.* **23** 1917–1928
[2] Wohlers T, Campbell I, Diegel O, Huff R and Kowen J 2020 *Wohlers Report 2020: 3D Printing and Additive Manufacturing Global State of the Industry* (Wohlers Associates)
[3] Bertoli U S, Wolfer A J, Matthews M J, Delplanque J-P R and Schoenung J M 2017 On the limitations of volumetric energy density as a design parameter for selective laser melting *Mater Des* **113** 331–40
[4] Charles A, Bayat M, Elkaseer A, Thijs L, Hattel J H and Scholz S 2022 Elucidation of dross formation in laser powder bed fusion at down-facing surfaces: Phenomenon-oriented multiphysics simulation and experimental validation *Addit. Manuf.* **50**
[5] Guaglione F, Caprio L, Previtali B and Demir A G 2021 Single point exposure LPBF for the production of biodegradable Zn-alloy lattice structures *Addit Manuf* **48** 102426
[6] Viale V, Stavridis J, Salmi A, Bondioli F and Saboori A 2022 Optimisation of downskin parameters to produce metallic parts via laser powder bed fusion process: an overview *The International Journal of Advanced Manufacturing Technology* **123** 2159–82
[7] Grossmann A, Felger J, Froelich T, Gosmann J and Mittelstedt C 2019 Melt pool controlled laser powder bed fusion for customised low-density lattice structures *Mater Des* **181** 108054
[8] Bajaj P, Hariharan A, Kini A, Kürnsteiner P, Raabe D and Jägle E A 2020 Steels in additive manufacturing: A review of their microstructure and properties *Mater. Sci. Eng. A.* **772** 138633
[9] Obeidi M A, Mhurchadha S M U, Raghavendra R, Conway A, Souto C, Tormey D, Ahad I U and Brabazon D 2021 Comparison of the porosity and mechanical performance of 316L stainless steel manufactured on different laser powder bed fusion metal additive manufacturing machines *J. Mater. Res. Technol.* **13** 2361–74
[10] Sanaei N, Fatemi A and Phan N 2019 Defect characteristics and analysis of their variability in metal L-PBF additive manufacturing *Mater. Des.* **182** 108091
[11] Kjer M B 2023 Systems Engineering in the Pursuit of Open Architecture Additive Manufacturing a study. *PhD thesis* Technical University of Denmark
[12] Demir A G and Previtali B 2017 Multi-material selective laser melting of Fe/Al-12Si components *Manuf Lett* **11** 8–11
[13] Wei C, Li L, Zhang X and Chueh Y-H 2018 3D printing of multiple metallic materials via modified selective laser melting *CIRP Annals* **67** 245–8
[14] Bodner S C, van de Vorst L T G, Zalesak J, Todt J, Keckes J F, Maier-Kiener V, Sartory B, Schell N, Hooijmans J W, Saurwalt J J, Saurwalt J J and Keckes J F 2020 Inconel-steel multilayers by liquid dispersed metal powder bed fusion: Microstructure, residual stress and property gradients *Addit Manuf* **32** 101027
[15] Li X, Li G, Zhang M-X and Zhu Q 2021 Novel approach to additively manufacture high-strength Al alloys by laser powder bed fusion through addition of hybrid grain refiners *Addit Manuf* **48** 102400
[16] Chen P, Yao X, Attallah M M and Yan M 2022 In-situ alloyed CoCrFeMnNi high entropy alloy: Microstructural development in laser powder bed fusion *J Mater Sci Technol* **123** 123–35

[17] Valente E H, Nadimpalli V K, Christiansen T L, Pedersen D B and Somers M A J 2021 In-situ interstitial alloying during laser powder bed fusion of AISI 316 for superior corrosion resistance *Additive Manufacturing Letters* **1** 100006

[18] Naebe M and Shirvanimoghaddam K 2016 Functionally graded materials: A review of fabrication and properties *Appl Mater Today* **5** 223–45

[19] Loh G H, Pei E, Harrison D and Monzón M D 2018 An overview of functionally graded additive manufacturing *Addit Manuf* **23** 34–44

[20] Nadimpalli V K and Pedersen D B 2021 Additive Manufacturing of Functional Metals *3D Printing for Energy Applications* 1–32

[21] Pan Z, Nadimpalli V K, Funch C V, Andersen S A, Zhou L, Kjer M B, Christiansen T L and Zhang Y 2024 Microstructural evolution of multilayered AISI 316L-440C steel composites manufactured by laser powder bed fusion *Mater Charact* **211** 113907

[22] Li X, Pan Z, Smolej L, Nadimpalli V K and Moshiri M 2024 Towards manufacturing intra-layer multi-material mould tools with vertical interfaces using laser-based powder bed fusion *Mater Des* **243** 113056

[23] DebRoy T, Wei H L, Zuback J S, Mukherjee T, Elmer J W, Milewski J O, Beese A M, Wilson-Heid A, De A and Zhang W 2018 Additive manufacturing of metallic components – Process, structure and properties *Prog Mater Sci* **92** 112–224

[24] Wang Y M, et al. 2018 Additively manufactured hierarchical stainless steels with high strength and ductility *Nat Mater* **17** 63-71

[25] Laleh M, Sadeghi E, Revilla R I, Chao Q, Haghdadi N, Hughes A E, Xu W, De Graeve I, Qian M, Gibson I and Tan M Y 2023 Heat treatment for metal additive manufacturing *Prog Mater Sci* **133** 101051

[26] Maleki E, Bagherifard S, Bandini M and Guagliano M 2021 Surface post-treatments for metal additive manufacturing: Progress, challenges, and opportunities *Addit Manuf* **37** 101619

[27] Zhang C, Ahmed S, Nadimpalli V K, Yu T and Juul Jensen D 2024 Recrystallization kinetics in 3D printed 316L stainless steel *IOP Conf Ser Mater Sci Eng* **this volume** submitted

[28] Wang X 2024 Cell structure in steels induced by additive manufacturing *Materials Science and Technology* accepted

[29] Gao S, Hu Z, Duchamp M, Krishnan P S S R, Tekumalla S, Song X and Seita M 2020 Recrystallization-based grain boundary engineering of 316L stainless steel produced via selective laser melting *Acta Mater* **200** 366–77

[30] Fan J, Zhu Y, Wang W, Chen K, Godfrey A, Yu T and Huang X 2024 Recovery of dislocation cell structures in 316L stainless steel manufactured by selective laser melting *Journal of Materials Research and Technology* **30** 9472–80

[31] Zhang C, Yadav V, Moelans N, Juul Jensen D and Yu T 2022 The effect of voids on boundary migration during recrystallization in additive manufactured samples—a phase field study *Scr Mater* **214** 114675

[32] Thorsteinsdóttir E G, Primdahl D S, Zhang Y B, Juul Jensen D and Yu T 2019 Aging of 3D-printed maraging steel *IOP Conf Ser Mater Sci Eng* **580** 012047

[33] Snopinski P, Yu T, Zhang X and Juul Jensen D 2024 Effect of multi-pass shot peening on the microstructure of LPBF AlSi10Mg alloy *IOP Conf Ser Mater Sci Eng* **this volume** submitted

[34] Zhang C, Juul Jensen D and Yu T 2023 Effects of initial 3D printed microstructures on subsequent microstructural evolution in 316L stainless steel *Acta Mater* **242** 118481

An Image-Driven Machine Learning Method for Microstructure Characterization in Metal Additive Manufacturing: Generative Adversarial Network

Z Cao, Y Liu, J J Kruzic and X Li*

School of Mechanical and Manufacturing Engineering, UNSW Sydney, Sydney, 2052, Australia

*E-mail: xiaopeng.li@unsw.edu.au

Abstract. The recent development of artificial intelligence especially machine learning technology has provided an emerging direction for solving microstructure representation and analysis in additive manufacturing. In this work, we introduce an advanced image-driven machine learning algorithm that offers an effective way to abstract the features in microstructure and generates high-resolution and large-size images that can represent the original counterparts. The evolution of the model and the potential application of the algorithm in material science are also discussed.

1. Introduction

The microstructure of additively manufactured metals, encompassing aspects like grain evaluation, texture changes, and defect generation, serves as a record of the metal 3D printing processing history [1]. These microstructural features heavily influence material performance and properties such as strength, toughness, ductility, and corrosion resistance [2]. Researchers aim to establish quantitative relationships between processing, microstructure, and properties to design materials with desired structures and performance characteristics. Microstructure visualization, observation, and modelling are integral steps in material science and engineering research. Both empirical and computer simulation methods [3] have proven effective in visualizing surface morphology and characterizing microstructure facilitating research and development in the field. Historically, computational and numerical simulation methods relied on specific physical formulas with fixed parameters, limiting their ability to capture the overall characteristics of microstructure [4]. They often fail to represent defects and subtle features accurately, resulting in unsatisfactory prediction accuracy, especially when facing the complex morphology of microstructure in additive manufacturing. Moreover, these methods had narrow applicability, being suitable only for specific materials or processing conditions.

With rapid advancements in computer and data science, artificial intelligence (AI) has emerged as a promising tool in material science and engineering [5]. Machine learning, a key component of AI, enables computers to learn from experience and efficiently tackle complex calculation and expression problems [6]. This technology holds immense potential for discovering new materials, predicting defects, evaluating performance, and interpreting morphology. Image-driven machine learning (IDML) is a key method for parsing feature information from image data, enabling advanced functions [7]. Microstructure images contain valuable information about structural characteristics, processing history, and material properties. Establishing a model that accurately represents image features and integrates with other parameters is a current research focus. While traditional methods rely on manual feature extraction and physical models, IDML offers a unique approach using human-like thinking and statistical learning to overcome limitations.

Among IDML techniques, the Generative Adversarial Network (GAN) [8] stands out for its ability to generate and identify images effectively. GAN's mutual game theory between the Generator and Discriminator models enables high-quality microstructure image generation, facilitating accurate representation of material features and design. This study briefly introduces GAN and its principle, highlighting their potential in microstructure characterization and materials design for additively manufactured alloys.

2. Method

2.1 Material fabrication and processing

Ti-6Al-4V alloy powders (grade 23) and CP-Ti powders (grade 2) with a particle size distribution of 15 ~ 45 μm and 20 ~ 55 μm are used in this study. A ProX DMP 320 metal 3D printer, equipped with a 400 W fiber laser is employed to print block samples (8 mm × 8 mm × 30 mm). The laser power is 300 W, the scanning speed is 800 mm/s, the layer thickness is 30 μm and the hatch space is 120 μm. The as-fabricated Ti-6Al-4V samples are annealed at the temperature of 950 °C and 1040 °C for 2 hours. The as-fabricated CP-Ti samples are annealed at 950 °C for 2 hours. The annealed samples are then polished, etched, and observed under an optical microscope (OM).

2.2 Machine learning model

The image-driven machine learning model Generative Adversarial Network (GAN) consists of a Generator (*G*) and a Discriminator (*D*) as shown in figure 1. *G* generates samples from latent variables *z* to mimic real data, while *D* evaluates the authenticity of these samples. Through competition, *G* aims to produce increasingly realistic samples to fool *D*, which in turn improves its ability to distinguish real from fake. The objective function (1) maximizes *D*'s ability to differentiate real from fake samples and *G*'s ability to deceive *D*. The latent variable space *z* and microstructure data space *x* are defined, with *G* mapping from *z* to *x*. The training aims to make *G*'s samples indistinguishable from real data and *D*'s discrimination as accurate as possible. The objective function of the GAN model in this study can be described as follows [8]:

$$L_D = IE_{x\sim P_{data}}[(D(x)] - IE_{x\sim P_g}[D(G(z))] + \varphi IE_{\hat{x}\sim P_z}[(||\nabla_{\hat{x}} D(\hat{x})\,||_2 - 1)^2] \tag{1}$$

Here, IE stands for the desired distribution, P_{data} means the distribution of training image, P_z is the random distribution, P_g is the generator distribution, $IE_{x\sim Pdata}[D(x)]$ is the loss score of real data samples (g-loss), - $IE_{x\sim Pg}[D(G(z))]$ is the loss score of generated data samples (d-loss), and the final term serves as a gradient penalty (gp). The gp encourages the gradients of the critic function to have a norm close to 1, thereby maintaining the Lipschitz continuity and ensuring the stable training and effective approximation of Wasserstein distance, leading to better generative performance.

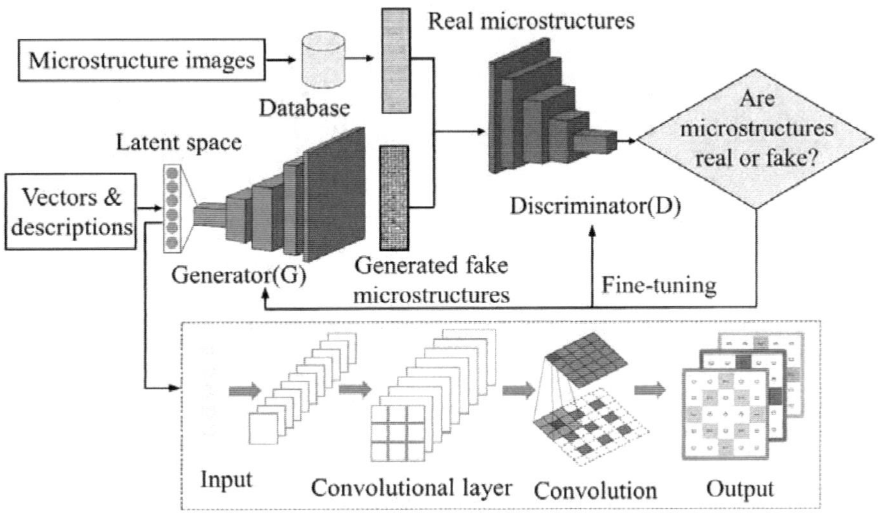

Figure 1. The structure of the GAN Model

3. Results

3.1 Microstructure digital characterization and model evaluation

The laser powder bed fusion (LPBF) fabricated Ti-6Al-4V samples were annealed at 1040 °C and held for 2 hours. Four different regions on the annealed sample surface were randomly selected to assess the homogeneity of the sample and were observed as shown in the figures 2(a), (b), (c), and (d) which consist of α and β phases. A dataset including 20,000 OM microstructure images was selected as training input. The proposed GAN model was trained with optimal parameters and generative results are shown in figures 2(e), (f), (g) and (h). The GAN-generated images closely resemble the original images capturing most features like grain shapes, distributions, and patterns. However, new images, while sharing common features, do not precisely match the originals. This is because the model learns internal regulations from general microstructural descriptors rather than mechanically replicating featural descriptions. Through continuous training, the discriminator assigns a score between 0 and 1, storing feature experiences as latent variables within the neural network. Given random latent input, the generator produces new images based on previous latent information, serving as digital representations of microstructural features and visual restoration.

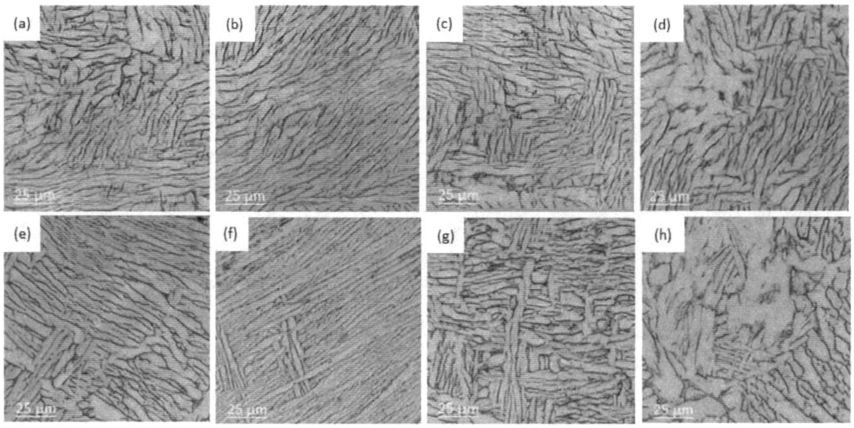

Figure 2. The microstructure images of original input (a)(b)(c)(d) and trained output (e)(f)(g)(h) for LPBF fabricated Ti-6Al-4V alloy annealing at 1040 °C/2h ($\alpha+\beta$ phase)

To quantitatively assess image generation effectiveness and output quality, a statistical perspective is adopted, proposing a loss function comprising Generator loss (g-loss), Discriminator loss (d-loss), and gradient penalty (gp). Figure 3 illustrates the relationship between loss values and training epochs. Initially, g-loss increases significantly, then gradually decreases, indicating avoidance of overfitting or collapse, and a shrinking distribution gap between real and predicted data. Frechet Inception Distance (FID) is a metric used to evaluate the quality of images generated by the GAN model by comparing the statistical characteristics of generated and real images. The continuous decrease of the Frechet Inception Distance (FID) score in figure 3(b) suggests synthesis image resolution and precision approach to real image quality. d-loss and gp curves remain stable, affirming their efficacy in data truth identification and gradient control.

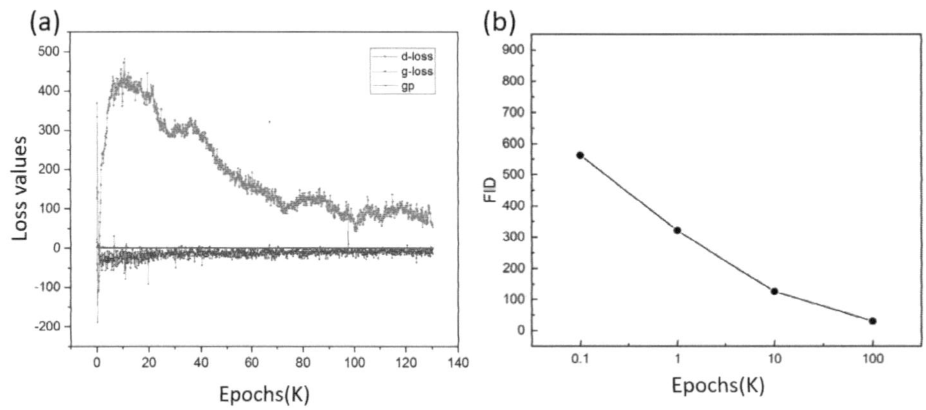

Figure 3. The loss values and FID scores for the GAN model in this study

3.2 Transferability and scalability

The well-trained latent within the network enables the generation of high-resolution microstructure images with similar features under specific conditions. We applied microstructure images under different post-heat treatment conditions for LPBF fabricated Ti-6Al-4V to train the same GAN model, showcasing its transferability. Figure 4 shows that the synthetic images accurately reflect the original morphology, capturing complex and disorganized shapes and elemental distributions at various scales. Furthermore, we validated the model's wide applicability by training it on microstructural distributions for LPBF fabricated CP-Ti alloy annealed at 950 °C/2h in figure 5, yielding synthetic images highly similar to real data.

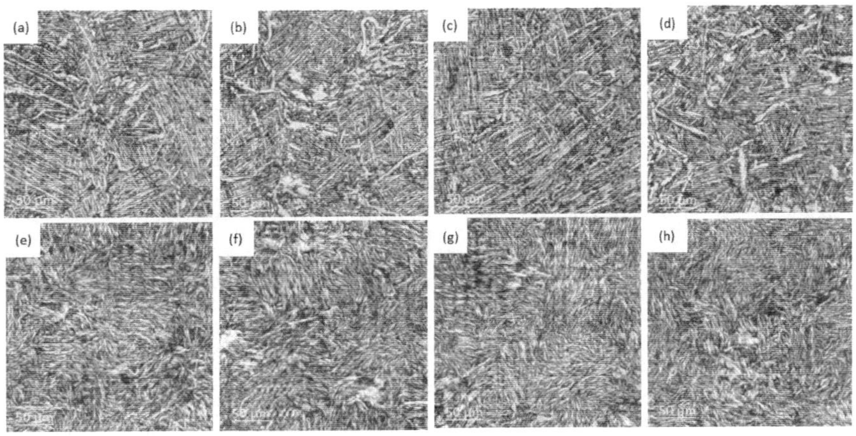

Figure 4. The microstructure images of original input (a)(b)(c)(d) and trained output (e)(f)(g)(h) for LPBF fabricated Ti-6Al-4V alloy annealed at 950 °C/2h (α+β phase)

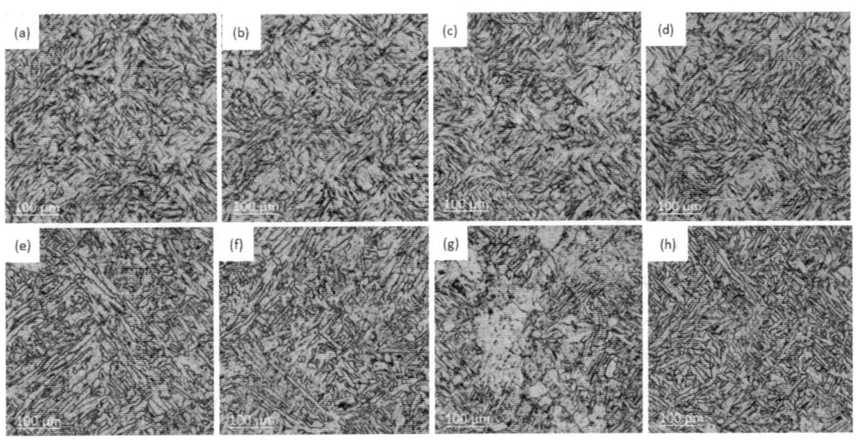

Figure 5. The microstructure images of original input (a)(b)(c)(d) and trained output (e)(f)(g)(h) for LPBF fabricated CP-Ti alloy annealed at 950 °C/2h (α' martensite phase)

The model's scalability, facilitated by its open-source and easily modifiable structure, allows for flexible input and output of randomly sized images. This flexibility is crucial for learning features from real data, particularly when original images lack detail due to their small size. Adjusting convolutional layer parameters like kernel size, stride, and padding enables the generation of desired image sizes for diverse analysis purposes. Additionally, as an image-driven machine learning network, the GAN model can perform image-to-image functions for microstructure representation and reconstruction. Furthermore, it has the potential to implement process-microstructure (label-to-image) or microstructure-property (image-to-text)

IOP Conf. Series: Materials Science and Engineering **1310** (2024) 012015 doi:10.1088/1757-899X/1310/1/012015

functions by incorporating labels as extra input variables to generate conditional images. This capability holds promise for quantitatively studying complex relationships between processing parameters, microstructure, and mechanical properties.

4. Discussions

An improved image-driven machine learning method, Generative Adversarial Network (GAN), is proposed for representing and redesigning the characteristics of microstructure for LPBF fabricated Ti-6Al-4V and CP-Ti alloys in this work. It achieved high-quality microstructure image generation with adjustable parameters and stable training. Future work includes using conditional GAN for text-to-image transformation and studying the physical meaning of latent weights, which can learn the complex relationships between process-microstructure-property for additively manufactured alloys. Similar to our recent work [9], this will further enhance our understanding of how material descriptions (such as processing parameters, mechanical properties, etc.) can be translated into visual microstructure representations.

References

[1] Prepared under the direction of the ASM International Handbook Committee 1985 Metallography and Microstructures *ASM Metals Handbook Ninth Edition* vol. **9** (Metals Park, OH: ASM International) 12
[2] Paulson N H, Priddy M W, McDowell D L and Kalidindi S R 2017 Reduced-order structure-property linkages for polycrystalline microstructures based on 2-point statistics *Acta Mater.* **129** 428-438
[3] Herzog D, Seyda V, Wycisk E and Emmelmann C 2016 Additive manufacturing of metals *Acta Mater.* **117** 371-392
[4] Cao Z, Liu Q, Liu Q C, Yu X, Kruzic J J and Li X 2023 A machine learning method to quantitatively predict alpha phase morphology in additively manufactured Ti-6Al-4V *npj Comput. Mater.* **9** 195
[5] DebRoy, T, Wei H L, Zuback J S, Mukherjee T, Elmer J W, Milewski J O, Beese A M, Wilson-Heid A, De A and Zhang W 2018 Additive manufacturing of metallic components-process, structure, and properties *Prog. Mater. Sci.* **92** 112-224
[6] Hart G L W, Mueller T, Toher C and Curtarolo S 2021 Machine learning for alloys *Nat. Rev. Mater* **6** 720-755
[7] Richman J M, Lookman T and Kalinin S V 2019 Materials informatics: From the atomic-level to the continuum *Acta Mater.* **168** 473-510
[8] Gulrajani I, Ahmed F, Arjovsky M, Dumoulin V and Courville A C 2017 Improved training of Wasserstein GANs *Proc. Adv. Neural Inf. Process. Syst.* **30**
[9] Liu Q, Chen W L, Yakubov V, Kruzic J J, Wang C H and Li X 2024 Interpretable machine learning approach for exploring process-structure-property relationships in metal additive manufacturing *Addit. Manuf.* **85** 104187

Inhomogeneous deformation in melt-pool structure of Al-Fe-Cu alloy manufactured by laser powder bed fusion

Y Cheng[1*], Y Otani[1], N Takata[1*], A Suzuki[1], M Kobashi[1] and M Kato[2]

[1] Department of Materials Process Engineering, Graduate School of Engineering, Nagoya University, Furo-cho, Chikusa-ku, Nagoya 464-8603, Japan
[2] Aichi Center for Industry and Science Technology, 1267-1 Akiai, Yakusa-cho, Toyota 470-0356, Japan

*E-mail: cheng.yue.i6@s.mail.nagoya-u.ac.jp, takata.naoki@material.nagoya-u.ac.jp

Abstract. The laser powder bed fusion (L-PBF) processed Al–2.5Fe–2Cu (mass%) alloy exhibited a high tensile strength above 340 MPa and pronounced directional dependence. The sample exhibited a characteristic inhomogeneous microstructure (in melt-pool structure) arising from the local melting and rapid solidification in the L-PBF process. The coarsened cellular structure localized along the melt pool boundary resulted in the local soft regions affected by the melt pool structure. The local vulnerability contributed to the direction dependence of the tensile ductility of the specimen.

1. Introduction

Laser powder bed fusion (L-PBF) is a metal additive manufacturing technology that uses a laser beam to selectively melt and fuse powder. Renowned for its high precision, the L-PBF manufacturing process facilitates the realization of intricate geometries and components with exceptional accuracy and resolution [1, 2, 3, 4]. Due to the local melting and rapid solidification during the L-PBF process, unique inhomogeneous microstructures are formed in the L-PBF-processed alloys. The unique microstructure enhances mechanical properties, surpassing those of components produced through traditional casting processes, with a notable feature being the presence of the melt-pool structure [1, 5]. The microstructural inhomogeneity within the melt-pool structure has been suggested to cause the directional dependence observed in mechanical performance [6, 7, 8, 9, 10].

L-PBF-processed Al–2.5Fe (mass%) binary alloy exhibited high tensile strength exceeding 300 MPa, attributed to the refinement of the Al_6Fe metastable phase [11], in contrast to the $Al_{13}Fe_4$ stable phase often formed in high-Fe contained Al alloys during the L-PBF process [12]. The formation of the $Al_{13}Fe_4$ stable phase is intentionally avoided in producing conventional Al alloys due to its coarsened morphology and brittleness [13]. Cu was chosen as the third added element to further strengthen the L-PBF-processed Al–2.5Fe alloy [14]. The replacement of Al atoms by Cu atoms transformed the Al_6Fe phase into the thermodynamically stable $Al_{23}CuFe_4$ phase [15, 16], promoting the formation of the refined Al_6Fe phase with Cu in solution and enhancing the strength of the alloy [14]. While high tensile strength was achieved through alloy design, pronounced directional dependence was also observed simultaneously.

This study investigates the relationship between the inhomogeneous microstructure of L-PBF-processed Al–2.5Fe–2Cu alloy and the directional dependence observed in its mechanical properties from both macro and micro perspectives.

IOP Conf. Series: Materials Science and Engineering **1310** (2024) 012016 doi:10.1088/1757-899X/1310/1/012016

2. Materials and Methods

The Al–2.5Fe–2Cu alloy powder with a mean particle size below 20 μm was prepared via gas atomization. The major compositions of the powders are described as follows; Fe: 2.7, Cu: 1.9, Si: 0.1 mass%. Bulk samples were manufactured at ambient temperature using a ProX200 3D printer (3D SYSTEMS, Rock Hill, SC, USA). The parameters were described in the literature [14]. The L-PBF-processed samples exhibited high relative densities above 99.7%.

Plate-shaped specimens for tensile tests were prepared through wire electrical discharge machining and then subjected to a uniaxial tensile test at a strain rate of 1.2×10^{-3} /s at room temperature. Each sample type underwent at least three tensile tests conducted both parallel and perpendicular to the build direction. The fracture surface of specimens was observed using an electron microscope (SEM; JSM-IT500 and JSM-6610A, JEOL Ltd).

To understand the local hardness of L-PBF-processed samples with inhomogeneous microstructure, the nanoindentation tests were performed by a Nanoindentation Tester (ENT-5, ELINOIX, JP) equipped with a Berkovich triangular pyramid indenter [17, 18] at a maximum load of 0.2 mN with a constant loading rate of 0.02 mN/s and a holding time of 1 s. All indents on the sample surface were observed by SEM. The nano-hardness map and the local indentation hardness (H_{IT}) distribution were generated using raw nanoindentation data in Python codes employing the numpy and matplotlib modules.

3. Results and discussion

3.1 Microstructure

Figure 1 shows the microstructural features of Al–2.5Fe–2Cu alloy samples at different scales, observed on the plane perpendicular to the build direction. The sample exhibited a characteristic inhomogeneous microstructure consisting of melt pools with fish-scale-like shapes (figure 1 (a)), corresponding to the repeated selective melting and rapid solidification during the L-PBF process as reported in many different sorts of L-PBF-processed Al alloys [19, 20, 21]. SEM observation with high magnification (figure 1 (b)) revealed the coarsened cellular structure localized along the melt-pool boundary, whereas numerous fine particles were distributed in the α-Al matrix. The local coarsening of microstructures created variations in image contrast, facilitating the visualization of melt-pool morphologies as observed in the optical micrographs. The bright-field transmission electron microscope (TEM) images (figure 1 (c)) showed the microstructure inside the melt pool, and numerous dispersoids sized several tens of nanometers were homogeneously distributed in the α-Al matrix inside the melt pool. These intermetallic phases appeared to be connected. Similar microstructure features were observed in the Al–2.5Fe (mass%) binary alloy [22] and Al–2.5Fe–2Mn (mass%) ternary alloy [23] samples. The presence of a diffraction ring in the selected area electron diffraction (SAED) pattern (figure 1 (d)) derived from the <222> plane of the orthorhombic Al_6Fe phase (*o*C28) was identified in previous research [16, 24].

Figure 2 presents the scanning transmission electron microscopy (STEM) image and corresponding energy-dispersive X-ray spectroscopy (EDS) analyses of the Al–2.5Fe–2Cu alloy sample. The EDS composition analyses (figure 2 (a)) revealed a relatively high concentration of Fe in the spherical intermetallic phases. The results revealed Cu enrichment in the Al_6Fe phase. The XRD profiles [14] suggested high-intensity peaks associated with the $Al_{23}CuFe_4$ (*o*C28) phase, indicating the formation of the $Al_{23}CuFe_4$ phase. The quantitative EDS point analyses revealed the Cu-enriched particles contained approximately 14.3 mol% Fe and 4.7 mol% Cu (figure 2 (b) [14]), corresponding well to the composition of the $Al_{23}CuFe_4$ phase. Consequently, observed nanoscale particles corresponded to the $Al_{23}CuFe_4$ (*o*C28) phase with several Cu atoms occupying Al sublattice sites in the Al_6Fe phase [16].

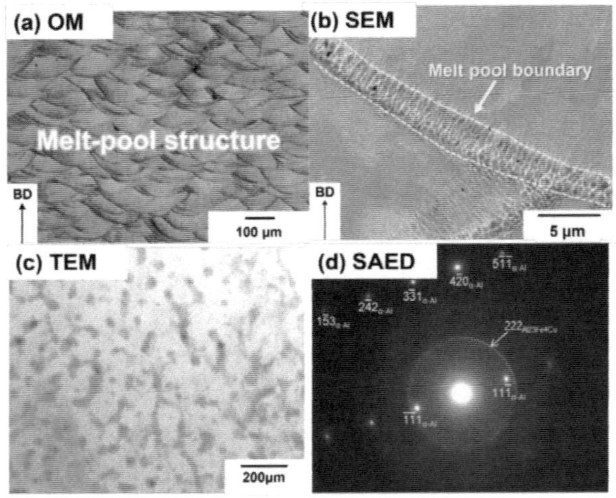

Figure 1. Multi-scale microstructural features of L-PBF-processed Al–2.5Fe–2Cu alloy samples: (a) optical micrograph and (b) SEM image showing the melt-pool structure; (c) TEM bright-field image showing fine intermetallic phases; (d) selected-area electron diffraction pattern [14].

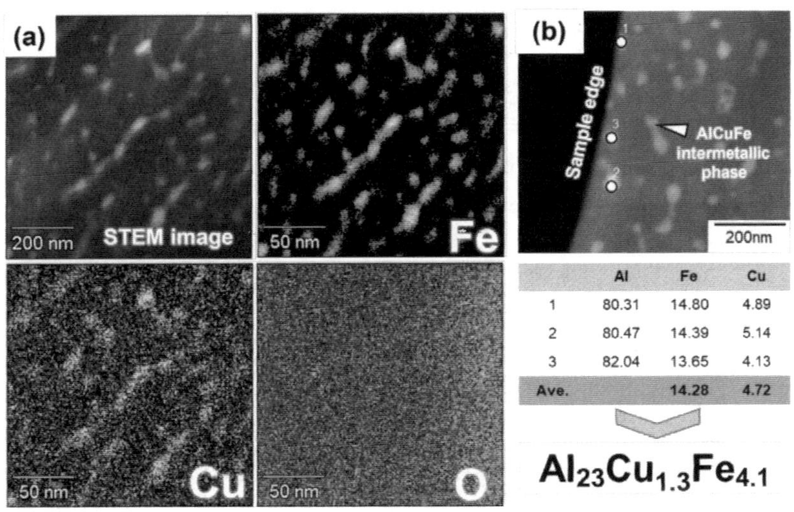

Figure 2. (a) EDS elemental map and (b) EDS point analyses for intermetallic phases [14] in L-PBF-processed Al–2.5Fe–2Cu alloy.

3.2. Mechanical property

Figure 3 shows the nominal stress-strain curves of specimens that were tensile-deformed either along or perpendicular to the build direction. The Al–2.5Fe–2Cu ternary alloy specimen exhibited a notable tensile strength, surpassing 360 MPa, marking a more than 15% increase compared to the Al–2.5Fe binary alloy [22]. This heightened strength is attributed to the augmented formation of numerous finely dispersed Al_6Fe phases, stabilized by the incorporation of Cu [14]. The tensile properties of Al–2.5Fe–2Cu alloy specimens exhibited a directional dependence, similar to other L-PBF-processed Al alloys [20, 23]. Specimen with the tensile direction perpendicular to the build direction, exhibited a high nominal stress of ~360 MPa and elongation of ~10%, while the specimen tensile along the build direction showed a nominal stress of ~347 MPa and elongation of ~7.5%. Specimens with the tensile direction perpendicular to the build direction often exhibit higher strength and ductility compared to those in the other direction. Such a direction dependence of the tensile properties was observed in Al-2.5Fe alloy [22] and Al-12Si alloy [7] manufactured by L-PBF.

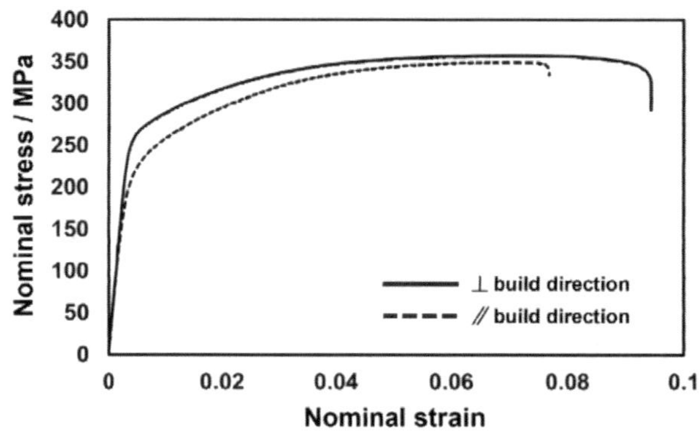

Figure 3. Nominal stress-strain curves of L-PBF-processed Al–2.5Fe–2Cu alloy specimens with tensile direction perpendicular or parallel to the build direction [14].

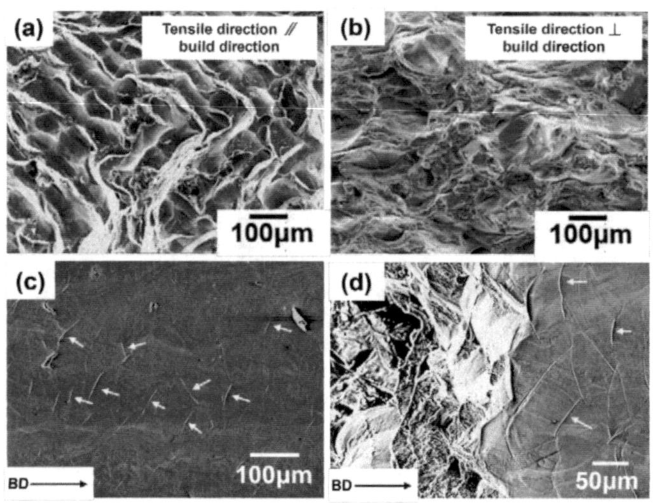

Figure 4. Fracture surface of L-PBF-processed Al–2.5Fe–2Cu alloy: (a) tensile direction parallel to build direction, (b) tensile direction perpendicular to build direction. SEM images (c,d) showing the fracture portion of specimen with the tensile direction parallel to build direction.

Figure 4 presents the SEM images displaying the fracture surface (figures 4 (a)(b)) of Al–2.5Fe–2Cu alloy specimens with different tensile directions: (a) parallel and (b) perpendicular to the build direction. It also shows the fracture portion (figure 4 (c)(d)) of the specimen with tensile direction along the build direction. Arrays of band-shaped and wavy steps forming terraces were observed on the fracture surface of the specimen with tensile direction along the build direction (figure 4 (a)). The macroscopic morphology aligned with the melt-pool morphology observed by optical microscopy (figure 1 (a)). The observed round surface indicated that fracture preferentially occurred along melt-pool boundaries. Similarly, a comparable round surface was observed for the specimen with the tensile direction perpendicular to the build direction. However, typical ductile fracture surfaces with numerous spherical dimples were often observed on the fracture surface (figure 4 (b)), similar to L-PBF-processed Al–Fe and Al–Fe–Mn alloy specimens [22, 23]. The fracture edges (figure 4 (d)) exhibit a relatively smooth appearance, corresponding to both the melt-pool structure (figure 1 (a)). The observations of the fracture surface (figure 4 (a)). Additionally, a large number of arcuate cracks can be observed on the polished specimen surface (indicated by bright contrast, highlighted with yellow arrows in figure 4 (c)), aligning with the morphology of the melt pool boundary. This suggests that localized fractures preferentially occur under loading at the local region of the melt pool boundary. Subsequently, these localized cracks expand and interconnect as the tensile test progresses, ultimately leading to specimen fracture.

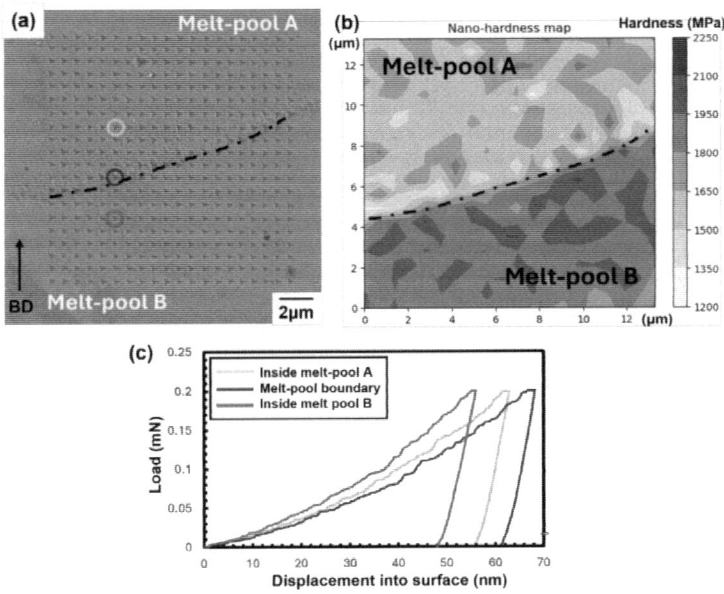

Figure 5. (a) SEM image and (b) nano-hardness maps of L-PBF-processed Al–2.5Fe–2Cu alloy sample. (c) Typical load-displacement curves measured on different regions (as indicated in (a)).

In the present study, the melt-pool structure is posited as a primary contributing factor to the direction dependence observed in the mechanical performance (particularly, tensile ductility) of L-PBF-processed Al alloy. To explore the relation of local mechanical property within the inhomogeneous microstructure, a micron-scale region of approximately 14×14 μm^2 containing the melt pool boundary was chosen for nanoindentation tests (figure 5 (a)). This region spans two melt pools and includes the interface between them (melt-pool boundary). For better illustration, the melt pool at the top of the region was designated as melt-pool A, and the one at the bottom as melt-pool B. To visually represent the nano-hardness (denoted as H_{IT}) distribution within the test region more intuitively [25], the nano-hardness map of the test region was prepared (figure 5 (b)). The summarized results are displayed in figure 5. Notably, a few indentation tests failed, likely due to the local surface roughness, resulting in larger-sized indents locally observed (figure 5 (a)). The indents on the melt pool boundary exhibited a displacement-load curve with a lower gradient and larger indent sizes compared to those within the melt pool (figure 5 (a)(c)). The nano-hardness map reveals an uneven distribution of H_{IT} within the test region, mirroring the inhomogeneous microstructure observed in this area. The region with the lowest H_{IT} (the lightest blue area in figure 5 (b)) appears in an arc shape, corresponding to the location of the melt-pool boundary. The coarse microstructures of the melt-pool boundary, as shown in figure 1 (b), lead to the low hardness in this region. Apart from localized softening attributed to the presence of melt pool boundaries, the remaining areas exhibit a relatively uniform hardness distribution, consistent with the uniformly dispersed fine particles, as shown in figure 1 (b). The melt pool boundary acted as the softest part of the L-PBF-processed Al–2.5Fe–2Cu alloy and the most vulnerable part during deformation, as corroborated by the fracture surface morphology (figures 4 (a)(b)) and fracture portion (figures 4 (c)(d)).

4. Summary

The L-PBF-processed Al–2.5Fe–2Cu alloy sample exhibited a characteristic inhomogeneous microstructure (in melt pool structure) arising from the local melting and rapid solidification during the L-PBF process. The coarsened cellular structure localized along the melt pool boundary resulted in the local soft regions affected by the melt pool structure. The local vulnerability contributed to the direction dependence of the tensile ductility.

Acknowledgments
The authors are grateful for the alloy powder preparation provided by Dr. Isao Murakami of Toyo Aluminum K.K. This work was supported by JST PRESTO Grant Number JPMJPR22Q4 (Japan), JSPS KAKENHI Grant Number 24H00378 (Japan), and the "Knowledge Hub Aichi" Priority Research Project of the Aichi Prefectural Government (Japan).

References

[1] Yadroitsev I, Yadroitsava I, Du Plessis A, and MacDonald E 2021 *Fundamentals of Laser Powder Bed Fusion of Metals* (Amsterdam: Elsevier) 349-376
[2] Herzog D, Seyda V, Wycisk E, and Emmelmann C 2016 *Acta Mater.* **117** 371–392
[3] Khorasani A, Gibson I, Awan U S, and Ghaderi A 2019 *Addit. Manuf.* **25** 176-186
[4] Abdulhameed O, Al-Ahmari A, Ameen W, and Mian S H 2019 *Adv. Mech. Eng.* **11(2)** 1687814018822880
[5] Takata N, Liu M, Kodaira H, Suzuki A, and Kobashi M 2020 *Addit. Manuf.* **33** 101152
[6] Zhao L, Santos Macías J G, Ding L, Idrissi H, and Simar A 2019 *Mater. Sci. Eng. A* **764** 138210
[7] Liu M, Wada T, Suzuki A, Takata N, Kobashi M, and Kato M 2020 *Crystals* **10** 1007
[8] Otani Y, Takata N, Suzuki A, Kobashi M, and Kato M 2023 *Scripta Mater.* **226** 115259
[9] Kotadia H R, Gibbons G, Das A, and Howes P D 2021 *Addit. Manuf.* **46** 102155
[10] Zhao L, Song L, Santos Macías J G, Zhu Y, Huang M, and Simar A, Li Z 2022 *Addit. Manuf.* **56** 102914
[11] Qi X, Takata N, Suzuki A, Kobashi M, and Kato M 2020 *Addit. Manuf.* **35** 101308
[12] Wang W, Takata N, Suzuki A, Kobashi M, and Kato M 2020 *Intermet.* **125** 106892
[13] Tsukahara T, Takata N, Kobayashi S, and Takeyama M 2016 *Tetsu-to-Hagane* **102** 89–95
[14] Cheng Y, Miyawaki T, Wang W, Takata N, Suzuki A, Kobashi M, Kato M 2024 *Addit. Manuf. Let.* **8** 100191
[15] Raghavan V *J. Phase Equilibria Diffus.* 2005 Al-Cu-Fe (aluminum-copper-iron), **26** 59–64
[16] Black P J, Edwards O S, Forsyth J B 1961 *Acta Cryst.* **14** 993–998
[17] *Nanoindentation Tester ENT-5* last viewed on 04.10.2023 available online: https://www.elionix.co.jp/english/keisoku_en/ent_5_en (04 October 2023)
[18] Fischer-Cripps A C 2010 *Nanoindentation* 2nd ed (New York Springer) 19-38
[19] Pauzon C, Buttard M, Després A, Chehab B, Blandin J J, Martin G 2022 *Scripta Mater.* **219** 114878
[20] Qi X, Takata N, Suzuki A, Kobashi M, Kato M 2022 *J. Mater. Sci. Technol.* **97** 38-53
[21] Sun S B, Zheng L J, Liu J H, Zhang H 2017 J. *Mater. Sci. Technol.* **33(4)** 389-396
[22] Qi X, Takata N, Suzuki A, Kobashi M, Kato M 2021 *Mater. Sci. Eng. A* **805** 140591
[23] Wang W, Takata N, Suzuki A, Kobashi M, Kato M 2023 *Addit. Manuf.* **68** 103524
[24] Walford L. K 1965 *Acta Cryst.* **18** 287-291
[25] Dhal A, Thapliyal S, Gaddam S, Agrawal P, Mishra R S 2022 *Scientific Reports* **12** 18344

Micro-macro relationship between microstructure and mechanical behavior of 316L stainless steel fabricated using L-PBF additive manufacturing

C Ozdogan[1], R A Yildiz [1], L Tavares[2], and M Malekan[1]

[1] Centre for Industrial Mechanics, Institute of Mechanical and Electrical Engineering, University of Southern Denmark, 6400 Sønderborg, Denmark

[2] Centre for Industrial Electronics, Institute of Mechanical and Electrical Engineering, University of Southern Denmark, 6400 Sønderborg, Denmark

*E-mail: malekan@sdu.edu

Abstract. Compared to traditional production techniques, additive manufacturing (AM) of metallic components has several benefits, mainly little material waste and more design freedom. AM process based on laser powder bed fusion has many key process parameters including scanning speed, layer thickness, build direction, and printing power. Each one of these parameters influences microstructure, and hence macro-mechanical behavior of the manufactured part, as the part microstructure plays a critical role in determining the mechanical properties. This work aims to address a relationship between micro-structure and macro-mechanical behavior of AM fabricated parts made of 316L Stainless Steel. Both as-built and heat-treated samples are being used for experimental testing and microstructure characterizations. Arcan fixture is used to evaluate the macro-mechanical fracture behavior of the material under mode-I, mode-II, and mixed-mode conditions. Microstructure evaluations of the fracture surfaces are done using scanning electron microscopy and X-Ray diffraction techniques. Finally, a correlation between micro-scale characteristics and macro-mechanical behavior is obtained together with different AM process parameters.

Keywords: Additive manufacturing, Laser powder bed fusion, 316L stainless steel, Fracture analyses, Arcan fixture.

1. Introduction

Additive manufacturing (AM) is becoming an alternative method to produce complex geometries without the macroscopic defects compared to the conventional methods [1]. There are more than hundred AM process parameters capable of affecting the part quality and the production process [2]. The most challenging process parameters can be listed as laser power [3, 4], scanning speed [3, 5], laser spot size, hatch spacing [5], layer thickness [6], point distance, and scan patterns [7]. A combination of laser power (P), scanning speed (V), layer thickness (T), and hatch spacing (H) define a parameter called volumetric energy density (VED), which is defined as $\mathrm{VED} = \mathrm{P}/\mathrm{V} \cdot \mathrm{T} \cdot \mathrm{H}$. One of the main material behaviors which can be evaluated based the VED is material fracture. Yakout et al. [8] classified three fracture types into brittle, brittle-to-ductile transition and ductile with respect to VED amount for L-FPB Invar 36 and SS 316L parts. The brittle fracture was detected at the lowest energy amount (41.7 J/mm3) due to the void formation while the brittle-to-ductile fracture was visible within transition period due to the less void formation at relatively higher energy density. Another parameter to be considered which can affect the anisotropy is the build orientation. All these parameters can potentially affect different material properties, such as yield stress, ultimate tensile strength, elastic modulus, elongation at break, and fracture toughness among others.

Fracture toughness is usually used as a generic term for measures of material resistance to extension of a crack [9]. This parameter is different depending on mode of fracture based on the type of loading applied so-called mode-I (opening mode), mode-II (in-plane shear mode), mode-III (out-of-plane shear mode), and mixed mode condition. In contrast to the well-established testing protocols such as ASTM E1820 [10] and ISO 12135 [11] that provide guidelines for measuring mode-I fracture toughness, determining mixed mode fracture toughness poses certain challenges. Different combinations of mixed mode loading have necessitated the development of numerous methodologies, such as asymmetric three-point bending specimen [12], compact tension shear fixture [13], all-fracture-mode test [14], and Arcan fixture [15]. The Arcan fixture is a method employed to examine isotropic or composite materials under various fracture mode conditions, including mode-I, mode-II, and mixed mode-I/II conditions. However, according to the best of authors' knowledge, there is limited effort on fracture behavior, mechanical response, and microstructure of the *316L SS* produced by laser powder bed fusion (L-PBF) under different modes of fracture. In this manner, the primary goal of this research is to get detailed information regarding the macro-mechanical fracture behavior of the material under mixed-mode conditions and to measure mode-I (0°), mode-II (90°) and mixed-mode fracture toughness values at the angle of 45°. In addition, heat treatment process is applied to assess its effects on the microstructure, mechanical properties, and fracture behavior of the samples and compare with the as-built ones.

2. Methodology

The 316L stainless steel specimen was manufactured using an L-PFB machine for Arcan testing. A total of 36 samples were produced with a build direction of 0°, a layer thickness of 30 μm, a laser power of 350 W, a hatch spacing of 50 μm, and a scanning speed of 1400 mm/s in the same batch. Figure 1 depicts the additively manufactured 316L SS specimen on the printing bed and the Arcan test setup. These 36 samples were divided into four sets: as-built (without heat treatment), heat-treated at 700°C, heat-treated at 900°C, and heat-treated at 1100°C. Each set sub-divided into three sub-sets to be tested at 0° (mode-I), 45° (mixed-mode), and 90° (mode-II). The heat treatment (annealing) was done using a furnace under argon gas for 2 hours each. This procedure entails subjecting the substance to a precise temperature and subsequently gradually reducing it. It aims to relieve residual stresses, improve ductility, and refine the microstructure. Furthermore, all the specimens were spray painted to be able to get displacement and strain formation during the testing via digital image correlation device.

The stainless steel Arcan fixture was specifically designed for conducting biaxial loading experiments on both metallic and non-metallic samples under strain control condition at room temperature. This fixture, as shown in figure 1b, was equipped with top and bottom gripping jaws that allow for the application of mixed mode loading at various angles, ranging from 0° (corresponding to pure tensile condition) to 90° (corresponding to pure shear loading), through strategically placed holes. The Arcan fixture was affixed to our Lloyd machine, which has an upper force limit of 30 kN. Throughout the Arcan testing process, a single CCD camera and a lighting device were utilized to capture high-quality images which were then analyzed and transformed into deformation data, including force, displacement, strain, and stress using a digital image correlation software.

IOP Conf. Series: Materials Science and Engineering **1310** (2024) 012017 doi:10.1088/1757-899X/1310/1/012017

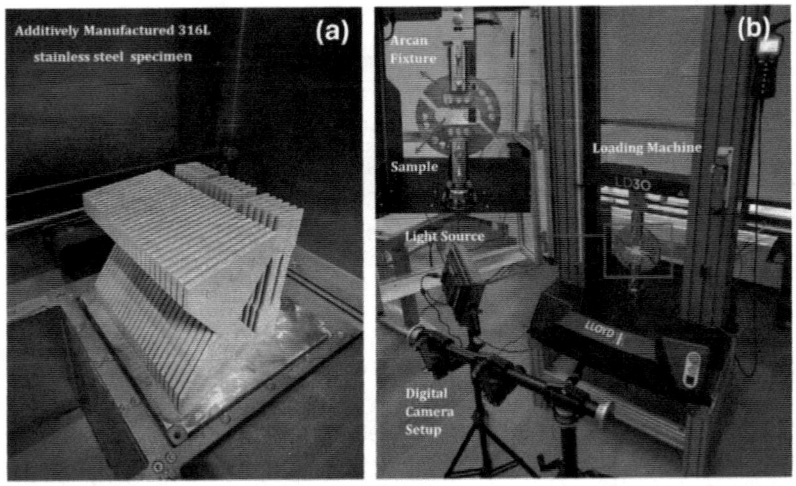

Figure 1. (a) AM produced 316L stainless steel samples, and (b) Arcan testing setup.

3. Results and Discussion

The applied loading in the 0° (LD_0), 45° (LD_45), and 90° (LD_90) directions for both as-built and heat-treated samples. Figure 2a illustrates that the as-built sample exhibits the highest force value when loaded at 0°, with the shortest elongation at fracture. As the loading angle increases, i.e., shifting from mode-I to mode-II, the fracture force decreases while the elongation at break significantly increases. Under pure tensile loading (mode-I), the specimen displays brittle behavior, with increased ductility observed at 45° and 90° loading angles. Moreover, failure is attributed to strain under mixed-mode loading conditions (45°). The samples are then subjected to a loading angle of 90° (mode-II) at room temperature, 700°C, 900°C, and 1100°C. Figure 2b demonstrates the impact of temperature on the shear behavior of the 316L SS samples. It indicates that both as-built and heat-treated samples at 700°C exhibit higher shear strength compared to heat-treated samples at 900 and 1100°C. Additionally, an increase in temperature resulted in reduced shear resistance.

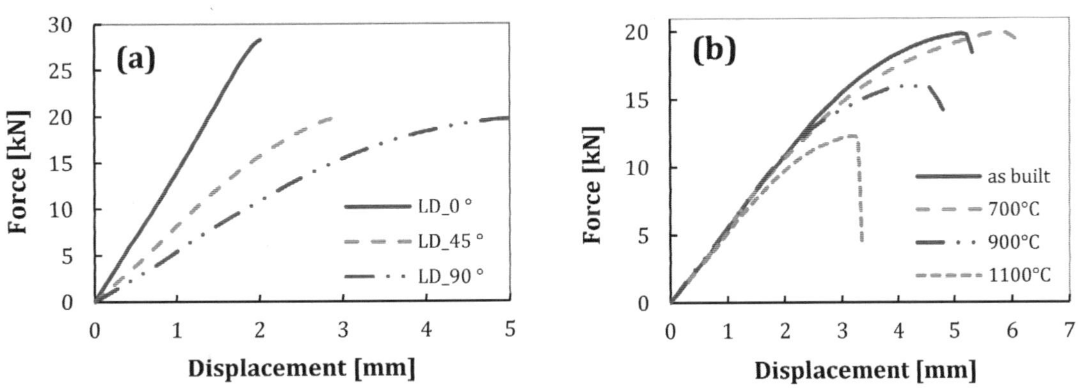

Figure 2. Force-displacement curves for: (a) as-built samples at different angles, and (b) as-built and heat-treated samples under pure shear loading (mode-II fracture) using Arcan testing.

The Hitachi S-4800 scanning electron microscope (SEM) was utilized to analyze the microstructure of both as-built and heat-treated samples. Operating at 10kV and 10μA, this device provided detailed microstructure images of the 316L SS samples. A comparison between as-built and heat-treated samples reveals that the former exhibits higher powder particle percentage and porosity, particularly at 1100°C. The as-built sample displays porous structures, lack of fusion, and clusters of powder particles with spherical shapes, predominantly in specific regions as depicted in figure 3. The presence of a brittle fracture zone is evident in both figure 3(a) and figure 4, with the as-built sample showing a less prominent brittle fracture zone compared to the sample heat-treated at 1100°C, which also affects the macro-mechanical behavior as shown in figure 2(b). This suggests that heat treatment significantly impacts the microstructure, similar to the findings reported in [16]. In Ozer's study, it was seen that heat treatments had a considerable effect on the AM 316L SS microstructure. Accordingly, when the AM 316L SS samples were heated up to 1150°C, the deterioration of microstructure was inspected and the rate of coarse and homogenous grains increased whereas the porosity was reduced significantly. Furthermore, this analysis signifies that the heat-treated sample at 1100°C had a higher ratio of brittle fracture zones, potentially indicating a decrease in force regarding this sample.

Figure 3. SEM images of the as-built sample fractured at loading angle of 90° (mode-II) at different magnifications.

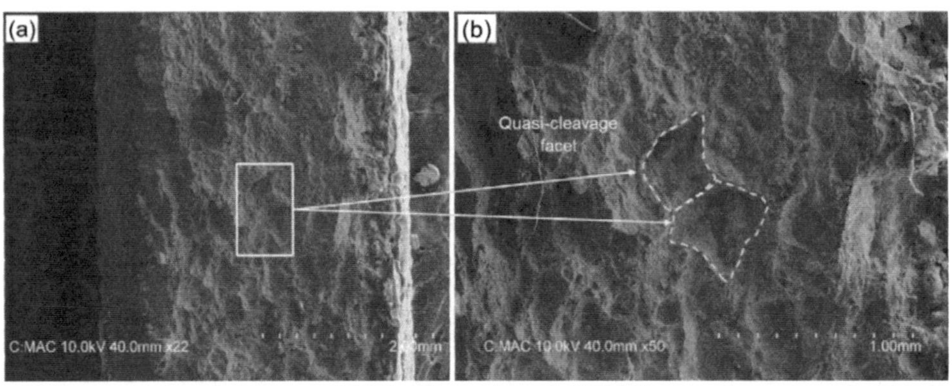

Figure 4. SEM images of the heat-treated 316L SS sample at 1100°C fractured under mode-I loading.

A Bruker D8 Advance X-ray diffractometer (XRD) with a 1600 W Cu K-α source was used for the analysis of chemical composition and crystallographic structure. The background subtraction and phase identification were done with the Bruker DIFFRAC.EVA software. Figure 5 shows that the as-build and 700°C heat-treated samples contain the same peaks at around 43.6°, 50.8°, 74.7°, and 90.7°, corresponding to a single-phase iron, space group Fm-3m (225). Also identified as face-centered-cubic (FCC) crystal structure, austenite, or γ-Fe [16, 17]. In addition to the peaks present in the as-build and 700°C samples, the sample treated at 900°C shows two new peaks at 33.2° and

35.7°, which were identified as Fe_2O_3 hematite, space group R-3c:R(167). The sample treated at 1100°C maintain the later mentioned peaks, while several more peaks occur that can also be indexed by the Fe_2O_3 hematite, space group R-3c:R(167). This might be because of oxidation due to exposure to air for long hours. These analyses revealed the absence of ferrite phase in the fabricated 316L SS samples, indicating that they were composed solely of the austenite phase. Despite efforts to detect the ferrite phase in both as-built and heat-treated samples, it was not possible due to limitations of the XRD device in measuring low content and small ferrite particles, similar to the reported conclusion by Ciu et al. [18].

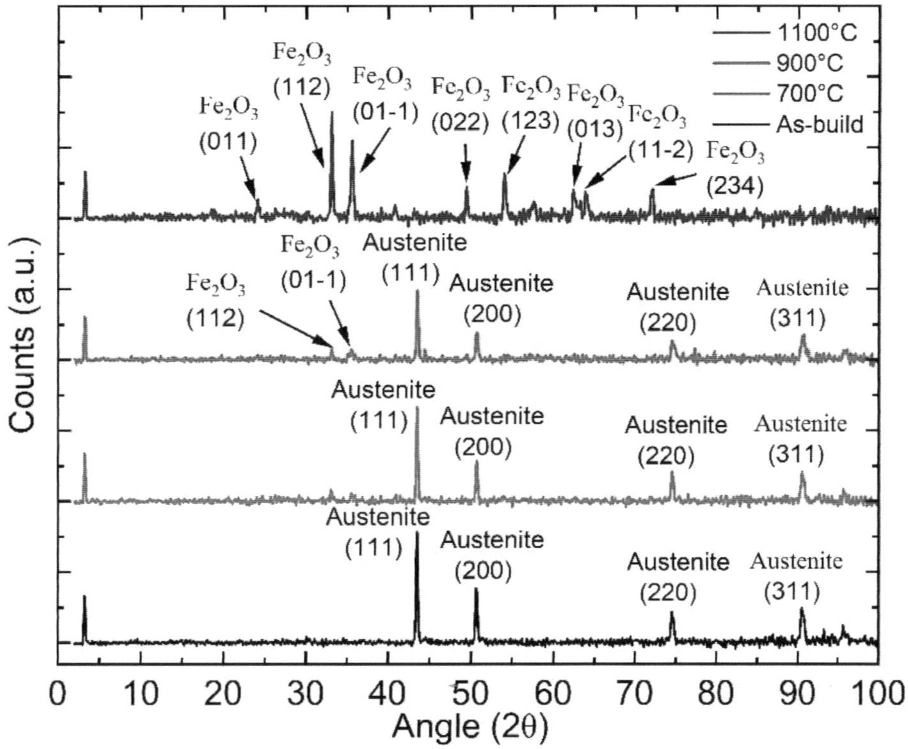

Figure 5. The XRD results of the AM produced as-built and heat-treated 316L stainless steel samples.

Conclusion

The main objective of this study was to investigate the impact of heat treatment on the microstructure analysis and fracture properties of additively manufactured (AM) 316L stainless steel specimens. Analysis of the force-displacement curve from Arcan testing revealed minimal differences between the as-built and heat-treated samples at 700°C in comparison to other heat treatment conditions. The as-built sample exhibited a higher percentage of powder particles and porosity compared to the heat-treated sample at 1100°C. Consequently, the heat-treated sample at 1100°C displayed a more dominant brittle fracture zone than the as-built sample. The X-ray diffractometer (XRD) analysis exhibits that the as-build and 700°C heat-treated samples contain equal peaks at approximately 43.6°, 50.8°, 74.7°, and 90.7°, corresponding to a single-phase iron, space group Fm-3m (225). This analysis revealed that the composition of the structure consisted of austenite phase, and it was found that the austenite microstructure remained unaffected by the heat treatments. Oxidation was predominantly seen in the heat-treated sample at 1100°C which could be because of long hour exposure to air.

Additional research and analysis are under progress to further explore the relationship between AM process parameters, heat treatment process, microstructure, and macro-mechanical behavior (including hardness) of the 316L stainless steel samples. This involves conducting further scanning electron microscopy (to characterize porosity and defect formations at the fracture surface) and electron backscatter diffraction (to quantify the geometrically necessary dislocation density in different alloys under different deformation states) analyses, as well as examining the crystallographic structure.

References

[1] Mahmoudi, M, Tapia, G, Franco B, Ma J, Arroyave R, Karaman I and Elwany A 2018 On the printability and transformation behavior of nickel-titanium shape memory alloys fabricated using laser powder-bed fusion additive manufacturing *J Manuf Proc* **35** 672–680

[2] Kurzynowski T, Chlebus E, Kuźnicka B, Reiner J 2012 Parameters in selective laser melting for processing metallic powders, Proc. SPIE 8239, High Power Laser Materials Processing: Lasers, Beam Delivery, Diagnostics, and Applications, 823914 (6 February 2012)

[3] Hao L, Wang W, Zeng J, Song M, Chang S and Zhu C 2023 Effect of scanning speed and laser power on formability, microstructure, and quality of 316L stainless steel prepared by selective laser melting *J Mater Res Technol* **25** 3189–3199

[4] Yildiz RA, Popa AA and Malekan M 2024 On the effect of small laser spot size on the mechanical behaviour of 316L stainless steel fabricated by L-PBF additive manufacturing *Mater Today Commun* **38** 108168

[5] Han X, Zhu H, Nie X, Wang G and Zeng X 2018 Investigation on selective laser melting AlSi10Mg cellular lattice strut: molten pool morphology, surface roughness and dimensional accuracy *Mater* **11** 3

[6] Nguyen QB, Luu DN, Nai S, Zhu Z, Chen Z and Wei J 2018 The role of powder layer thickness on the quality of SLM printed parts *Arch Civil Mech Eng* **18** 3 948–955

[7] Ahmed N, Barsoum I, Haidemenopoulos G and Abu Al-Rub RK 2022 Process parameter selection and optimization of laser powder bed fusion for 316L stainless steel: A review *J Manuf Proc* **75** 415–434

[8] Yakout M, Elbestawi MA and Veldhuis SC 2019 Density and mechanical properties in selective laser melting of Invar 36 and stainless steel 316L *J Mater Proc Technol* **266** 397–420

[9] Zhu XK and Joyce JA 2012 Review of fracture toughness (G, K, J, CTOD, CTOA) testing and standardization *Eng Fract Mech* **85** 1–46

[10] ASTM-E1820 2018 Standard test method for measurement of fracture toughness. West Conshohocken, PA: ASTM International

[11] ISO-12135 2016 Metallic materials—unified method of test for the determination of quasistatic fracture toughness. The International Organization for Standardization

[12] Keiichiro T and Hitoshi I 1992 Elastic-plastic fracture toughness test under mixed mode I-II loading *Eng Fract Mech* **41** 4 529–540

[13] Kamat S and Hirth J 1996 Mixed mode I/II fracture toughness of 2034 aluminum alloys *Acta Mater* **44** 201–208

[14] Richard HA and Kuna M 1990 Theoretical and experimental study of superimposed fracture modes I, II and III *Eng Fract Mech* **35** 6 949–960

[15] Voloshin A and Arcan M 1998 Failure of unidirectional fiber-reinforced materials - new methodology and results *Exp Mech* **20** 8 280–284

[16] Özer G 2023 Heat treatment, microstructure and corrosion relationship on 316L SS produced by AM *Mater Sci Technol* **9** 6 671–682

[17] Zhou C, Wang J, Hu S, Tao H, Fang B, Li L, Zheng J and Zhang L 2020 Enhanced corrosion resistance of additively manufactured 316L stainless steel after heat treatment *J Electrochem Soc* **167** 14 141504

[18] Cui L, Jiang S, Xu J, Peng RL, Mousavian RT and Moverare J 2021 Revealing relationships between microstructure and hardening nature of additively manufactured 316L stainless steel *Mater Des* **198** 109385

Improved corrosion resistance of powder manufactured AISI 904L parts by hot isostatic pressing post treatment

A Sorea[1*], P Valler[2], P Kjeldsteen[3] and P Hjelmeborn[4]

[1] Development and Project Engineer, Sintex, A/S, Hobro, Denmark
[2] Head of Development & Project, Sintex, A/S Hobro, Denmark
[3] Senior Business Development Manager, Sintex A/S, Hobro, Denmark
[4] Senior Innovation Specialist, Grundfos China Holding Co.Ltd., Suzhou, China
E-mail: ALES@Sintex.com

Abstract Metal binder jetting (MBJ) of AISI 904L super austenitic stainless steel makes it possible to produce complex components with a higher corrosion resistance compared to austenitic stainless steels such as AISI 304L and AISI 316L. The initial sintering trials resulted in a porous part with low corrosion resistance. As AISI 904L is a steel with austenitic phase throughout the entire sintering window, densification during sintering was inhibited which resulted in the reduced corrosion resistance due to open porosities. The trials were made with AISI 904L powder suited for MBJ but shaped through metal powder extrusion due to lack of printer availability. This paper will show how to enhance densification and utilize pickling in order to improve the corrosion resistance closer to the expected level comparable to cast and rolled material but with the shaping possibilities of MBJ.

1. Introduction

Austenitic stainless steel is a widely used industrial material due to its excellent corrosion resistance, formability, and weldability. The microstructure is controlled by the chemical composition and the manufacturing routes from melt to final components. With the implementation of the additive manufacturing (AM) technology based of sintering of metal powder, metal binder jetting (MBJ), it is possible to obtain different microstructures and thereby properties compared to the wrought and cast alloys. As it was not a possibility to change to 904L powder and run in the powder in the MBJ printer at the time of the trials it was investigated how to obtain components as close to MBJ as possible. The metal powder extrusion (MPE) technology was chosen as it enabled the use of a powder intended for MBJ, a relatively low amount of binder and was readily available to easily make small test samples.

A prior study highlights that the corrosion resistance of sintered stainless steel is expected to be lower compared to cast and wrought equivalents due to crevice corrosion mechanism taking place in the pores [1]. The authors emphasize the importance of controlling the density by adjusting the sintering temperature, sintering time, and precipitation of other unwanted phases during cooling that can result in lower corrosion resistance [1].

Other studies [2] have shown that the formation of delta-ferrites during high temperature sintering enhances the shrinkage rate as the coefficient of diffusion of iron atoms is 1-2 order of magnitudes higher in delta-ferrite than in the austenitic phase. A ThermoCalc simulation of AISI 904L shows that the alloy is stable in the austenitic phase in the entire sintering window as shown in figure 1. Therefore, to increase the density and the corrosion properties it was chosen to densify the samples by introducing hot isostatic pressing (HIP) after sintering.

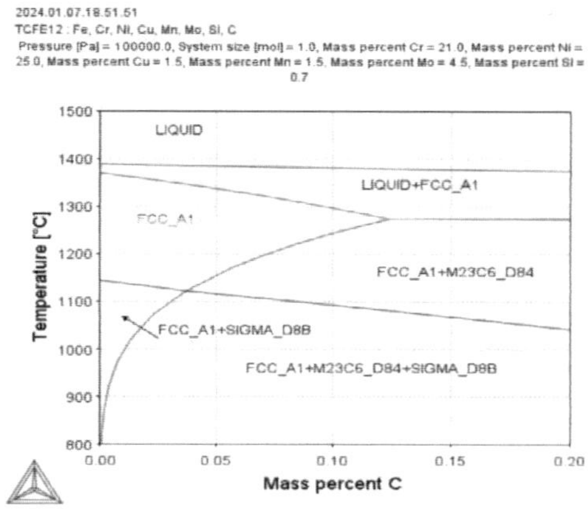

Figure 1. Phase diagram showing the stability of the austenitic phase during sintering in the sintering window (marked yellow).

In this work, the microstructure of AISI 904L stainless steel powder extruded and sintered has been analysed and correlated to the corrosion performance. The aim is to test the corrosion resistance in high saline environments and to understand the corrosion nucleation mechanism in the samples and compare it with wrought stainless-steel equivalents.

2. Materials and Methods

Tests and analysis were performed on AISI 904L powder extruded and sintered material as well as wrought EN 1.4539, cast EN 1.4584 used for benchmarking.

The AISI 904L samples were manufactured by MPE by mixing Osprey® 904L -22 µm metal powder with a binder and a solvent which is kneaded using a high shear mixer and then extruded through a die with a dimension of 34.15 x 1.70 mm and later cut to a length of 85 mm. The extrusion, solvent removal, and air debinding were performed according to the methodology described in a previous article [3] The components were sintered in a Nabertherm VHT 40-16 batch sintering furnace at a temperature of 1330°C for 3 hours in a pure hydrogen atmosphere. After sintering some of the samples were densified using a HIP cycle at 1200°C for 3 hours at 105 MPa pressure in an argon atmosphere. Some of components that received HIP treatment were later pickled in 60% nitric acid for 70 minutes at room temperature.

The wrought test samples were cut out of a cold-rolled coil with a thickness of 1 mm delivered with a 2E surface finish (Cold rolled, heat treated, mechanically descaled) and the cast EN 1.4584 samples were cut out from a product component.

The chemical composition of all samples was measured by XRF using a Thermo Fischer Niton XL3t Gold portable analyzer by measuring 3 areas per plate sample run twice as batch measurement.

The particle size and particle size distribution of the AISI 904L powder was measured with a Malvern Mastersizer 3000 diffraction particle size analyzer. The percentile values used were d10, d50 and d90.

A phase diagram was generated based on the chemical composition of the raw powder material using ThermoCalc 2022a with the TCFE11 database. The diagram, in combination with carbon measurements from an ELTRA Elementrac CS-i measurement, was used to identify the sintering window as shown in figure 1.

2.1. Sample examinations

Samples for metallographic examination were cold moulded in epoxy, grinded to 220 SiC paper and polished with diamond suspension (9 µm, 3 µm, 1 µm). The samples were cleaned in an ultrasonic bath for 5 minutes in acetone and 5 minutes in ethanol. The microstructural analysis was carried out on a Zeiss Axio Observer microscope.

The density and both closed and open porosities of the components were determined by the Archimedes Principle (DIN ISO 3369).

Surface roughness of the samples were measure in accordance with ISO 16610-21 with a Marwin with a SD26 driver unit equipped with a BFW A 10-45-2/90 6852403 probe. The travel length was 4 mm with a speed of 0.5 mm/2 giving a total of 8000 measuring points.

2.2. Corrosion experiments

The test environment used was a 6 wt% $FeCl_3$ solution with a pH of 1.2. The solution was prepared according to ASTM G48. The test duration was 24 hours, and the temperature was selected to 25°C. Evaluation of the corrosion resistance was done by comparing weight loss/surface area (g/m^2).

The Cyclic Potentiodynamic Polarization (CYP) were all performed using a Gamry Reference 600 potentiostat, with a standard three electrode and Avesta cell setup [4]. The solution media for the Avesta cell consist of 250 ml of artificial seawater (ASW) produced according to ASTM D1141-98 although not including the stock solution No. 3 and pH adjustment steps. The chlorinity of the ASW is 19.4 g. The solution temperature is regulated using a temperature-controlled water bath. All steel samples were prepared by grinding the surface using SiC paper with a grit size of 600 and cleaned in acetone and ethanol for five minutes, using an ultrasonic bath, before being mounted to the bottom of the Avesta cell and tightened with a torque of 20 Nm exposing a sample area of 1 cm^2. To ensure pitting corrosion conditions, demineralized water is used to flush a filter paper ring, with a 1 cm^2 hole, located between the sample and Avesta cell. The demineralized water flow is set to 0.4-0.5 ml/min using a peristaltic pump. The addition of demineralized water is counterweighted by the addition of double concentrated ASW using the same peristaltic pump. Another peristaltic pump is used to extract solution keeping the solution volume constant at 250 ml. The exposed sample acts as the working electrode, a Pt mesh is used as the counter electrode, while a saturated calomel electrode (SCE) is used as the reference electrode. The CYP measurements (T = 60°C) started by measuring the open circuit potential (OCP) for 1 h before conducting a cyclic sweep from −0.25 V vs. OCP to 1V vs the SCE with a scan rate of 0.167 mV/s. The scan is set to reverse when the current density reaches 5 mA/cm^2 or the potential reach 1 V vs. the SCE. The reverse scan rate was 0.167 mV/s back to −0.25 V vs. OCP. The breakdown potential (E_{break}) is approximated as the potential when reaching a current density of 0.01 mA/cm^2.

3. Results and discussion

The particle size distribution of the AISI 904L powder used for the MPE samples was measured to a d10 of 6.0 µm, a d50 of 12.3 µm and a d90 of 20.7 µm which correlates well with the powder certificate that states 90% -22 µm.

Figure 2 shows the MPE produced components after sintering (1), sintering + HIP (2) where a dark oxide layer is present on the surface due to overlapping of several samples during the HIP treatment. The dark oxide layer has later been removed on some of the samples by pickling (3).

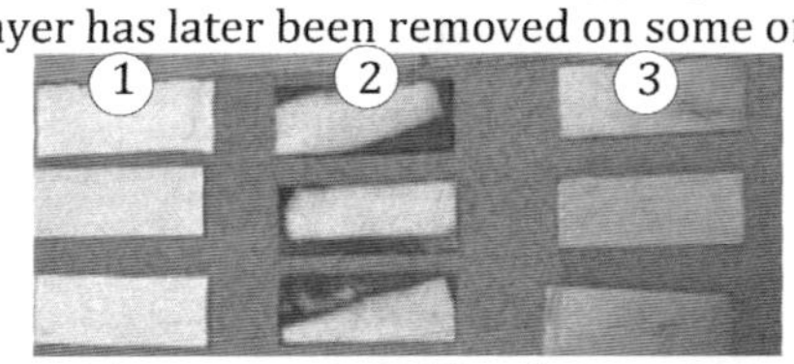

Figure 2 Components after sintering (1), sintering + HIP (2) and sintered + HIP and pickling (3).

The measured chemical composition showed that MPE AISI 904L, wrought EN 1.4539, and cast EN 1.4584 were in accordance with the standard (ISO/EN10088-2 and ASTM A240/240M), showing similar compositions of the main elements Cr, Ni and Mo, see table 1.

Table 1. Chemical composition in wt% of wrought EN 1.4539, cast EN 1.4584, Extruded AISI 904L and 904L metal powder.

Elements	C	Si	Mn	P	Cr	Mo	Ni	Cu	Fe
Osprey® 904L	0.01	0.5	2	0.023	20.3	4.2	24.8	1.38	Bal.
MPE Sintered + (HIP) AISI 904L	0.06	0.40	1.78	0.07	20.11	4.15	25.46	1.31	Bal.
Wrought EN1.4539	0.02	0.24	1.56	0.07	19.8	4.31	24.79	1.50	Bal.
Cast EN1.4584	0.02	0.53	0.91	0.08	19.07	4.48	25.18	1.48	Bal.

The carbon after sintering of the MPE samples was measured to 0.06% and the difference from the powder to the sintered components can be explained due to residuals from binder removal during debinding.

The pitting resistance equivalent number (PREN) is indicating a steels corrosion resistance towards localized corrosion based on the chemical composition. The number can be calculated by the simplified equation $PREN = \%Cr + 3.3 * \%Mo + 16 * \%N$. table 2 shows the calculated PREN number.

Table 2. Pitting resistance equivalent number

MPE Sintered AISI 904L	MPE Sint. + HIP AISI 904L	Wrought EN 1.4539	Cast EN 1.4584
33.85	33.81	34.02	33.85

The surface roughness of the extruded 904L components was measured to an average Ra of 1.3 µm and an average Rz of 9.1 µm, this was compared to a 316L component produced by binder jetting where the roughness had an average Ra value of 3.7 µm and an average Rz value of 21.7 µm.

3.1 *Microstructure characterization*

Figure shows the micrographs of the unetched samples which shows a large number of porosities in the sintered AISI 904L sample while the closed porosities of similar sample of sintered AISI 904L has been eliminated with HIP. The two sintered samples both show quite a lot of open porosities in the surface. Even though the cast EN 1.4584 shows a few porosities both this and the wrought EN 1.4539 are comparable to the sintered AISI 904L + HIP material but without the same amount of surface porosities.

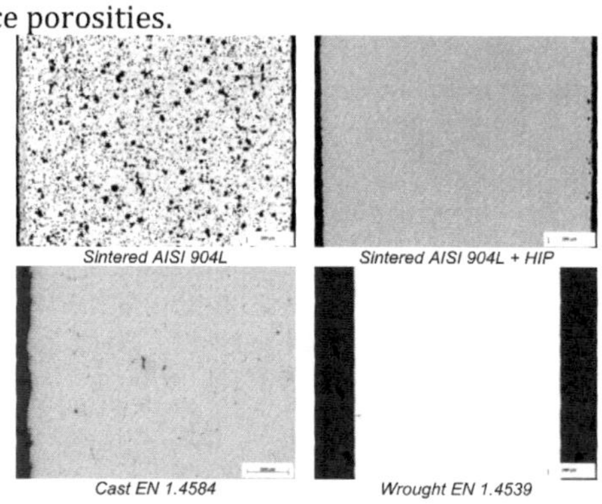

Figure 3. Metallographic observations of sintered, sintered + HIP, cast and wrought samples.

3.2 *Density testing*

The Archimedes measurements presented in table 3 show that the sintered components have a density of 7.13 g/cm^3 and a total porosity of 10.8% out of which 0.4% are open and 10.4% are closed porosities. The HIP treatment, as expected, removes all the closed porosities, and increases

IOP Conf. Series: Materials Science and Engineering **1310** (2024) 012018 doi:10.1088/1757-899X/1310/1/012018

the density to 8 g/cm³ which is close to the maximum value. The subsequent pickling has not provided any changes in the density or porosities.

Table 3 Density, open and closed porosities.

	MPE AISI 904L Sintered	MPE AISI 904L Sintered + HIP	MPE AISI 904L Sintered + HIP + Pickling
Density [g/cm³]	7.13	8.00	8.00
Open Porosities [%]	0.4%	0.4%	0.4%
Closed Porosities [%]	10.4%	0.0%	0.0%

3.3 *Corrosion testing*

The weight loss after a 6% $FeCl_3$ test at 25°C of MPE AISI 904L as-sintered, MPE AISI 904L + HIP and MPE AISI 904L + HIP and pickled wrought EN1.4539 E2 and cast EN 1.4584 as-cast are shown in table 4. The experimental results shows that the corrosion is more severe on MPE samples compared to wrought and cast samples despite the post treatment. More so, it shows that the HIP treated samples have the highest weight loss with an average weight loss of 241 g/m² which is evaluated to be due to oxidation of the surface, which is verified by the higher corrosion resistance of the HIP treated samples after pickling with a weight loss of 146 g/m².

Table 4 Weight loss after Ferric chloride test

	MPE AISI 904L	MPE AISI 904L HIP	MPE HIP and pickled	Wrought EN 1.4539 E2	Cast EN 1.4584
Weight Loss [g/m²]	187.04	240.94	146.26	0.08	76.92

The cyclic potentiodynamic polarization curves of MPE AISI 904L SiC600, MPE AISI 904L + HIP SiC600, wrought EN 1.4539 SiC600 and cast EN 1.4584 SiC600 are shown in Figure 4. All samples experience self-passivation characteristics, meaning that they spontaneously form a Cr_2O_3 layer at the surface [5]. More so, figure 4 shows a successful re-passivation when reversing the potential. The pitting potential was 0.01 mA, showing pitting potential of the following samples: wrought EN1.4539 0.909 V ± 0.01 (n=2), cast EN 1.4584 0.359 V ± 0.13 (n=2), MPE AISI 904L 0.050 V ± 0.04 (n=2), MPE AISI 904L HIP 0.219 V ± 0.13 (n=2). Wrought EN1.4539, cast EN 1.4584 and MPE AISI 904L + HIP shows a stable passive region with few metastable pitting, while MPE AISI 904L indicate more crevice like behavior without any clear passive behavior, see figure 4.

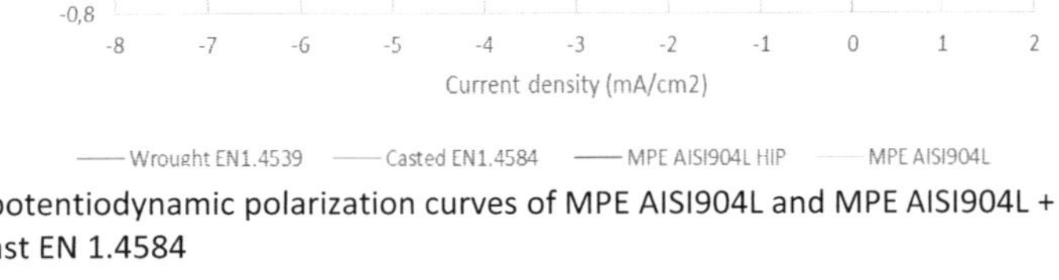

Figure 4. Cyclic potentiodynamic polarization curves of MPE AISI904L and MPE AISI904L + HIP, wrought EN1.4539 and cast EN 1.4584

The results show a significant higher corrosion resistance of wrought EN 1.4539 compared to the other samples. More so, the results highlight that the densification of MPE samples by HIP has a positive impact on the corrosion resistance, highlighting the significant impact porosities have due to local crevice environment, which is formed resulting in limiting oxygen supply and the acidification due to the present of Cl. However, the results shows that the corrosion resistance of MPE samples without porosities still are significant lower compared to equivalent wrought and cast samples. Further analysis of the microstructure using SEM, XRD or similar technics should be conducted to evaluate phases, segregations, and inclusions to further understand the lower corrosion resistance.

4. Conclusion

A detailed analysis of the porosities impact on the corrosion resistance of MPE AISI 904L has been conducted. While the measured surface roughness of the MBJ printed and extruded component were not directly comparable we assume that the surface porosities we see with MPE will most likely also in some form be present in a printed MBJ component. So, the presented work was able to reveal the following information:

1. The tested samples MPE AISI 904L Sintered, MPE AISI 904L Sintered + HIP, Wrought EN 1.4539 and Cast EN 1.4584 has similar chemical composition, resulting in similar PREN number in the range of 33.85 - 34.02.
2. The microstructure of MPE AISI 904L Sintered contains a large number of porosities homogeneous distributed in the cross section. HIP treatment significantly improves the density of the bulk material achieving a density of 8.0 g/cm^3, but still with open surface porosities. region.
3. The $FeCl_3$ corrosion test shows that the weight loss of MPE AISI 904L Sintered + HIP (240.94 g/m^2) was higher compared to MPE AISI 904L Sintered (187.04 g/m^2), despite the lower number of porosities. The higher weight loss of the HIP treated samples is a result of oxides on the surface. It has been further evaluated by performing post pickling after HIP (146.26 g/m^2). The wrought EN 1.4539 has the lowest weight loss of the tested samples with an average value of 0.08 g/m^2.
4. Densification by HIP treatment shows to be an efficient way to improve the corrosion resistance of MPE AISI 904L. The samples need to undergo mechanical removal of the surface layer to remove open porosities [6]. More so, despite conducting a HIP treatment did the result show that MPE AISI 904L has a lower corrosion resistance compared to wrought EN 1.4539 and cast EN 1.4584

Acknowledgements

The authors would like to extend a special thanks to the following colleagues for their contribution; Jimmi Ammitzbøll, Allan Lind Sørensen and Christian Lyck from Sintex as well as Espen Zink Eikeland and John Julius Flounders from Grundfos.

References

[1] Mirzababaei S and Pasebani S 2019 A Review on Binder Jet Additive Manufacturing of 316L Stainless Steel *J. Manuf. Mater. Process* ***3*** 82.

[2] Y Wu R M German D Blaine B Marx and C Schlaefer 2002 Effect of residual carbon content on sintering shrinkage microstructure and mechanical properties of injection molded 17-4 PH stainless steel *J. Mater. Scie.* ***37*** 3573-3583.

[3] A Sorea C Lyck and P Valler 2022 Metal Powder Extrusion Of Large, Thin-Walled Stainless Steel Structures At Room Temperature *World PM 2022 Cong. Proc 978-189907255-2.*

[4] R Qvarfort 1988 New Electrochemical Cell for Pitting Corrosion Testing *J. Corr. Scie.* ***28*** 131-140.

[5] P H Kaae and E Z Eikeland 2022 Corrosion Performance of Additively Manufactured Stainless Steel by Binder Jetting *Conf. Ser.: Mater. Sci. Eng.* ***1249*** 012051.

[6] T Mathiesen 1993 Corrosion Properties of Sintered Stainless Steels *Ph.d.-thesis.*

Additive Manufacturing of 7xxx Aluminium Alloys by Laser Powder Bed Fusion

K Zhang*, S Wenner[1], C D Marioara[1], E W Hovig[1], Q Du[1], M Onsøien[1], K Marthinsen[2]

[1] SINTEF Industry, Trondheim, Norway
[2] Department of Materials Science and Engineering, NTNU, Trondheim, Norway

*E-mail: kai.zhang@sintef.no

Abstract. Additive manufacturing of high strength wrought grade aluminium alloys is of great industrial and academic interests, but still difficult in practice due to the tendency of crack formations in the microstructure. This work aimed at producing dense crack-free samples of AA7075-type aluminium alloys by the laser powder bed fusion (LPBF) technology, and to investigate the microstructure. It was found that LPBF of pure AA7075 powder gave micro-cracks for all tested 25 processing parameters which covered a wide range of laser power and speed. In order to avoid the cracks, different methods were tried including high preheating temperature, introducing nano-particles, and adding Si into the alloys. Introducing ZrH_2 nano-particles into AA7075 by mechanical blending could successfully produce crack-free samples using LPBF. Adding 5% Si into the AA7075 powder by mechanical blending could also give crack-free samples with high relative density. It was also found that the as-built samples with preheating at 250°C had hardness as-low-as 70HV. Solutionizing at 480°C with water quenching followed by artificial aging at 120°C can bring the hardness to 140HV and 120HV, for the AA7075 variants with ZrH_2 and with addition of Si, respectively. Detailed characterizations revealed clear influences of both preheating temperature and alloying on the microstructure. It is concluded that preheating temperature should be carefully selected for AM of high strength precipitation hardening Al alloys.

1. Introduction

Metal additive manufacturing (AM) has developed significantly in the past decades and has moved from lab development to wide industrial applications nowadays. There are many different AM technologies available for metals, depending on feedstock nature, feeding method, energy source, needs of post-processing (de-binding and sintering) etc., for example laser powder bed fusion (LPBF, or PBF-LB/M), wire-arc AM (WAAM), direct energy deposition (DED), binder jetting (BJ) etc. Each AM technology has its own advantages and disadvantages, and hence different optimal application scenarios. LPBF is one of the most applied AM technologies in industry [1].

By LPBF, the metal powder particles are melted by a laser beam to form the part in a layer-by-layer manner. The cooling rates in the melting pool is very high, of 10^4 K/s magnitude according to some estimations [2]. The solidified melt pool will also be partially remelted and solidified by laser input sideways and from upper layers. These unique solidification features together with other contributions from, for example, powder feedstock quality, process parameters etc., can lead to various defects in the parts processed by LPBF. Typical LPBF defects include lack of fusion, porosity, solidification cracking, stress cracking, rough surface quality etc. [3] Research and development aimed at eliminating those defects to improve productivity and reduce costs are crucial for a wider application of the LPBF, and AM in general.

Aluminium alloys are widely applied in various industries and have very good recyclability. There are many well developed and approved Al alloys for different applications [4]. However, only very limited number of Al alloys are commercially available for AM and LPBF. The most common Al alloys for LPBF are AlSi10Mg and AlSi7Mg, which both are derived from casting aluminium alloys. Wrought aluminium alloys, especially mid-to-high strength 6xxx Al-Si-Mg(-Cu) and 7xxx Al-Zn-Mg(-Cu) systems are scarcely combined with LPBF. The low processability, especially a tendency of solidification cracks along columnar grain boundaries, is one of the main reasons [5,6].

This work aims to produce crack-free dense AA7075 type alloys by LPBF. Inspired by other works, microalloying methods by introducing nanoparticles [5], adding Si into the alloys [7], and combining with elevated preheating temperatures [8] have been applied. Detailed microstructure investigations are performed to understand the hardness results in the as-built and heat-treated states.

2. Experimental

2.1 Powder, LPBF printing and post-AM heat treatment

The LPBF printer was a model SLM280 from Nikon SLM Solutions AG. A high temperature preheating chamber was used. The metal powder was gas atomized AA7075 supplied by ECKART TLS GmbH. The AA7075 (7075 in short) powder has a chemical composition of Al-5.44wt%Zn-2.56wt%Mg-1.61wt%Cu with the additions of 0.12wt%Si, 0.22wt% Fe and 0.21wt% Cr. The powder particles are well sphere-shaped with a size range of 20-63 μm. In addition to the pure 7075 powder, two other alloy variants were prepared by mechanical mixing. The '7075+ 1.6wt%ZrH_2' variant was added 1.6wt% zirconium hydride (ZrH_2) nano-powder which was of 99.9% purity and of average particle size 80-100nm, supplied by Nanoshel UK. The '7075+5wt%Si' variant was produced by mixing with gas atomized hypereutectic AlSi powder (Si 40wt%, size range 20-63 μm) supplied by Elkem, such that the final Si content is about 5wt%. A TURBULA® 3D shaker mixer was used for mixing powders, with a speed of 23rpm and mixing time ~9 hours for each alloy variant.

The processing parameters for LPBF were varied in order to find optimal processing parameters. The laser powder (P) was varied between 175 to 425W; the laser speed (v) 950 to 2020mm/s; hatching (h) in the range of 105-170um; layer thickness (t) 30 or 60um. In total 25 different processing parameters were used, and each of them were assigned to a different cylindrical sample (diameter 10mm and height 10mm) for the printing job. The base plate preheating temperature was either room temperature (RT) or 250°C. Nitrogen was used as the protective gas.

Some of printed cylindrical samples were heat treated. The heat treatment procedure consisted of solutionizing 1h@480°C, water quench and artificial aging 24h@120°C.

2.2 Material characterization

Different characterization methods were employed. Light optical microscopy (LOM) was mainly used for revealing any printing defects, especially solidification cracking here. The polarized LOM and EBSD were used for revealing the grain structure. Scanning electron microscopy (SEM) and transmission electron microscopy (TEM) were used to investigate the second-phase particles and their distributions. The SEM imaging was carried out with a Hitachi S3400-N microscope. The TEM was a Jeol 2100 equipped with an LaB_6 filament and operated at 200 kV.

Vickers hardness was performed as for mechanical testing.

3. Results and discussions

None of the as-built samples had cracks visible to the naked eye on the exterior surface, i.e. no macro-scale cracks, independent of the laser parameters, preheating temperatures, and alloying variants.

a) 7075, cross diameter section LOM b) 7075+1.6wt%ZrH_2 c) 7075+5wt%Si

d) 7075, grain structure e) 7075+1.6wt%ZrH_2 grain structure f) 7075+5wt%Si grain structure

Figure 1. (a-c) Typical LOM images of polished cross sections; (d-f) typical grain structure by anodising and polarized LOM, for three different alloy variants, i.e. pure 7075, 7075+1.6wt%ZrH_2 and 7075+5wtSi. BD – build direction. Printing preheating temperature: RT for (d), 250°C for all others.

However, cracks were observed by LOM for all printed cylindrical samples with pure 7075 powder on the polished cross section, independent of processing parameters or preheating. A typical such crack is shown in figure 1(a), where the crack is along the build direction (BD). The grain structure shows clear columnar grains which grow along the BD across melting pools due to high temperature gradients along this direction, see figure 1(d). The cracks follow the columnar grain boundaries, as indicated by white arrows in the figure. The micro-cracks, called hot cracking / solidification cracking in the literature, can be attributed to factors such as lack of melt feeding at the end of solidification, thermal stresses exerted on the mushy zones, and / or high misorientation angles at the grain boundaries (GBs) etc. [9] The current results also imply that preheating, at least up to 250°C, cannot effectively prevent micro-cracking by LPBF with pure 7075 feedstock, even though it has been reported that elevated temperature preheating can effectively reduce the thermal stress in LPBF [8].

IOP Conf. Series: Materials Science and Engineering **1310** (2024) 012019 doi:10.1088/1757-899X/1310/1/012019

By adding ZrH_2nano-particles or Si into 7075 powder, crack-free samples were successfully produced, as shown in figure 1(b) and (c) as two examples. Very high relative density can be achieved especially for the Si alloying variant, >99.9% based on image analysis. Despite the relatively low resolution of figure 1(e) and (f), a fine and more equiaxed grain structure can be observed in the added alloy variants as compared to the pure 7075 counterpart. A more detailed SEM image, using EBSD, of grain structure for the Si alloying variant printed samples can found elsewhere [10], which supports the above statement regarding grain structure refinement.

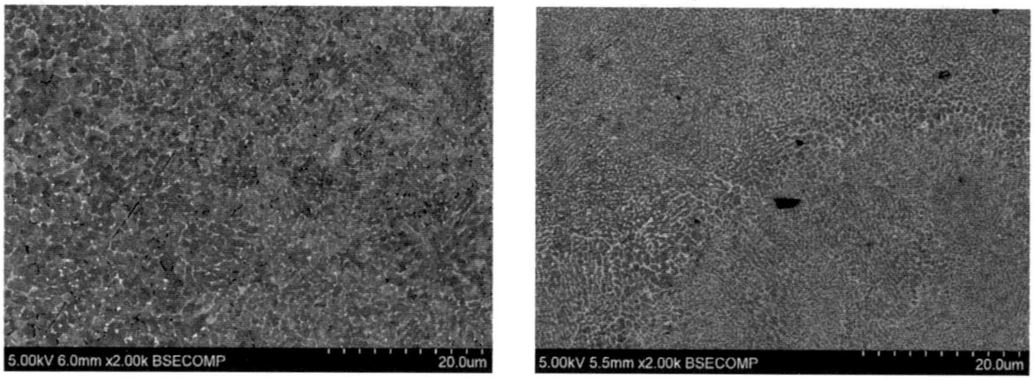

a) 7075+5wt%Si, as-built, preheating: 250°C b) 7075+5wt%Si, as-built, preheating: RT

Figure 2. Typical microstructure by SEM for (a) 7075+5wt%Si, printed with 250°C preheating and (b) 7075+5wt%Si, printed with RT preheating. Image perpendicular to BD.

Typical SEM microstructures of 7075+5wt%Si printed with 250°C and with RT preheating are shown in figure 2. A fine cell structure at the melting pool boundary and slightly elongated cellular / columnar structure into the melting can be observed in figure 2(b), with RT preheating, which is a typical microstructure for metals by LPBF [11]. The cell size here is roughly half a micron. According to EDS mapping (not shown here) the alloying elements segregated into the cell structures are mainly Si and Mg, with little Cu and Zn being detected. The 250°C preheating counterpart features a coarser, more indistinct structure, due to an incomplete network of particles and segregations as shown in figure 2(a). The typical cell size in this case is about 2 microns, four time larger than in the case with RT preheating. The alloying elements segregated into the cell structures with 250°C preheating were mainly Si and Mg with low levels of Cu and Zn, similar to the RT preheating case.

The hardness results for different alloying variants and tempers are compiled in table1. Typical Vickers hardness for samples printed with 250°C preheating was about 75-80HV, for both 7075+1.6wt%ZrH_2 and 7075+5wt%Si. However, the hardness was about 140HV for 7075+5wt%Si with RT preheating. The preheating temperature dependent hardness for 7075+5wt%Si may be due to the differences into the microstructure, as illustrated above and in figure 2: higher preheating temperatures lead to coarser cellular structures at the melt pools and coarser particles and alloying element segregation, hence lower hardness than the RT preheating condition.

The heat treatment, i.e. solutionizing 1h@480°C, water quench and aging 24h@120°C, can dramatically change the hardness, as given in table 1. The hardness improved significantly for those with samples printed with 250°C preheating. However, the 7075+1.6wt%ZrH_2 one ended

IOP Conf. Series: Materials Science and Engineering **1310** (2024) 012019
doi:10.1088/1757-899X/1310/1/012019

with higher hardness than the 7075+ 5wt%Si counterpart, which again should originate from the microstructure differences.

Table 1. Typical Vickers hardness for LPBF fabricated samples with different 7075 alloy variants and tempers (heat treatment: solutionizing 1h@480°C, water quench and aging 24h@120°C). All samples listed here were printed with 250°C preheating.

Alloy variant	Temper	Hardness (HV)
7075+1.6wt%ZrH_2	as-built	75
7075+1.6wt%ZrH_2	after heat treatment	145
7075+5wt%Si	as-built	80
7075+5wt%Si	after heat treatment	120

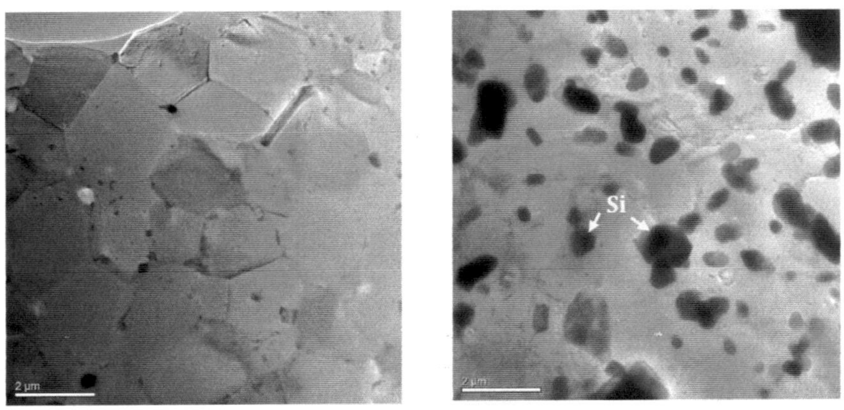

a) 7075+1.6wt%ZrH_2, after heat treatment b) 7075+5wt%Si, after heat treatment

Figure 3. Typical microstructure by TEM for LPBF samples after heat treatment of two alloying variants, (a) 7075+1.6wtZrH_2 and (b) 7075+5wtSi. TEM sample perpendicular to the BD.

TEM images for the two modified 7075 variants after heat treatment are given in figure 3. Looking along the BD, clear faceted grains are resolved in the 7075+1.6wt%ZrH_2 condition, with typical grain size about 2 microns. At higher magnifications, fine η' and η_2 precipitates with {111}Al habit planes can be resolved by TEM, similar to what is observed in 7075 alloys manufactured via conventional routes. The 7075+5wt%Si variant after heat treatment has a quite distinct microstructure compared to as-built temper and to the ZrH_2 sample after heat treatment, due to the newly formed and dispersed fine Si particles of typical sub-micron size. At higher magnifications TEM reveals the presence of short needles elongated in <100>Al directions, typically found in Al-Mg-Si system alloys, and the absence of η type precipitates of the Al-Mg-Zn system. This could be an effect of excess Si in solid solution binding Mg during artificial aging, preventing the precipitation of Mg-Zn phases. The 7075+5wt%Si variant has lower hardness than the 7075+1.6wt%ZrH_2 variant after the same heat treatment, which could also be attributed to absent of η type precipitates. Detailed TEM investigations will be reported elsewhere.

4. Conclusions

The current study demonstrates that preheating at 250°C does not prevent the formation of micro-cracks by LPBF using pure AA7075 powder. Instead, microalloying by adding ZrH_2 nanoparticles as grain refiner, or alloying with additions of Si are two effective ways to avoid crack formation, due to refined grain structures in these cases. Preheating at 250°C leads to low as-built hardness due to a coarse cell structure and particles. A conventional heat treatment consisting of solution heat treatment followed by artificial aging can restore the hardness of the 7075+1.6wt%ZrH_2 variant at levels comparable to conventionally processed 7075 alloys. However, the 7075+5wt%Si variant gave a lower hardness after the heat treatment, due to a different precipitation behaviour.

Acknowledgement

This research project is funded by the SFI PhysMet 'Center for sustainable and competitive metallurgical and manufacturing industry' under the grant number of 309584/F40, and by the project 'EO-SEP Fast-track development of AM materials' with support from the Research Council of Norway. Elkem is acknowledged for providing the hypereutectic AlSi powder used in this study.

References

[1] Armstrong M, Mehrabi H and Naveed N 2022 An overview of modern metal additive manufacturing technology *J Manuf Process* **84** 1001–29

[2] Kosiba K, Gustmann T, Kim J T, Seok J, Jung J, Beyer L, Scudino S, Giebeler L, Han J and Hufenbach J K 2023 Experimental cooling rates during high-power laser powder bed fusion at varying processing conditions *J Alloys Compd* **967** 171773

[3] Brennan M C, Keist J S, Palmer T A, Bourell D L, Frazier W, Kuhn H and Seifi M 2021 Defects in Metal Additive Manufacturing Processes *J Mater Eng Perform* **24**

[4] Raabe D, Ponge D, Uggowitzer P J, Roscher M, Paolantonio M, Liu C, Antrekowitsch H, Kozeschnik E, Seidmann D, Gault B, De Geuser F, Deschamps A, Hutchinson C, Liu C, Li Z, Prangnell P, Robson J, Shanthraj P, Vakili S, Sinclair C, Bourgeois L and Pogatscher S 2022 Making sustainable aluminum by recycling scrap: The science of "dirty" alloys

[5] Martin J H, Yahata B D, Hundley J M, Mayer J A, Schaedler T A and Pollock T M 2017 3D printing of high-strength aluminium alloys *Nature* **549** 365–9

[6] Zafar F, Reis A, Vieira M, Emadinia O, Zafar F, Reis A, Vieira M and Emadinia O 2023 Additively Manufactured High-Strength Aluminum Alloys: A Review *Recent Advancements in Aluminum Alloys*

[7] Montero Sistiaga M L, Mertens R, Vrancken B, Wang X, Van Hooreweder B, Kruth J P and Van Humbeeck J 2016 Changing the alloy composition of Al7075 for better processability by selective laser melting *J Mater Process Technol* **238** 437–45

[8] Hovig E W, Azar A S, Mhamdi M and Sørby K 2020 Mechanical Properties of AlSi10Mg Processed by Laser Powder Bed Fusion at Elevated Temperature *Minerals, Metals and Materials Series* 395–404

[9] Phan M A L, Fraser D, Gulizia S and Chen Z W 2021 Mechanism of hot crack propagation and prevention of crack formation during electron beam powder bed fusion of a difficult-to-weld Co-Cr-Ni-W superalloy *J Mater Process Technol* **293** 117088

[10] Du Q, M'hamdi M, Reiersen M, Hovig W and Zhang K 2023 A fully kinetic phase diagram-coupled multicomponent columnar-to-equiaxed grain transition model with an application to additive manufacturing

[11] Kempf A and Hilgenberg K 2020 Influence of sub-cell structure on the mechanical properties of AlSi10Mg manufactured by laser powder bed fusion

On the chemical composition, microstructure and mechanical properties of a Nitrogen-contaminated Ti-6Al-4V component built by Wire-Arc Additive Manufacturing

D Hu[1]*, R Biswal[2], V K Sahu[1,3], J W Fellowes[4], A Zadehkabir[2], S W Williams[2] and A E Davis[1]

[1] Department of Materials, The University of Manchester, Manchester, M13 9PL, UK
[2] Welding and Additive Manufacturing Centre, Cranfield University, Bedfordshire, MK43 0PL, UK
[3] Now at Mechanical Engineering, University of North Texas, Denton, 76201, US
[4] Department of Earth and Environmental Sciences, The University of Manchester, M13 9PL, UK

*E-mail: dongchen.hu@manchester.ac.uk

Abstract. Additive manufacturing (AM) using recycled Ti-6Al-4V (Ti64) feedstock material from wrought waste streams is a novel process that can reduce the overall energy cost and carbon (CO_2) footprint when compared to primary-production routes. The potential contamination of recycled feedstock material (e.g. C, O, N and Fe) can affect the microstructure and mechanical properties of the component. In this work, a Ti64 test wall built using wire arc AM (WAAM) was studied, where the top half only was contaminated by N through the shielding gas during deposition. This allowed a direct comparison of Ti64 WAAM material with high and low N content, deposited under otherwise identical conditions, to replicate the worst-case scenario of N contamination from using recycled swarf. The hardness of the N-contaminated section was found to be 25% higher than the uncontaminated section of the wall, demonstrating the N solid solution strengthening in Ti64. The room temperature transformed microstructure was found to have a 25% coarser α-lath thickness, which was proposed to be an effect of the AM cyclical heating and increasing of the β-transus temperature due to a higher level of N. Additionally, the outer layer of the N-contaminated sample section was found to have a refined parent β grain structure.

1. Introduction

Titanium alloys are embraced in the aerospace industry due to their high specific strength, fracture resistance and corrosion resistance [1, 2]. Additive Manufacturing (AM) methods have the advantages of high material utilisation, flexibility for complex shapes and the ability to manufacture near net-shape components [3, 4]. Wire-Arc Additive Manufacturing [4] (WAAM) is one suitable method for fabricating large-scale parts with 'kg/h' high deposition rates, which makes them suitable for production of large aerospace components [5].

A high level of material is wasted in conventional wrought processes during machining [5]. Recycling and reusing the swarf as feedstock materials for AM processes can reduce energy consumption and CO_2 emissions. However, using recycled Ti can introduce contaminant elements like C, N, O and Fe. N is an α-stabilisers [2], which means that a higher level of N could affect the transformation microstructure of an AM as-built component, especially if the contaminant distribution is not homogeneous. However, it is inherently challenging to detect the presence of

IOP Conf. Series: Materials Science and Engineering **1310** (2024) 012020 doi:10.1088/1757-899X/1310/1/012020

light elements (like O and N) in Ti microstructures, and the metallurgical effects of N in particular are rarely studied [6]. To accelerate the adoption of recycled Ti into AM aerospace component production, it is important to evaluate the effect of N contamination in Ti64 (and indeed other potential contaminants) on microstructure and resulting mechanical properties, to highlight potential risks for safety-critical parts.

In this study, a Ti64 test WAAM wall was built, with the top half contaminated by N, to directly compare the influence of N on the macro and microstructure. Multiple chemical analyses were conducted to characterise the chemical distribution and microstructure, to discern the metallurgical effects of N contamination in WAAM Ti64.

2. Materials and experiments

A WAAM wall specimen was produced using a 1.2 mm diameter Ti64 wire (wire feed speed: 2.5m/min) with an EWM® Plasma Torch with a travel speed of 5 mm/s. The deposition was performed over a 6.3 mm thick hot rolled Ti64 substrate. A local shielding device with Pureshield® Argon was used to maintain the inert atmosphere around the melt pool during the deposition process. The inert atmosphere was maintained until the temperature of the solidified melt pool dropped to 500°C, as it can be assumed that below this temperature the O_2 and N_2 ingress into Ti64 is negligible. Any changes made in the gas composition supplied to the local shielding device can directly impact the quality of deposition. Therefore, the first 15 layers were deposited using the Pureshield Argon and followed by another 20 layers with a controlled amount of N_2 being added to the local shielding gas mixture. The volumetric ratio of the N_2 to Pureshield Argon was estimated to be 0.11. Since the local shielding device consists of multiple honeycomb flow channels to establish a laminar flow, this also helps in homogenising the gas mixture. The reference frame for the images relative to the WAAM build is defined as: heat source travel direction parallel to the wall or 'welding' direction as WD, the transverse direction as TD and, build height direction normal to each layer as ND (see figure 1a). After deposition, the WAAM wall was removed from the substrate, and several ND-TD cross sections were cut and polished to a 0.04 micron OPS finish.

One ND-TD cross section covering the uncontaminated/contaminated interface region was etched with Kroll's reagent and then a Zeiss AxioCam was used to produce large area stitched optical micrographs. Scanning Electron Microscopy (SEM)-Backscattered Electron (BSE) images were acquired using an FEI Magellan Field Emission Gun (FEG)-SEM to investigate the transformation microstructure. Micro-hardness mapping was performed on another cross section using a Struers Durascan 80 hardness tester. Large area Electron Backscattered Diffraction (EBSD) mapping of both cross sections was performed on a Thermo-Fisher Apreo FEG-SEM equipped with an Oxford Instrument 'Symmetry 2' EBSD detector, operating at an accelerating voltage of 20kV. The EBSD data was further analysed using commercial software AZtecCrystal provided by Oxford Instruments. The β-phase solidification microstructure was reconstructed based on the Burgers orientation relationship between the parent β phase and the transformation α using the parent grain reconstruction capability in AZtecCrystal. A novel Electron Probe Microanalysis with Soft X-ray Emission Spectroscopy (EPMA-SXES) technique was employed to obtain the spatial distribution and quantified concentration of C, N and O, since the characteristic peaks of these light elements are extremely difficult to resolve reliably in a conventional SEM-EDS X-ray spectrum. A more detailed description of the technique will be included in a future publication.

3. Results

The hardness profile shown in figure 1(a) was obtained from micro-hardness mapping by averaging 25 indents in TD, which indicates a significant increase in hardness in the contaminated part (top, ~445HV) than the clean part (bottom, ~ 354 HV). A large area optical image of the etched cross section is presented in figure 1(b), and reveals that the dissimilar-colour interface between the uncontaminated Ti64 material (referred to as 'clean part') and the first contaminated layer (black arrows) is not flat, but instead curved, matching the bead profile of similar Ti64 WAAM deposits in this plane [7]. HAZ bands are also clear in figure 1(b). However, the HAZ bands adjacent to the interface are curved and irregular (red arrows), unlike the parallel HAZ bands in the clean bottom part and those in the top contaminated part but located farther from the interface (orange arrows). The β-reconstructed EBSD map in figure 1(c) shows clear grain refinement in the outer layer of the contaminated part, which differs from the highly textured (<001>//ND) coarse columnar β within the central region and the bottom clean part of the wall. figure 1(d) is a quantified EPMA map showing the N level across the interface. The concentration of N across the interface changes with a stepped profile with a sharp transition between 0.03 wt.% and 0.67 wt.% followed by a gradual gradient, with a final discontinuous transition to ~ 1.1 wt.%, matching the bead profile of several deposited layers. This 'transition gradient' is similar to those seen in WAAM dissimilar-alloy builds where the alloy feedstock is switched mid-deposition [8]. Averaged N concentrations measured by EPMA in the clean/contaminated sections are given in table 1. Similar analysis has also been applied to C and O but show a uniform distribution throughout the deposition.

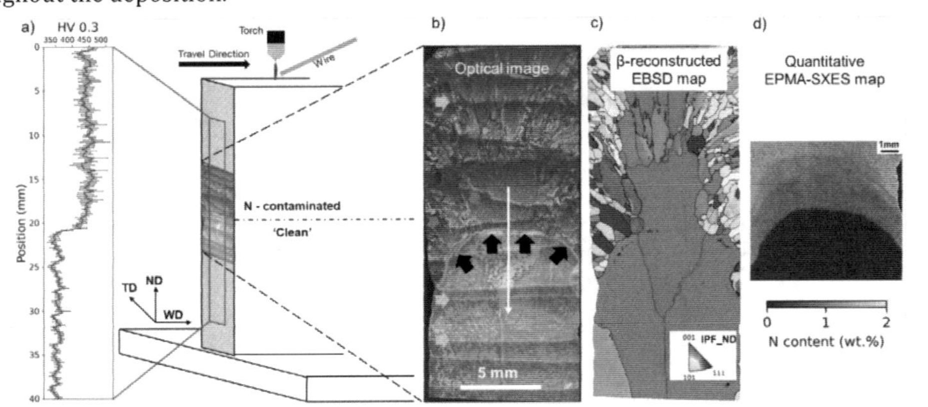

Figure 1. (a) N-contamination WAAM wall schematic diagram and reference frame, alongside a hardness profile on the ND across the interface. (b) An optical image taken from the interface. Fusion boundary between clean and contaminated part is indicated by black arrows. Some HAZ bands are highlighted by red and orange arrows. (c) A β-reconstructed EBSD map of the same area. (d) Quantified EPMA-SXES map showing N level at the interface. The white line on the optical image indicates where the SEM-BSE scanning was performed (figure 2).

The bulk chemical compositions of the contaminated and clean parts were analysed by Inductive Coupled Plasma mass spectroscopy (ICP), and the results are presented in table 1. The ICP results revealed that the top part is only contaminated with N without any significant changes in the level of other elements. Although a bulk concentration of ~0.8 wt.% N was recorded, the unusually high level in the sample was unfortunately outside the instrument calibration range and so can only be considered an estimate. The concentrations of other elements were well within calibration ranges and are therefore considered much more reliable.

Overall, the transformation microstructure consisted of typical α + β basketweave lamellae [7]. Inter-lamellar spacing (ILS) measurements were performed across the contamination interface using the method described by Zhao et al. [9]. ILS measurement was only performed on a single line to avoid the effect of the irregular (non-flat) HAZ bands adjacent to the interface in TD. Across

Table 1. ICP measurement of bulk sample compositions in wt.% taken from the contaminated top and clean bottom of the WAAM wall. * 0.805 wt.% is outside the calibration of the ICP measurement, and can only be taken as an estimate.

Element	Al	V	Fe	C	H	O	N (ICP)	N (EPMA)
Contaminated	6.34	3.83	0.18	0.023	0.0015	0.17	0.805*	1.18
Clean	6.24	3.86	0.17	0.023	0.0011	0.17	0.013	0.03

the HAZ bands a characteristic 'saw tooth', repeating pattern is observed, where the ILS increases as it approaches the bottom of the HAZ band and subsequently drops rapidly at the top of the HAZ band [7]. The abrupt change in ILS is linked to the effect of the $\alpha \rightarrow \beta$ phase transformation on transformation microstructure as shown in figure 2, which is consistent with previous work [7].

More importantly, the ILS measurements show that the average ILS is ~25% larger in the contaminated part compared to the clean part. This discrepancy suggests that a higher level of N has a more pronounced effect than the heat accumulation on the α-lath coarsening.

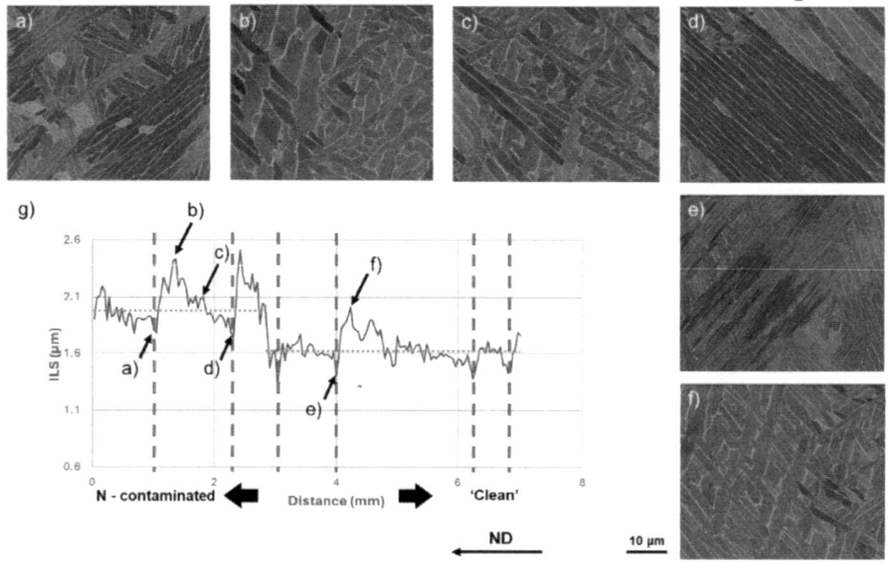

Figure 2. (a-d) Sequential SEM-BSE images showing the microstructure from one HAZ band to another. (e) and (f) are two SEM-BSE images showing the microstructure across one HAZ band from the clean section. (g) Linear intercept α-lamellar spacing (ILS) measurement across 6 HAZ bands across the interface taken from the position highlighted in figure 1(b) by the white arrow.

4. Discussion

The hardness measurement showed that there is a ~25% increase in the contaminated part compared to the clean part, which is indicative of an increase in the yield strength as has been demonstrated previously (e.g. [10]). This increase in hardness is likely attributed to N solid solute strengthening effect, as previously reported in pure Ti [11, 12] but conversely reduces the ductility.

The transformation microstructure analysis reveals a ~25% coarsening of the lath thickness in the contaminated part. Coarsening of the transformation microstructure of Ti64 leads to a higher ductility [13], which could cause the WAAM-built Ti64 component to exhibit an increased strength without compromising the ductility drastically. O was found to have a similar effect by Caballero [14], where a 0.25% addition of O demonstrated an increase in the strength while retaining similar ductility in uniaxial tensile tests, which could be linked to transformation microstructure coarsening. This could be an advantage of AM processing over conventional casting/wrought processing, due to the high cooling rate that produces a very fine as-built transformation microstructure that can tolerate such coarsening and maintain beneficial mechanical properties.

Nitrogen is an α stabiliser in Ti [2] and is known to have a similar effect as O in increasing the β-transus temperature (T_β) [15]. This is an important consideration in AM Ti64 microstructures with variable N concentration because of the thermal cycling that occurs below the T_β causing partial α→β phase transformation, leading to the formation of HAZ bands. In order to predict the effect of N on the T_β in Ti64, phase transformation temperature calculations of Ti64 were performed using the 'Thermo-calc' software with the ICP measured composition of the contaminated wall, except for the N where the averaged EPMA value (1.18 wt.%) was used that was considered more reliable (see figure 3a). The significant increase in the T_β may explain the coarsening of the transformation microstructure on the N-contaminated side. During the thermal cycling below T_β, the α regions are subjected to partial α→β phase transformation leading to transformation microstructure coarsening [7]. A higher N content in the contaminated section is proposed to accelerate this coarsening kinetic as the partial phase transformation occurs at a higher temperature and leads to a coarser transformation microstructure than in the clean section.

However, the spatial distribution of N, as revealed by the EPMA map in figure 3(d), is heterogeneous along the TD. The EPMA map shows a sharp N gradient directed towards the 'C' shaped interface, which is consistent with the bead profile. As illustrated by the N profile along the ND in figure 3(b), there is a sharp transition at the interface followed by a gradual gradient which results from the remelting and liquid mixing of the previous layer during deposition. After depositing another 2 layers above the interface (approximately 4 mm), the N content reaches a steady level. Figure 3(c) is a schematical illustration of the N spatial distribution of a ND-TD cross section. Red, yellow and green dashed lines along the TD were plotted on figure 3(c) to represent locations that experienced the same thermal history and align with the HAZ bands at the build centreline. Corresponding T_β schematic profiles (based on the N EPMA map and phase diagram trends) that match the same centreline positions have been added in figure 3(d) and are proposed to correspond to the curvature of the irregular HAZ bands shown in figure 1(b). Initially, there is a high N gradient when crossing the fusion boundary (that is aligned with the bead profile) in TD between layers, which results in high HAZ band curvature. However, as the N gradient diminishes moving further from the interface, the curvature of the T_β profile and HAZ bands decrease.

Furthermore, the outer layer of the N-contaminated section was found to have a refined parent β grain structure as illustrated in figure 1(c). This observation is consistent with previous findings in dual phase Ti alloys with wider mushy zones [8]. It is suggested that the outer layer refinement is caused by the maximum thermal gradient tilting away from the ND during solidification, following the normal of the curved bead profile. This allows the dendrites growing away from the ND to gain a growth advantage over the <001> // ND columnar β grains and block their growth. This effect is explained in more detail in reference [16]. The thermal gradient is more likely to affect the N contaminated section, since N is known to increase the width of the mushy zone. In addition, due to the relatively high level of N in the sample, nitrides may form in the melt pool, which could act as heterogeneous nucleation sites and contribute to the β grain refinement during solidification, which has been demonstrated with artificial nucleants introduced to the melt pool [17].

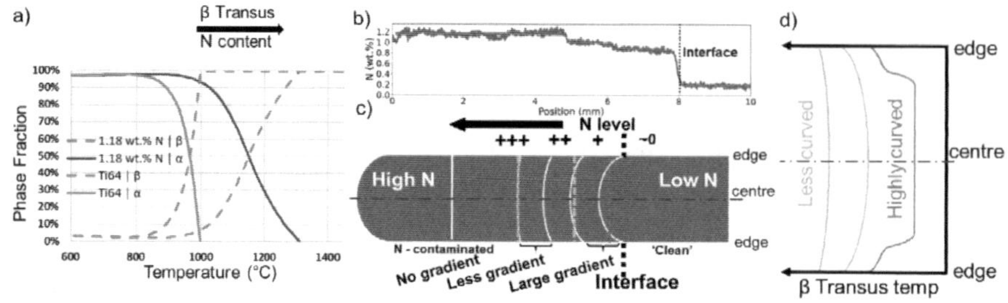

Figure 3. (a) A diagram showing the β-transus temperature of Ti64 (no contaminants) and Ti64 with higher N content (1.18 wt.% as given by EPMA). (b) A line profile of a longer higher spatial resolution EPMA map to show the N content on each part and the gradient along ND. (c) A schematic illustration of the distribution of the N gradients along the ND and (d) a schematic diagram showing T_β profiles relating to HAZ band behaviour.

5. Conclusions

- A higher level of N has an effect of refining of the parent β grains at the outer layer of the WAAM wall.
- Hardness measurements showed an increase in solid solution strengthening by ~25% due to the addition of ~1.2 wt.% N.
- Our results also show that the transformation microstructure is coarsened by 25% in the N-contaminated section.
- The room-temperature transformed microstructure appeared to be affected by the inhomogeneous spatial distribution of the N. The N concentration gradient was suggested to change the position of HAZ bands accordingly due to its influence on the β transus temperature.

Acknowledgements

The authors are appreciative of the EPSRC program grant Sustainable Additive Manufacturing (EP/W01906X/1) for supporting this research. The authors acknowledge the use of equipment within the Advanced Metals Processing and Characterisation themes of the Henry Royce Institute for Advanced Materials, funded through EPSRC grants EP/R00661X/1, EP/S019367/1, EP/P025021/1 and EP/P025498/1.

Reference

[1] Ezugwu E and Wang Z 1997 *J. Mater. Process. Technol.* **68** 262-74
[2] Welsch G, Boyer R and Collings E 1993 *Materials properties handbook: titanium alloys*: ASM international)
[3] Gong G, Ye J, Chi Y, Zhao Z, Wang Z, Xia G, Du X, Tian H, Yu H and Chen C 2021 *J. Mater. Res. Technol.* **15** 855-84
[4] Williams S W, Martina F, Addison A C, Ding J, Pardal G and Colegrove P 2016 *Mater. Sci. Technol.* **32** 641-7
[5] Allen J 2006 *Cost Effective Manufacture via Net-Shape Processing* **17** 1-17
[6] Gardner H, Gopon P, Magazzeni C, Radecka A, Fox K, Rugg D, Wade J, Armstrong D, Moody M and Bagot P 2021 *J. Mater. Res.* 1-16
[7] Ho A, Zhao H, Fellowes J W, Martina F, Davis A E and Prangnell P B 2019 *Acta Mater.* **166** 306-23
[8] Kennedy J, Davis A, Caballero A, White M, Fellowes J, Pickering E and Prangnell P 2021 *Mater. Charact.* **182** 111577
[9] Zhao H, Ho A, Davis A, Antonysamy A and Prangnell P 2019 *Mater. Charact.* **147** 131-45
[10] Keist J S and Palmer T A 2017 *Mater. Sci. Eng., A.* **693** 214-24
[11] Chong Y, Tsuru T, Guo B, Gholizadeh R, Inoue K and Tsuji N 2022 *Acta Mater.* **240** 118356
[12] Issariyapat A, Visuttipitukul P, Umeda J and Kondoh K 2020 *Addit. Manuf.* **36** 101537
[13] Lütjering G and Williams J C 2007 Engineering materials and processes: titanium *New York* 203-50
[14] Caballero A 2023 Effect of different shielding conditions, thermal cycles and post-deposition treatments on melting behaviour and mechanical properties of additively built components. PhD Thesis, Cranfield University
[15] Wriedt H and Murray J 1987 *Bull. Alloy. Phase. Diagr.* **8** 378-88
[16] Davis A, Wainwright J, Sahu V, Dreelan D, Chen X, Ding J, Flint T, Williams S and Prangnell P 2024 *Metall. Mater. Trans. A.* 1-23
[17] Kennedy J R, Davis A E, Caballero A, Williams S, Pickering E and Prangnell P 2021 *Addit. Manuf.* **40** 101928

Processability of K340 Cold Work Tool Steel by Directed Energy Deposition Technique

MS Kenevisi[1,2,*], PA Martelli[1,2], FS Gobber[1,2], D Ugues[1,2] and S Biamino[1,2]

[1] Department of Applied Science and Technology (DISAT), Politecnico di Torino, Italy
[2] Integrated Additive Manufacturing iAM@PoliTO Center, Politecnico di Torino, Italy

[*]E-mail: saleh.kenevisi@polito.it

Abstract. Directed Energy Deposition (DED) is an additive manufacturing process which can be used to repair defected components, such as blanking dies made of K340 tool steel. In this work, double tracks of K340 steel were deposited using DED process to study the processability of the alloy, and the tracks were characterized by light optical microscopy (LOM), scanning electron microscopy (SEM) and microhardness test. The results showed that near full-dense deposits can be made. However, the thermal cycle imposed by the process alters the microstructure of the material. Further investigation is required to make it possible to achieve a more homogeneous microstructure.

1. Introduction

Directed Energy Deposition (DED) is an additive manufacturing (AM) process that involves accurately depositing material onto a substrate by using a focused energy source, such as electron beam or laser. The process involves melting or sintering a feedstock material, which may be in the form of powder or wire, onto the substrate's surface layer-by-layer [1]. This process has been used both for fabricating new components and repairing damaged parts [2,3]. The processing parameters in each AM process have a significant impact on the quality of the fabricated components. Chen et al. [4] and Cho et al. [5] investigated how processing parameters can affect the deposition process and the microstructure of the material. Additionally, different studies have been conducted to investigate and to improve the effectiveness of the DED repairing approach for various alloys, such as steels [6], Ti [7], and Ni [8] alloys.

K340 tool steel, also known as Böhler K340 ISODUR, is a high-carbon and high-chromium tool steel widely used for cold work applications because of its well-balanced properties. It possesses high hardness and wear resistance, improved toughness, and dimensional stability [9]. It is widely used as blanking tools and dies. However, with time, defects such as wear or cracking may occur on the dies.

The use of the powder-fed DED process for depositing K340 steel provides the capability to repair damaged parts. However, the DED repair process may modify the microstructure of the component. Generally, high-carbon steels are not considered as suitable alloys for AM processes because of their high susceptibility to cracking [10]. This study investigates the DED processability of the alloy, which, to the authors' knowledge, has not been previously examined in the literature. The study examines how processing parameters impact the geometry and the microstructure of the deposits.

2. Materials and Methods

The material used in this study was BÖHLER K340 Isodur powder particles. As the powder particles of this material is not commercially available, HERMIGA 100/10 VI gas atomizer by PSI was used to produce the powder. The obtained powder was then sieved to obtain a particle size range of 45-125 μm with D50 = 87.8 μm. In terms of the powder characteristics, the particles were generally spherical with some satellites attached to them, as can be seen in figure 1(a). The cross-section analysis of the particles (figure 1(b)) also did not reveal high amounts of internal gas pores.

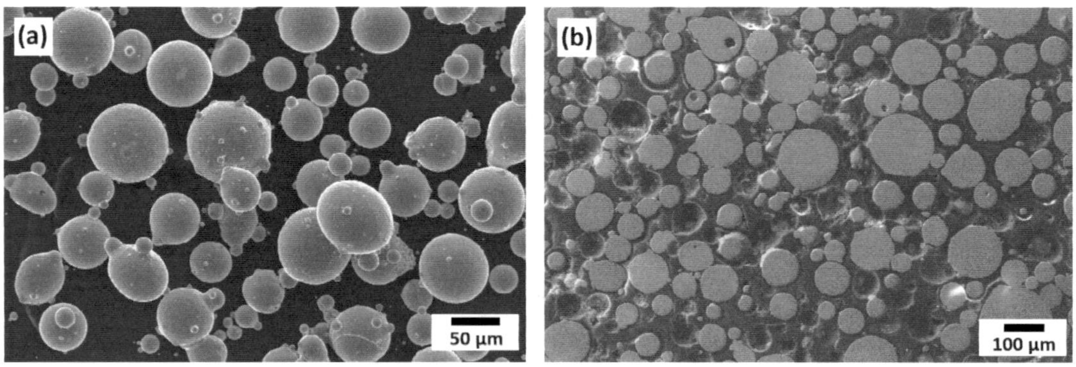

Figure 1. SEM micrographs of a) particles and b) cross-section of the powder particles used in this work.

The chemical composition of the powder was analyzed by LECO ONH836 and CS844 instruments and a ZEISS EVO 15 scanning electron microscope equipped with Ultim Max EDS detector by Oxford Instruments. The results of the alloying elements are provided in table 1, which is comparable to the nominal composition provided by Böhler.

Table 1. Chemical composition (wt.%) of the produced powder.

Element	C	Cr	Mo	V	Si	Mn	Al	Nb
Powder	1.12	8.05	1.91	0.42	0.81	0.46	0.89	0.16

Fifteen double-tracks with different sets of processing parameters (nozzle travel speed from 400 to 700 mm/min, laser powder from 500 to 900 W, and carrier gas flow rate from 4.5 to 7.5 l/min) were deposited on a K340 platform by Laserdyne 430 laser metal deposition machine by Prima Additive equipped with 1 kW Yb laser with a spot size of 2 mm.

The cross-sections of the deposits were studied by SEM. Moreover, the hardness measurement was done by using VMHT 1200 Leica Vickers microhardness tester and applying 300 g force with a dwell time of 15 s.

3. Results and Discussion

3.1 Single Tracks Morphology

The morphology of different deposits was studied, and the results for three samples are illustrated in figure 2. It can be seen that by increasing the energy density both the width and the height of the deposit increases. At S = 700 mm/min and P = 500 W (sample A) the average height and the width of the deposit are 214.6 μm and 2.47 mm, respectively. By decreasing the nozzle speed to 550 mm/min and increasing laser power to 700 W (sample B), an increase in the height of the track to 300.6 μm and width of it to 2.84 mm was observed. Further increasing the energy density (S = 400 mm/min and P = 900 W) resulted in a track with 682.5 μm of height and 2.89 mm of width (sample C).

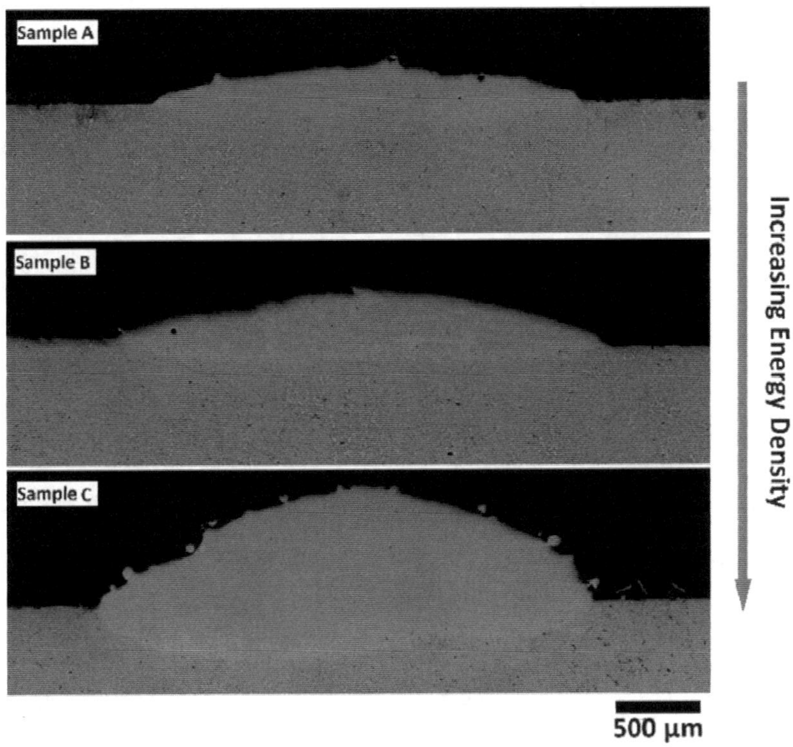

Figure 2. LOM micrographs of the deposits illustrating the tracks' morphology.

This is because by lowering the travel speed and increasing the laser power, the temperature of the melt pool increases and more powder particles are deposited and incorporated into the melt pool [11,12]. It was also observed that the sample with the highest energy density had the biggest deposit. Moreover, it showed the highest value of dilution which is 26%. However, on the other side, the dilution of the track deposited with the highest travel speed and lowest laser power was only 5%. Deeper penetration of the material can be achieved by the elevated melt pool temperature at high laser powers [13]. The statistical analysis showed that the laser powder was the most significant factor among the three parameters that affects the dilution. Generally, dilution values between 10 – 30 % are desired in the literature [14].

3.2 Defects and Microstructural Evolution

The porosity content of the deposits was measured from different cross-sections of each sample and the average value for samples A, B, and C were 0.5, 0.3, and 0.1%, respectively. The cross-section analysis did not disclose any lack of fusion (LOF) defect in all the samples. The decling trend in the porosity content by increasing the energy density the melt pool is remained at higher temperature and longer times before solidification. Therefore, more gases, which were within the powder particles or generated due to the evaporation of elements, that are entrapped in the melt pool can escape and reach the surface.

Regarding the microstructural evolution, depositing the material caused a modification in the microstructure of the material, which occured in all depositions. To represent this modification, figure 3 depicts the SEM micrograph of sample C. As can be seen, the microstructure of the deposit differes from that of the substrate and there is a heat affected zone (HAZ) as well. The microstructure of the deposited material is characterized by a fine equiaxed cellular structure. EDS analysis suggested the segregation of elements such as C, Cr, Mo, and V at cell boundaries as well as the presence of austenite.

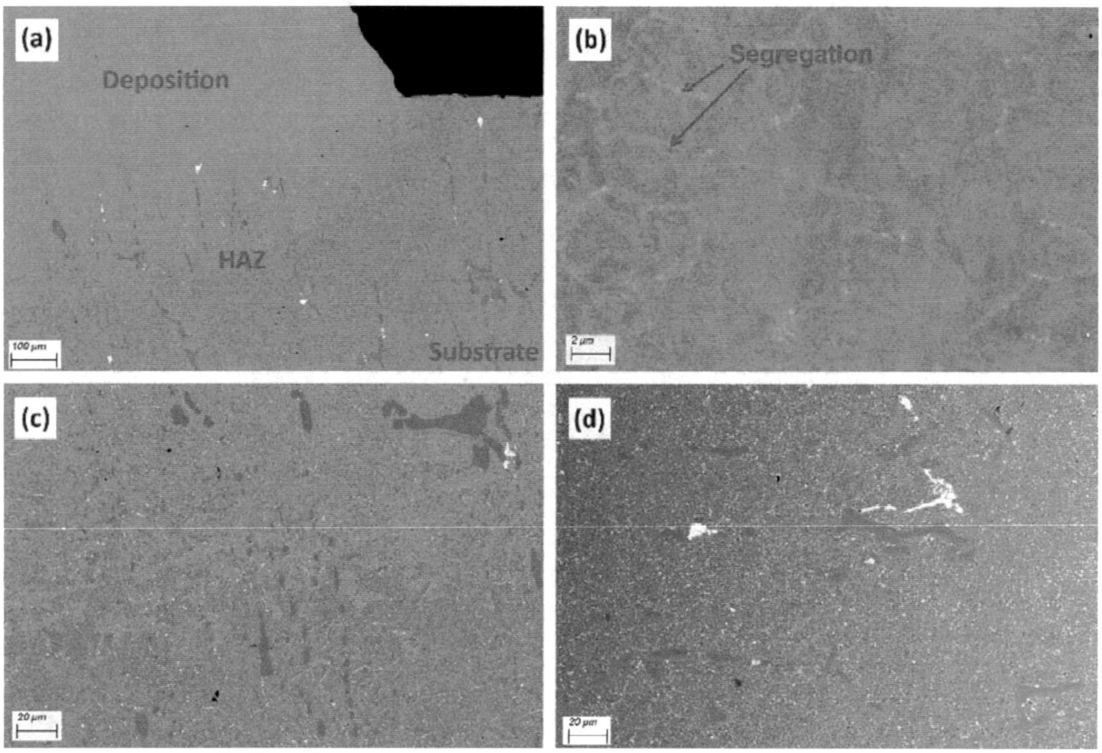

Figure 3. SEM micrographs from the cross-sections of sample C illustrating the microstructure of a) all regions at low magnification, b) deposit, c) HAZ, and d) substrate material.

On the other side, the substrate was used in an annealed condition and both spherodized and elongated carbides were disperesed in a ferritic matrix. The big blocky bright regions are also carbides rich in Nb. Due to the applied thermal cycle that the material close to the deposition has experienced, an HAZ was identified. In this region, fine carbides are precipitates along the grain boundaries. These carbides were rich in Cr and Mo. Compared to the microstructure of the substrate, there was no evidence of fine spherical carbides (gray carbides sized 1-3 µm). This is due to the dissolution of these carbides at relatively high temperatures imposed to this region.

3.3 Hardness Measurement

Vickers Microhardness test was performed on a 1200 × 1200 μm^2 area consisting all different regions. The hardness map was produced and illustrated in figure 4. It shows that the HAZ close to the deposition possesses the highest hardness value. The hardness in this region reaches up to 690 HV, while the corresponding values for the base metal and the deposit are within the range of 240-300 and 390-440 HV, respectively.

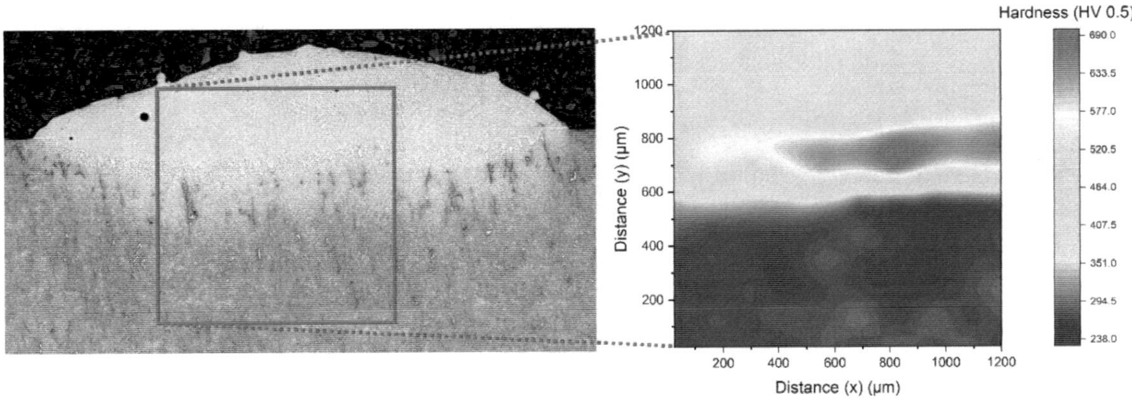

Figure 4. Microhardness map of the selected region of sample C.

The high hardness value of the HAZ close to the deposit suggests that the material in this region may reach the austenitizing temperature during the deposition process. In the cooling stage, the microstructure transformed to martensite and the primary carbides, which are responsible for the increased hardness [15].

4. Conclusions

In this study, a preliminary investigation of the processability of K340 cold work tool steel by DED process was conducted. Double tracks were deposited with different sets of processing parameters. The results showed that by increasing laser power and decreasing nozzle travel speed, in other words increasing the energy density, both height and width of the deposits increased. Laser power was shown to be the most significant factor affecting the morphology of the tracks. Microstructural analysis revealed the presence of retained austenite and segregation of alloying elements in the deposited material. Moreover, the applied thermal cycle introduced some HAZs with modified microstructures. Hardness measurement also confirmed the different microstructures of the substrate, HAZ, and the deposited material. More in-depth microstructural analysis is going to be performed in future works.

References

[1] Vafadar A, Guzzomi F, Rassau A and Hayward K 2021 *Appl Sci* **11** 1–33
[2] Kim T G and Shim D S 2021 *Mater Sci Eng A* **828** 142004
[3] Piscopo G and Iuliano L 2022 *Int J Adv Manuf Tech* **119** 6893–6917
[4] Chen B, Huang Y, Gu T, Tan C and Feng J 2018 *Rapid Prototyp J* **24** 964–972
[5] Cho K T, Nunez L, Shelton J and Sciammarella F 2023 *J Manuf Mater Process* **7** 105
[6] Simoneau L, Bois-Brochu A and Blais C 2020 *J Mater Eng Perform* **29** 6139–6146
[7] Rauch M, Hascoet J-Y and Mallaiah M 2020 *MATEC Web Conf.* **321** 3017
[8] Wilson J M, Piya.C, Shin Y C, Zhao F and Ramani K 2014 *J Clean Prod* **80** 170–178
[9] Jovicevic-Klug P, Toth L and Podgornik B 2022 *Coatings* **12** 1296
[10] Vincic J, Aversa A, Lombardi M and Manfredi D 2023 *Met Mater Int* **30** *501-516*
[11] Peyre P, Aubry P, Fabbro R, Neveu R and Longuet A 2008 *J Phys D Appl Phys* **41** 025403
[12] Hua T, Jing C, Xin L, Fengying Z and Weidong H 2008 *J Mater Process Tech* **198** 454–462

[13] Félix-Martínez C, Ibarra-Medina J, Fernández-Benavides D A, Cáceres-Díaz L A and Alvarado-Orozco J M 2021 *Int J Adv Manuf Tech* **115** 3999–4020
[14] Ahn D G 2021 *Int J Pr Eng Man-Gt* **8** 703–742
[15] Zhang X, Liu W, Sun D and Li Y 2007 *Metall Mater Trans A* **38** 499–505

Microstructure Evolution and Precipitation Strengthening Behaviour of Additively Manufactured High-speed Steels

H Zhang[1*], D A Venero[2], J Park[3], S V Petegem[4], A Özsoy[4], G Soundarapandiyan[4], S Robertson[5], X Zhang[6] and B Chen[1,7]

[1] School of Engineering, University of Leicester, Leicester, UK
[2] ISIS Neutron and Muon Facility, Rutherford Appleton Laboratory, Didcot, UK
[3] X-ray Science Division, Argonne National Laboratory, Lemont, USA
[4] Structure and Mechanics of Advanced Materials, Paul Scherrer Institut, Villigen, CH
[5] Materials Characterization Centre, Loughborough University, Leicestershire, UK
[6] Nuclear Science and Engineering Division, Argonne National Laboratory, Lemont, USA
[7] Department of Mechanical Engineering, University of Southampton, Southampton, UK

*Email: hz202@leicester.ac.uk

Abstract Additively manufactured (AM) high-speed steels were investigated, focusing specifically on the microstructure evolution during post-treatment in S390 steel and the rapid solidification process in M50 steel. An improved understanding of the processing-microstructure-property relationship for AM high-speed steel is achieved through a combination of post-mortem microstructure characterisation on precipitates and in-situ tracking of phase evolution. Quantitative characterisation of primary carbides and nanoprecipitates highlights the strengthening through nanoprecipitates that contribute to the exceedingly high hardness of 921 HV. Phase evolution during tempering was examined through in-situ synchrotron diffraction and ex-situ small-angle neutron scattering, revealing primary carbide growth by 60 nm within 2 minutes and nanoparticle precipitation with a size of 1.4 nm after 60-minute tempering. Additionally, the microstructure evolution of AM M50 steel was investigated by operando synchrotron diffraction, unveiling cooling rates in the order of 10^5 K/s during liquid-solid transformation. After printing, the carbon content of 0.47 wt.% in the matrix was derived from the martensite tetragonality. The insights gained serve as a valuable guide for designing future steel groups and developing heat treatment procedures tailored for the AM process.

1. Introduction

Compared to conventional means, additive manufacturing (AM) has demonstrated its reliability in producing various grades of steel exhibiting refined microstructure and improved performance, e.g. stainless steel [1-3] and maraging steel [4, 5]. Important progress has been made in understanding the microstructural features of AM high alloy steels, e.g. dendritic grain microstructure [6, 7] and processing defects [8]. It is recognised that high-speed steels with high contents of carbon and alloying elements undergo intricate phase transformations due to the complex thermal histories encountered during AM processing, leading to intricate microstructures with multiple precipitates that are crucial in strengthening steels.

Greenen et al. [7] found that the hardness level of AM M3:2 high-speed steel was similar to that processed by hot-isostatic pressing and casting ones, ranging from 920-945 HV. Sander et al. [9] observed a high hardness of 900 HV in AM FeCrMoVC tool steel and related the property enhancement to the discontinuous network of carbides with a homogeneous distribution in the steel matrix. Saewe et al. [10] studied the processability of AM HS6-5-3-8 high-speed steel and they claimed that hardness decreased from 650 HV to 570 HV despite more carbide precipitation as a result of increased carbon content. In addition, several studies focused on the microstructure evolution in AM tool steel [11-13] but were primarily concerned with primary carbides with a size of about 0.7-1 μm during heat treatment [14]. Overall, there remains a significant gap in the understanding of how the microstructure evolves and its influence on the properties of AM high-speed steels.

This study aims to elucidate the microstructure evolution and underlying strengthening mechanisms in high-speed steel during AM and post-treatment processes. Our goal is achieved through a combination of techniques, including post-mortem characterisation, ex-situ small-angle neutron scattering and in-situ synchrotron diffraction, spanning all length scales down to nanometres.

2. Methods

2.1 Materials

S390 high-speed steel (1.48C-10.34W-5.08V-4.82Cr-2.07Mo, wt.%) was processed by electron-beam powder-bed-fusion using a power of 330 W, scanning speed of 700 mm/s and hatch offset of 0.15 mm. The AM samples were heat-treated by austenitisation (1230 °C, 4 mins) and triple tempering (560 °C, 120 mins×3). M50 high-speed steel powders with nominal compositions of 0.79C-4.36Cr-4.19Mo-1.15V, wt.% were produced by vacuum induction melting inter-gas atomisation with a D50 value of 33.65 μm. This was then used for in-situ laser printing study.

2.2 In-situ synchrotron diffraction and ex-situ small-angle neutron scattering experiments

In-situ measurements of phase evolution during the tempering process were performed at APS 1-ID using a beam energy of 76.11 keV. The AM S390 sample after austenitisation with a size of 5×5×1 mm^3 was tempered at 560 °C, 120 mins×3 with heating and cooling rates of 200 °C/min. The Debye-Scherrer rings were continuously collected every 30 seconds in transmission mode. The integrated 1D X-ray diffraction patterns were analysed by the Rietveld method using GSAS II.

In-situ synchrotron measurements during laser printing with a miniaturised laser-beam powder-bed-fusion facility [15, 16] were performed at SLS MicroXAS with an energy of 17.18 keV. Steel powders spread on the baseplate were melted with a laser power of 150 W, speed of 500 mm/s and hatch offset of 0.05 mm. The diffraction images were collected on the top-layer surface in reflection mode. The integrated diffraction patterns were analysed by Matlab to track the peak intensities and positions.

Small-angle neutron scattering for nanoprecipitate characterisation was performed at the ISIS ZOOM beamline. AM S390 samples with a thickness of 0.3 mm were placed in a magnetic field of 1.5 T in the horizontal direction to saturate the steel matrix. Details can be found elsewhere [17]. The collected scattering data, covering a q range of 0.004-0.5 $Å^{-1}$, was reduced by Mantid software [18] and analysed using the software SASfit [19].

2.3 Statistical analysis of precipitates

The primary carbides in AM S390 steel were characterised using focused-ion-beam imaging combined with gas XeF_2 etching. The micrographs used for statistical analysis were processed by ImageJ [20] to distinguish different types of carbides. The nanoprecipitates in steel were quantitatively analysed by fitting the obtained magnetic neutron scattering data to the scattering intensity model [17] based on the assumption that they are spherical following a lognormal size [21], from which the particle volume fraction and size distribution were determined.

3. Results

3.1 Precipitation strengthening in heat-treated condition

The primary carbides in AM S390 steel are mainly V-rich MC and W-rich M_6C types and nano-precipitates are V, Cr-rich particles. Figures 1a and 1b present the size distribution of M_6C and MC carbides before and after heat treatment, respectively, with lognormal plots displayed for the statistical interpretation of the data. After heat treatment, both types of carbides exhibit a decreasing trend in percentage within the smaller size range but with increasing proportion within the larger size range. This indicates the coarsening behaviour of primary carbide particles, as evidenced by the increased mean values from 351±202 nm to 443±274 nm for MC carbides and from 396±241 nm to 411±210 nm for M_6C carbides. Additionally, all plots showed the presence of positive skewness in the particle size and shifted to the right after heat treatment with the increase of mean and median values for both types of carbides. The peak of MC carbides shifted to the left after heat treatment with the mode value decreasing slightly from 168 nm to 146 nm, while that of M_6C carbides shifted to the right with the mode value increasing from 168 nm to 202 nm. It is noteworthy that using a single parameter to describe the particle size is not appropriate given the significant difference among the three values of mean, median and mode.

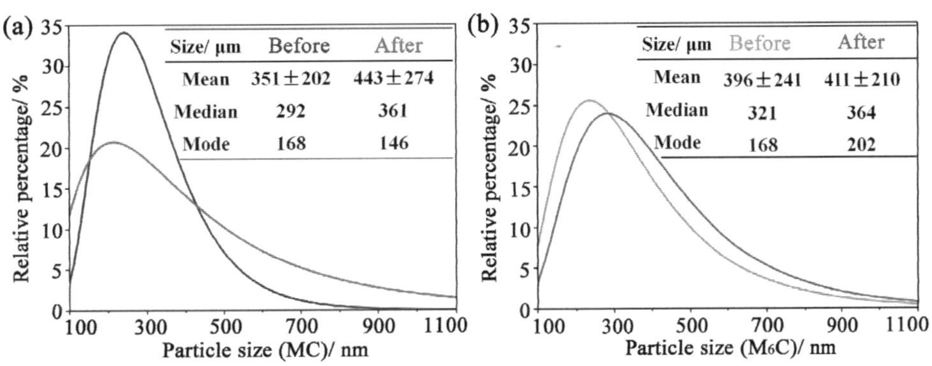

Figure 1. Particle size distribution histograms for (a) MC carbides and (b) M_6C carbides for the conditions of before and after heat treatment.

The statistical analysis of nanoprecipitates in AM S390 steel before and after the heat treatment was obtained from the small-angle scattering results. Compared to the AM sample before heat treatment, the determined mean radius of nanoprecipitates in the AM steel sample before heat treatment was 7.6 nm and the volume fraction was 1.6 %. After heat treatment, the nanoprecipitates exhibited the feature with a volume fraction of 2.3 % and a much smaller size of 1.0 nm. This indicates the precipitation of a larger quantity of tiny particles during heat treatment. These nanoprecipitates dispersed in the matrix as observed in the heat-treated condition contributed to the steel strengthening which will be assessed as follows.

3.2 Tempering behaviour

The phase evolution in AM S390 steel during triple tempering was tracked by in-situ synchrotron diffraction (figure 2) with the dashed lines separating three cycles, each including the heating, holding at 560 °C and cooling stages. Significant changes in phase fraction occurred during cooling to ambient temperature and reheating to the tempering temperature while remaining relatively stable during the 120-minute tempering periods. Focusing only on phase fractions after each tempering cycle, the most notable change in matrix fraction occurred after the first tempering, with an increase in ferrite fraction from 0.70 to 0.75 and a decrease in austenite fraction from 0.07 to 0.005. Carbide fractions remained relatively stable, with MC at about 0.193 and M_6C ranging from 0.046 to 0.053. Subsequent second and third tempering cycles resulted in marginal changes in phase fraction. Multiple tempering, as performed in this study, is commonly employed in conventional steel heat treatment aiming to effectively reduce the retained austenite [22].

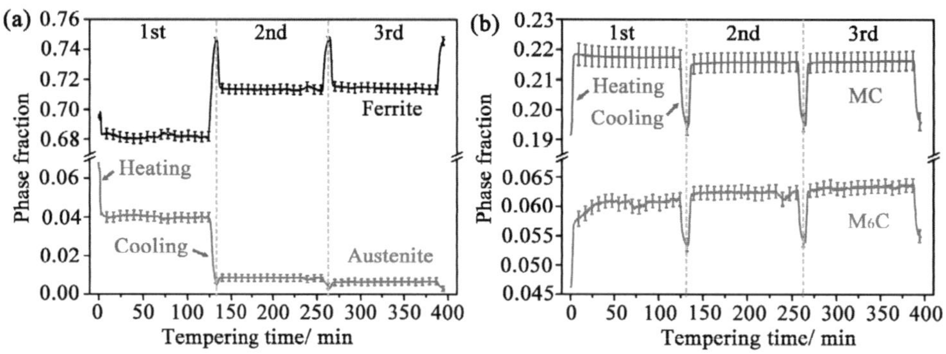

Figure 2. Phase fraction evolution of (a) matrix (ferrite and austenite) and (b) carbides (MC and M_6C) in AM S390 steel during triple tempering processes, each including heating, holding at 560 °C and cooling stages as indicated in the figures.

Representative samples after tempering for 2, 10, 60 and 120 minutes were then characterised to determine the size and volume fraction of primary carbides and nanoprecipitates together with hardness tested to evaluate the mechanical property for understanding the precipitate evolution during the first tempering cycle. The average particle size of MC primary carbides increased from 150 nm to 234 nm during tempering, with significant growth observed to 212 nm after 2 minutes of tempering. Following this, the size slightly increased to 217 nm after 10 minutes and reached 237 nm after 60 minutes. For M_6C carbides, the average size increased significantly within the initial 2 minutes, from 180 nm to 240 nm and slightly decreased to 217 nm after 120-minute tempering. No nanoprecipitates were detected within the initial 60 minutes possibly due to the limited size resolution, until the nanoprecipitates with an average size of about 1.4 nm were detected with a volume fraction of about 1.9 %. Consequently, the hardness of AM steel decreased by 30 HV from its initial value of 860 HV after 2 minutes of tempering, attributed to the coarsening of primary carbides. Subsequently, the hardness increased to 864 HV within a 10-minute tempering period and further increased to the highest hardness of 921 HV after 60-minute tempering potentially due to the strengthening effect of nanoprecipitate formation. Although the primary carbides inevitably coarsened with significantly increased sizes, the nanoprecipitates formed during heat treatment played an important role in steel strengthening.

3.3 Microstructure evolution in high-speed steel during the rapid solidification

The diffraction patterns collected during laser printing over time are presented in figure 3a, illustrating the phase evolution of α′-martensite and γ-austenite during laser melting. The powders underwent heating as the laser beam approached the probed region at around 180 ms, with the disappearance of martensite peaks. The inserted enlarged view of the melting process is highlighted by the red box. It can be seen that the austenite peaks shifted to lower angles due to thermal expansion during heating, disappeared as melting occurred and returned due to thermal contraction during cooling. The repeated peak shifting observed in the enlarged view was caused by the line scanning strategy, wherein the laser beam passed through the probed region multiple times. Martensite peaks reappeared with further cooling as the laser beam moved far away.

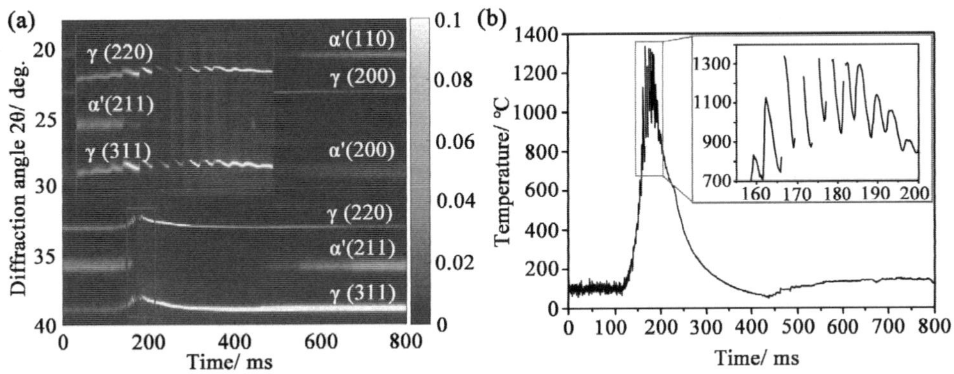

Figure 3. (a) Evolution of diffraction patterns versus time during laser scanning (laser power: 150 W, speed: 500 mm/s, hatch offset: 0.05 mm) with an enlarged view of the melting process. (b) calculated temperature profile based on the lattice change of γ(311).

Based on the peak shifting of γ(311) that can represent bulky strain, the temperature profile was calculated and presented in figure 3b. Austenite was initially tracked after laser melting (at 166 ms) at the calculated temperature of about 1340 °C. The highest cooling rate, determined from the derivative of the temperature profile, was found to be 4.7×10^5 °C/s. Hooper [23] measured the cooling rates in the order of 10^6 °C/s during laser printing of Ti6Al4V. Mazumder et al. [24] reported that the cooling rates for the AM H13 tool by direct metal deposition were in the order of 10^5 °C/s. The cooling rate determined in this study agrees well with the reported cooling rates for AM metals.

During rapid solidification, carbon atoms that are unable to diffuse become trapped within the martensite lattice, thereby strengthening the steel. The carbon content in martensite is found to be related to martensite tetragonality. The diffraction pattern collected after printing, with the split of α′(002) and α′(200) peaks, was fitted using the Pseudo-Voigt function to calculate the martensite lattice parameters. The martensite tetragonality was evaluated by the c/a ratio of 1.021 based on the determination of c=2.91 Å and a=2.85 Å. Thus, the carbon content in martensite was calculated to be 0.47 wt.%. by the equation $c/a=1+0.045\times C$ (wt.%) [25].

4. Conclusion

The microstructure evolution of AM high-speed steels during rapid solidification and heat treatment was investigated to help understand the processing-microstructure-property relationship of AM steel. The growth of primary carbides and the formation of nanoprecipitates were observed after heat treatment. During tempering, retained austenite was effectively optimised with a fraction decrease of 0.065. Nanoprecipitates with an average size of 1.4 nm were tracked to appear after tempering at 560 °C for 60 minutes, achieving a high hardness of 921 HV despite the coarsening of primary carbides within the initial 2-minute tempering. In addition, the rapid cooling rate was evaluated to be 4.7×10^5 °C/s, leading to the formation of martensite and retained austenite. The carbon content was measured from martensite tetragonality, estimated to be 0.47 wt.% in the AM steel after printing.

Acknowledgement

The authors acknowledge financial support from the UK's Engineering and Physical Sciences Research Council EPSRC grant (EP/P025978/1 and EP/R043973/1,) and China Scholarship Council (No. 202006020022). We thank the Advanced Photon Source for awarding synchrotron beamtime (GUP 77461) based on research supported by the U.S. DOE Office of Science-Basic Energy Sciences under Contract No. DE-AC02-06CH11357, Paul Scherrer Institute for awarding synchrotron beamtime (No. 20220055) and the ISIS Neutron and Muon Source for awarding SANS beamtime (No. 2220379).

References

[1] Wang Z, Palmer T A and Beese A M, 2016, Acta Mater., 110, 226-235
[2] Wang X, Zheng B, Yu K, Jiang S, Lavernia E J and Schoenung J M, 2021, Addit. Manuf., 46, 102154
[3] Boes J, Röttger A, Theisen W, Cui C, Uhlenwinkel V, Schulz A, Zoch H W, Stern F, Tenkamp J and Walther F, 2020, Addit. Manuf., 34, 101379
[4] Sun Z et al., 2024, Acta Mater., 263, 119534
[5] Kürnsteiner P, Barriobero-Vila P, Bajaj P, De Geuser F, Wilms M B, Jägle E A and Raabe D, 2023, Addit. Manuf., 73, 103647
[6] Fonseca E B, Gabriel A H G, Araújo L C, Santos P L L, Campo K N and Lopes E S N, 2020, Addit. Manuf., 34, 101250
[7] Geenen K, Röttger A, Feld F and Theisen W, 2019, Addit. Manuf., 28, 585-599
[8] Großwendt F, Röttger A, Strauch A, Chehreh A, Uhlenwinkel V, Fechte-Heinen R, Walther F, Weber S and Theisen W, 2021, Mater. Sci. Eng. A, 827, 142038
[9] Sander J, Hufenbach J, Giebeler L, Wendrock H, Kühn U and Eckert J, 2016, Mater. Design, 89, 335-341
[10] Saewe J, Carstensen N, Kürnsteiner P, Jägle E A and Schleifenbaum J H, 2021, Addit. Manuf., 46, 102125
[11] Saewe J, Gayer C, Vogelpoth A and Schleifenbaum J H, 2019, BHM Berg- und Hüttenmännische Monatshefte, 164, 101-107
[12] Kunz J, Saewe J, Herzog S, Kaletsch A, Schleifenbaum J H and Broeckmann C, 2020, Steel Res. Int., 91, 1900562
[13] Qin S, Herzog S, Kaletsch A and Broeckmann C, 2021, Metals, 11, 596
[14] Botero C, Ramsperger M, Selte A, Åsvik K, Koptyug A, Skoglund P, Roos S, Rännar L-E and Bäckström M, 2020, Steel Res. Int., 91, 1900448
[15] Hocine S, Van Swygenhoven H, Van Petegem S, Chang C S T, Maimaitiyili T, Tinti G, Ferreira Sanchez D, Grolimund D and Casati N, 2020, Materials Today, 34, 30-40
[16] Hocine S, Van Petegem S, Frommherz U, Tinti G, Casati N, Grolimund D and Van Swygenhoven H, 2020, Addit. Manuf., 34, 101194
[17] Zhang H, Peng H, Bagot P A J, Wang Y, Venero D A, Robertson S, El-Fallah G M A M, Guo H, Moody M P and Chen B, 2024, Acta Mater., 270, 119834
[18] Arnold O et al., 2014, Nucl. Instrum. Methods Phys. Res., Sect. A, 764, 156-166
[19] Bressler I, Kohlbrecher J and Thunemann A F, 2015, J. Appl. Crystallogr., 48, 1587-1598
[20] Schneider C A, Rasband W S and Eliceiri K W, 2012, Nat. Methods, 9, 671-675
[21] Sharma H, Parfitt D, Syed A K, Wimpenny D, Muzangaza E, Baxter G and Chen B, 2019, Mater. Sci. Eng. A, 744, 182-194
[22] Bryson W E, 2015, The Heat-Treating Processes Step 4: Tempering, Heat Treatment, 108-111.
[23] Hooper P A, 2018, Addit. Manuf., 22, 548-559
[24] Mazumder J, Schifferer A and Choi J, 1999, Material Research Innovations, 3, 118-131
[25] Bhadeshia H and Honeycombe R, 2017, Steels: microstructure and properties, Butterworth-Heinemann

Challenges in characterizing additively manufactured AlSi10Mg using X-ray Laue micro-beam diffraction

Y Zhang[1*], M Defer[1], W Liu[2], E F F Knipschildt-Okkels[1], J Oddershede[3], A Slyamov[3], F Bachmann[3], E Lauridsen[3], and D Juul Jensen[1]

[1] Department of Civil and Mechanical Engineering, Technical University of Denmark, Kgs. Lyngby, DK-2800, Denmark
[2] Advanced Photon Source, Argonne National Laboratory, Lemont, IL 60439-4800, USA
[3] Xnovo Technology ApS, Køge, DK-4600, Denmark

*E-mail: yubz@dtu.dk

Abstract. Additive manufacturing of metals using for example laser powder bed fusion systems generally results in grains of complex shapes with cellular structure of submicron sizes, accompanied by a high dislocation density. This paper presents preliminary results from characterizing an AlSi10Mg alloy manufactured by L-PBF using non-destructive three-dimensional X-ray Laue micro-beam diffraction. Both synchrotron and laboratory X-ray methods are used. The aim is to identify challenges in characterizing these microstructural features and to propose future research directions to address them.

1. Introduction

Additive manufacturing (AM), for example through laser powder bed fusion (L-PBF), offers distinct advantages in net-shape manufacturing, which eliminates the need for extensive post-processing. However, the inherent rapid cooling rates associated with L-PBF can lead to the development of complex microstructures, predominantly characterized by grains with complex branching morphology and cellular or cellular-dendritic structures. The complexity is further amplified by the laser scanning routines employed during manufacturing, where the patterns of laser movement not only influence the melt pool geometry but also have profound effects on the resulting microstructure and texture [1]. To effectively analyze the evolution of microstructure during L-PBF processes and optimize printing parameters to enhance material properties, comprehensive three-dimensional (3D) microstructure characterization is of key importance.

Though 3D characterization techniques using e.g. serial sectioning and electron microscopy provide a high spatial resolution [2,3], their destructive nature limits their use for dynamic studies. Over the last two to three decades, several non-destructive X-ray based characterization techniques have been developed, offering new insights into the internal microstructures of crystalline materials. These techniques include X-ray Laue micro-beam diffraction (μXRD) [4], three-dimensional X-ray diffraction microscopy (also known as high-energy X-ray diffraction microscopy, HEDM) [5,6], and dark field X-ray microscopy (DFXM) [7]. However, due to the complex microstructures in AM materials the applications of these sophisticated techniques have been limited. The few examples include: Liu et al. recently employed HEDM to monitor the recrystallization behavior of AM 316L stainless steel during in-situ annealing at 1250 °C, mapping only recrystallized grains [8]. Using DFXM, Chen et al. characterized the lattice parameters and orientations of 5 μm-sized cells in direct energy deposited nickel superalloy [9]. Lynch et al.

employed laboratory-based μXRD with a focused polychromatic X-ray beam of ~20 μm to perform 2D mapping of well-recovered grains larger than 25 μm in AM Ti55531 thin-wall specimens [10]. To date, 3D characterization of microstructures in AM samples using μXRD remains unexplored.

This work aims to provide results from the characterization of L-PBF processed AlSi10Mg alloy using μXRD and to identify related challenges. Both synchrotron and cutting-edge laboratory-based μXRD systems were used; the latter was established recently using state-of-the-art Pt-coated glass capillary focusing optics, which enable sub-10 μm focusing of X-rays [11]. AlSi10Mg was chosen as a model material due to its dual-phase structure, which introduces more generality, thereby providing a unique perspective on the future development of μXRD techniques for characterizing metals processed by AM.

2. Experimental

2.1 Materials

An AlSi10Mg sample manufactured using L-PBF was used for this study. Details about L-PBF processing can be found in reference [12]. The samples for the synchrotron experiment were polished with 3 μm diamond suspension and OPU. For the laboratory μXRD experiment, a cylindrical rod with a diameter of 0.5 mm was cut using electron discharge machining followed by electropolishing in two stages: initially in a 10% solution of perchloric acid in ethanol at a voltage of 18 V, and subsequently in a 2% solution at 10 V, to better control the thinning of the tip. Liquid nitrogen was used to cool down the solutions. The obtained sample had a diameter of less than 50 μm within a 2 mm range from the tip. Heat treatment was performed at 520 °C for either 5 min or 2 h using a Nabertherm L 1/12/R6 Mini Muffle Furnace, followed by air cooling.

2.2 Synchrotron μXRD

The synchrotron μXRD experiments were conducted at beam line 34-ID-E at the Advanced Photon Source (APS), Argonne National Laboratory. A focused polychromatic micro-beam with a full width at half maximum of ~0.3 μm was used to determine the crystallographic orientations of the grains by indexing the Laue diffraction patterns. The sample was mounted on a holder inclined at a 45° angle to the incoming beam. A flat panel detector with 2048 × 2048 pixels and a 200-μm pixel size, positioned in a 90° reflection geometry 510.3 mm above the specimen, was utilized to record the Laue diffraction patterns. For 3D characterization, a Pt knife-edge was applied as a differential aperture, scanning along the sample surface at a distance of 250 μm to resolve diffraction patterns from different sample depths penetrated by the micro-beam. An exposure time of 0.25 s and a step size of 0.75 μm were chosen for the 3D mapping. The Laue patterns from each depth were reconstructed and indexed using the LaueGo software, available at beamline 34-ID-E. Further details about the experimental setup can be found in reference [4].

Both Al and Si phases were used to index the depth-resolved diffraction patterns. However, because both phases are cubic, many voxels were indexed as both phases. Considering that Si generally leads to more diffraction spots than Al and exhibits lower internal misorientation, the number of indexed spots and grain internal misorientation were utilized as additional criteria to distinguish between the two phases. Si particles were defined as clusters of more than three neighboring voxels with an internal misorientation below 0.1° from the dataset indexed with the Si phase. Since grain boundary regions typically yield fewer diffraction spots than grain interiors [13], a simple threshold on number of indexed spots proved ineffective. Consequently, clusters of Si were considered as particles if they additionally contained at least one voxel with more than 15 indexed spots. The Al matrix was then identified by excluding the voxels classified as Si particles from the dataset indexed with the Al phase. Voxels with diffraction patterns not indexed as Al but indexed using the Si phase and not assigned to a Si particle were left as not-indexed.

2.3 Laboratory μXRD

A recently established laboratory μXRD system, equipped with a microfocus X-ray source (maximum 160 keV and 10 W) and Pt-coated twin paraboloidal X-ray focusing optics was used to characterize the needle AlSi10Mg sample. Details of the optics characteristics can be found in reference [11]. A schematic of the setup is shown in figure 1. In this setup, a motorized five-axis stage was used for optics alignment. A 400 μm upstream aperture and an additional 2 mm aperture downstream of the optics were used to eliminate stray radiation. A photon counting detector (Advacam s.r.o.) was used to enhance the signal-to-noise ratio and thus reduce the exposure time. This detector, consisting of 256×256 pixels with a physical pixel size of 55 μm, was mounted in a transmission geometry 15.36 mm from the specimen. A 2 mm thick tungsten beamstop was mounted in the center of the detector to block the direct beam and improve the diffracted signal contrast. The detector geometry was calibrated using a single crystal Si of 100 μm. For the Si single crystal, an exposure time of 1 s was used. At this first attempt, a single diffraction image was taken from the as-printed sample with an exposure time of 20 s.

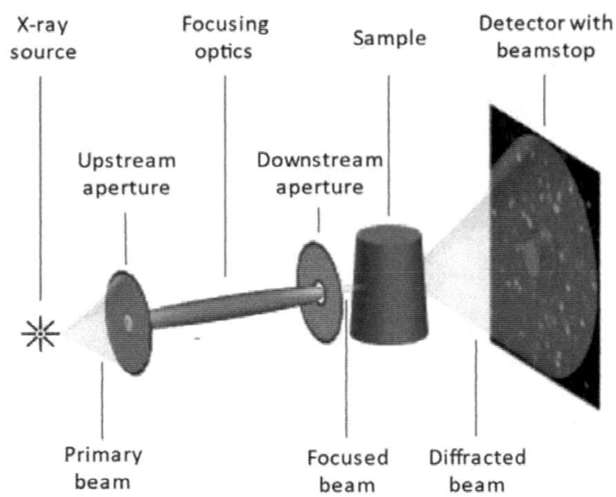

Figure 1. Schematic of the laboratory-based μXRD setup.

3. Results and discussion

3.1 Synchrotron μXRD results

The as-printed sample exhibits a typical cellular structure with cell sizes ranging from 0.55 to 0.7 μm [14]. To determine the feasibility of 3D volume mapping of the samples, diffraction patterns were collected without a wire scan, representing an integrated pattern along the whole depth illuminated by the focused beam. The results for the two heat-treated samples are shown in figure 2. It is seen that after heat treatment at 520 °C for 5 min, the diffraction spots are still quite streaking, whereas reasonably sharp spots are observed for the sample annealed for 2h. Based on this screening, it was concluded that the diffraction signals from the as-printed sample, which likely contains an even higher density of dislocations, would be weaker than those shown in figure 2(a); thus, it is not suitable for the time-consuming volume mapping. Consequently, only the sample annealed for 2h was characterized by full 3D volume mapping.

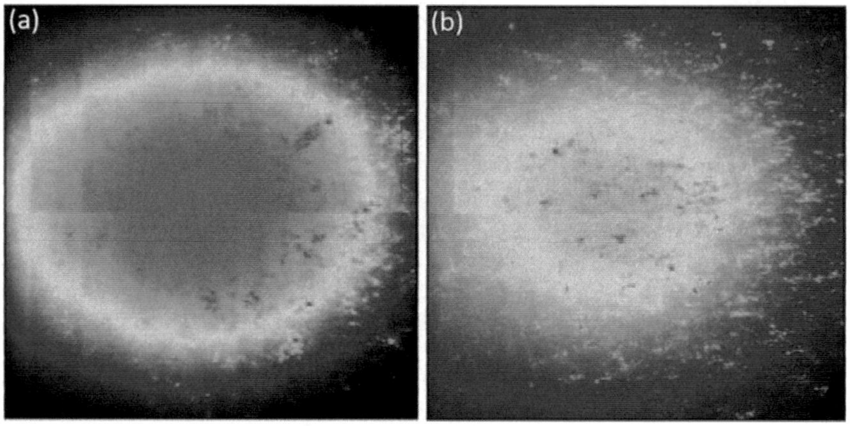

Figure 2. Diffraction patterns collected for the two annealed samples using a synchrotron micro-beam and a 90° reflection detector geometry: (a) 520°C for 5 min, and (b) 520 °C for 2h. Only screenshots of these images were saved; hence, the actual intensities could not be quantified. However, the precise intensity values are not critical for evaluating the sample state.

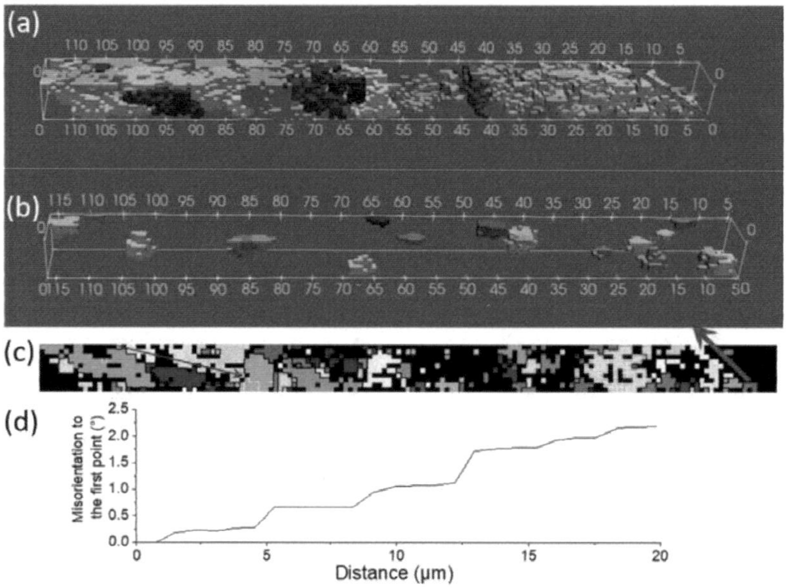

Figure 3. Synchrotron μXRD results for AM AlSi10Mg heat-treated at 520 °C for 2h. 3D microstructure of the (a) Al matrix and (b) Si particles. (c) Cross section of the Al matrix and (d) misorientation profile along the thin red line marked in (c). The thick arrows mark the building direction. Units in (a) and (b) are μm.

The 3D volumes of the Al and Si phases are shown in figure 3. It is seen that in agreement with previous EBSD results [14], the Al matrix has not recrystallized even after annealing at 520°C for 2h but is only extensively recovered. The misorientation accumulated within a grain can be up to 2° (see figure 3(d)). The Si particles are distributed reasonably uniformly within the mapped volume, with sizes ranging from 1.5 to 5 μm. Compared to the results characterized by optical microscopy and nano-tomography in reference [12], it is evident that many smaller Si particles

IOP Conf. Series: Materials Science and Engineering **1310** (2024) 012023 doi:10.1088/1757-899X/1310/1/012023

are missing in the present measurements. This is likely due to the relatively large step size used and the weak diffraction signals from small Si particles. The fraction of indexed voxels for the whole volume is about 50%, demonstrating the challenges in full 3D microstructural mapping for this material. Using a longer exposure time and smaller step size may result in a higher indexing rate, though the maximum allowable exposure time for the present experimental setup is 2 s.

3.2 Laboratory μXRD results

A diffraction pattern of the as-printed sample is shown in figure 4, alongside a diffraction pattern from a single-crystal Si for reference. Unlike the circular spots from Si, streaking diffraction spots are observed for the as-printed AlSi10Mg sample, attributable to the orientation variations between cells within individual grains (see also figure 3(d)). Note that the different appearance of the diffraction spots in figure 4(a) compared to those in figure 2 is due to the different detector geometries used. A large number of diffraction spots are visible in figure 4(a) from multiple illuminated grains. The intensity variation between spots typically results from the combined effects of grain size, X-ray energy within the spectrum range, lattice structure factor, and the detector quantum efficiency [15]. While the orientation of the single-crystal Si can be readily indexed, no orientation was indexed for the as-printed AlSi10Mg sample using a recently developed dictionary-based branch-and-bound indexing algorithm [16], which was designed for indexing multiple orientations in a single Laue diffraction pattern. This is likely due to the streaking of the spots from this sample, which introduces large errors in determining the center of mass of the spots. However, it cannot be ruled out that the number of available spots for each illuminated grain is not sufficient for indexing (i.e. less than 4), possibly because of the comparatively weak beam. A longer exposure time should be considered to possibly enable a successful orientation indexing.

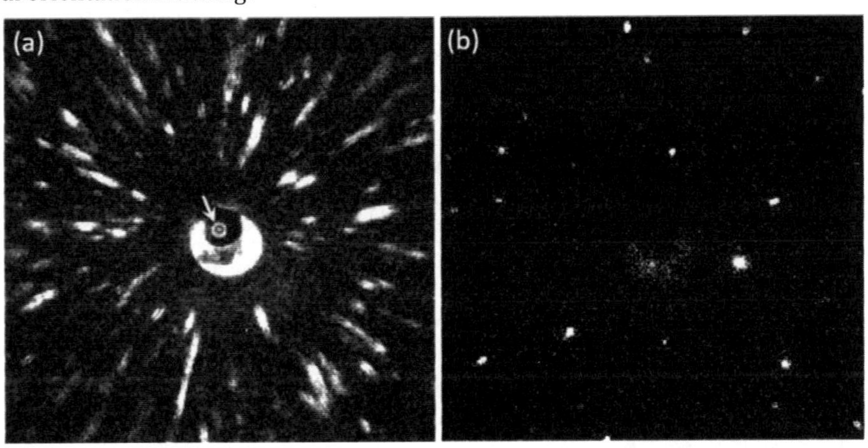

Figure 4. Laboratory μXRD diffraction images from (a) as-printed AlSi10Mg alloy and (b) Si single crystal. The ring feature marked by the yellow arrow in (a) is the footprint of the direct focused beam penetrating the beamstop, while the large bright area at the central region is from a leakage of the direct beam through the pinhole in the shielding. The higher background level in (a) than (b) is the result of longer exposure time, 20 s versus 1 s.

4. Summary and outlook

The microstructure of an AlSi10Mg alloy processed by L-PBF was characterized using both synchrotron and laboratory μXRD. Due to the fine scale of the cellular structure within individual grains, characterizing as-printed materials that contain a high density of dislocations remains challenging, even using X-rays from a third-generation synchrotron source. Only after extended recovery by annealing at 520°C for 2 h the Al matrix and some comparatively large Si particles can be characterized with reasonable success.

Following the recent X-ray source upgrade at APS, the X-ray flux is expected to be enhanced by a factor of 10-100, while the beam size is reduced by the same factor. The resulting intense, small beam of 50 nm, in combination with an advanced photon-counting detector, will likely allow fast 3D mapping of even the as-printed materials, providing a unique possibility for studying microstructural evolutions. A combination of μXRD with nano-tomography as described in reference [12] is suggested for a more precise separation of the Al and Si phases.

For laboratory μXRD, future directions include advancements in software development that can accommodate orientation spreads, and data collection routines designed to include diffraction spots from multiple incident angles [17], the combination of which may enable reasonable microstructural characterization of AM metals. Despite the comparatively long exposure time of 20 s and large beam size of ~10 μm, the unlimited access to such an instrument makes laboratory μXRD a compelling tool for 3D microstructural characterization, once its full potential is realized.

Acknowledgements

Use of the Advanced Photon Source was supported by the U. S. Department of Energy, Office of Science, Office of Basic Energy Sciences, under Contract No. DE-AC02-06CH11357.

Funding

Y Zhang, M Defer, and D Juul Jensen acknowledge the support by a research grant (VIL54495) from the VILLUM FONDEN.

References

[1] Aboulkhair N T, Simonelli M, Parry L, Ashcroft I, Tuck C and Hague R 2019 *Prog. Mater. Sci.* **106** 100578
[2] Rowenhorst D J, Nguyen L, Murphy-Leonard A D and Fonda R W 2020 *Curr. Opin. Solid State Mater. Sci.* **24** 100819
[3] Fonda R W and Rowenhorst D J 2022 *IOP Conf. Ser. Mater. Sci. Eng.* **1249** 012007
[4] Larson B C and Levine L E 2013 *J. Appl. Crystallogr.* **46** 153–64
[5] Juul Jensen D and Poulsen H F 2012 *Mater. Charact.* **72** 1–7
[6] Li S F and Suter R M 2013 *J. Appl. Crystallogr.* **46** 512–24
[7] Simons H, King A, Ludwig W, Detlefs C, Pantleon W, Schmidt S, Snigireva I, Snigirev A and Poulsen H F 2015 *Nat. Commun.* **6** 1–6
[8] Liu H, Zhang Y, Wilkin M, Park J-S, Kenesei P, Rollett A D and Suter R M 2022 *IOP Conf. Ser. Mater. Sci. Eng.* **1249** 012054
[9] Chen Y et al. 2023 *Scr. Mater.* **234** 115579
[10] Lynch P A, Stevenson W, Fox D, Ghaderi A, Xie L and Barnett M R 2019 *SSRN Electron. J.* 0–27
[11] Seret A, Kehres J, Gundlach C, Olsen U L, Poulsen F H, Juul Jensen D, Cordier M, Stripe B, Yun W and Zhang Y 2023 *Opt. Express* **31** 38840-38853
[12] Defer M, Dasgupta S, Shahini A J, Xiao X, Juul Jensen D and Zhang Y 2024 *IOP Conf. Ser. Mater. Sci. Eng.* **this volume** submitted
[13] Lindkvist A, Liu W, Juul Jensen D and Zhang Y 2023 *Mater. Res. Lett.* **11** 942–8
[14] Defer M 2023 *Master thesis* Technical University of Denmark
[15] Fang H, Juul Jensen D and Zhang Y 2020 *Acta Crystallogr. Sect. A* **76A** 652–663
[16] Seret A, Gao W, Juul Jensen D, Godfrey A and Zhang Y 2022 *J. Appl. Crystallogr.* **55** 1085–1096
[17] Zhang Y and Juul Jensen D 2020 WO2022013127

IOP Conf. Series: Materials Science and Engineering **1310** (2024) 012024 doi:10.1088/1757-899X/1310/1/012024

The effect of processing parameters on dislocation density and tensile properties in laser powder bed fusion 316L

M Schreiber[1], C Brice[2], K Findley[1], J Klemm-Toole[1], and J Gockel[2,*]
[1]Department of Metallurgy and Materials Engineering, Colorado School of Mines, Golden, Colorado, USA
[2]Department of Mechanical Engineering, Colorado School of Mines, Golden, Colorado, USA

*E-mail: joygockel@mines.edu

Abstract. The processing-structure-properties relationships in laser beam powder bed fusion (PBF-LB) additive manufacturing (AM) are complex with multiple aspects of the processing impacting the microstructure and mechanical properties. Though, the influences of process parameters on strengthening mechanisms are less clear. In this work, laser power, scanning velocity, and hatch spacing were varied to promote PBF-LB 316L microstructures with distinctive thermal histories to vary microstructures and tensile properties. Tensile data were collected for over 100 different processing parameters on a single PBF-LB platform. Across the process parameter matrix yield strength, work hardening behavior, and ductility varied considerably. In this work, the effect of process parameters on initial dislocation density was studied. By quantifying the dislocation density with X-ray diffraction and line profile analysis, a relationship between processing parameters and initial dislocation densities was established. The contribution of dislocation density and other strengthening mechanisms to the yield strength is discussed.

1. Introduction

Additive manufacturing (AM), specifically laser beam powder bed fusion (PBF-LB), is a processing technique that enables the fabrication of geometrically complex components. Studies that correlate the effects of process parameters (e.g. laser power, scanning speed, hatch spacing, scan strategy, etc.) to porosity, microstructure defects, and mechanical properties are readily available [1-5].

The generation of a dislocation structure in 316L during PBF-LB has been explained in detail using high resolution characterization and modelling [6-9]. The dislocation structure develops from thermal distortions driven by high temperature gradients and resultant high thermal strain rates in PBF-LB 316L [6]. Increasing the part volume, decreasing substrate temperature, or increasing part to melt pool volume ratio can increase cooling rates, constraint, and subsequently the dislocation density [6]. Cyclic heating and cooling in subsequent deposition passes decrease the dislocation density via recovery mechanisms [9, 10]. A lower initial dislocation density will decrease strength and increase work hardening behavior in post-build testing [11-13].

In the following work, differences in tensile data across a matrix of 316L manufactured by PBF-LB 316L with different process parameters were explained in terms of their as-built dislocation densities. Dislocation densities were calculated by analyzing peak widths of X-ray data using two line profile analysis (LPA) techniques. Electron backscattered diffraction (EBSD) grain boundary analysis was utilized to support conclusions drawn from tensile data.

2. Materials, methods, and models

2.1 Material and part fabrication

316L round tensile bars were directly prepared with a 3D Systems DMP Flex 350 PBF-LB platform in an argon atmosphere. Each tensile bar was built with unique parameters across four builds with laser powers ranging from 110-400 W, scanning speeds ranging from 700-1500 mm/s, hatch spacings from 50-150 μm, layer heights ranging from 30-60 μm, and a beam spot size of ~ 84 μm. No thermal post-processing or machining was performed on the tensile bars. Four samples were down selected based on tensile results and their processing parameters and volumetric energy densities are described in table 1. Volumetric energy density was calculated as the power divided by the product of the scanning speed, hatch spacing, and layer height.

Table 1. Representative 316L print parameters to generate different microstructures.

Sample	Power, P (W)	Scanning Speed, v (mm/s)	Hatch Spacing (μm)	Layer Height (μm)	Volumetric Energy Density (J/mm^3)
A	362	849	115	60	62
B	114	997	91	60	26
C	255	1100	100	60	39
D	400	700	50	60	190

2.2 Tensile property testing

Tensile tests were performed using an MTS Landmark servo-hydraulic load frame with a 25 mm gauge length extensometer. Tests were performed with a strain rate of 0.005 s^{-1}. Tensile coupon geometry followed ASTM E8/E8M-16a guidelines, with the exception of the rough as-built surface, for circular samples with a nominal diameter of 6 mm and gauge length of 25 mm.

2.3 Material characterization

Metallographic coupons were sectioned out of the tensile grip ends after testing. Standard metallographic procedure was followed to prepare coupons for analysis. EBSD was performed with 0.15-0.6 μm step sizes depending on magnification (500-2000x). EBSD was performed to measure average misorientation between pixels in regions of the as-built 316L. High angle grain boundaries (HAGB) were separated by regions of the microstructure with greater than 15° misorientation and low angle grain boundaries (LAGB) were separated by regions of the microstructure between 1 and 15° misorientation. Kernel average misorientation (KAM) analysis was performed on all micrographs using a constant kernel size of 6 μm. No image post-processing was performed to remove or clean-up EBSD data. XRD profiles were collected on metallographic coupons using a Cu Kα anode (λ = 0.154 nm), a step size of 0.02°, dwell time of 20 s, and a scan range of 40-100°. Dislocation density was calculated with Convolutional Multiple Whole Profile (CMWP) and the modified Williamson Hall (mWH) line profile analyses (LPA) [14-17].

3. Results & discussion

3.1 Tensile properties

The engineering stress strain data for the entire sample matrix are plotted as gray lines in figure 1 and the tensile properties of a representative subset are shown in table 2. Nominal tensile properties of PBF-LB 316L processed by 3D Systems using optimal parameters were also included for comparison. The tensile data demonstrated that changing laser power, scanning velocity, and hatch spacing will influence the yield strength, total elongation, and work hardening. A 5-fold variation in yield strength, measured with the 0.2% offset method, and 10-fold variation in total elongation were observed. Most of the stress strain curves exhibited a similar elastic to plastic transition, albeit with different yield strengths, followed by low work hardening, then necking to failure or abrupt failure. Similar engineering stress-strain curve shapes have been reported in tensile samples with artificially introduced porosity [3] which suggested that porosity covering 16% of the cross-sectional diameter will influence total elongation, yield and ultimate tensile strength. Interestingly, there is one tensile sample (Sample D) with lower strength values and nominally high total elongation (45%) that exhibited atypically high work hardening behavior. Low yield strength, high work hardening behavior, and high total elongation would suggest that the process parameters influenced microstructure features, excluding porosity.

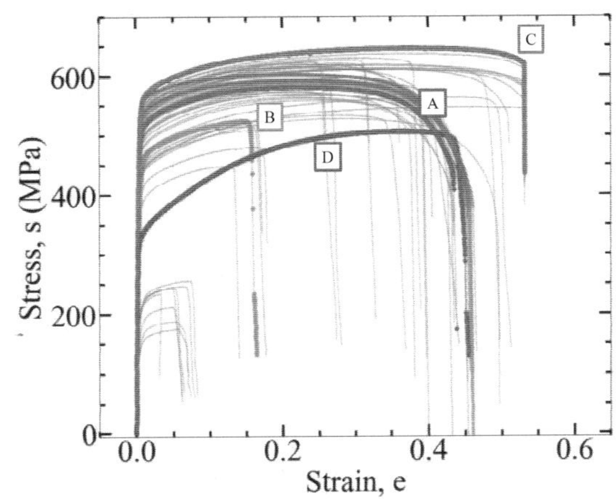

Figure 1. Engineering stress-strain curves for PBF-LB 316L, each processed with unique parameters. Samples selected for metallographic characterization are colored and labelled.

Table 2. Summarized tensile properties and porosity for PBF-LB 316L.

Sample	Yield Strength (MPa)	Ultimate Tensile Strength (MPa)	Total Elongation (%)	Area Percent Porosity (%)
Nominal	440±20	570±30	49±5	NA
A	490	582	44	0.021±0.01
B	405	524	16	6.7±3
C	505	646	53	0.039±0.02
D	325	506	45	1.57±0.1

The area percent porosity of two different surfaces within the grip end were also included in table 2. Due to the 7 % area porosity in Sample B, it was expected that the load carrying capacity not the yield strength was lower than the other samples. In fact, when adjusting the cross-sectional diameter by 7 % to calculate engineering stress of Sample B, the tensile curve shifts to a region between Sample A and Sample C. The parameters used to process Sample D (table 1) also generated moderate porosity, but the present porosity did not appear to influence total elongation. Dislocation densities and other strengthening mechanisms are discussed to conclude whether porosity and/or other microstructure features are responsible for the decreased yield strength of Sample D.

3.2 Dislocation density

The total dislocation densities from CMWP and mWH are shown in table 3. The dislocation densities determined with both LPA methods suggest that Samples A, B, and C have similar dislocation densities around 3×10^{15} (CMWP) or 6×10^{14} m^{-2} (mWH). The changes to processing parameters for Samples A, B, and C likely influence the strengthening contribution from other microstructure features. A lower laser power and faster scanning speed are expected to decrease melt pool size and consequently grain size [5, 18], increasing yield strength. Solid solution strengthening was estimated with empirical formulae available in the literature [19] and may be responsible for a portion of the yield strength difference. Composition variation in Sample B, specifically 0.30 and 0.15 weight percent differences in manganese and silicon, after melting and solidification can change strengthening by 5-30 MPa based on optical emission spectroscopy data collected according to ASTM A751-21 for Samples A, B, and C. Precipitation strengthening is not expected for PBF-LB 316L but nano-oxides observed with transmission electron microscopy (TEM) are reported to influence yield strength after annealing [20]. Therefore, the tensile behavior of Sample B was likely influenced by a finer grain size, composition variation, and the diminished load carrying capacity due to 7 % area porosity.

Table 3. Summary of XRD (dislocation density) and EBSD (strain and boundary density) results.

Sample	Dislocation Density, CMWP (10^{14} m^{-2})	Dislocation Density, mWH (10^{14} m^{-2})	KAM Peak Mode (°)	HAGB/area (mm/mm^2)
A	28	5.8	0.53±0.1	12±1
B	27	10	0.5±0.09	54±10[a]
C	21	7.22	0.57±0.07	30±12
D	16	2.2	0.59±0.1	19±10[a]

[a] Inflated due to porosity.

LPA determined that Sample D had a lower dislocation density of 2×10^{15} (CMWP) or 2×10^{14} m^{-2} (mWH). The lower dislocation density of Sample D agrees with the yield strength trend of figure 1, table 2, and the high work hardening rate. The relatively low dislocation density of Sample D suggests that the mechanisms responsible for forming dislocations and those responsible for annihilating dislocations during thermal cycling of the PBF process changed with the process parameters of Sample D. Considering that part volume and substrate temperature were similar across all samples and builds, it is unlikely that either variable would influence initial dislocation density values, as identified in the literature [6]. Solid solution strengthening decreased in Sample D which is believed to be driven by nitrogen content differences in the samples [21]. Nitrogen content, measured with combustion according to ASTM E1086-14, ranged from 0.074 and 0.1 weight percent across Samples A-D. The 0.074

weight percent nitrogen content in Sample D was estimated to decrease yield strength by up to 25 MPa compared to the other samples [19]. The higher laser power and lower scanning speed should increase melt pool size and grain size in the as-built microstructures [18]. The low yield strength of Sample D was attributed to larger grains, differences in composition (largely the nitrogen content), and lower dislocation density. Based on estimates of the cooling rate from primary dendrite arm spacing (PDAS) [22], the cooling rates experienced by Sample D were lower than the other samples which would influence strain gradients and the generation of dislocations during processing [6].

3.3 Electron backscattered diffraction

EBSD was performed to quantify the strain with KAM and HAGB line length per area to understand the influence of processing parameters on the strain and boundary density. Grain boundary maps (HAGB in black, LAGB in green) are shown for the PBF-LB 316L in figure 2 and a summary of KAM peak mode and HAGB line lengths per area are summarized in table 3. KAM peak mode has been shown to increase with increasing plastic strain [23, 24]. The KAM peak modes were statistically similar within the sample matrix, therefore EBSD was not able to distinguish differences in the present microstructures. Boundary line length per area trends were assumed to be correlated to grain size, i.e. a larger line length per area corresponded to a smaller grain size. The HAGB/area for Sample B was measured to be 54 mm/mm^2, the largest in the matrix, which agrees with the qualitative strengthening analysis above. Sample A and Sample D had similar HAGB/area at ~20 mm/mm^2 which suggests larger grains in these two as-built microstructures. Qualitatively, the strengthening contribution from grain boundaries in the finer grain microstructures (Samples B and C) may be 2-3× larger than the strengthening contribution of grain boundaries in the coarser microstructures (Samples A and D). Quantitative estimates of the magnitude of strengthening will depend on the Hall-Petch locking parameter which has been shown to be a function of grain size [10, 25].

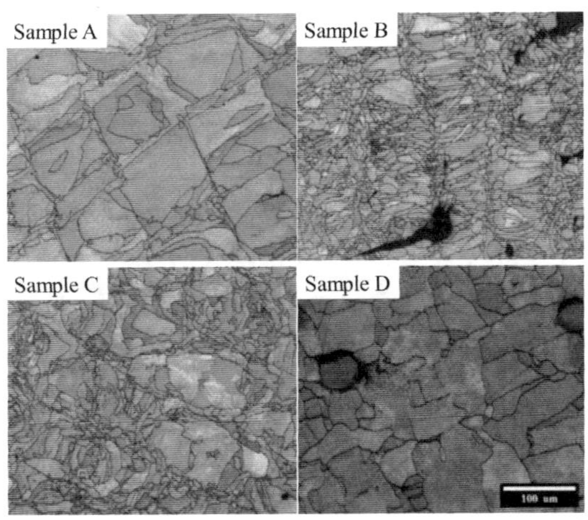

Figure 2. EBSD high (black) and low (green) angle grain boundary maps for PBF-LB 316L.

4. Conclusions

Dislocation densities were calculated with line profile analysis to explain the variation in tensile behavior of PBF-LB 316L manufactured with a range of processing parameters. Despite differences in yielding behavior by up to 100 MPa, dislocation densities of Samples A, B, and C were similar. The decreased yield strength of Sample B was attributed to a decreased load carrying capacity from porosity measured from 2D cross-sections and to a lesser extent composition variation and grain size. An EBSD boundary analysis confirmed expected trends in grain size for Sample B. Sample D, processed with the highest laser power and slowest scanning speed, exhibited the lowest yield strength, dislocation density, and highest work hardening rate in the present work. The lower dislocation density was attributed to a change in solidification conditions from higher laser power, lower scanning speed, and low hatch spacing. Therefore, PBF-LB 316L manufactured with higher laser power and lower scanning speed will generate microstructures with lower initial dislocation densities.

5. Acknowledgements

DISTRIBUTION STATEMENT A. Approved for public release: distribution is unlimited. OPSEC#8605. This material is based on research sponsored by the U.S. Army DEVCOM Ground Vehicle Systems Center and Air Force Research Laboratory under agreement number FA8650-20-2-5700.
Some of the work was performed in the following core facility, which is a part of Colorado School of Mines' Shared Instrumentation Facility: Electron Microscopy (RRID:SCR_022048) and Mechanical Testing (RRID:SCR_022051).
The authors would like to recognize Alex Gonzalez for conceptualization of the design of experiments matrix and Andrew Jones for tensile testing the sample matrix.

6. References

[1] Fayazfar H, Salarian M, Rogalsky A, Sarker D, Russo P, Paserin V and Toyserkani E 2018 Mater. Des. **144** 98-128
[2] Afkhami S, Dabiri M, Piili H and Bjork T 2020 Mater. Sci. Eng. A **802** 1-21
[3] Wilson-Heid A, Novak T and Beese A 2019 Exp. Mech. **59** 793-804
[4] Saeidi K, Gao X, Lofaj F, Kvetková L and Shen Z 2015 J. Alloys Compd. **633** 463-469
[5] Evans R, Walker J, Middendorf J and Gockel J 2020 Metall. Mater. Trans. A **51** 4123-4129
[6] Bertsch K, Meric de Bellefon G, Kuehl B and Thoma D 2020 Acta Mater. **199** 19-33
[7] Voisin T, Forien J, Perron A, Aubry S, Bertin N, Samanta A, Baker A and Wang Y 2021 Acta Mater. **203** 1-17
[8] Riabov D, Leicht A, Ahlström J and Hryha E 2021 Mater. Sci. Eng. A **822** 1-10
[9] Sabzi H and Rivera-Díaz-del-Castillo P 2020 Mater. Des. **195** 1-11
[10] Sabzi H, Hernandez-Nava E, Li X, Fu H, San-Martín D and Rivera-Díaz-del-Castillo P 2021 Mater. Des. **212** 1-17
[11] Carrington W, Hale K and McLean D 1960 Proc. Roy. Soc. A **259** 203-227
[12] Hasegawa T and Kocks U 1979 Acta Metall. **27** 1705-1716
[13] Humphreys F J and Hatherly M 2004 Recrystallization and Related Annealing Phenomena 2nd ed. (Oxford: Elsevier)
[14] Ribárik G and Ungár T 2010 Mater. Sci. Eng. A **528** 112-121
[15] Ribárik G, Jóni B and Ungár T 2020 Cryst. **10** 1-19
[16] Ungár T 2001 Mater. Sci. Eng. A **309-310** 14-22
[17] Shamsujjoha M, Agnew S, Fitz-Gerald J, Moore W and Newman T 2018 Metall. Mater. Trans. A **49** 3011-3027
[18] Gockel J, Beuth J and Taminger K 2014 Addit. Manuf. **1-4** 119-126
[19] Eliasson J and Sandström R 2000 Steel Res. **71** 249-254
[20] Chen S, Ma G, Wu G, Godfrey A, Huang T and Huang X 2022 Mater. Sci. Eng. A **832** 1-9
[21] Smith T, Sugar J, San Marchi C and Schoenung J 2019 Acta Mater. **164** 728-740
[22] Elmer J, Allen S and Eager T 1989 Metall. Trans. A **20A** 2117-2131
[23] Wright S, Nowell M and Field D 2011 Microsc. Microanal. **17** 316-329
[24] Im Y, Kim K, Jung K, Lee Y and Song K 2019 Metall. Mater. Trans. A **50A** 2014-2021
[25] Singh K, Sangal S and Murty G 2002 Mater. Sci. Tech. **18** 165-172

Understanding fatigue crack propagation pathways in Additively Manufactured AlSi10Mg

S Rangaraj[1*], S S I Ahmed[1], A Davis[1], P J Withers[1] and A Gholinia[1]

[1] Henry Royce Institute, Department of Materials, The University of Manchester, Manchester M13 9PL, United Kingdom

*E-mail: sethupathi.rangaraj@manchester.ac.uk

Abstract. Alloys produced through additive manufacturing (AM) offer substantial advantages, particularly in controlling material utilisation and precisely manipulating processing parameters, resulting in finely tuned material properties. However, the grain structure of AM material is typically complex, influenced by factors such as solidification dynamics, processing parameters, thermal gradients, and residual stress. Fatigue analysis shows considerable scatter due to entrained defects which limits their use as structural components. In this study, fatigue-failed samples from selective laser melted (SLM) AlSi10Mg alloy, oriented horizontal and vertical to the build direction were analysed to understand crack propagation paths. Here X-ray Computed Tomography (CT) was used to examine internal porosity from which fatigue cracks initiate, complemented by electron backscattered diffraction (EBSD) mapping. This enabled us to recognize the crucial role of the complex grain microstructure in controlling fatigue crack propagation.

1. Introduction

Selective Laser Melting is a powder-based metal additive manufacturing technique with potential applications across various industries, from biomedical to aerospace sectors. The technique is popular for its ability to manufacture intricate geometries with high materials utilisation compared to other traditional subtractive manufacturing methods. SLM can produce complex geometries offering drastic weight reduction while ensuring structural integrity [1]. The performance of the part is dependent on the complex anisotropic 3D microstructure, which can be controlled by the processing parameters (heat input and scanning configuration), build orientation, powder particle size, and post-processing treatments.

Aluminium alloys, particularly AlSi10Mg, are extensively employed for their high strength-to-weight ratio, corrosion resistance, and weldability. The specific strength of AlSi10Mg is enhanced through precipitation hardening of the metastable phase. The rapid solidification conditions inherent in SLM of AlSi10Mg powders result in directional solidification, leading to an anisotropic microstructure. This structure arises from the successive formation of melt-pools, within which complex grain morphologies are observed. This consists of α-Al cells surrounded by a eutectic-Si fibrous network [2]. During the solidification of AlSi10Mg various manufacturing defects, complex variations in grain size, and the arrangement of secondary phase particles can emerge, as a result of the complex solidification cycles. High thermal radiation and gas absorption in the surrounding environment can induce defects, notably gas porosity [3]. Additionally, insufficient energy input may lead to incomplete bonding between material layers, resulting in lack of fusion (LOF) defects.

These defects significantly compromise fatigue resistance, serving as preferred sites for crack initiation, particularly when they are irregularly shaped and positioned near the surface [4].

The aim of this paper is to examine the impact of the complex 3D defect and grain microstructure, on the initiation and propagation of fatigue cracks. To achieve this, high-resolution X-ray computed tomography (CT) is employed to characterise the entrained manufacturing defects in 3D. Subsequently, Electron Backscatter Diffraction (EBSD) mapping is performed to better understand the influence of the anisotropic microstructure on subsequent fatigue crack propagation, depending on whether the crack is growing parallel or perpendicular to the build direction.

2. Methodology

2.1 Sample Fabrication

The samples were resourced from Southwest Jiaotong University from pre-alloyed powders with particle sizes ranging from 15-40 μm through Selective Laser Melting. The AlSi10Mg composed of around 88wt% of Al, 9-10% of Si, 0.4-0.6% of Mg, and other minor elements. The fabrication process is described elsewhere [5] and was carried out on a BLT S310SLM using a laser beam power of 360-400W, a layer thickness of 0.05mm, a laser scan speed between 1200-1500 mm/s, and a laser scan-line spacing between 0.13-0.16mm, all within an argon atmosphere. To achieve optimal powder melting, each deposited track was laid down at 67° relative to the previous layer. Residual stresses generated during fabrication were relieved by treatment at a temperature of 300°C for 2 hours.

2.2 Fatigue Testing

Cylindrical fatigue specimens, following an M12 standard geometry, were extracted from the fabricated block, with a gauge diameter of 5mm. These fatigue specimens were cut with their tensile axis (TA) perpendicular and parallel to the build direction (BD). Fatigue testing of the samples was conducted using an Instron 8802K7237 100kN S1/2 T350 machine, with a maximum cyclic load of 140MPa and 170MPa in perpendicular and parallel to BD, respectively. The testing was carried out at a frequency of 40 Hz and an R ratio of 0.1.

2.3 X-ray CT and EBSD setup

Manufacturing defects were characterized by X-ray CT at the National X-ray Computed Tomography facility, utilizing the Nikon High Flux Bay scanner using an accelerating voltage of 100kV and current of 80A. A voxel size of 15μm and a spatial resolution of ~20μm was achieved. The CT scans were analysed using AVIZO 2019.3 software and the voids segmented manually using a grey-scale thresholding approach.

Crack growth within the grain microstructure was investigated by EBSD analysis on the TESCAN Mira3 LC FEG and Thermo Scientific Apreo 2 Scanning Electron Microscopes, equipped with Oxford Instruments' Symmetry EBSD detectors. The fatigue samples were cut in cross-section along the slicing plane. They were mounted with phenolic edge retention resin without compromising the fracture surface grains. The samples were polished with standard mechanical polishing method down to the oxide polishing suspension (OPS), to prepare the sample damage free for the EBSD analysis. For optical microscopy samples were etched in Keller's reagent. Mapping was conducted at an accelerating voltage of 20kV and varying currents between 13nA and 53nA. Oxford Instruments' AZtecCrystal software was used to post-process the data.

IOP Conf. Series: Materials Science and Engineering **1310** (2024) 012025 doi:10.1088/1757-899X/1310/1/012025

3. Results

3.1 As-built Microstructure

The as-built (Stress relieved) microstructure is shown in figure 1. Figure 1a shows the microstructure in the direction perpendicular to the BD where the solidification of the tracks is evident, as each layer is laid down at 67° to the previous layer. The microstructure in the direction parallel to BD in figure1b shows the typical fish-scale morphology caused by the melt pool boundaries, with a width of approximately 150-180 μm and a depth of roughly 100-120 μm.

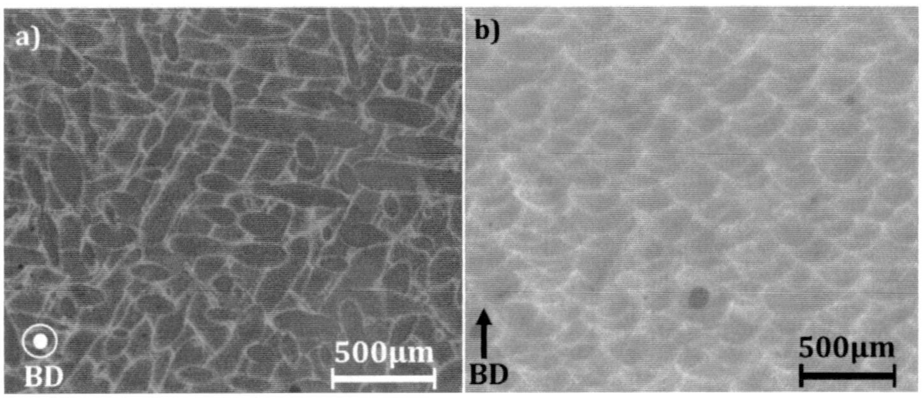

Figure 1. Optical images of the melt pool morphology (a) perpendicular and b) parallel to the BD.

The fish scale appearance is evident in the sample parallel to BD in the EBSD map in figure 2b, where near the melt pool boundary (MPB), the grains are fine equiaxed (<5 μm), and randomly oriented, as indicated by their random colours and depicted with dotted lines. By contrast, the grains within each fish-scale interior are larger, longer (~150mm), and columnar that grow away from the MPB. The columnar grains grow preferentially along the <001> direction [6,7] and are seen in red where they are aligned in the BD. For the sample perpendicular to BD (figure 2a) the larger columnar grains are viewed edge-on and grains delineate the laser scan tracks and MPB as shown with dotted lines, as can also be seen in figure 1a.

3.2 *Defect distribution*

There are two types of defects: small spherical gas pores and larger pancake shaped LOF defects. These manufacturing defects serve as stress concentration sites, particularly under fatigue loading conditions, and can significantly influence the initiation and propagation of cracks within the samples [5]. X-ray CT imaging reveals a population of defects of various sizes, as evident from the defect distributions shown in figure 3. One of the most important factors determining the fatigue performance of these samples, is the projected area of the defects. In order to understand the effect of defects tested in different directions relative to the build directions (BD), it is helpful to consider the defect distribution qualitatively. It can be observed that when tested with TA parallel to BD (figure 3b) the large irregularly shaped defects have a high projected area many of which are located close to the edge of the sample. These defects are oriented in-plane making it more likely they will cause early crack initiation. For tests TA perpendicular to BD (figure 3a), the large irregularly shaped defects are edge-on and have a low projected area making them less favourable for early crack initiation, due to lower stress concentration. This means that both types of samples show a large degree of scatter in fatigue life, but the samples tested TA perpendicular to the BD show a longer life because the defects are less potent in this orientation [8].

Figure 2. EBSD maps showing the grains, where the MPB regions are depicted with dotted lines viewed (a) perpendicular, and b) parallel, to the BD. In a) IPF-Z and b) IPF-Y are both parallel to BD.

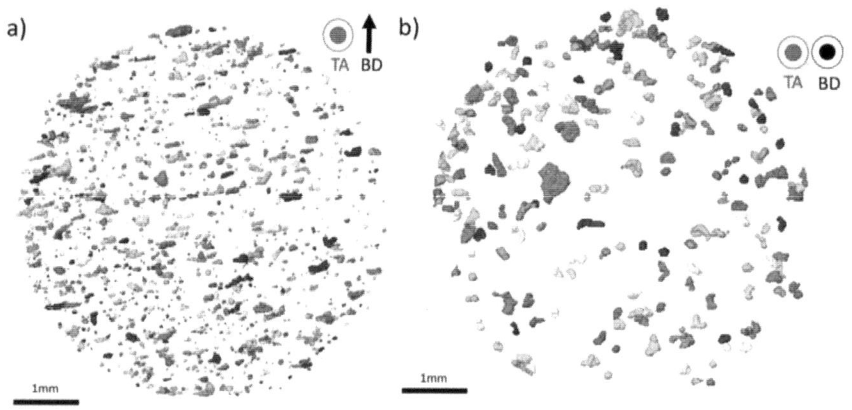

Figure 3. X-ray CT volume rendering of defects for fatigue samples cut with the TA a) perpendicular and b) parallel to the BD (matrix rendered transparent, voids arbitrarily coloured).

3.3 Microstructural Analysis of the Fatigue Fracture Surface

Figure 4 shows the secondary electron (SE) images of the fractured surface of samples, subjected to fatigue TA parallel and perpendicular to BD, featuring the crack initiation region, crack propagation zone, and fast fracture zone. For the sample TA parallel to BD (figure 4b), the critical defect displays a larger projected area and a more semicircular shape, compared to that perpendicular to BD (figure 4a). The fatigue crack region is also relatively flat.

The EBSD maps of the fracture surfaces in cross-section for both specimens are shown in figure 5, having been cut along the slicing plane indicated by broken dotted lines in figure 4. The crack path for the sample with TA perpendicular to BD is quite smooth (figure 5a), possibly due to the fact that there are many elongated grains perpendicular to the crack growth direction. While in samples with TA parallel to BD (figure 5b), it is somewhat more tortuous, which can be associated with the elongated LOF defects randomly distributed in the crack path. The crack propagation path appears to be predominantly trans-granular in both specimens, with some intergranular aspects observed in some regions, especially where the grains are fine.

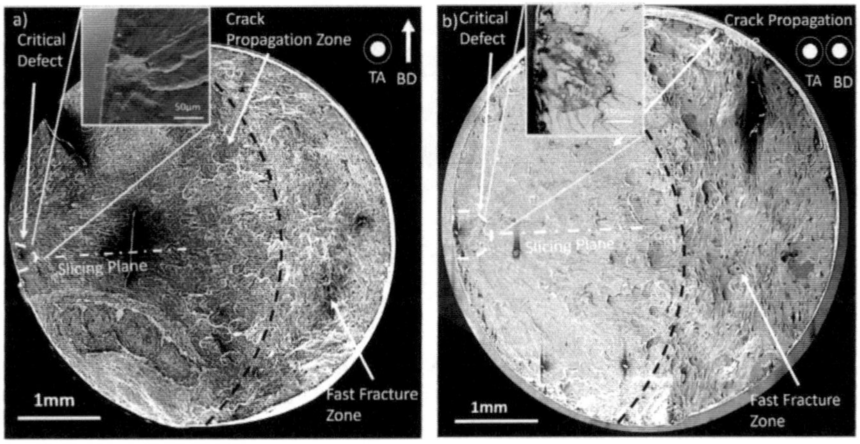

Figure 4. SE images showing the fatigue crack surface, depicting the top-down view, slicing plane and critical defect region for fatigue fractured sample with TA a) perpendicular to and b) parallel to the BD with subsets of the critical defect in the yellow box.

In figure 5, it is observed that the critical defect is associated with MPBs, where the grain size and shape are fine and equiaxed. These defects occur due to incomplete fusion or LOF, which tends to form at the MPBs [9]. There is some evidence that intergranular cracks propagate through these fine grains. It should be borne in mind that a) this is a 2D visualisation of crack propagation and 3D analysis is really required for such anisotropic microstructures and b) that a complete understanding also needs to consider the effect of the sub-micron eutectic Si structure, which may also influence the direction of crack propagation [10].

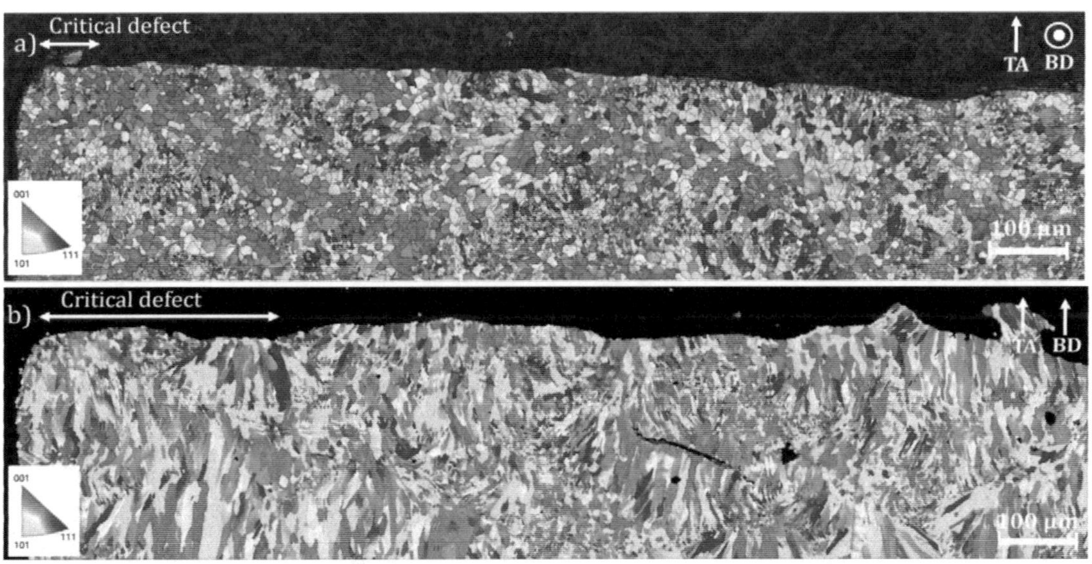

Figure 5. Cross-section large area EBSD maps, including the fracture surface for fatigue with TA a) perpendicular to and b) parallel to the BD, which are cut along the slicing planes indicated by broken dotted lines in figure 4. In a) IPF-Z and b) IPF-Y are both parallel to BD.

4. Conclusions and future work

In this study, samples of SLM AlSi10Mg alloy were subjected to fatigue loading with tensile axis (TA), oriented perpendicular and parallel to the build direction (BD). X-ray CT analysis identified spherical gas pores and pancake shaped LOF defects. Microscopy revealed the formation of fish scale patterns by the solidification of the melt pools. EBSD imaging revealed an anisotropic microstructure characterized by a wide range of grain sizes, ranging from fine equiaxed grains (<5 μm) to elongated columnar grains (150 μm) within the melt pool interiors. Fatigue cracks initiate from lack of fusion (LOF) defects located near the sample surface, more likely at the melt pool boundaries (MPBs) where the grains are small. Notably, EBSD analysis indicated that fatigue crack propagation initially occurred through the MPBs and further grows through the sample via a combination of transgranular and intergranular propagation. In the sample with TA parallel to the BD, the crack grows perpendicular to the elongated columnar grains and seems to be deflected by the grain structure. Conversely, in the sample with TA perpendicular to the BD, the elongated grains lie within the crack plane, resulting in less deflection. However, further 3D analysis is required to comprehensively characterise the crack path along with a consideration of the inter-dendritic silicon distribution.

Acknowledgements

The authors are grateful for the resources provided by the Electron Microscopy Centre funded by the Henry Royce Institute EP/X527257/1 and the Henry Moseley X-ray Imaging Facility funded through EP/T02593X/1. SR appreciates funding received from the University of Manchester's Dean's Doctoral Scholarship and SSAI acknowledges the TWI via National Structural Integrity Research Centre (NSIRC). We are grateful to Prof. Shengchuan Wu of Southwest Jiaotong University, China for providing the samples. We also acknowledge the support of the Collaborative Computational Project on Tomographic Imaging (CCPI) funded by the EPSRC through EP/T026677/1.

References

[1] Aboulkhair N T, Everitt N M, Ashcroft I and Tuck C 2014 Reducing porosity in AlSi10Mg parts processed by selective laser melting *Addit. Manuf.* **1-4** 77–86
[2] Di Egidio G, Martini C, Börjesson J, Ghassemali E, Ceschini L and Morri A 2023 Influence of Microstructure on Fracture Mechanisms of the Heat-Treated AlSi10Mg Alloy Produced by Laser-Based Powder Bed Fusion *Materials* **16** 2006
[3] Raja A, Cheethirala S R, Gupta P, Vasa N J and Jayaganthan R 2022 A review on the fatigue behaviour of AlSi10Mg alloy fabricated using laser powder bed fusion technique *J. Mater. Res. Technol.* **17** 1013–1029
[4] Bao H, Wu S, Wu Z, Kang G, Peng X and Withers P J 2021 A machine-learning fatigue life prediction approach of additively manufactured metals *Eng. Fract. Mech.* **242** 107508
[5] Wu Z, Wu S, Bao J, Qian W, Karabal S, Sun W and Withers P J 2021 The effect of defect population on the anisotropic fatigue resistance of AlSi10Mg alloy fabricated by laser powder bed fusion *Int. J. Fatigue* **151** 106317
[6] Thijs L, Kempen K, Kruth J P and Van Humbeeck J 2013 Fine-structured aluminium products with controllable texture by selective laser melting of pre-alloyed AlSi10Mg powder *Acta Mater.* **61** 1809–1819
[7] DebRoy T, Wei H L, Zuback J S, Mukherjee T, Elmer J W, Milewski J O, Beese A M, Wilson-Heid A, De A and Zhang W 2018 Additive manufacturing of metallic components – Process, structure and properties *Prog. Mater. Sci.* **92** 112–224
[8] Wu Z, Wu S, Gao X, Lin Y, Xue Y and Withers P J 2023 The role of internal defects on anisotropic tensile failure of L-PBF AlSi10Mg alloys *Sci. Rep.* **13** 14681
[9] Gui Y, Aoyagi K, Bian H and Chiba A 2022 Detection, classification and prediction of internal defects from surface morphology data of metal parts fabricated by powder bed fusion type additive manufacturing using an electron beam *Addit. Manuf.* **54** 102736
[10] Ren P, Withers P J, Zuo Z, Feng H and Huang W 2023 Microstructural effects on short fatigue crack propagation in cast Al-Si alloys *Int. J. Fatigue* **175** 107788–107788

Effect of hot isostatic pressing on the microstructure of laser powder bed fused A20X™ alloy

J Barode[1*], E Bassini[1], A Aversa[1,2], D Manfredi[1,2], D Ugues[1,2], S Biamino[1,2], M Lombardi[1,2], and P Fino[1,2]

[1] Department of Applied Science and Technology, Politecnico Di Torino, Corso Duca degli Abruzzi 24, Torino 10129, Italy
[2] Center of Integrated Additive Manufacturing (IAM), Politecnico di Torino, Corso Castelfidardo, 51, 10138 Torino, Italy

*E-mail: jayant.barode@polito.it

Abstract. In the present work, an attempt has been made to do a combined hot isostatic pressing (HIP) & solution treatment in a single-step and the effect of different cooling rates (quenching and annealing) on the microstructure was examined. For a comparison, solution treatment (without HIP) was also analysed. It was observed that HIP treatment was successful in reducing the macro-porosities of the L-PBF part. Although, it was not an efficient treatment for the dissolution of the θ-Al_2Cu phase. Furthermore, in both HIP-Quenched and HIP-Annealed treatments a sign of incipient melting was observed along the grain boundaries. Solution treatment without HIP did not show incipient melting.

1. Introduction

A20X™ is a newly developed high strength Al alloy (Al-Cu-Mg-Ag-Ti-B) which has overcome the limitations of hot cracking susceptibility and high fibre-laser reflectivity inherent to the high strength Al alloys thanks to the presence of pre-embedded TiB_2 particles [1]. Being a precipitation hardening alloy, much attention has been given to the post-processing T6 and T7 heat treatments i.e., solution treatment followed by artificial aging [2,3]. The alloy gains it strength by the precipitation of plate-type coherent-Ω and semi-coherent θ' phase during artificial aging. Recently, the response of the alloy towards T4 (natural aging) has also been explored [4].

It was well acknowledged that L-PBF processing of A20X™ alloy induces defects such as lack of fusion or gas induced porosities [5]. Such defects possess a detrimental effect on the mechanical properties, degrading the in-service life of the component. To mitigate such defects, a post-processing treatment of Hot Isostatic Pressing (HIP) has been found to be highly beneficial [6]. Four primary mechanisms are responsible for a pore elimination through HIP: plastic flow, power-law creep, Coble creep (grain boundary diffusion), and Nabarro–Herring creep (lattice diffusion) [7]. HIP conventionally offers slow cooling rate at the end of the cycle, however, with an integration of a new quenching module called uniform rapid cooling (URC), it is now possible to combine HIP & solution treatment in a single-step to reduce the time and energy of post-processing treatments [8]. HIP of L-PBF A20X™ alloy is still a less touched domain. To the authors knowledge, till now there has been only one work demonstrating the micro-mechanical response of the HIPed parts [6]. In the present work, an attempt has been made to observe the effect of combined post-processing heat treatment i.e., HIP & solution treatment on the microstructure developed. At the end of the HIP treatment, two cooling rates have been explored (i) quenching (~770°C/min) and (ii) annealing (~ 20°C/min). A comparison has also been made with a non-HIPed solution treatment.

2. Experimental

A commercial gas atomized A20X™ powder was purchased from ECKART GmbH. The powder morphology is shown in figure 1(a). The particle size distribution and nominal chemical composition are presented in reference [3]. L-PBF processing of cubic (10x10x10 mm^3) samples was performed in an EOS M270 dual mode system by an optimized set of processing parameters as described in our previous work [4]. HIP was performed in a Quintus QIH 15 L equipped with a molybdenum furnace in an inert Ar atmosphere. HIP cycle was comprised of heating from room temperature to 530°C with a ~7°C/min, holding for 1 h at 530°C under a pressure of 100 MPa followed by (i) quenching and (ii) annealing. Quenching was performed in a uniform rapid cooling (URC) module and annealing was done by a conventional process of furnace cooling (~20°C/min). Table 1 presents the nomenclature of various heat-treated conditions. Metallographic sample preparation was performed by grinding with SiC papers up to 4000 grit followed by colloidal silica polishing. Samples were not etched. Microstructure was examined using a Phenom XL table-top scanning electron microscope (SEM) equipped with secondary electron (SE), back scattered electron (BSE) detectors and an energy dispersive spectroscopy (EDS). Porosity content was measured from the SEM-BSE micrographs and analysed by ImageJ software. X-ray diffraction (XRD) was performed in a Panalytical Empyrean using a monochromatic Cu-Kα radiation of 1.54 Å operated at 40 kV and 40 mA with a step size of 0.013° and 25 s per step. Differential scanning calorimetry was done on the as-built (AB) samples in a NETZSCH 214 Polyma. Samples were heated from 40°C to 570°C and cooled to 40°C at 20°C/min in a N_2 atmosphere.

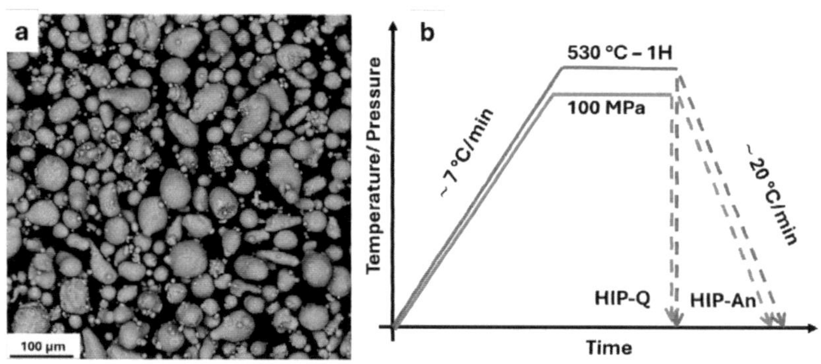

Figure 1. (a) SEM-SE micrograph of A20X™ alloy powder and (b) schematic of a HIP cycles.

Table 1. Nomenclature of various heat-treated conditions

Heat treatment	Temperature, pressure, and time
ST	530°C for 1 h + Quenched
R-ST	530°C as reached + Quenched
HIP-Q	530°C and 100 MPa for 1 h +Quenched
HIP-An	530°C and 100 MPa for 1 h + Annealed

3 Results

3.1 Macro-porosity

Figure 2 displays the low-magnification SEM-BSE micrographs of the AB and HIPed parts i.e., HIP-Q and HIP-An. Based on the morphology, porosities in the AB state were identified as gas induced porosities (spherical) and lack of fusion (high aspect ratio) (figure 2(a)). Post to the HIP treatment, macro-porosities reduced (figure 2(b) and 2(c)). The porosity content was reduced from AB (0.23±0.21%) to the HIPed parts (HIP-Q (0.06±0.02%) and HIP-An (0.05 ±0.03%)).

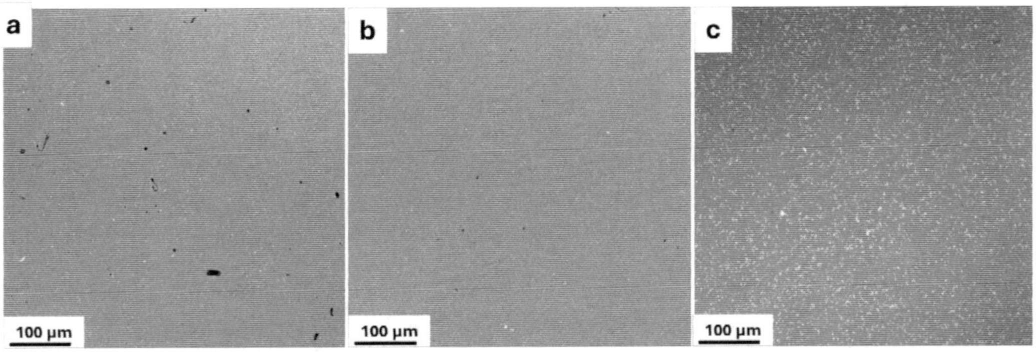

Figure 2. (a) Low-Mag SEM-BSE micrographs of (a) AB, (b) HIP-Q and (c) HIP-An

3.2 Phase and thermal analysis

The DSC profile of the AB state during heating and subsequent cooling is shown in figure 3(a). During heating a broad exothermic peak from 200-360°C was observed. It is a peak representing precipitation of various types of Al_2Cu phase such as Ω, θ' and θ. Precipitation peak was followed by a broad endothermic peak range from 400-530°C, representing the dissolution of all the forms of Al_2Cu [3]. The θ dissolution completes at ~530°C and θ starts to re-precipitate at ~502°C during cooling. The XRD patterns of the HIPed and non-HIPed conditions are shown in figure 3(b). The phases identified were Al, TiB_2 and θ-Al_2Cu. In the HIP-Q condition there was still a presence of θ peak. The peak intensity was much higher in the case of HIP-An. Whereas no peak was observed in the non-HIPed conditions i.e., ST and R-ST (see inset figure 3(b')).

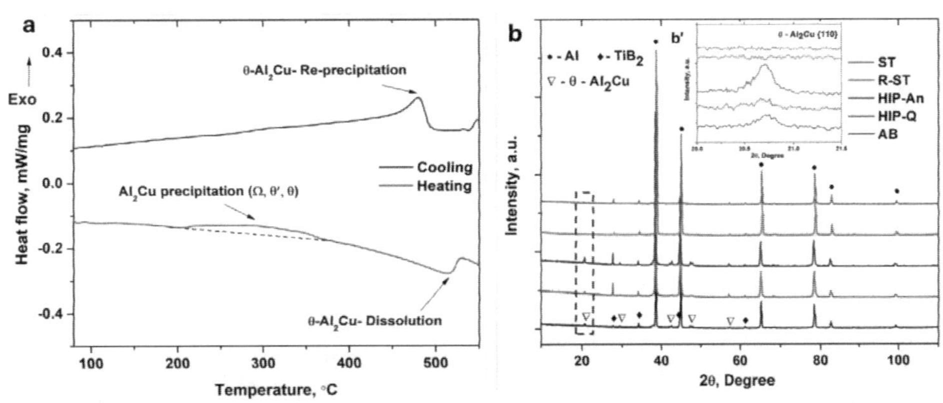

Figure 3. (a) DSC profile of AB condition, (b) XRD whole line profiles and (b') individual peak of θ-Al_2Cu {110} of HIPed and non-HIPed conditions.

3.3 Microstructure

Figure 4 shows the SEM-BSE micrographs of the AB, ST, HIP-Q and HIP-An conditions. The AB state showed a typical fine cellular microstructure with eutectic-θ as cell boundaries (figure 4(a)). In ST, the cellular microstructure vanished (figure 4(b)). Cellular microstructure also vanished after HIP-Q treatment, although, grain boundaries and grain triple junctions were decorated by fine precipitates (figure 4(c)). These precipitates are likely to be of an equilibrium θ phase.

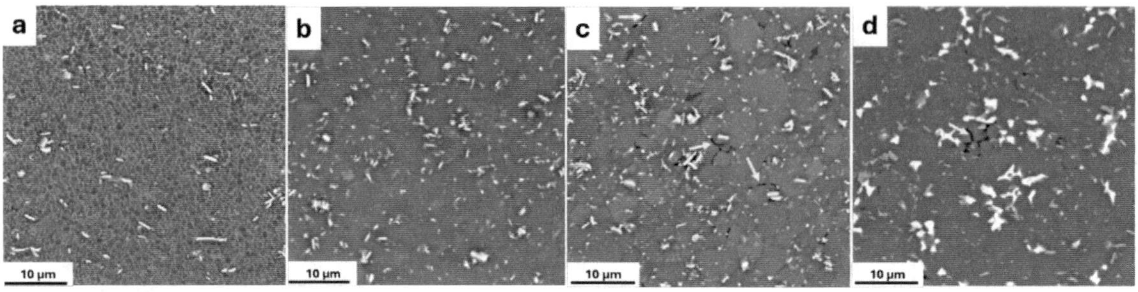

Figure 4. SEM-BSE micrograph of (a) AB, (b) ST, (c) HIP-Q and (d) HIP-An *(yellow arrow showing incipient melted region and purple arrow showing spheroidal blocky-θ inside grains).*

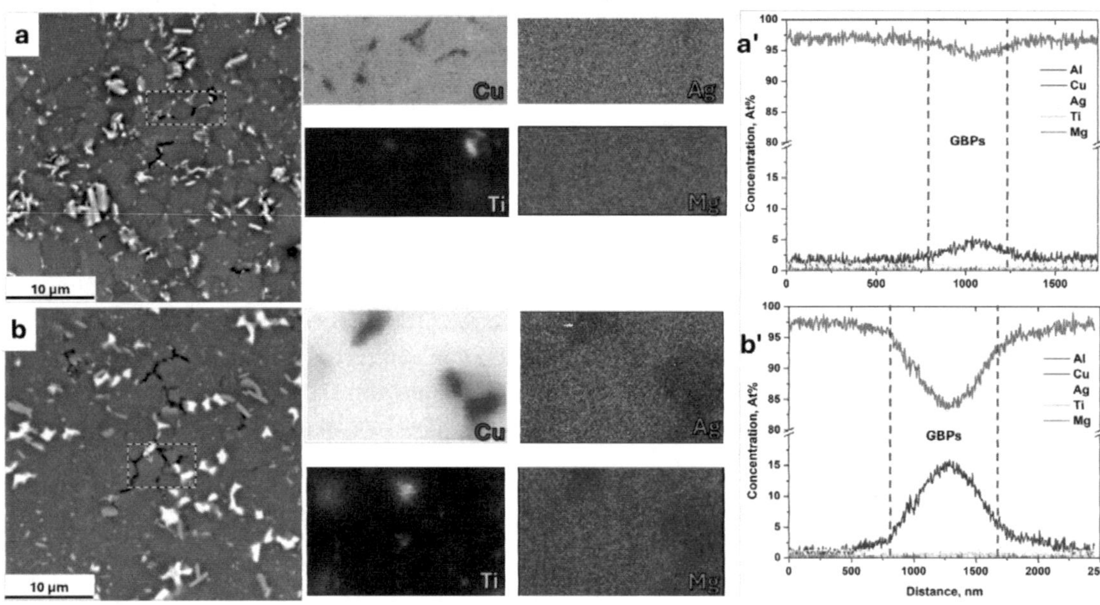

Figure 5. SEM-BSE micrograph of HIPed LPBF A20X™ alloy showing the EDS mapping on the selected region of incipient melting (a and b) and elemental line scan across a grain boundary precipitate (orange arrow) (a' and b') in (a-a') HIP-Q and (b-b') HIP-An sample.

In an HIP-An state, coarser θ precipitates were present (figure 4(d)). This observation was also supported by the XRD spectrum, see figure 3(b'). Surprisingly, in both the HIP-Q and HIP-An conditions, grain boundaries were also accompanied by elongated pores and in certain boundaries there were discontinuous pores (figure 4(c) and 4(d)). By looking at the morphology of these porosities, they are resembling of an incipient melting and not of cracks generated by de-cohesion during plastic deformation. Incipient melting is a localized melting of secondary phases primarily located at the grain boundaries [9]. Such porosities due to incipient melting was not observed in ST. The elemental mapping around the incipient melted region and line scan across the grain boundary precipitate (GBP) of the HIPed samples is presented in figure 5. In both the HIP-Q and HIP-An conditions, it was observed that GBPs are Cu-rich (figure 5(a) and 5(b)). The same was also suggested by the line scan across the precipitate (figure 5(a') and 5(b')). The concentration profile of other solutes (Ag and Mg) remains more or less the same. This indicates that these GBPs in the HIPed conditions are only of an equilibrium θ-Al_2Cu phase. It was also revealed that the grain boundary pores were accompanied by fine particles of TiB_2.

4 Discussion

4.1 Incomplete dissolution

An incomplete dissolution of θ was witnessed in HIP-Q. Whereas a more complete dissolution was observed in the non-HIPed solution treatments performed at the same temperature i.e., ST (figure 3(b')). Interestingly, no θ peak was observed even in R-ST, indicating that most of the θ phase has been dissolved during the ramping stage itself. It was speculated that the incomplete dissolution in the HIP-Q was associated to diffusional creep that assists in the densification during HIP. During diffusional creep, solute atoms migrate along the grain boundaries (Coble creep) and towards the grain boundaries through lattice diffusion (Nabarro-Herring creep) [10]. Thus, the presence of θ phase along grain boundaries is likely to be a result of a precipitation by diffusional creep during densification. These GBPs are of blocky-θ phase and possess much slower dissolution characteristics [11]. In the case of ST and R-ST, during ramping the eutectic-θ cell boundaries initially breakdown into fine spheroidal particles and hence would accelerate the dissolution. Thus, in a ST and R-ST condition, the finer spheroidal eutectic-θ dissolves rather than forming blocky-θ. A more detailed analysis of the θ dissolution was reported in our previous work [3]. In HIP-An, the blocky-θ phase re-precipitated and coarsened due to much slower cooling rate (figure 4(d) and 5(b)).

4.2 Incipient melting

The HIPed samples (HIP-Q and HIP-An) showed the sign of incipient melting along grain boundaries (figure 5). Whereas no incipient melting was observed in non-HIPed conditions. The spheroidal blocky-θ present inside the grains in both HIPed and non-HIPed conditions doesn't melt (figure 4). We suspect that there could be two possible reasons for the formation of incipient melting in an A20X™ alloy:

(i) Deformation/strain induced GBPs of low-melting phases such as S-Al_2CuMg and Q-$Al_7Cu_3Mg_6$
(ii) Morphology assisted pre-melting of a grain boundary θ-Al_2Cu precipitate.

The melting point of S-Al_2CuMg and Q-$Al_7Cu_3Mg_6$ is 495°C and 513°C, respectively, which is much lower than the solutioning temperature (530°C). However, we did not find these phases either by XRD (figure 3(b)), or in the EDS (figure 5(a') and 5(b')). The EDS line scan across GBPs showed only Cu segregation. Thus, we suspect that the incipient melting was due to the morphology assisted pre-melting of θ-Al_2Cu. The melting point of a θ phase is influenced by its morphology [9]. The higher interfacial area of the elongated θ GBPs would favour pre-melting with respect to the spheroidal θ phase present inside the grains.

5 Conclusion

Microstructure of the L-PBF A20X™ alloy was investigated in two HIPed treatments (i) quenched (HIP-Q) and (ii) annealed (HIP-An). The conclusions drawn are the following: (i) Macro-porosity was reduced after the HIP treatment, (ii) HIP treatment was inefficient to dissolve θ-Al_2Cu, whereas solution treatment without HIP was more efficient, (iii) in HIP-Q sample, θ phase was precipitated along the grain boundaries, whereas coarser θ phase was observed in HIP-An, and (iv) incipient melting along grain boundaries was observed in the HIPed samples. Morphological assisted pre-melting of grain boundary θ-Al_2Cu phase could be the possible reason. An optimized HIP & solution treatment parameters (temperature, pressure and cooling rate) should be further investigated to get a desirable microstructure i.e., no incipient melting and a more complete dissolution of θ-Al_2Cu.

Acknowledgement

The study was carried out within the MOST – Sustainable Mobility National Research Center and received funding from the European Union Next-GenerationEU (Piano nazionale di ripresa e resilienza (PNRR) – missione 4 componente 2, investimento 1.4 – D.D. 1033 17/06/2022, CN00000023). This manuscript reflects only the authors' views and opinions, neither the European Union nor the European Commission can be considered responsible for them.

References

[1] Butler S A 2003 An aluminium-copper alloy UK Patent: GB2334966A

[2] Jiang F, Tang L, Li S, Ye H, Attallah M M and Yang Z 2023 Achieving strength-ductility balance in a laser powder bed fusion fabricated TiB2/Al–Cu-Mg-Ag alloy *J. Alloys Compd.* **945** 169311

[3] Barode J, Vayyala A, Virgillito E, Aversa A, Mayer J and Fino P 2023 Revisiting heat treatments for additive manufactured parts : A case study of A20X alloy *Mater. Des.* **225** 111566

[4] Barode J, Vayyala A, Aversa A, Yang L, Mayer J, Fino P and Lombardi M 2024 Natural and artificial aging behaviour of Al-Cu-Mg-Ag-Ti-B (A205) alloy processed by laser powder bed fusion: Strengthening mechanisms and failure analysis *Mater. Today Commun.* **39** 108978

[5] Ghasri-Khouzani M, Karimialavijeh H, Pröbstle M, Batmaz R, Muhammad W, Chakraborty A, Sabiston T D, Harvey J P and Martin É 2023 Processability and characterization of A20X aluminum alloy fabricated by laser powder bed fusion *Mater. Today Commun.* **35** 105555

[6] Shakil S I, Beamer C, Attallah M M and Haghshenas M 2024 The Effect of Heat Treatment on Local Mechanical Properties of Laser Powder Bed Fused and Hot Isostatically Pressed Al-Cu-Mg-Ag-TiB2 (A20X) Aluminum Alloy *JOM* **76** 31–41

[7] Atkinson H V and Davies S 2000 Fundamental aspects of hot isostatic pressing: an overview *Metall. Mater. Trans. A* **31** 2981–3000

[8] Mujica Roncery L, Lopez-Galilea I, Ruttert B, Huth S and Theisen W 2016 Influence of temperature, pressure, and cooling rate during hot isostatic pressing on the microstructure of an SX Ni-base superalloy *Mater. Des.* **97** 544–52

[9] Lombardi A, Mu W, Ravindran C, Dogan N and Barati M 2018 Influence of Al2Cu morphology on the incipient melting characteristics in B206 Al alloy *J. Alloys Compd.* **747** 131–9

[10] Hondros E D and Henderson P J 1983 Role of grain boundary segregation in diffusional creep *Metall. Trans. A* **14** 521–30

[11] Han Y M, Samuel A M, Samuel F H and Doty H W 2008 Dissolution of Al2Cu phase in non-modified and Sr modified 319 type alloys *Int. J. Cast Met. Res.* **21** 387–93

Characterization of Si particles in additively manufactured AlSi10Mg using synchrotron transmission X-ray nanotomography

M Defer[1], S Dasgupta[2], A J Shahani[2], X Xiao[3], D Juul Jensen[1] and Y Zhang[1]

[1] Department of Civil and Mechanical Engineering, Technical University of Denmark, Kgs. Lyngby, Denmark
[2] Materials Science and Engineering, University of Michigan, Ann Arbor, USA
[3] National Synchrotron Light Source II, Brookhaven National Laboratory, Upton, USA

E-mail: mcade@dtu.dk

Abstract. In AlSi10Mg samples manufactured by Laser Powder Bed Fusion, distinguishing the Si eutectic network/Si particles from the Al matrix by X-ray imaging is challenging due to the low absorption contrast between the Al and Si. This work investigates the possibility of overcoming this obstacle in synchrotron transmission X-ray microscopy. Effects of both different defocusing conditions and X-ray beam energies are evaluated and optimal conditions are identified for imaging a sample annealed post-print for 2h at 520°C. It is shown that both large particles (e.g. 4μm) and particles as small as 0.5 μm, can be imaged with reasonable precision in 3D non-destructively.

1. Introduction

AlSi10Mg has good mechanical properties and corrosion resistance while being relatively cheap and light weight. When processed through additive manufacturing, specifically laser powder bed fusion (LPBF), excellent mechanical properties are obtained, typically better than for samples produced by traditional manufacturing methods, like casting [1]. This improvement is attributed to the rapid cooling rates (reaching 10^6 K/s [2]) during LPBF, resulting in a fine microstructure composed of submicrometer Al cells surrounded by a 10^2 nm thick eutectic network.

While the focus on AlSi10Mg manufactured by LPBF has been on optimizing the process to minimize porosity and defects, there has been limited emphasis on high-magnification microstructural characterization, especially after post-LPBF heat treatments. Notably, there is a scarcity of three-dimensional (3D) nano-scale data regarding the morphology and topology of phases and, to our knowledge, no experiments have been conducted to monitor the 3D microstructural evolution in-situ during heat treatment. This may be due to the multiple challenges associated with 3D non-destructive characterization of this material. A main one relates to the length-scale of the eutectic network and Si particles, which requires imaging with a spatial resolution below 50 nm. This resolution can be achieved by synchrotron transmission X-ray microscopy (TXM) experiments, where the minimum effective pixel size reaches 20-30 nm [3]. A second challenge arises from the very similar X-ray linear absorption coefficients of Al and Si leading to a low absorption contrast in nano-tomography.

The present work aims to investigate the possibility of overcoming these challenges using TXM. For this first attempt, an annealed LPBF AlSi10Mg sample with comparatively large Si particles, in the range of 0.5-4 μm, was used. To enhance the contrast between Si and Al phases, TXM scans were conducted at different defocusing conditions, introducing phase contrast. Five focusing distances and three beam energies were employed to assess their impact on the reconstructions, aiming to determine an optimal solution.

2. Material and experimental procedures

2.1 Material

The material studied in this work is an AlSi10Mg alloy, and the sample was manufactured using LPBF at the Danish Technical Institute. The printing parameters fell within the typical ranges utilized for this alloy, including a hatch spacing of 170 μm, a layer thickness of 60 μm, a laser speed of 1850 mms^{-1}, and a laser power of 650 W. A cylindrical rod, 0.5 mm in diameter and 10 mm in height, was cut by electron discharge machining from the printed bulk sample. It was then annealed for 2 hours at 520 °C in a Nabertherm L 1/12/R6 Mini Muffle Furnace and air-cooled. After heat treatment, samples are expected to contain Si particles with an average equivalent diameter of 0.5 μm as measured by light optical microscopy (LOM). The alloy Si wt% suggests a maximum area fraction of 11% for Si particles. The annealed rod was electropolished using a 10% solution of perchloric acid in ethanol at a voltage of 18 V followed by a 2% solution at 10 V. Liquid nitrogen was used to cool down the solution before electropolishing. The obtained needle-shaped sample had a diameter below 40 μm along the whole gauge volume.

2.2 TXM experiment

The nano-tomography experiment was performed at the 18-ID Full-field X-ray Imaging beamline of the National Synchrotron Light Source II, USA. The set-up consists of an upstream hutch housing the beamline optics (collimating mirror, monochromator and toroidal mirror) and an experimental hutch. In the experimental hutch, a condenser focused the beam onto the sample while the transmitted X-rays exiting the sample were magnified by a zone plate onto a 20 μm thick CsI scintillator to be converted to visible light on the detector [4]. For the experiment, the beam energy was kept between 8 and 10 keV. The needle-shaped sample was rotated 180° at a speed of 3°/s and the TXM scan was performed concurrently with an exposure time of 0.1 s. A projection was recorded at each rotation angle for a total of 600 projections per dataset. A pixel size of 22.54 nm × 22.54 nm with an overall magnification of 288x was obtained for a field of view of 57.70 × 57.70 μm^2. In this investigation, the zone plate was incrementally shifted in steps of 25 μm from its initial position to a total 100 μm offset towards the detector.

2.3 Data analysis

The datasets were processed using already available functions in TomoPy [5]; a Python package designed for tomographic data processing and image reconstruction. Figure 1 presents the reconstructed images. Following a standard intensity normalization, a gridrec algorithm was used to reconstruct the volumes [6, 7]. This algorithm, originally proposed by Dowd et al. [6], utilizes the Fourier transform method. A low-pass Parzen filter is then applied to smooth the Fourier transform of the projection data and thereby reduce the noise. The obtained reconstructed slice displays very strong ring artefacts (figure 1(a)). A ring removal function based on a combination of Vo et al.'s algorithms [8] was applied to eliminate stripes from the sinogram, thus effectively removing the rings in the reconstructed slice, as depicted in figure 1(b). To enhance the contrast between the Al matrix and the Si particles but also reduce the noise, a phase retrieval algorithm utilizing Paganin's method [9] was executed, transforming the phase-contrast effects between Al and Si into an absorption contrast image with an enhanced signal-to-noise ratio (see figure 1(c)). The few remaining ring artefacts, which can obstruct clear identification of particles, are removed at the last step using a 1D median filter on reconstructed slices in polar coordinates [10]. Almost all remaining ring artefacts are eliminated with the second ring removal algorithm. However, it may introduce void-center artefacts [8] as a trade-off, see figure 1(d). Other artefacts such as shadowing are still evident at the sample edges and around Fe particles. Median and Gaussian filters were tested to reduce the background influence but resulted in more blurring and were thus not utilized. The same data processing routine as illustrated in figure 1(d) is applied to all subsequent reconstructed data.

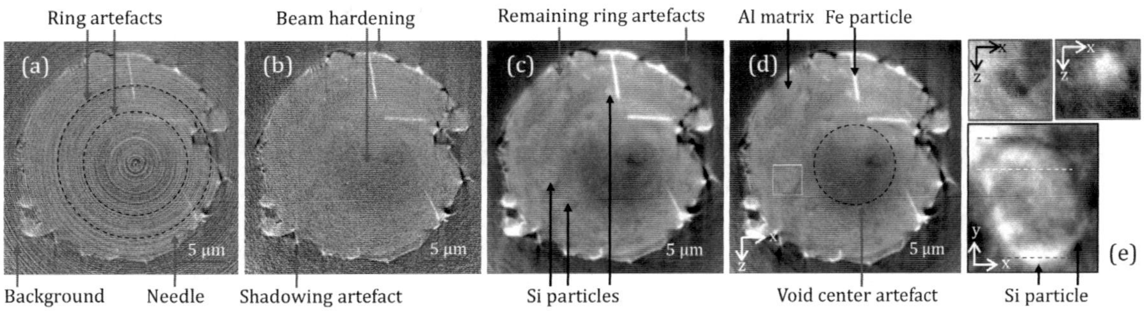

Figure 1. Slice of the sample post-LPBF heat treated for 2h at 520°C, acquired by synchrotron TXM at 9 keV and with 50 µm defocusing of the zone plate. The result of each step in the data processing is illustrated: (a) normalisation and reconstruction, (b) ring removal algorithm, (c) phase retrieval and (d) ring removal on the reconstructed data. (e) shows 2D cuts of a Si particle in the xy plane along with top and bottom cross sections in zx plane indicating that, at its edges, the particle appears totally black or white.

The LPBF AlSi10Mg sample, post-heat treated for 2 hours at 520°C as shown in figure 1(d), consists of needle-shaped Fe-containing particles and faceted Si particles embedded in the Al matrix. The significantly higher linear attenuation coefficient of Fe compared to Al or Si provides a good contrast between the Fe particles and the matrix, facilitating their identification. However, due to minimal absorption contrast between Al and Si, Si particles are primarily visible through propagation-based phase contrast (at 50 µm defocusing of the zone plate). This phase contrast causes the edges of Si particles to appear either brighter or darker than the matrix, as illustrated in figure 1(e). Consequently, if the slice studied intersects a Si particle at an edge (e.g. top and bottom in figure 1(e)), it can appear completely dark or bright. Applying a stronger phase retrieval algorithm can better homogenize the contrast but at the expense of the spatial resolution. It is worth noting that phase retrieval algorithms are inherently low-pass filters [11].

3. Results and discussion

3.1 Effects of X-ray energy

At a fixed defocusing distance of 50 µm, the effect of different energies—8 keV, 9 keV, and 10 keV—was investigated. This range is expected to allow sufficient absorption (75-85%) while remaining within the capacity constraints of the beamline. The optimal energy was determined based on the quality of the beam illumination and the ability to detect Si particles.

In figure 2, the same slice of the sample is shown for each energy value. A non-uniform intensity distribution, in particular at higher energies, is readily observed in the reconstructed slices. This is attributed to both temporal and spatial fluctuations in beam illumination. As shown in figure 2(a3), the beam intensity on the flat field projection at 8 keV exhibits a distinct and sharp peak. At 9 keV, it appears broader, displaying a plateau-like maximum with slight variations, while at 10 keV the sample illumination is highly inhomogeneous presenting a darker centre, see figure

IOP Conf. Series: Materials Science and Engineering **1310** (2024) 012027 doi:10.1088/1757-899X/1310/1/012027

2(b3) -(c3). When comparing projections across the whole scan, it is found that the beam presents greater temporal fluctuations in average intensity, shape and position at 10 keV than at 8 and 9 keV. The beam spatial variation can therefore not be eliminated completely by intensity normalization. Besides intensity fluctuation, the beam is also likely to exhibit residual wavefront phase variations, which locally influence the phase contrast. These observations highlight the critical need for microscope calibration when characterizing this challenging sample.

Considering now the ability to detect the Si particles, at 8 keV (figure 2(a)), the lower energy leads to slightly higher absorption contrast between the Al and Si phases. This absorption contrast facilitates the identification of large Si particles as shown in the 3D intensity plot of a large particle, figure 2(a1), where the Si particle appears noticeably brighter than the surrounding matrix (even without the phase retrieval algorithm). However, smaller particles are more difficult to discern, as indicated by the surface plot in figure 2(a2). The more uniform beam intensity along with a lower phase contrast at 8 keV led to an enhanced smoothing during phase retrieval when using the same parameters as for 9 keV and 10 keV.

At 9 and 10 keV (figure 2(b)-(c)), there is a weaker absorption contrast between the Si particles and Al matrix but the good phase contrast (edge enhancement) still allows to identify the Si particles, including smaller particles. Using a beam energy of 9 keV proves to be an advantageous compromise, offering a balance between good illumination and phase contrast. This configuration enables the observation of both large and smaller particles, making it a favourable choice for quantitative investigations that rely on a segmentation of particles.

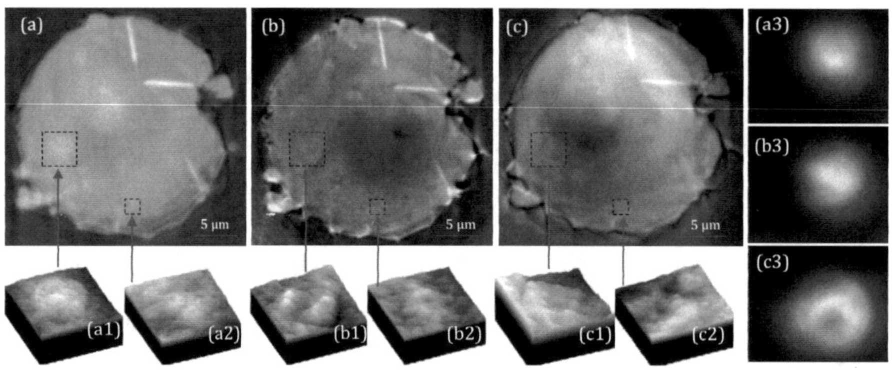

Figure 2. Effects of the X-ray beam energy: (a) 8 keV, (b) 9 keV and (c) 10 keV. All measurements were performed at a defocussing distance of 50 µm. Intensity surface plots of the same two particles are presented for each condition: a large particle (a1), (b1) and (c1) and a smaller one (a2), (b2) and (c2). Flat field projections for each energy are shown in (a3), (b3) and (c3).

3.2 Effects of defocusing distance

To further optimize the measurements, different defocusing distances were considered for a fixed beam energy of 9 keV. Figure 3 presents the same reconstructed slice at each defocusing distance. For no or minimal defocusing (figure 3(a)-(b)) the particles cannot be clearly identified. Further increasing the defocusing distance results in enhanced phase contrast and thereby improved edge detection. As shown in the 3D intensity plots (figure 3(a2)-(e2)), a relatively large particle, which is barely noticeable at low defocusing, becomes increasingly apparent with greater distance. However, comparing the 3D intensity plots of smaller particles (figure 3(a1)-(e1)), the blurring caused by defocusing is predominant at 100 µm, preventing the detection of some small particles. From figure 3, a good balance between phase contrast and blurring is found at 50 and 75 µm, where the highest number of visible particles is observed. After analysing various slices, a defocusing distance of 50 µm emerged as the most optimal choice.

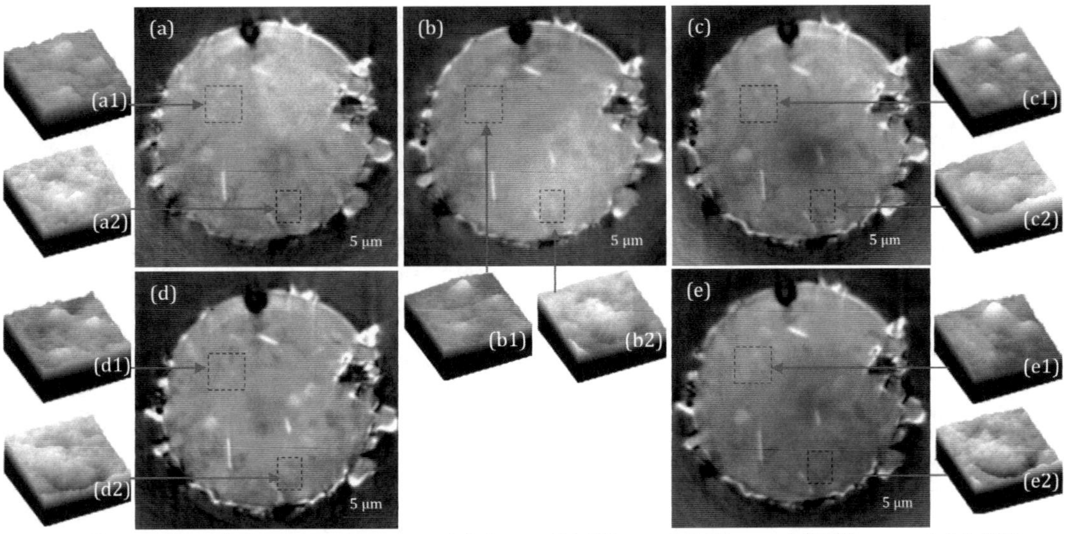

Figure 3. Effects of the defocusing distance: (a) 0 µm, (b) 25 µm, (c) 50 µm, (d) 75 µm and (e) 100 µm. All TXM measurements were performed with a beam energy of 9 keV. Along with each slice, the intensity surface plots of the same block of three small particles (a1)-(e1) and a larger particle (a2)-(e2) are shown.

3.3 Comparison with light microscopy

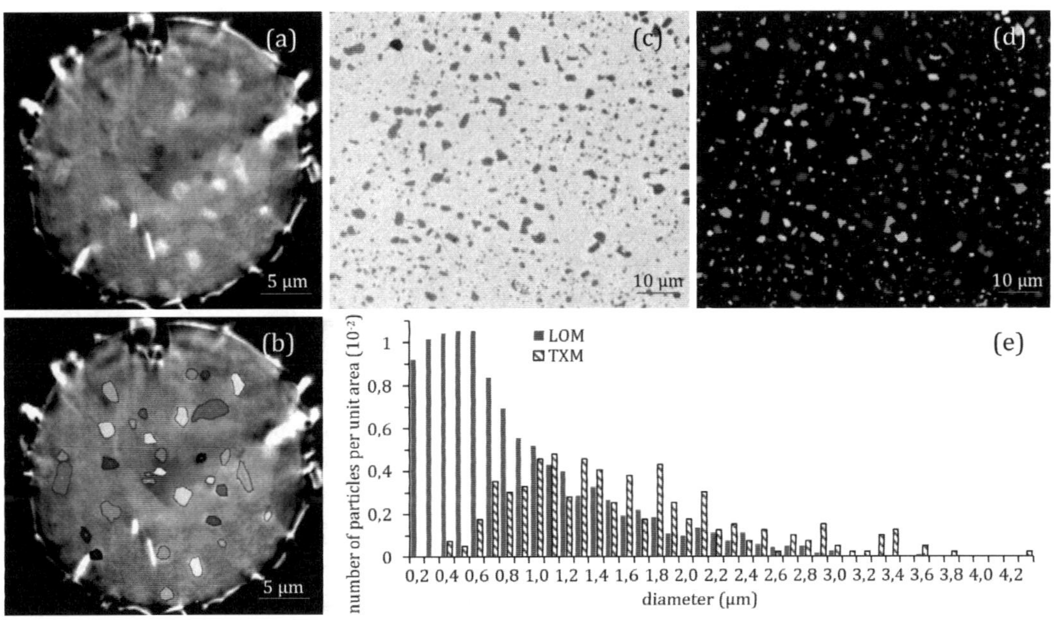

Figure 4. Si particles in LPBF AlSi10Mg after 2 h post-heat treatment at 520 °C were imaged using (a) TXM at 9 keV with 50 µm defocusing distance and were segmented in (b). For comparison, LOM (c) imaging and segmentation (d) were performed. (e) shows their particle size distributions (equivalent diameter).

A total of 261 Si particles were segmented manually from 10 well-separated TXM slices at optimal contrast conditions and covering a total area of 4140 μm^2. To further understand the effects of defocusing and estimate the effective spatial resolution, the resulting particle size distribution from TXM was compared to that of Si particles, segmented from high-magnification LOM micrographs (see figure 4). The high contrast between Al and Si in LOM enables the identification of small particles of sub-100 nm diameter within a relatively large area of 35 875 μm^2, thus providing a reasonable benchmark for the comparison. When comparing the two size distributions, it appears that the largest particles identified from TXM are about 10% larger than those observed in LOM, possibly due to defocusing-induced blurring effects. Moreover, figure 4 indicates that a large fraction of smaller particles below 0.5 µm are missing in the segmentation TXM. This is primarily attributed to the low absorption contrast, variations in beam intensity, and various artefacts (e.g. shadowing and scattering), hindering the visual segmentation of these small particles. Interestingly, the total area fraction of Si particles is about the same, 9.9%, measured using both LOM and TXM, with standard deviations of 0.8% and 1.8%, respectively. The Si particles artificial coarsening in TXM is likely compensating for the missing area covered by small particles.

4. Summary and outlook

The analysis of effects of beam energy and defocusing distance in synchrotron X-ray nano-tomography of post-LPBF heat treated AlSi10Mg highlights that lower X-ray energy levels offer improved beam illumination but lower phase contrast. Increasing the defocussing distance enhances phase contrast but induces blurring. The best compromise of these effects for the present sample is found for a defocusing distance of 50 µm, coupled with a beam energy of 8 to 9 keV. Using these parameters, particles larger than 0.5 µm in diameter can be segmented manually.

In future work, improved particle segmentation will be sought through machine learning using an artificial neural network (as the random forest approach did not provide an acceptable outcome), where the manually segmented data will be set as training data. The 3D distribution of the Si particles will then be visualized. Combined with the microstructure characterized by micro-beam X-ray diffraction [12] and TXM measurement during in-situ annealing, a comprehensive 4D (3D space- and time-resolved) study will be possible to track the evolution of the Si network and enable quantification of the Si particles distribution in the Al matrix grain structure.

Acknowledgements

Use of the Advanced Photon Source was supported by the U. S. Department of Energy, Office of Science, Office of Basic Energy Sciences, under Contract No. DE-AC02-06CH11357. The sample was printed at the Danish Technological Institute, and we thank N. Kjelgaard Vedel-Smith for providing it for this investigation. M Defer, D Juul Jensen and Y Zhang acknowledge the financial support by a research grant (MicroAM - VIL54495) from VILLUM FONDEN. S.D. and A.J.S. thank the U.S. Department of Energy, Office of Science, Office of Basic Energy Sciences (DE-SC0023147).

References

[1] Girelli L, Tocci M, Gelfi M and Pola A 2019 *J. mater. sci. eng. A* **739** 317-328
[2] Tang M, Pistorius P C, Narra S and Beuth J L 2016 *JOM* **68** 960-966
[3] Shahani A J, Xiao X, Lauridsen E M and Voorhees P W 2020 *Mater. Res. Lett.* **8** 462–476
[4] Cobrun D S et al. 2019 *Rev. Sci. Instrum.* **90** 053701
[5] Gürsoy D, De Carlo F, Xiao X and Jacobsen C 2014 *J. Synchrotron Radiat.* **21** 1188–1193
[6] Dowd B, Campbell G H, Marr R B, Nagarkar V, Tipnis S, Axe L and Siddons D P 1999 *Proc. SPIE* **3772** 224-236
[7] Rivers M L 2012 *Proc. SPIE* **8506** 85060U
[8] Vo N T, Atwood R C and Drakopoulos M 2018 *Opt. Express* **26** 28396-28412
[9] Paganin D M, Mayo S C, Gureyev T E, Miller P R and Wilkins S W 2002. *J.* Microsc. ***206*** 33-40
[10] Prell D, Kyriakou Y and Kalender W A 2009 *Phys. Med. Biol.* **54** 3881
[11] Shahani A J, Gulsoy E B, Gibbs J W, Fife J L and Voorhees P W 2014 *Opt. Express* **20** 24606-24621
[12] Zhang Y, Defer M, Liu W, Knipschildt-Okkels E F F, Oddershede J, Slyamov A, Bachmann F, Lauridsen E and Juul Jensen D 2024 *IOP Conf. Ser. Mater. Sci. Eng.* **this volume** submitted

Grain size assessment using EBSD on heterogeneous additively manufactured microstructures

J del Gaudio[1*], J Donoghue[1], P J Withers[1], A Garner[1,2], K Mingard[3], M Gee[3] and A Gholinia[1]

[1] Henry Royce Institute, Department of Materials, University of Manchester, Manchester, M13 9PL, United Kingdom

[2] Jacobs, Warrington, United Kingdom

[3] National Physical Laboratory, Teddington, United Kingdom

*E-mail: jacopo.delgaudio@manchester.ac.uk

Abstract. Grain microstructures are important for a whole range of materials properties. With the widespread use of additive manufacturing (AM), current standards for grain boundary evaluation, such as ASTM E2627-13, may not be fit for purpose due to their microstructural complexity. They are notoriously anisotropic across a range of length scales due to the non-equilibrium nature of the solidification process and the tracking of the heat source. In this work we examine the grain microstructure of a nickel superalloy produced by laser powder bed AM by electron back scatter diffraction. We find that the guidelines provided by the ASTM E2627-13 are not suitable for such heterogeneous and anisotropic materials and some modified guidelines are provided on the field of view and EBSD step size that should be used to characterise such materials.

1. Introduction

Grain size measurement is widely undertaken in industry and research and is a fundamental property that can significantly affect the overall material behaviour. The influence of grain size on the deformation and failure of polycrystalline metal is well documented, and for many models that have been developed to predict the performance of the materials, such as the Hall-Petch equation [1] for predicting yield strength.

The measurement procedure for grain size by electron backscatter diffraction (EBSD) is standardized by the ASTM E2627-13(2019) [2], which is an extension of the ASTM E112 (2021) [3], that is based on stereological principles [4]. It states that in order to perform an EBSD grain size measurement in compliance with the standard some requirements need to be met:

- No more than 10% of the points should be unindexed.
- The grains should be defined with grain boundary (GB) misorientation above 5°.
- Multiple maps can be used with minimum of 50 grains in each map until at least 500 grains have been measured.
- The average grain should contain about 500 pixels in the EBSD map.
- All grains must contain more than 100 pixels.

Following these guidelines the average area of the grains is calculated by $\bar{A} = \frac{1}{n}\sum_i^n A_i$ where A_i is the area of the grain i. The standard deviation is calculated on the average area of the grain $\bar{A}$. Finally, the ASTM grain size number "G" is calculated by using the formula $G = (-3.3223 \cdot \log \bar{A}) - 2.995\bar{A}$, where $\bar{A}$ is in mm². This procedure is based on stereological principles that assume the "independent uniform and random" (IUR) [4] selection of the grains on each map and the Gaussian distribution of the resulting grain size population.

Additive manufactured (AM) materials are built-up layer by layer in the "build direction" (BD) by melting and rapidly solidifying (cooling rate of 10^4-10^6 K/s [5]) the raw material in the form of powder or wire. The resulting microstructure is highly anisotropic and it exhibits a substantial divergence from a Gaussian distribution arrangement [6] due to the complex thermal history as well as a very wide range of grain sizes.

The aims of this work are to examine the extent to which the ASTM standard is suitable for the measurement of the grain size in AM materials, what parameters can be used to define the microstructure and to measure the grain size, and what EBSD conditions should be used in order obtain reliable and representative analyses.

2. Experimental details

2.1 Materials

The exemplar material studied is AM Hastelloy X produced on an EOS M400-4 laser powder bed system with a sinusoidal grid raster, using a powder with a nominal composition: Ni 47 wt.%, Cr 22 wt.%, Fe 18 wt.%, Mo 9 wt.%, Co 1.5 wt.%, W 0.6 wt.%, C 0.1 wt.%, Mn 1 wt.%, Si 1 wt.%, B 0.008 wt.%, Nb 0.5 wt.%, Al 0.5 wt.%, Ti 0.15 wt.%.

2.2 EBSD measurement and post-processing

The AM block was cut parallel to the BD. The samples were polished using standard mechanical procedure. The EBSD maps were acquired on a Zeiss Auriga 60 with an Oxford Instruments Symmetry 2 detector, at electron beam power of 30 kV and current of 10 nA. A small area map of 766 μm x 480 μm with a total area of 0.37 mm^2 was acquired at processing 1076 patterns/s with step size of 0.5 μm. A larger area map of 4293 μm x 1306 μm with a total area of 8.76 mm^2 (A_0) was acquired with the detector processing 2311 patterns/s and a step size of 1.25 μm. Each map was postprocessed using AZtecCrystal 2.2. The clean-up procedure consisted of "Auto Clean-Up", "Remove Wild Spikes", "Zero Solution Removal" with the level of "5-Five Neighbourhood" and 10 iterations on the small area map and the same procedure but with level of the zero solution removal at "4 -Four Neighbourhood" 10 iterations on the large area map. A total of 221813 pixels for the small area map (7%) and 505311 pixels for the large area map (9%) have been replaced, (i.e. within the ASTM guidelines).

3. Results and discussion

3.1 Setting the grain boundary criteria

The first step is to define what represents a grain boundary (GB). Figure 1 shows the highly anisotropic structure, both in terms of grain shape, with small equiaxed grains at melt pool boundaries and large elongated grains along the BD direction and a variety of differently shaped small grains, from equiaxed to elongated. The 5 and 10° GB criteria are highlighted by the white and black lines respectively. From this it is clear that the white lines represent sub-grain boundaries (21% of the total) and the black predominantly high angle GBs (79%). Clearly for this AM system the 5° misorientation proposed by ASTM is not appropriate, indeed it could be argued whether 10° is sufficient misorientation to correctly segment the grains. Throughout this paper 10° is used as the GB criteria.

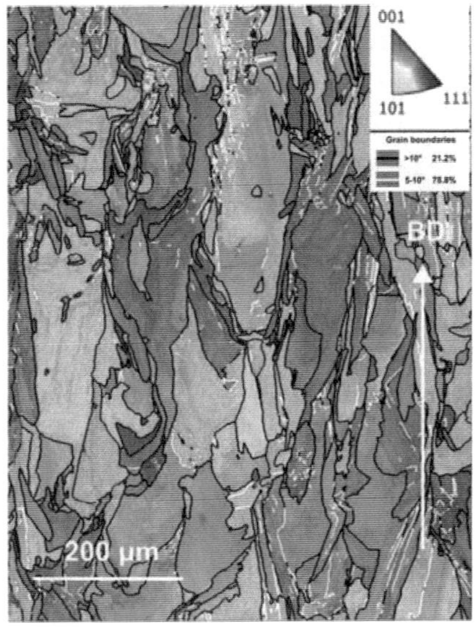

Figure 1. Small area EBSD map of Hastelloy X AM, in IPF-Z colours, step size 0.5 μm, 1533x480 pixels with the BD vertical. The 5° criteria is illustrated by the white lines, the 10° misorientation by the black lines.

3.2 Area required for grain size analysis

Collecting large area maps is time consuming. In this section we examine the distribution of grain sizes to examine whether it conforms to the Gaussian distribution and look at the dimensions necessary that should be used or exceeded to define a statistically representative grain size. We divided the large area (A_0) map in figure 2 into six progressively smaller subset sizes ($A_0/2$, $A_0/4$, $A_0/8$, $A_0/16$, $A_0/32$, $A_0/64$). This is done by progressively halving perpendicular to the BD (e.g. dashed white lines) or parallel to the BD (the white solid lines). In each subset size, 8 patches of the microstructure are taken at different locations. The area and equivalent circle diameter of the grains are measured, including only the grains that are wholly within the patch. Over the whole area (A_0) a total of 4522 grains have been detected with an average equivalent circle diameter of 36.7 μm and area weighted equivalent circle diameter of 94.5 μm.

The average equivalent circle diameters measured from the six subsets parallel to the BD and the six subsets perpendicular to the BD are compared in figure 3. For the subsets parallel to the BD the results show similar mean values for the four larger fields of view with the standard

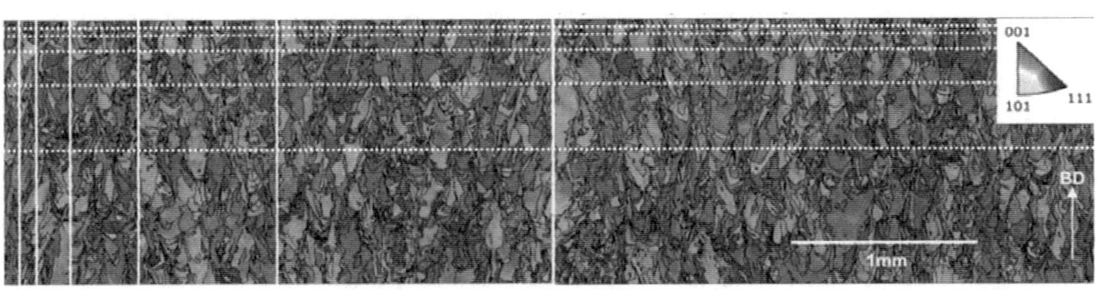

Figure 2. Large area EBSD map (A_0) in IPF-Z colours with the strategy to divide into subsets parallel to the BD in solid white lines and perpendicular to BD in dotted lines.

IOP Conf. Series: Materials Science and Engineering **1310** (2024) 012028 doi:10.1088/1757-899X/1310/1/012028

deviation increasing as the field of view decreases. This might suggest that a subset area ¼ or even ⅛ as wide as A_0 might suffice. By contrast for the subset areas perpendicular to the BD the results only begin to plateau for the $A_0/2$ subset area and this suggests that an area equal to the full height of the image in figure 2 is required. This result is of course related to the elongated shape of the larger grains and consequently the likelihood of cutting very large grains which, due to their orientation, are unlikely to be fully included in the field of view of the subset perpendicular to the BD. In both cases the average tends to smaller grain sizes as the sampled area becomes smaller because few larger grains lie within the subset area, because grains at the boarder are excluded. Therefore, similar number of grains need to be covered in either direction by matching the aspect ratio of the larger grains and the EBSD map.

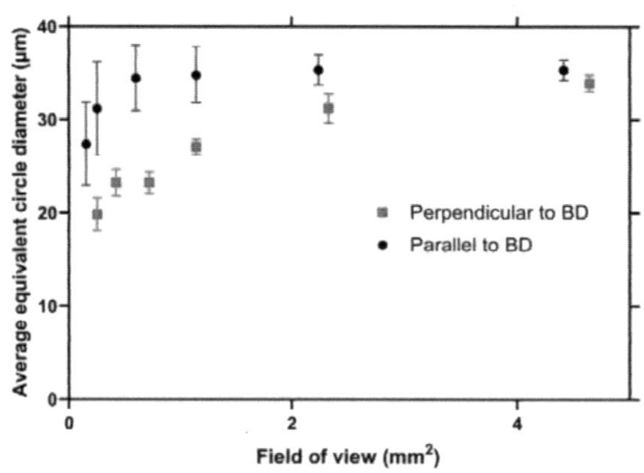

Figure 3. Measurement of the average grain equivalent circle diameter for the differently sized subsets from the large area map (A0/2, A0/4, A0/8, A0/16, Ao/32, A0/64) and 8 patches for each subset.

3.3 Step size required for grain size analysis

As well as considering the size of the patch sampled it is important to consider the sampling step size. Using a small step size over a large area is very cost and time inefficient while sampling at a coarse step size will exclude the smaller grains. In order to examine this the grain size (equivalent circle diameter) population was assessed for the largest subset ($A_o/2$) from the large area map using a step size of 1.25 μm and 2.5 μm by sub-sampling by 2x2. These populations are shown in figure 4(a) as histogram with bin width calculated with the Freedman-Diaconis rule [7] of 3.1μm. The resulting distribution is evidently highly non-Gaussian (contravening the ASTM requirement) and shows that over 40% of grains lie within the first three bins of the distribution for the 1.25 μm map. The distribution of the 2.5 μm map shows essentially the same distribution but with a different cut off due to the threshold on the minimum number of pixels (100) defining the grain in the grain detection procedure which determines the smallest recordable diameter for a grain. As a result the number fractions change significantly because >40 % of the grains recorded for the smaller step size are unrecorded such that the grain average increases from 39 μm for 1.25μm step size to 50μm at 2.5μm. Such a skewed population distribution means that the finer the step size the smaller the average measured grain size. However, as one would expect, it is clear from the grain area weighted population distribution in figure 4(b) that these smaller grains exert a smaller and smaller influence on the area weighted grain size. As a result the area weighted average grain size is much less dependent on the step size being 96 and 97 for the 1.25 and 2.5μm step sizes respectively.

The implications of this wide and highly skewed grain size distribution on the average grain size is shown in figure 5 as a function of the different sampling patch areas. For the reasons noted above, the change in step size introduces a significant shift towards a larger average grain size as

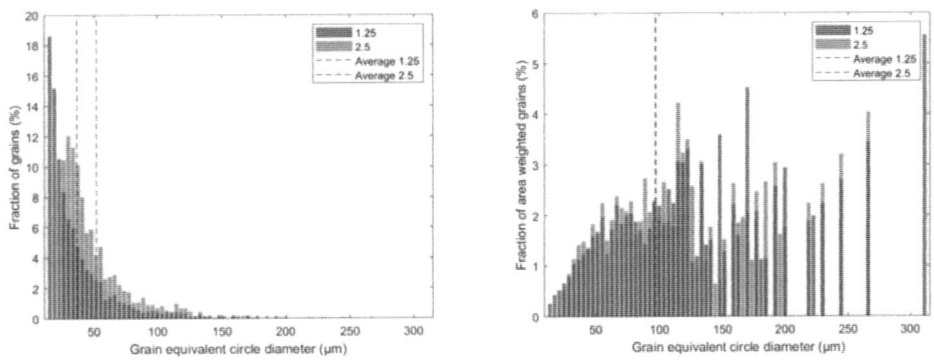

(a) (b)

Figure 4. (a) Distribution of the grain equivalent circle diameter for the largest subset ($A_0/2$) parallel to the BD with step size 1.25 μm (grey) and 2.5 μm (red). (b) Distribution of the grain area weighted diameter over the grain equivalent circle diameter for the largest subset parallel to the BD with step size 1.25 μm (grey) and 2.5 μm (red).

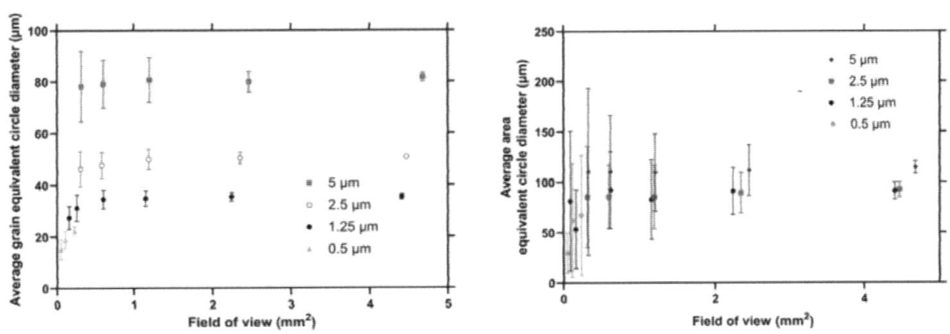

(a) (b)

Figure 5. Measurement of the a) average equivalent circle diameter and b) average area equivalent circle diameter for the differently sized subsets with 8 patches from each subset, which are sampled from the large area map as a function of the EBSD step size.

the step size is increased. By contrast the area average grain size is much less affected by the step size although the scatter increases as the size of the sampling subset is decreased.

4. Conclusions

In this paper we have looked at the suitability of the ASTM standard for grain size measurement for a typical AM microstructure. The grain structure is highly anisotropic and the grain size varies greatly from very small grains and very large columnar grains. The following conclusions are made:

1. 10° misorientation is a more appropriate grain boundary criterion than 5° because of the highly developed sub grain structure.
2. The grain size distribution in this AM microstructure is not Gaussian as required in ASTM standard, therefore it is not possible to follow the ASTM standard.
3. The average equivalent circle diameter reduces with reduction in step size, which is due to the possibility to capture finer grains at finer step sizes.
4. Because of the large tail of the fraction of very small grains, the average equivalent circle diameter is affected by the cut off or threshold used for allowable minimum grain diameter. However the average equivalent circle diameter is less affected by the threshold used for minimum grain diameter when considering the area weighted grain fraction.
5. The area or field of view required to achieve good grain size statistics in an EBSD map needs to have similar number of grains perpendicular and parallel to BD. Therefore the aspect ratio of an EBSD map for these types of AM microstructures will the aspect ratio of the grains.
6. The average weighted area fraction is less dependent on step size, which is a more robust method to capture grain size in such microstructures.

In an ideal world it would be preferable to measure the grain size in 3D using plasma dual beam or laser tribeam microscopy [8]. Future work is underway to determine the minimum block size required for characterisation of AM 3D complex grain structure. This work can provide guidelines on the dimensions of the sampled area and the in-plane step size that should be employed.

Acknowledgements

The authors are grateful for the resources provided by the Electron Microscopy Centre funded by the Henry Royce Institute (EP/X527257/1) for providing the microscopy capability. JdG is grateful for the sample and the funding provided by the National Physical laboratory and Jacobs.

References

[1] Hall E O 1951 The deformation and ageing of mild steel: III discussion of results *Proc. Phys. Soc. B* **64** 747
[2] Standard practice for determining average grain size using electron backscatter diffraction (EBSD) in fully recrystallized polycrystalline materials 2019 **ASTM E2627-13**
[3] Standard test methods for determining average grain size 2021 **ASTM E112-13**
[4] Wojnar L 2020 Stereology *Encyclopedia of Mathematical Geosciences* Springer international publishing ed B S Daya Sagar et al Cham: Springer International Publishing 1–12
[5] Mostafaei A, Ghiaasiaan R, Ho I, Strayer S, Chang K, Shamsaei N, Shao S, Paul S, Yeh A, Tin S and To A C 2023 Additive manufacturing of nickel-based superalloys: A state-of-the-art review on process-structure-defect-property relationship *Prog. Mater. Sci.* **136** 101108
[6] Zhang LC, Liu Y, Li S and Hao Y 2017 Additive Manufacturing of Titanium Alloys by Electron Beam Melting: A Review *Adv. Eng. Mater.* **20** 1700842
[7] Freedman D and Diaconis P 1981 On the histogram as a density estimator: L2 theory *Probab Theory Relat Fields* **57** 453
[8] Gholinia A, Donoghue J, Garner A, Curd M, Lawson M J, Winiarski B, Geurts R, Withers P J and Burnett T L 2024 Exploration of fs-laser ablation parameter space for 2D/3D imaging of soft and hard materials by tri-beam microscopy *Ultramicroscopy* **257** 113903

IOP Conf. Series: Materials Science and Engineering **1310** (2024) 012029 doi:10.1088/1757-899X/1310/1/012029

Applying lab-based DCT to reveal and quantify the 3D grain structure of a miniature chess rook produced by binder jetting

J Sun[1], F Bachmann[1], J Oddershede[1*] and E Lauridsen[1]

[1] Xnovo Technology ApS, Køge, Denmark

[*]E-mail: joddershede@xnovotech.com

Abstract. Imaging the 3D grain microstructure of additively manufactured (AM) metal samples non-destructively is key to a better understanding of the material performance. Recent advances of lab-based diffraction contrast tomography (DCT) enable non-destructively mapping the spatial distribution of crystallographic orientations in the bulk of certain types of metal AM parts. Combining these 3D texture measurements with traditional absorption contrast tomography (ACT) gives unprecedented insights into materials structure, such as the spatial distribution of porosities and the relationship between microstructural anisotropies and the build direction. Here, the combination of ACT and DCT is exemplified for a metal AM sample produced by binder jetting. This application highlights some of the recent technical progress enabling 3D mapping of metal AM structures and leads to a discussion of remaining challenges for lab-based DCT in the metal AM field.

1. Introduction

Producing metal components and parts by AM in a layer-by-layer fashion have many benefits when it comes to achieving intricate, customized geometries fast and at a low cost. However, metal AM parts still face a number of challenges, such as void formations, residual stresses and anisotropies in mechanical properties and/or microstructures. Integrating sophisticated, non-destructive 3D characterization techniques in the process design and quality control would be a valuable step towards overcoming some of these challenges.

A first important step towards 3D characterization of metal AM samples is the combination of EBSD and automated serial sectioning, see e.g. [1]. Although this approach is destructive, limited to relatively small sample volumes and may require specimens with intricate geometries to be cast into epoxy before mapping the grain structure, it benefits from a very high spatial resolution, and it also works for samples with residual stresses. More recently, results from serial sectioning of a laser powder bed fusion (PBF) Inconel 625 alloy sample were compared to non-destructive 3D data from a high-energy diffraction microscopy (HEDM) experiment [2]. There, the overall grain structure was found to be the same, with noticeable differences in the number and spatial arrangement of annealing twins. Later, HEDM was also used to characterize recrystallization during annealing of laser powder bed 316L stainless steel sample [3].

Because near-field HEDM synchrotron experiments are performed with a line-focused beam in a layer-by-layer manner, the resulting grain maps typically have a relatively high resolution, both spatially and angularly. Synchrotron DCT [4,5], applying a box beam, is a significantly faster grain mapping alternative for samples with limited levels of deformation. When DCT migrated to the lab-based X-rays sources [6] and was commercialized as the LabDCT™ solution for ZEISS Xradia Versa and CrystalCT™ X-ray microscopes [7–9] it became possible to perform non-destructive 3D grain mapping of similar quality in the users own home lab, with the lower brilliance of the lab-based X-ray source limiting only the data collection speed and the minimum grain size.

An example of successfully applying lab-based DCT to obtain a 3D map of 100,000+ grains in a metal AM sample with an intricate morphology will be outlined in Section 2. This experiment has been enabled by recent technical advances, namely the introduction of advanced acquisition strategies for seam-less collection and reconstruction of data from longer, larger, high-aspect ratio samples [10], along with developments to boost the reconstruction algorithm, GrainMapper3D™ [11]. Examples of quantifying structural anisotropies with respect to the build direction from the 3D grain map will also be given in Section 2, before rounding up with a discussion of what works and why, and which challenges still remain for lab-based DCT of metal AM samples in Section 3 and conclusions in Section 4.

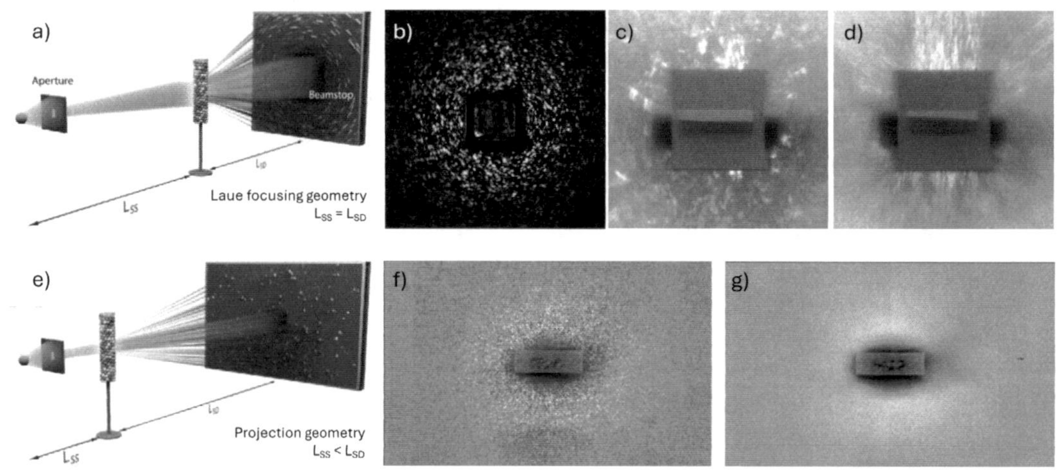

Figure 1. DCT setups and corresponding diffraction contrast patterns (DCPs). a) Laue focusing geometry and DCPs from b) the binder-jetted chess rook and c)-d) different parts (with varying amounts of residual stresses) of a Ni-based superalloy IN718 structure in the form of Mona Lisa produced by electron beam PBF [16]. e) Projection geometry and DCPs from HIAM [17] honeycomb structures of f) coarse-grained Fe and g) fine-grained CuNiFe.

2. Example of applying lab-based DCT to a metal AM miniature chess rook

2.1 Chess rook – binder jetting

The chess rook is made of 316L stainless steel and produced by binder jetting [12]. After the inherent sintering step, it was further annealed. It has a total heigh of 8.2 mm along the vertical cylinder and build direction, and max. and min. diameters of 4.8 mm and 2.9 mm at the base and in the central part, respectively. The central part has a wall thickness of 270 μm and an inner winding staircase with a step height of 360 μm.

2.2 3D experiment

The experimental conditions for acquiring the ACT data of the full chess rook can be found in Table 1 along with select settings for the DCT experiment. The DCT data of the central part (2.6

IOP Conf. Series: Materials Science and Engineering **1310** (2024) 012029 doi:10.1088/1757-899X/1310/1/012029

mm in height) was collected in Laue focusing geometry (figure 1(a), $L_{SS} = L_{SD} = 15$ mm) using the Helical Phyllotaxis Raster advanced acquisition strategy [10]. An example diffraction pattern can be seen in figure 1(b). The grain map reconstruction was performed in GrainMapper3D v.3.2 using three hkl families and a voxel size of 6 μm. Adjacent regions in space were assigned to the same grain if the voxel-to-voxel misorientation was less than 0.3°, yielding just above 110,000 grains in the mapped volume.

Table 1. Experimental settings

Column heading	Source Voltage	Filter or Aperture	Detector (binning)	Number of Projections	Exposure Time	Voxel Size
Chess rook, ACT	160 kV	-	0.4X (1)	2,401	2 s	4.78 μm
Chess rook, DCT	100 kV	375x375	DCT4X (2)	4,505	45 s	6 μm

2.3 Quantification of the grain structure

Figure 2 shows the full 3D morphology of the chess rook from the ACT reconstruction overlaid with the DCT grain map of the central part colored by IPF. From figure 2(c) and the pole figures in figure 3(f)-3(h) (note that min. and max. are 0.9 and 1.1 mrds, respectively) it is clear that the grain orientations are not textured with respect to the build direction. Figure 3(a) shows the grain size distribution, given as equivalent sphere diameters (ESDs), which is found to follow a standard lognormal distribution with a mean around 30 μm. Further analysis of the spatial distribution of different grain sizes reveals no correlation between grain size and location within the sample. A closer look at the ACT reconstruction shows 15-25 μm porosities at some of the grain boundaries. Here it should be noted that the smallest detectable void size is given by the current voxel size. Furthermore, the surface roughness is found from the ACT reconstruction to be 25-40 μm, which is essentially one grain diameter, c.f. figure 3(a)).

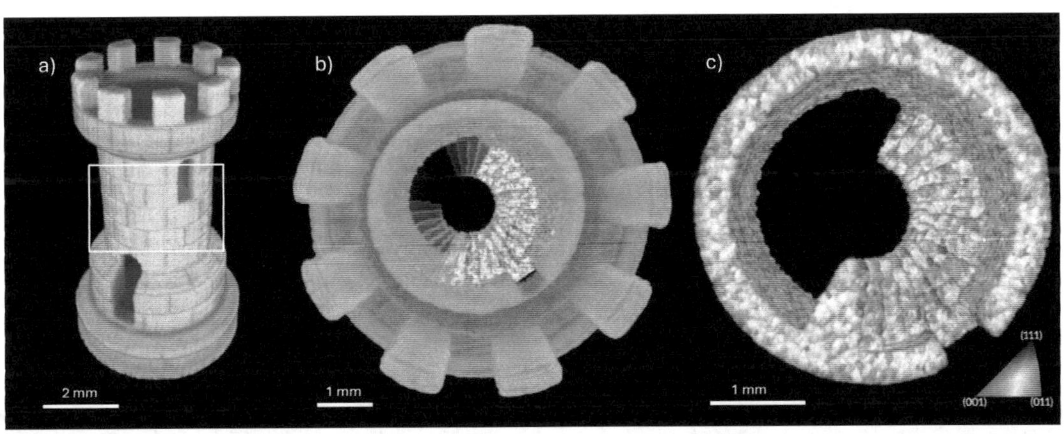

Figure 2. a) 3D shape of the chess rook from ACT seen from the side with the central part mapped by DCT boxed in. b) Top view showing the inner spiral staircase in an overlay of the ACT (grey) and the DCT grain map (IPF color along vertical cylinder and build axis). c) Zoom of grain map (IPF).

In order to investigate whether the grain shapes are irregular with respect to the build direction, the spatial distribution of voxels assigned to each grain were approximated by ellipsoids with radii $a \leq b \leq c$ as exemplified in figure 3(d) and 3(e) where the voxels belonging to two different grains are shown as dots. Defining the ellipsoid factor [13], $EF = a/b - b/c$, where $EF < 0$ corresponds to oblate and $EF > 0$ to prolate grains, respectively, the ellipsoid factor versus grain size can be plotted in figure 3(b). Here the ellipsoid factor of the prolate grain in figure 3(d) is marked by a red cross, while EF for the oblate grain in figure 3(e) is marked by a blue cross. Based on this we conclude that, although we see a slightly larger spread in EF for smaller grains, there is no preferred sign of EF, hence a certain grain shape (*e.g.* oblate or prolate along any direction) is not dominating. Finally, figure 3(c) is a so-called Flinn diagram, originally used to describe strain anisotropy in geology [14] and later adapted to quantify changes in the micro-architecture of trabecular bone caused by osteoporosis [15]. It plots a/b versus b/c for each ellipsoid, and the closer the observations are to the upper right corner at (1,1), the more spherical the objects are. The Flinn diagram confirms that the predominant grain morphology is close-to-spherical with a symmetric spread of oblate and prolate shapes. Had we found a preferred grain shape, the 3D data would have allowed us to correlate the axis of elongation (corresponding to c, when $a \approx b$) or contraction (corresponding to a, when $b \approx c$) with any direction in the sample, particularly with the build direction.

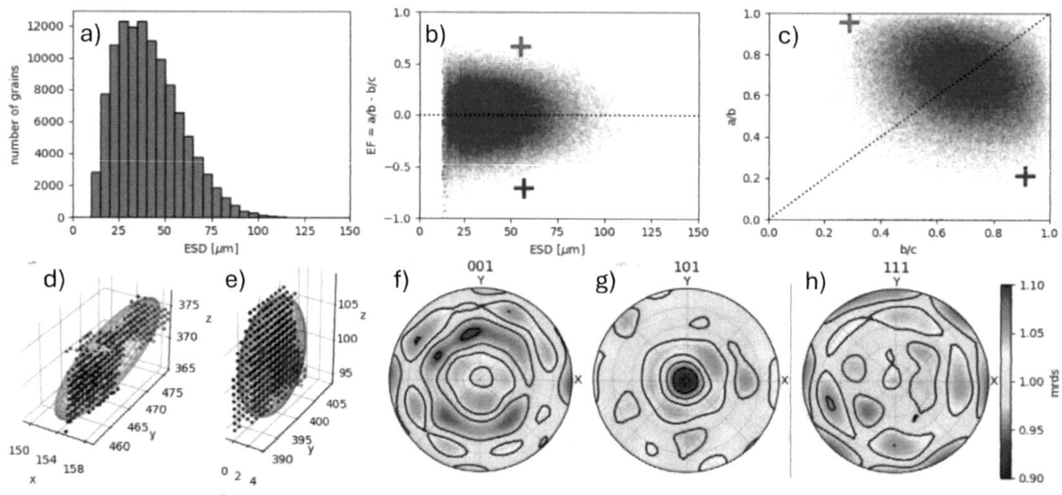

Figure 3. Grain size, shape and texture quantification for the chess rook. a) Grain size distribution. b) Ellipsoid factor versus grain size. c) Flinn diagram. Ellipsoid fits to the voxels (represented by dots) assigned to d) a prolate (EF=0.67) bulk grain and e) an oblate (EF=−0.70) surface grain, respectively, marked with the same color crosses in b) and c). f)-h) Pole figures along the vertical cylinder and build axis.

3. Discussion

3.1 Remaining challenges for lab-based DCT mapping of AM samples

Some of the inherent microstructural features of metal AM microstructures are currently still challenging for 3D grain mapping with lab-based DCT, and to some degree also for synchrotron DCT and HEDM. The first main challenge is diffraction spots that smear radially as exemplified in figure 1(c) and 1(d), which were collected at the top and bottom, respectively, of a Ni-based superalloy IN718 structure in the form of Mona Lisa produced by electron beam PBF [16]. The typical causes for smearing of diffraction spot are either large grain mosaicities (intragranular orientation gradients) or residual stresses in as-printed samples. The second main challenge is structures with grain sizes that fall below the detection limit due to the limited flux of the lab-based X-ray source. This is exemplified in figure 1(f) and 1(g), which were collected on coarse-grained Fe and fine-grained CuNiFe honeycomb structures produced by hydrogel infusion additive manufacturing (HIAM) [17], respectively. These data are collected in projection geometry (figure 1(e)) as opposed to the Laue focusing data in figure 1(b)-1(d).

The obvious way forward to address these challenges would be moving towards a more focused beam [18,19], which should both improve the signal-to-noise ratio for small grains and make the illuminated volume small enough that peak overlap due to deformation smearing is minimal. The scanning protocols are already in place in terms of the advanced acquisition strategies, while the reconstruction algorithm may need some updating to handle the variations in lattice constants across the sample.

3.2 Lab-based DCT on AM samples, what works and why?

The abovementioned example of a metal AM structure successfully characterized by lab-based DCT was performed on a sample produced by binder jetting. Also, the coarse-grained Fe HIAM honeycomb structure in figure 1(f) seems to produce reconstructable DCT data, hence it would be natural to look at the similarities of these two processes in relation to some of the other AM methods that have resulted in samples unsuitable for lab-based DCT. Both binder-jetting and HIAM works with metal particles of a composition and size that can be chosen by the operator. These particles are bound together, either by adhesive droplets applied by a print head (binder-jetting) or by infusing them into a hydrogel (HIAM). At the end, the AM structure is cured and further densified by sintering/calcination. Essentially both processes to a large degree preserve the original grain structure of the metal powders (no melting) and the final heat treatment, besides densifying the structure, may also help to relieve any potential residual stresses from the build process. As opposed to this, many other metal AM processes are based on (partially) melting the powder precursor material such that the particles bind together during solidification. The electron beam PBF technique used to produce the IN718 Mona Lisa sample [16] is one such example. The corresponding diffraction contrast patterns with large, smeary spots in figures 1(c) and 1(d) clearly illustrate that the resulting structure does not bear much resemblance of the precursor powder particles, but rather has notably larger sample regions of similar orientation with significant residual stresses.

4. Conclusion

Lab-based DCT holds great promise for mapping the 3D grain structure of metal AM parts, especially if these are produced from precursor particles larger than the grain detection limit (currently 10-20 μm) [20]. Good results have been achieved for different ways of adjoining the particles, and although it appears preferable if the particles are adjoined without a melting and solidification step, this is by no means a requirement. Finally, it seems to be beneficial if the printing process includes an inherent additional heat treatment to relieve residual stresses. Developments are on the way to handle smaller particles and more deformed structures. Finally, it is worth mentioning that already now more than 100,000 grains inside metal AM parts with large, intricate geometries can be mapped within 2-3 days to yield quantitative 3D information about the relationship between microstructural anisotropies and the build direction.

Acknowledgements

The authors thank Digital Metal (now Markforged Sweden) for providing the chess rook sample, Alex Plotkowski, Oak Ridge National Laboratory, for providing the Mona Lisa sample and Sammy Shaker, Caltech, for providing the honeycomb structures.

References

[1] Rowenhorst D J, Nguyen L, Murphy-Leonard A D and Fonda R W 2020 *Curr. Opin. Solid State Mater. Sci.* **24**

[2] Chapman M G, Shah M N, Donegan S P, Scott J M, Shade P A, Menasche D and Uchic M D 2021 *Integr. Mater. Manuf. Innov.* **10** 129–41

[3] Liu H, Zhang Y, Wilkin M, Park J-S, Kenesei P, Rollett A D and Suter R M 2022 *IOP Conf. Ser. Mater. Sci. Eng.* **1249** 012054

[4] King A, Johnson G, Engelberg D, Ludwig W and Marrow J 2008 *Science (80-.).* **321** 382–5

[5] Ludwig W, Reischig P, King A, Herbig M, Lauridsen E M, Johnson G, Marrow T J and Buffière J-Y 2009 *Rev. Sci. Instrum.* **80** 033905

[6] King A, Reischig P, Adrien J, Peetermans S and Ludwig W 2014 *Mater. Charact.* **97** 1–10

[7] McDonald S A, Reischig P, Holzner C, Lauridsen E M, Withers P J, Merkle A P and Feser M 2015 *Sci. Rep.* **5** 14665

[8] Holzner C, Lavery L, Bale H, Merkle A, McDonald S, Withers P, Zhang Y, Jensen D J, Kimura M, Lyckegaard A, Reischig P and Lauridsen E M 2016 *Micros. Today* **24** 34–43

[9] Bachmann F, Bale H, Gueninchault N, Holzner C and Lauridsen E M 2019 *J. Appl. Crystallogr.* **52** 643–51

[10] Oddershede J, Bachmann F, Sun J and Lauridsen E 2022 *Integr. Mater. Manuf. Innov.* **11** 1–12

[11] Sun J, Bachmann F, Oddershede J and Lauridsen E 2022 *IOP Conf. Ser. Mater. Sci. Eng.* **1249** 012045

[12] Digital Metal (now Markforged Sweden) *https://markforged.com/3d-printers/px100*

[13] Doube M 2015 *Front. Endocrinol. (Lausanne).* **6** 1–5

[14] Flinn D 1962 *Q. J. Geol. Soc. London* **118**

[15] Felder A A, Monzem S, De Souza R, Javaheri B, Mills D, Boyde A and Doube M 2021 *R. Soc. Open Sci.* **8**

[16] Plotkowski A, Ferguson J, Stump B, Halsey W, Paquit V, Joslin C, Babu S S, Marquez Rossy A, Kirka M M and Dehoff R R 2021 *Addit. Manuf.* **46** 102092

[17] Saccone M A, Gallivan R A, Narita K, Yee D W and Greer J R 2022 *Nature* **612** 685–90

[18] Lynch P A, Stevenson W, Fox D, Ghaderi A, Xie L and Barnett M R 2019 *SSRN Electron. J.* 0–27

[19] Seret A, Kehres J, Gundlach C, Lund Olsen U, Friis Poulsen H, Juul Jensen D, Cordier M, Stripe B, Yun W and Zhang Y 2023 *Opt. Express* **31** 38840

[20] Oddershede J, Sun J, Gueninchault N, Bachmann F, Bale H, Holzner C and Lauridsen E 2019 *Integr. Mater. Manuf. Innov.* **8** 217–25

Influence of printing parameters on part density in L-BPF of Ti-6Al-4V and correlation with static mechanical properties measured using indentation testing

J Schulz[1*], E C Santos[1], A Kalliath[2], B Schmaling[2], P Zok[2], S Siegert[2], I S Dandu[2], KH Lindner [2], E Bruder[3] and K Durst[3]

[1] IQS, Carl Zeiss IMT GmbH, Oberkochen, Germany
[2] Imprintec GmbH, Bochum, Germany
[3] Department of Physical Metallurgy, TU Darmstadt, Germany

*E-mail: julian.schulz@zeiss.com

Abstract. A complete workflow of parameter development and process qualification of a part manufactured using laser powder bed fusion (L-PBF) AM technique requires data of porosity, pore morphology, distortion and static mechanical properties. Using a combination of two non-destructive techniques: Computed tomography and indentation testing, the influence of printing parameters on defect types and part density in L-PBF of Ti-6Al-4V is described. In addition, the results using the combined method are compared to conventional testing methods.

1. Introduction

Possible applications of Additive manufacturing (AM) of metal parts are e.g. re-printing legacy parts, production of functional prototypes for mechanical engineering or production of highly complex geometries where other fabrication methods might be replaced in the future [1].

One of the most used techniques for metal AM today is laser powder bed fusion (L-PBF). This technique uses a laser paired with a scanning system traces the area in each layer that needs to be melted to make up the final part. The main parameters in the laser exposure, that affect the part quality in terms of porosity are laser power, scanning speed and hatching distance (distance between the laser tracks). Together with the layer height, these four parameters determine the volumetric energy density (VED) of the printing process, defined by equation (1).

$$\text{VED} = \frac{\text{laser power}}{\text{scanning speed} \cdot \text{layer height} \cdot \text{hatch distance}} \tag{1}$$

The volumetric energy density highly impacts the likelihood for two common but detrimental kinds of defects in additive manufacturing, namely lack of fusion (LoF) and keyhole (KH) porosity. Typically, with a low VED the powder scanned by the laser is not fully melted, resulting in the neighbouring tracks not being fused together thoroughly, which will cause lack of fusion type porosity in the final part. In contrast a high VED leads to the pool of molten metal being too deep and liquid metal being ejected from it, forming a defect that when observed after cutting a sample sideways looks similar to a traditional keyhole. The consecutive layers above partially close this keyhole, leaving behind a spherical keyhole pore [2].

Traditionally AM parameters are developed by printing test specimens and determining their density either using Archimedes principle or light microscopy of the cross-sections. Archimedes only gives an average absolute density for the whole part, so the information about pore distribution (borders vs. centre of part) and their shape is not captured. Cross-sections do give information about pore distribution and do not rely on a reference material density, but they are only giving two-dimensional information and sample preparation is slow when compared to Archimedes or computed tomography (CT).

Computed tomography solves the short comings of the previously mentioned analysis methods for porosity in AM parts, by giving three-dimensional information about individual pore locations, their shape (allowing differentiation between LoF and KH) and size (relevant for fatigue), while still being a non-destructive testing (NDT) method like Archimedes. At the same time geometrical deviations from a parts CAD model can also be measured from the CT data. In terms of time effort, novel reconstruction techniques as published by [3] combined with the automated evaluation software Zeiss VERCES enable CT scanning of a 15mm test specimen (ZTO) at 17μm resolution in 15 minutes. This resolution is the output of optimizing between detectability of pores, that are relevant for mechanical properties, and scanning time. Reference scans at higher resolutions showed that smaller pores do not significantly affect the determined porosity.

Traditional mechanical testing requires testing sample of a specific geometry (e.g. tensile samples or fatigue samples) to be produced, which increases time and cost for developing new materials, while still not giving information on the exact manufactured part, but just on the printing process of the specific alloy.

Indentation testing is an excellent method to measure local static mechanical properties directly on the printed part as it is quasi-non-destructive and does not require expensive and time-consuming sample preparation. The i3D technology developed at Imprintec uses a sphero-conical indenter tip and residual imprint geometry to determine the static mechanical properties. As the indenter tip penetrates the material with a certain force, the resulting plastic deformation causes vertical displacement or pile-up of the surrounding material. This is known as the residual imprint. The three-dimensional geometry of the residual imprint is captured using a white light interferometer. A numerical finite element model is set up based on the experiment. The imprint profiles (vertical displacement of the material surface) resulting from the numerical computation and experiment are compared, and an optimisation algorithm varies the material parameters of the simulation until a satisfying convergence of the parameters. The converged parameters from the inverse analysis effectively represent the material parameters [4].

In this paper the influence of LoF and keyhole porosity on the static mechanical properties, when measured using either traditional uniaxial tensile testing or novel indentation techniques, is compared, showcasing the possibilities of new material characterisation and testing approaches.

2. Materials and Methods

In this study AM samples made from Ti-6Al-4V (Ti64) were manufactured using the laser powder bed fusion process (L-PBF) using 3 different parameters aiming for samples with high density, artificial lack of fusion and keyhole porosity, respectively. Ti-6Al-4V is an alpha-beta alloy, with the alpha phase being a hexagonal-closed-packed crystal structure and the beta phase being body-centred-cubic. Ti-6Al-4V offers excellent mechanical properties, especially in relation to its

IOP Conf. Series: Materials Science and Engineering **1310** (2024) 012030 doi:10.1088/1757-899X/1310/1/012030

specific weight, making it ideal for applications in the medical (e.g. hip or dental implants) or aerospace field.

2.1 Geometry of samples
The selected parameters were tested with a geometry called "Zeiss Test Object" (ZTO) including features like overhangs, walls of different thickness and rods (cylinders) of different diameters on top of the bulk base. Figure 1 shows a CAD model of the ZTO (a) and real-world image of the printed part (b).

For the uniaxial tensile testing samples with a gauge diameter of 5mm, gauge length of 25mm and M8 thread for mounting have been produced. Samples were printed with a little excess material in the radial direction, to machine them down to tolerance and eliminate impact from AM surface roughness.

2.2 L-PBF parameters and post-processing
The parts were printed in an EOS M290 using the L-PBF process with the parameters in table 1. The #LoF parameter features the lowest VED provoking lack-of-fusion porosity. The #KH parameter is scanning the laser path too slowly for its given power, hereby provoking keyhole porosity. The #Opt parameter is selected to obtain very high material density. After printing the parts were heat-treated at 850°C for 2 hours under argon atmosphere with successive quenching in water. This results in a similar microstructure for all four conditions after the heat-treatment, so only the different types of porosity are affecting the mechanical properties.

For the samples used in uniaxial tensile testing the post-processing consists of machining the gauge region to attain a consistent diameter with a low roughness surface finish and cutting threads into the top and bottom of the samples to allow mounting in the tensile machine.

Since the indentation test is a quasi-non-destructive testing method done on the surface of a test sample (ZTO), the material surface quality greatly affects the overall measurement. Therefore, it is vital to have the surface adequately prepared. The test surface of the ZTO is prepared using abrasive papers mounted on rotary disc grinding machines. The ZTO is held in contact parallel to a wet rotating disc, thus removing small amounts of material by abrasion. In the beginning, abrasive paper of coarse grit size (400/600) is used, and progressively, finer grit paper (2500) is used to ensure a better surface finish.

2.3 Non-destructive testing using CT
In the CT investigation, the samples were mounted in a designated sample holder to allow for up to 10 samples to be scanned successively. This ensures that the samples are centred to the rotation axis of the CT system, and they are slightly tilted to reduce CT artefacts. For shortened CT scans,

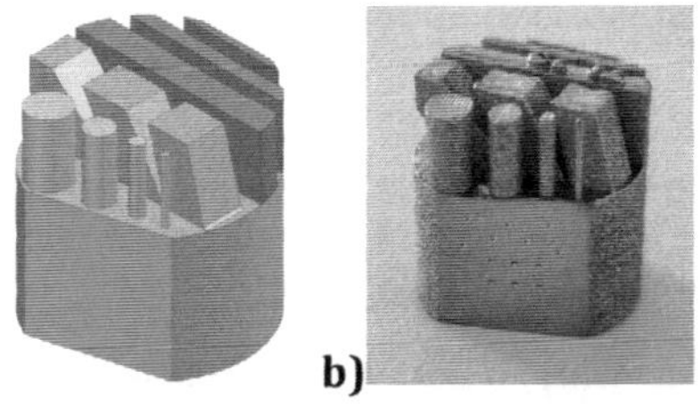

a) b)

Figure 1. a) CAD model of the ZTO testing geometry. The cylindrical base is 15mm in diameter and 10mm in height. Diameter of the rods is 3mm/2mm/1mm/0.5mm, wall thicknesses are 4mm/2mm/0.5mm and overhang angles are 15°/30°/45°. b) Printed ZTO with indentations from i3D testing.

IOP Conf. Series: Materials Science and Engineering **1310** (2024) 012030 doi:10.1088/1757-899X/1310/1/012030

Table 1. Printing parameters used in the production of the ZTO samples and test geometries for uniaxial tensile testing. The layer height was 60µm for all parameters.

Process parameter	Laser power	Scanning speed	Hatch spacing	VED
#LoF	370 W	2000 mm/s	120 µm	25.7 J/mm³
#Opt	370 W	1250 mm/s	120 µm	41.1 J/mm³
#KH	200 W	600 mm/s	120 µm	46.3 J/mm³

the output data was enhanced using the techniques described in [3]. The segmentation of the volumes was done using Zeiss Inspect X-Ray in conjunction with a specifically developed add-on package, that allows for repeatable batch processing. Simultaneously, geometrical properties of the pores allow differentiation of different types of defects.

2.4 Static mechanical properties from indentation testing using i3D®
For the indentation testing, the prepared ZTO is held in a V-shape sample holder. This ensures that sample under investigation is securely held during the entire procedure. A special diamond sphero-conical indenter with opening angle of 90° and tip radius of 350µm (referred as Typ-A indenter) is selected. The indentation is performed with a load of 588.6N and the entire procedure is performed in accordance with DIN SPEC 4864:2019-11 [5]. The 3D residual imprint geometry is measured using an optical white light interferometer (WLI). i3D ® algorithm calculates the plastic flow curve from each indentation using the geometrical data from the residual imprint. For statistical accuracy, an averaged plastic flow curve from 5-6 indents is considered. The entire procedure is fully automated and requires less sample preparation compared to the tensile test.

3. Results and Discussion

3.1 Porosity analysis by CT
For the CT analysis, ZTO samples with identical parameters to those used in indentation testing have been manufactured and scanned. The results of these investigations are summarized in table 2. The trends in the table clearly indicate, that the optimal parameter performed the best with less than 0.1% detected porosity in the samples. The lack-of-fusion parameter also showed lower porosity than the keyhole parameter. The samples without heat-treatment showed overall lower porosity than the heat-treated ones, however those samples were printed, and CT scanned in a different batch, which can account for those fluctuations.

3.2 Light microscopy of microstructures
Additionally, to the CT investigations, metallographic samples were produced and etched with Kroll's reagent for analysis using light microscopy. Figure 2 shows images of the microstructure of all printing parameters used in this paper. The microstructure in both cases shows fine acicular martensitic structures, which are on the order of 100µm in diameter. At the same time a clear differentiation between the pores caused by lack of fusion and keyholing mechanism can be made.

3.3 Indentation testing (I3D technology)
The results from indentation testing include 5-6 indents on one specimen per condition (table 3). The specimens labelled as #LoF and #KH show low strength for both conditions as-built and heat-

Table 2. Porosity in the ZTO test samples, determined by CT analysis using Zeiss Inspect X-Ray for evaluation of the volumetric data. The data for all parameters (except the one noted separately) was averaged from two samples. The as-built and heat-treated samples where printed and analysed in separate batches.

Process parameter	Heat-treatment	CT porosity	Heat-treatment	CT porosity
#LoF	none	0.138 %	HT 850°/2h	0.489 %
#Opt	none	0.039 % [a]	HT 850°/2h	0.080 %
#KH	none	0.978 %	HT 850°/2h	1.371 %

[a] Porosity value was averaged from only one ZTO specimen.

treated compared to the specimen with optimal parameters (#Opt). The agreement between the uniaxial tensile test and the determination of the mechanical properties by indentation is on average 4.4% for the yield strength and 1.6% for UTS. The maximum deviation is 9.5% for the yield strength and 5.4% for UTS. The calculation of the mechanical properties is based on an isotropic material model. If there is an anisotropic material behaviour due to the L-PBF process, this can influence the results. Other factors should be considered when comparing with the tensile test: The indentation test was not carried out on the same specimens and was indented perpendicular to the build direction. The tensile tests, on the other hand, were performed in the build direction.

3.4 Uniaxial tensile testing

The uniaxial tensile testing experiments used 3 specimens for each condition, averaging the calculated properties like yield strength (YS) and ultimate tensile strength (UTS). The static mechanical properties do not vary significantly between the different process parameters and respectively levels and type of porosity. There is a significant difference between the samples, that underwent heat-treatment and the ones left in as-built condition. This can be explained by the coarsening of the fine-grained microstructure from the AM process during the annealing. The difference between the yield strength and UTS is consistent among all samples indicating a similar impact of work hardening regardless of the heat-treatment used.

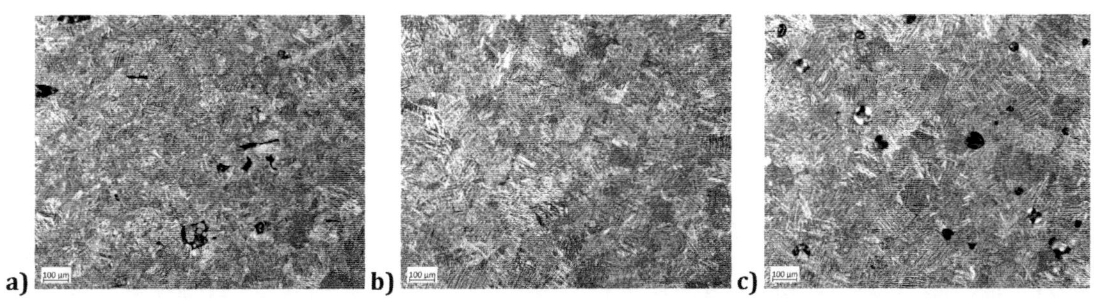

Figure 2. Light microscopy of metallographic surfaces etched with Kroll's reagent using polarized light to enhance contrast. All samples were heat-treated at 850°C for 2h and built with a) #LoF parameter, b) #Opt parameter, c) #KH parameter.

Table 3. Static mechanical properties determined by uniaxial tensile experiments (columns 3-4) and i3D indentation technology (columns 5-6). The uniaxial tensile data was averaged from 3 individual specimens per condition.

Process parameter	Heat-treatment	YS $R_{p0.2}$ (uniaxial)	UTS R_m (uniaxial)	YS $R^I_{p0.2}$ (indent.)	UTS R^I_m (indent.)
#LoF	as-built	1116 MPa	1156 MPa	1016 MPa	1140 MPa
#Opt	as-built	1107 MPa	1158 MPa	1076 MPa	1174 MPa
#KH	as-built	1119 MPa	1154 MPa	1013 MPa	1099 MPa
#LoF	HT 850°/2h	960 MPa	1016 MPa	999 MPa	1015 MPa
#Opt	HT 850°/2h	955 MPa	1033 MPa	1041 MPa	1089 MPa
#KH	HT 850°/2h	970 MPa	1020 MPa	980 MPa	1020 MPa

4. Conclusion

Development of parameters for new alloy along with process qualification for additive manufacturing using L-PBF requires fast and reliable testing of porosity and mechanical properties. While most conventional methods do not provide a fast way of doing so and lack information about mechanical properties (e.g. cross sections are time consuming to prepare and Archimedes averages over volume), the combination of CT using ZEISS VERCES and indentation testing using Imprintec i3D overcomes those limitations by providing complete test information, quickly reliably and quasi-non-destructively. The influence of both LoF and keyhole porosity, at the level produced in this work, on static mechanical properties is low, when measured with uniaxial tensile testing. Pores, which are localized to the indentation sites cause a higher impact of porosity on the properties determined from the i3D technology, due to high susceptibility to local changes in material properties (e.g. irregularities in the microstructure).

Deviation between uniaxial tensile testing and indentation testing will be investigated further to increase alignment of the two measurement techniques. Since the entire indentation procedure can be performed in 20 minutes including sample preparation, this is considerably lower in time and effort when compared to the uniaxial tensile test, where sample preparation by machining can take easily multiple hours. This clearly gives opportunities where this technology is advantageous over traditional testing methods.

References

[1] Montero J, Weber S, Bleckmann M and Paetzold K 2020 A methodology for the decentralised design and production of additive manufactured spare parts *Prod. Manuf. Res.* **8** 313–334
[2] Du C, Zhao Y, Jiang J, Wang Q, Wang H, Li N and Sun J 2023 Pore defects in Laser Powder Bed Fusion: Formation mechanism, control method, and perspectives *J. Alloy. Compd.* **944** 169215
[3] Ziabari A et al. 2023 Enabling rapid X-ray CT characterisation for additive manufacturing using CAD models and deep learning-based reconstruction *Npj Comput. Mater.* **9** 91
[4] Schmaling B and Hartmaier A 2012 Determination of plastic material properties by analysis of residual imprint geometry of indentation *J. Mater. Res.* **27** 2167-2177
[5] Deutsches Institut für Normung 2019 DIN SPEC 4864:2019-11. Test method for the determination of flow curves and benchmark characteristic values for tensile testing by means of minor destructive indentation, 3D measurement and finite-element material models (Berlin: Beuth Verlag)

Cracking Mechanism in E-Beam 3D-Printed DZ125 Ni-based Superalloys

Z Wang, Y Lin, Y Zhao, F Shangguan and K Chen

[1] Center for Advancing Materials Performance from the Nanoscale (CAMP-Nano), State Key Laboratory for Mechanical Behaviour of Materials, Xi'an Jiaotong University, Xi'an, Shaanxi 710049, P. R. China

*E-mail: kchenlbl@gmail.com

Abstract. Directionally solidified Ni-based superalloys are extensively employed to manufacture turbine blades due to their outstanding high-temperature mechanical properties. To reduce overall costs, repairing techniques are highly demanded to restored the shape and properties of damaged turbine blades. In this study, as a surrogate for the repair process, DZ125 Ni-based superalloys were grown epitaxially with the electron beam powder bed fusion 3D-printing method on a base metal with the same chemical composition. Cracks are detected within the printed part, always along the high-angle grain boundaries and roughly parallel to the building direction. The cracks are identified to be liquation cracks, and the thermal cycling effects are proved to play an important role in crack initiation and propagation. The knowledge gained from this work provides valuable insights towards 3D-printing strategy development to obtain crack-free directionally solidified superalloys.

1. Introduction

Directionally solidified (DS) Ni-based superalloys play a vital role in manufacturing the hot sections of gas turbine engines, showcasing significantly improved high-temperature performance comparing to their polycrystalline counterparts. The absence of transverse grain boundaries and the substantial presence of coherent intermetallic precipitates, such as γ'-$Ni_3(Al, Ti)$, impart exceptional resistance to creep [1] and fatigue [2] in these materials. Nonetheless, the extensive applications in harsh environment frequently result in damages of DS superalloy blades. To prolong the service life and lower the overall costs, repairing techniques are highly demanded to restore the shape, microstructure, and performance of damaged superalloy parts [3]. The recently emerged additive manufacturing (AM) process, also known to public as 3D-printing, especially electron beam powder bed fusion (EB-PBF), holds great promise towards this goal [4]. This layer-by-layer fusion-based process exhibits high design flexibility to produce DS superalloys with high geometry complexity; however, cracking is still the first of all challenge to overcome before the application of 3D-printed superalloys.

Although 3D-printing exhibits remarkable advantages and feasibility in processing condition variation [5], alloy composition redesign [6] and grain refinement [7], which can be used to alleviate crack formation to some extent, these approaches are frequently inadequate to fabricate crack-free non-printable superalloys, such as CM247LC, DZ125, etc. Furthermore, repairing damaged DS superalloys has more restrictions compared to manufacturing a new one. Generally, the repaired parts are required to possess similar composition, microstructure, and corresponding properties to the alloy in the damaged parts. Therefore, composition modification is usually not a choice, and epitaxial growth is necessary to maintain the columnar grain structure. Moreover, although it has been demonstrated that either refined [6] or single-crystalline [8] grain structure favours the suppression of cracks, this is not a possible solution when it comes to repairing a DS superalloy. All of these restrictions narrow the process window and hinder the easy access to a successful AM-repair on non-printable DS superalloys.

On the other hand, due to the high solidification rate and the cyclic thermal effect accompanying the repeated layer-by-layer deposition, 3D-printing produces finer dendrite and precipitations [9], higher defect concentration [10], less element segregation [11], and different stress distribution [12]. A comprehensive measurement of elemental segregation, phase distribution and grain orientation is thus needed to understand the crack formation mechanism and optimize the repairing process. In this study, we employed an EB-PBF method to 3D print DZ125 DS superalloy on a base metal with the identical composition. Liquation cracking and a strong correlation with the thermal cycling effect during the printing process are evidenced. The mechanism of crack initiation and propagation is proposed, which provides a valuable guideline to pre-empt hot cracks in EB-PBF manufactured and/or repaired DS superalloys.

2. Experimental

2.1 Material and EB-PBF process

The chemical composition of DZ125 powders and the base metal was listed in Table 1. A SAILONG Y-150 system was utilized to conduct the EB-PBF process, employing an acceleration voltage of 60 kV and maintaining a controlled helium pressure of 5×10^{-3} bar. The pre-alloyed powders (70 ~ 150 μm) were manufactured using the Plasma Rotating Electrode Process with Ar and deposited by 70 μm thick layers during the printing process. The as-cast DZ125 base metal was placed within perforated SS316 stainless steel plates to create a composite structure, which was subsequently positioned on the platform to finalize the printing process (figure 1(a)).

The powders underwent consolidation using a defocused beam prior to each single-layer process to prevent the occurrence of the "smoke" phenomenon [13] and ensure the stability of the powder bed. To maintain the high preheating temperature during the whole process, an identical consolidation procedure was employed subsequently after each individual layer printing (figure 1(b)). Finally, cylindrical samples with dimensions of $\Phi 10 \times 10$ mm were fabricated by a scanning speed of 600 mm/s, beam power of 240 W, rotation angle of 90° and line offset of 0.10 mm (figure 1 (c)-(d)). The chosen building direction (BD) was parallel to the normal of the (001) crystal plane.

Table 1. The nominal composition of the DZ125 base metal and powder (wt. %).

Material	Co	Cr	W	Al	Ti	Ta	Mo	Hf	C	B	Ni
Powder	10.10	8.97	6.79	5.14	0.99	3.93	1.95	1.43	0.094	0.015	Bal.
Base metal	10.03	9.02	6.99	5.26	0.98	3.91	2.05	1.54	0.100	0.012	Bal.

2.2 Microstructural characterization

As shown in figure 1 (c)-(d), we denoted longitudinal (i.e. parallel to the electron beam scanning direction), transverse, and building directions as X, Y, and Z respectively. Slices were cut along the Y-Z plane, mechanically ground with SiC papers ranging from 60 to 2000 grit, and then electro-polished using a solution of 10% $HClO_4$ and 90% ethanol. The cracked grain boundaries were investigated under scanning electron microscopy (SEM, Hitachi SU8230) in either secondary electron (SE) or backscattering electron (BSE) mode. To analyse the microstructure features of various precipitates, the other polished specimens were electro-etched in aqueous phosphoric acid (30vol%) at 5V for 5s.

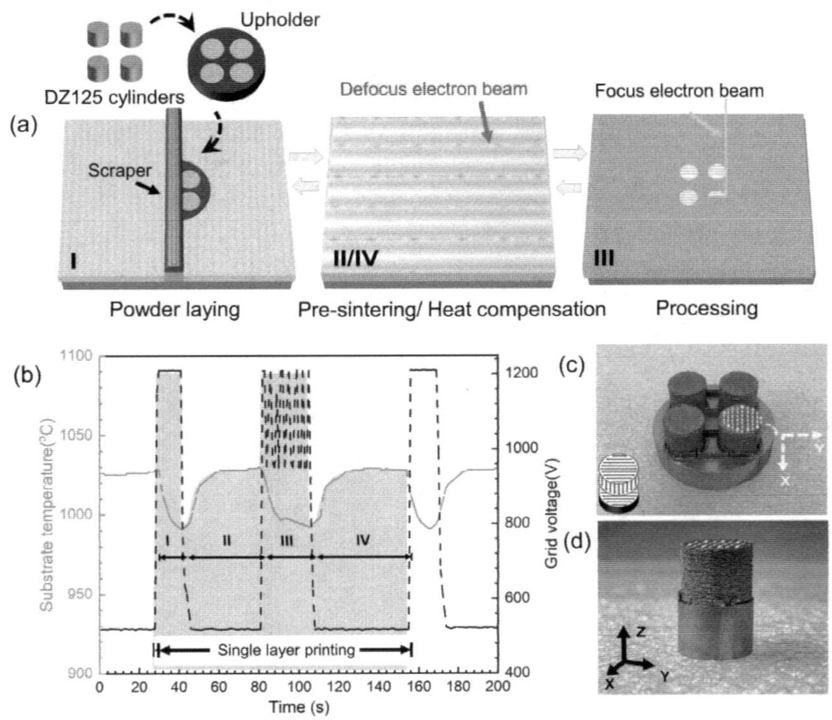

Figure 1. (a) Schematic of the composite base metal and single-layer-printing process; (b) curves showing the change of temperature and grid voltage during the process; (c-d) as-printed samples with labelled coordinate axis.

3. Results

3.1 Cracking behaviour

Figure 2 illustrates the cracking behaviour in the printed samples. With process optimization, nearly epitaxial fabrication is achieved for the entire build, except for a few stray grains within the bottom printed area (figure 2(a)). A significant contrast is evident between the uppermost and the rest of the printed area, which is manifested as 'bright strips'. This contrast is attributed to the distinct size and morphology of γ' particles, which is explained subsequently. As the site of crack initiation, the grain boundary (GB) within the heat affected zone (HAZ) exhibits severe cracking (figure 2(b)) compared to other printed areas (figure 2 (c)-(d)), which can be attributed to the tortuous morphology of grain boundaries [14], the liquation induced by γ/γ' eutectic remelting [15], and the exacerbating effect on tearing caused by irregular MC carbides [16]. The mechanism of HAZ cracking in 3D-printed superalloys can be found in our previous work [15]. As the printing progresses, the GB crack propagates upward and ultimately terminates within the sample, resulting in a crack-free zone of approximately 250 microns in depth (figure 2(d)).

The existence of the crack-free zone has rarely been reported in previous research on EB-PBFed superalloys. To clarify the critical factor that triggers crack termination, three samples with different heights are printed using similar EB-PBF parameters (details in section 2.1). The coverage (height) of cracked GBs and crack-free zone is quantified and summarized in figure 2(e). It is worth noting that a long-term solution treatment (1230 °C, 20h) of the as-cast substrate is carried out before the printing process. The purpose is to eliminate the adverse impact of element segregation on cracking. Consequently, with the new substrate, cracking in both the HAZ and the initial printed area is less pronounced. From figure 2(e), there exists a strong correlation between the coverage of cracked GBs and the printing height. For all the samples, the average coverage span of cracked GBs rises proportionally with the increase of the printing height. But in contrast, the extent of crack-free zone remained consistent regardless of the printing height. Moreover, the crack-free zones are exclusively distributed within the uppermost region of the printed sample, exhibiting an average coverage of approximately 250 μm. This suggests that the formation of cracks is closely related to the thermal cycling effect during the EB-PBF process.

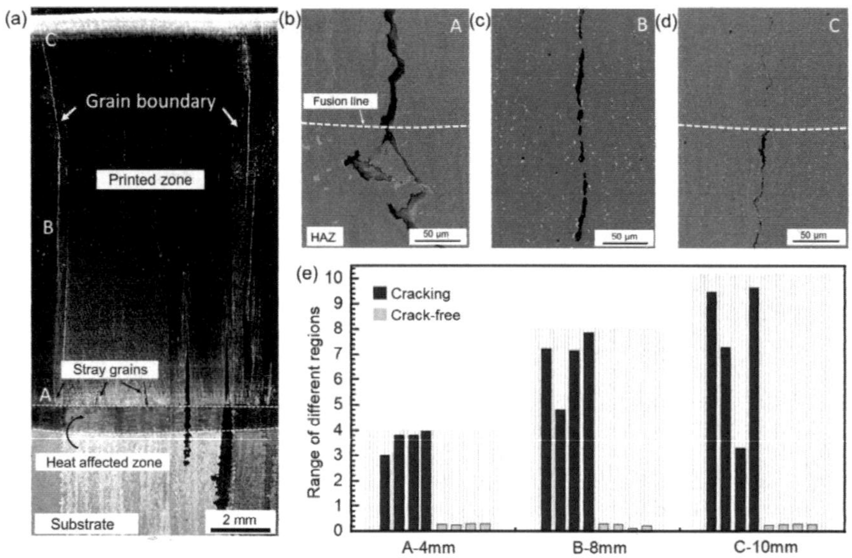

Figure 2. Cracking behaviour of the EB-PBFed DZ125 superalloy. (a) stereo-microscope image showing the substrate, HAZ and as-printed part; (b) crack morphology in the bottom, (c) middle and (d) uppermost regions of the printed sample marked in (a); (e) the diagram demonstrating the variation of cracked and crack-free regions according to the different printing heights.

3.2 Crack termination at last deposited layer

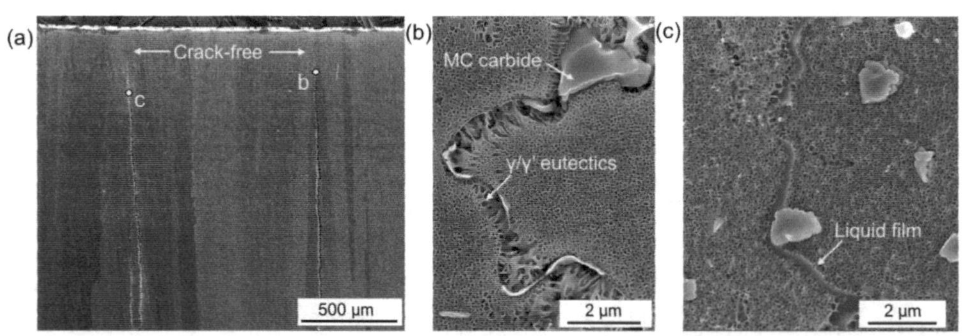

Figure 3. (a) SEM-image showing the uppermost printing layers which including cracked and crack-free grain boundaries; (b) γ/γ′ eutectics and (c) liquid film observed in the crack tip marked in (a).

The presence of the uppermost crack-free zone confirms a strong correlation between crack termination and the final solidification process. Figure 3(a) illustrates the microstructure of the uppermost printing layers which including cracked and crack-free GBs. In the crack-free zone, SEM images reveal the presence of comb-like γ/γ' eutectic phases and blocky MC carbides (figure 3(b)). Nevertheless, the embedded γ/γ' eutectic phases undergo transformation into liquid films along at a lower location (figure 3(c)).

4. Discussion

Considering the presence of the "bright strips" in figure 2(a) and the abrupt termination of cracks, a closer microstructural analysis is conducted on the uppermost printed area. Figure 4 reveals distinct features of γ' particles within the "bright strip". The uppermost region I contains small and spherical γ' particles (~100 nm) within the γ matrix due to the rapid cooling rate of the EB-PBF process [17]. The γ/γ' eutectic phases are exclusively observed in the interdendritic regions (IDRs) and GBs. Serving as the last printed layer, the region I retains the most inherent solidification characteristics of AM without cycling thermal effect. However, at a deeper position, the coarse γ/γ' eutectic phases disappear, leaving behind only spherical γ' particles and MC carbides (figure 4(b)). Notably, liquid films appear concurrently within this region (figure 3(c)). It is thus inferred that the transformation from γ/γ' eutectic phases to liquid films can be clearly attributed to the heat generated during the final layer printing. In region III, the γ' particles within the IDRs underwent significant coarsening due to prolonged reheating, ultimately transforming into regular cubic shapes (figure 4(c)). The proportion of enlarged γ' particles will steadily increase with greater depth. Until it reaches a depth of about 1mm (according to the border of the prominent bright strip), the regular cubic γ' particles (~200 nm) are uniformly distributed within the dendrite cores (DCs) and IDRs (figure 4(d)), indicating that the thermal effect produced by the subsequent printing is stabilized.

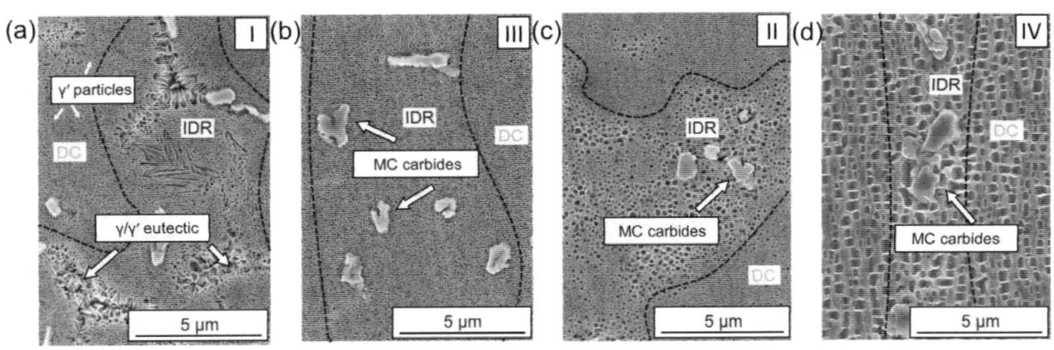

Figure 4. The typical morphology and size of γ' at the different locations within the topmost printed region.

Based on the hierarchical structure present within the topmost region, the actual cracking type in the printed DZ125 superalloy can be inferred. Firstly, as common types of solid-state cracking, the ductility-dip cracking or strain-age cracking typically occur within the temperature range of 0.5-0.8 T_m [18] or during post-welding heat treatment [19]. Given that the printing temperature is maintained at around 1000 °C during the whole process (figure 1(b)), it is suggested that the cracks in this study are most likely hot crack, a.k.a. solidification or liquation crack. Furthermore, there is no crack occurs in the uppermost as-solidified region, thus the possibility of solidification cracking is very low. The presence of γ/γ' eutectic phases in region I provides further evidence on the cracking type. As the typical low-melting-point phases, these eutectic phases are easily melted and transform into liquid films during cyclic reheating (figure 3(c)). All of the aforementioned evidence illustrates that the observed cracks in the printed samples are liquation cracks. And also, the cracking mechanism is proposed as follows.

Firstly, as the newly-formed printed layer, the uppermost region preserves the most intrinsic solidification characteristics of AM without thermal cycling effect. Due to element segregation at the GBs and interdendritic regions, low-melting-point phases such as γ/γ' eutectics are formed during the final stage of solidification. Fortunately, thanks to the high preheating temperature during the EB-PBF processing, the thermal stress level is mitigated and the susceptibility of solidification cracking is reduced. Therefore, a crack-free region can be observed in the vicinity of the top surface (figure 3(a)). As the printing progresses, near the fusion line, a.k.a. melt pool boundary, the material is subjected to rapid reheating to a temperature approaching T_m, resulting in the dissolution of γ' particles into the γ matrix. Concurrently, the low-melting-point eutectics along the GBs undergo incipient melting, leading to the formation of liquation films (figure 3(c)). During the subsequent cooling stage, the vulnerable liquefied GBs are torn due to the combined effect of thermal stress and phase transition stress induced by γ' re-precipitation. Each deposited layer experiences at least one such thermal cycle process, except for the uppermost one. Therefore, liquation cracking occurs and persists throughout the printing process.

5. Conclusions

In this study, we employed an EB-PBF method to epitaxially fabricate the DS superalloy DZ125 from the cast base metal. The printed alloy demonstrates a typical liquation cracking behaviour. Within the uppermost printing layer, γ/γ' eutectic precipitation occurs along the grain boundaries and no crack occurs. However, for the previous layers according to a lower position, the persistent thermal effect leads to the γ/γ' eutectics remelting and stress accumulation, resulting in liquation cracking. As the printing progresses, the successive liquation of γ/γ' eutectics persists along the GB, which leads to a long and straight cracked channel in the printed samples. This liquation cracking behaviour proposed in this study firstly emphasizes the influence of thermal cycling effect during the EB-PBF process. The knowledge acquired from these phenomena will directly contribute to a more comprehensive comprehension of microstructure evolution and cracking mechanism during the EB-PBF process, thereby providing insights for solution optimization in additive manufacturing of DS superalloys.

References

[1] Zhang H, Li C, Guo Q, Ma Z, Li H and Liu Y 2019 *Scr. Mat.* **164** 66-70.
[2] Suzuki S, Sakaguchi M, Domen M, Karato T and Suzuki K 2022 *Acta Mater.* **240** 118288.
[3] Rajkumar V, Nagesha B K, Tigga A K, Barad S and Suresh T N 2021 *Mater. Today Proc.* **45** 5395–5399.
[4] Körner C 2016 *Int. Mater. Rev.* **61**(5) 361–377.
[5] Ramsperger M, Singer R F and Körner C 2016 *Metall. Mater. Trans. A* **47**(3) 1469–1480.
[6] Griffiths S, Ghasemi Tabasi H, Ivas T, Maeder X and De Luca A 2020 *Addit. Manuf.* **36** 101443.
[7] Kontis P, Chauvet E, Peng Z, He J, da Silva A K and Raabe D 2019 *Acta Mater.* **177** 209–221.
[8] Li Y, Yu YF, Wang Z B, Liang X Y, Kan W B and Lin F 2022 *Acta Metall. Sin.* **35**(3) 369–374.
[9] Gäumann M, Bezençon C, Canalis P and Kurz W 2021 *Acta Mater.* **49**(6) 1051–1062.
[10] Xue J, Zhang A, Li Y, Qian D, Wan J and Qi B 2015 *Sci. Rep.* **5**(1) 14903.
[11] Zhong M, Sun H, Liu W, Zhu X and He J 2005 *Scr. Mat.* **53**(2) 159–164.
[12] Li Y, Qian D, Xue J, Wan J, Zhang A and Tamura N 2015 *Appl. Phys. Lett.* **107**(18) 181902
[13] Cordero Z C, Meyer H M, Nandwana P and Dehoff R R 2017 *Acta Mater.* **124** 437–445.
[14] Zhang X, Chen H, Xu L, Xu J, Ren X and Chen X 2019 *Mater. Des.* **183** 08105.
[15] Li Y, Chen K and Tamura N, 2018 *Mater. Des.* **150** 171–181.
[16] Chen Q Z, Jones C N and Knowles D M 2004 *Mater. Sci. Eng. A* **385**(1) 402–418.
[17] Sames W J, Unocic K A, Dehoff R R, Lolla T and Babu S S 2014 *J. Mater. Res.* **29**(17) 1920–1930.
[18] Qian D, Xue J, Zhang A, Li Y, Tamura N and Song Z 2017 *Sci. Rep.* **7**(1) 2859.
[19] Kayacan R, Varol R and Kimilli O, 2004 *Mater. Res. Bull.* **39**(14) 2171–2186.

Revisiting the Theory of Spinodal Strengthening: Implications for Inhomogeneous Additive and Irradiated Structural Alloys

Riccardo Civiero[1] and Anter El-Azab[1*]

[1] School of Materials Engineering, Purdue University, West Lafayette, IN, USA

*E-mail: aelazab@purdue.edu

Abstract. Composition fluctuations are commonly observed in additive manufactured alloys, high entropy alloys and irradiated alloys. These fluctuations, which arise due to thermodynamic and/or kinetic factors, influence the mechanical strength of alloys. The classical theory of spinodal strengthening in alloys with coherency stress arising due to the spinodal long-range composition fluctuations has been developed by Cahn, Kato and others. We present the motivations for revisiting Cahn's theory to account for the concurrent solid solution strengthening in spinodally strengthened alloys by taking into consideration the effect of the solute friction along with the internal coherency stress. We briefly expose the model we developed to achieve the aforementioned aim and discuss some findings.

1. Introduction

A common feature among additive manufactured (AM) alloys, high-entropy alloys (HEAs) and irradiated alloys is compositional inhomogeneity, which affects the mechanical strength of these alloys. Increase of yield and tensile strengths and total tensile elongation to failure, or embrittlement and loss of ductility are features that have been frequently reported in literature for these kinds of materials [1–11]. Compositional inhomogeneities refer to mesoscale undulations (or fluctuations) of the elemental composition, resulting in the formation of composition-rich and composition-deficient regions and spinodal/spinodal-like decomposition with wavelengths that range from tens to hundreds of nanometers [2,4,7].

In recent years, there has been a surge in investigating additive manufactured AISI 316 and 304 stainless and maraging steels. These studies explore the unique microstructural and mechanical properties that can be achieved on components with complex geometries by additive manufacturing techniques, such as Laser-Powder Bed Fusion (L-PBF), Selective Laser Melting (SLM) or Electron Beam Melting (EBM). Transmission Electron Microscopy (TEM) analysis coupled with Energy-Dispersive Spectroscopy (TEM-EDS) maps of components fabricated from austenitic stainless steels, as AISI 316 and 304, by additive manufacturing revealed a fine cellular structure with dimensions on the order of hundreds of nanometers [12,13]. This microstructural morphology is characterized by a high dislocation density ($\rho_{disl} \approx 10^{15}\ m/m^3$) along the cellular walls [12,14,15], which likely comes from the stresses arising due to the rapid thermal contraction [15]. The cellular walls consist of dislocation tangles of different Burgers vectors and stacking faults [14,15]. From the EDS composition maps, researchers have also observed the enrichment of heavy elements, namely, Cr, Mo, Ni and Mn, along the cellular walls [12–15]. This segregation significantly impacts the formation energy and strength of dislocation-dislocation junctions [12], leads to dislocation trapping and twinning [10,11,14]. Also, the precipitates that are commonly

IOP Conf. Series: Materials Science and Engineering **1310** (2024) 012032 doi:10.1088/1757-899X/1310/1/012032

formed in these regions may exert a pinning effect on dislocation motion [14,16]. Turning attention to maraging steels, as PH 13-8 or 17-4 steels, morphologies and segregations similar to the ones characterizing the austenitic stainless steels have been reported: Cr, Mo and Mn enrichments has been found in inter-columnar regions and in lath boundaries, with high dislocation densities, especially in the martensitic grains [17,18]. Researchers have also applied heat treatments, as recrystallization annealing, to additive manufactured stainless steels: when temperature is kept above 700°C, the element segregation is eliminated and dislocation tangles rearrange towards higher misorientation [11,13,19].

In a different context, the mechanical properties of irradiated Fe-Cr, Fr-Cr-Al and Fe-Cr-Ni alloys have been widely assessed, for their potential for applications in light water nuclear reactors [20–22]. Specifically, hardening is induced in these materials from the interaction of dislocations with radiation-induced defects and composition fluctuations [2,10,23]. Type a/2<111> and a<100> dislocation loops are frequently observed defects in Fe-Cr alloys due to irradiation [3,24–26]. When the Cr concentration in the alloy rises, it results in a reduction in size of these loops, while simultaneously leading to an increase in loop density [27,28] . In Fe-Cr alloys, α' phase appears in the form of precipitates or via spinodal decomposition depending on the Cr concentration in the alloy [3]. In irradiated Fe-Cr, irradiation causes spinodal-like decomposition to take place, which results in Cr fluctuations in the matrix with amplitudes not generally consistent with the α' phase composition [29,30]. This impacts the dislocation mobility and in turn the alloy plasticity [3,31]. Composition inhomogeneities manifest themselves also as enrichment and segregation of heavy elements, like Cr, Ni and Mo, at the grain boundaries and defect sinks [24,32]. For example, Cr has been observed to segregate at the dislocation loops, limiting their mobility [26,27].

From the above studies, it appears that a strong correlation exists between compositional and microstructural features, such as segregation of heavy elements, precipitates and irradiation-induced defects, and the behaviour of dislocations. The interplay between these characteristics results in alteration of the alloy mechanical properties relative to their homogeneous counterparts. In this regard, dislocation dynamics may serve as an effective tool that can capture the effect of composition inhomogeneity on alloy deformation.

2. Classical models for hardening via composition fluctuations

We are primarily interested on the impact of composition fluctuations on the strength of alloys, a question which was tackled in the literature through the theory of spinodal hardening of alloys. As argued in section 3, spinodal hardening is not completely independent of solution hardening, as was assumed in the classical literature. Solid solution strengthening models are categorized into two types. The models developed by Fleischer [33,34], Friedel [35] and Suzuki [36] are referred to as the strong pinning models. These models have been developed for dilute alloys and consider solute atoms as obstacles that pin the dislocations while the latter bow out in between. The models developed by Mott [37], Nabarro [38] and Labusch [39] are referred as weak pinning models. They model concentrated alloys where solute atoms are treated as a static ensemble generating static fields which dislocation interact with. The primary issues associated with these models are that (i) complex systems where the solute hardening mode changes between the two categories cannot be treated correctly with a single model, (ii) the effect of the solute atoms on the dislocation core structure is not taken into account, and that (iii) these models are valid for a random distribution of solute atoms, but not when the composition shows short- or long-range order, as in the case of segregation in additive manufactured or irradiated metals.

Cahn [40] and Kato [41,42] (see also relevant work in [43]) developed models that take into consideration the influence on strengthening due to internal stresses arising from long-range composition fluctuations along <100> directions in FCC crystals. In this model, the stress field is modulated in such a way that positive and negative resolved shear stress arise over the {111} glide plane. They computed the strengthening associated with such stress undulations on this plane by setting up a force equilibrium of dislocation using string model,

$$\gamma \frac{d^2y}{dx^2} \Big/ \left[1 + \left(\frac{dy}{dx}\right)^2\right]^{3/2} + b\tau_I + \tau_a b = 0 \qquad (1)$$

where the first term multiplies the line tension γ and dislocation curvature, τ_I and τ_a the stresses arising on the dislocation line due to internal and applied stresses respectively. Equation (1) governs the profile y(x) of the dislocation line that extends along the x coordinate. This equation was solved by Cahn by means of a Galerkin-Ritz method. The largest applied stress for which the above equation has a solution, which is called the critical resolved shear stress (CRSS), has been derived for screw and edge line dislocations to be $\tau_a = A^2\eta Y^2 b/3\sqrt{6}\beta\gamma$ and $\tau_a = A^2\eta Y^2 b/\sqrt{2}\beta\gamma$, respectively [40], where A is the amplitude of the composition fluctuation, $\eta = \partial \ln(a)/\partial c$, a is the lattice parameter, c is the composition, $\beta = 2\pi/\lambda$, λ is the wavelength of the composition fluctuations, b the magnitude of Burgers vector and $Y = (c_{11} + 2c)(c_{11} - c_{12})/c_{11}$. Kato, however, noted that mixed dislocations with slope corresponding to $\sqrt{3}$ show the highest resolved shear stress for the composition modulation [41]. Following the same mathematical methodology, he derived for mixed dislocations the CRSS to be $\tau_a = A\eta Y/\sqrt{6}$. The main aspect that differentiates the results of Cahn and Kato is that a linear dependence on the composition amplitude is derived, instead of a quadratic one, and that it is independent on the composition wavelength and fluctuation, showing excellent agreement with measured values for different kind of alloys [43]. Cahn and Kato models are suitable for the early stages of spinodal decomposition, when the composition modulation follows a sinusoidal pattern. At more advanced stages, however, no analytical solution for the force balance exists, so the equation is solved numerically. Cahn and Kato models neglect chemical strengthening or modulus effects in the alloys as second order effects, sometimes unpredicting the experimental observations [42,44].

3. A suggested dislocation dynamics framework

Cahn [40] and Kato's [41,42] works consider the hardening in spinodal alloys to be the part due to the effect of coherency stress associated with the composition undulations. The literature is not clear on whether that strengthening should be added to the solid solution strengthening at, say, average composition, or whether there is interference between the two mechanisms. We have worked out this problem by defining the spinodal strengthening to be that strengthening gained (or lost) by introducing composition fluctuations over the average composition of the alloy. We have also adopted dislocation dynamics to solve this problem for single and multiple dislocation. We highlight here only the case of a single dislocation, which is counterpart to Cahn's model. We did so for a FeCrAl BCC system, which can turn inhomogeneous via irradiation [45]. A 3D single sinusoidal mode composition field is generated along the [120], $[\bar{2}10]$, and [001] crystal orientations, with expressions following the works from Cahn and Kato. The lattice friction stress and dislocation mobility previously obtained by molecular dynamics (MD) simulations for all composition values dictated by this construction were used [46]. Of these MD simulations we are particularly interested in the lattice friction stress due to solute, which was added to equation (1) to couple solute friction to coherency stress, resulting in a modified Cahn's theory. We then define the hardening due to spinodal decomposition as an incremental change in CRSS, denoted by ΔCRSS, due to introducing composition undulations in the homogeneous alloy. That is,

$$\Delta CRSS = CRSS_S - CRSS_H \qquad (2)$$

In the above, $CRSS_S$ and $CRSS_H$ are CRSS in decomposed and the homogeneous cases. The results differ from Cahn/Kato's theory for spinodal hardening in many respects. First, it was found that the dislocation takes on a different equilibrium profile both in the unloaded and loaded states, see figure 1 (a). This difference results from the imposition of the local friction stress on dislocation equilibrium expressed by equation (1). Locally, such friction stress depends on the concentration of solute. It was also found that ΔCRSS due to spinodal decomposition is a function of dislocation line orientation, which was expected from the classical work. However, the coupling with solute friction brings in some dependence on the local composition and that, in turn, makes the dependence on the dislocation line orientation different from that reported by Cahn and Kato. In particular, we found that the spinodal decomposition may lead to softening (relative to the strength of the alloy at average composition) in certain cases, e.g., screw dislocations on $(110)[\bar{1}11]$ slip system as shown in figure 1 (b). Finally, the induced effect was found to depend on the average alloy composition, not just the amplitude of the composition modulation.

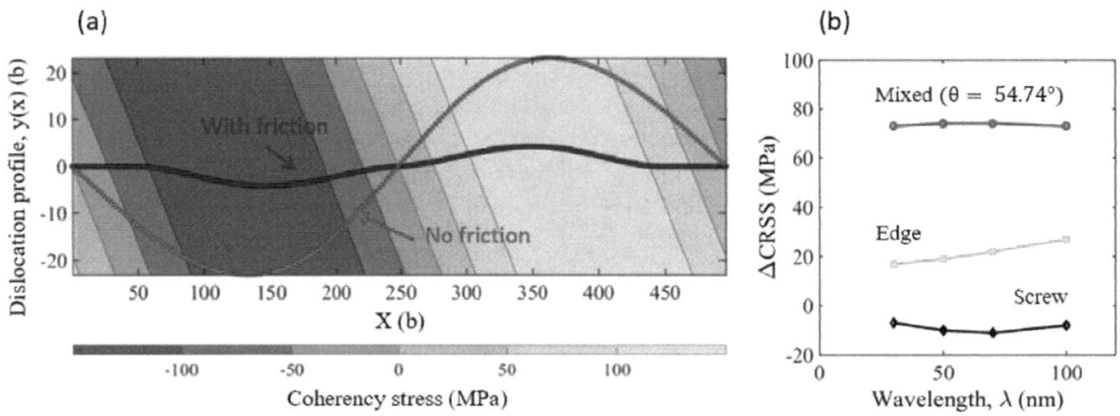

Fig. 1 (a) Equilibrium dislocation profile obtained at no applied stress when friction stress arising due to composition modulations in the material is introduced in the model (blue line) compared with the case of no friction stress (red line). The contour plot shows the resolved coherency stress on the dislocation glide plane. (b) Incremental change in the CRSS due to spinodal decomposition as a function of wavelength for screw, edge and mixed dislocation with line direction making an angle of 54.74° with the slip direction.

4. Concluding remarks

Composition inhomogeneities are commonly found in both additive manufactured and irradiated alloys due to thermodynamic or kinetic factors. This composition inhomogeneity alters the mechanical response of alloys. The use of dislocation dynamics models is crucial in studying the mechanical response of these alloys, as it allows for the 3D representations of the composition inhomogeneity and the resulting coherency stresses and, therefore, leads to accurate prediction of material behaviour. Establishing such a link between the composition state of alloys and their mechanical response facilitates the design of the alloys. Preliminary use of dislocation dynamics to investigate the composition fluctuations in a sinusoidal composition field gave results that are different from the classical results of Cahn and Kato's for the CRSS. Further investigation of this problem using 3D dislocation dynamics simulations in the post-yielding regime is underway.

Acknowledgement

This work was supported by the US Department of Energy, Office of Fusion Energy Sciences via award number DE-SC0024585 at Purdue University.

References

[1] Massey CP, Zhang D, Briggs SA, et al. 2021 *Journal of Nuclear Materials* **549**:152804. https://doi.org/10.1016/J.JNUCMAT.2021.152804.

[2] Brager HR, Garner FA, Hamilton ML. 1985 *Journal of Nuclear Materials* **133–134**:594–8. https://doi.org/10.1016/0022-3115(85)90217-X.

[3] Field KG, Hu X, Littrell KC, et al. 2015 *Journal of Nuclear Materials* **465**:746–55. https://doi.org/10.1016/J.JNUCMAT.2015.06.023.

[4] Garner FA, Brager HR, Dodd RA, et al. 1986 *Nucl Instrum Methods Phys Res B* **16**:244–50. https://doi.org/10.1016/0168-583X(86)90020-0.

[5] Garner FA, McCarthy JM, Russell KC, et al. 1993 *Journal of Nuclear Materials* **205**:411–25. https://doi.org/10.1016/0022-3115(93)90105-8.

[6] Field KG, Snead MA, Yamamoto Y, et al. 2018 *ORNL/SPR-2018/905*. https://doi.org/10.2172/1474581.

[7] Nakai K, Kinoshita C, Nishimura N. 1991 *Journal of Nuclear Materials* **179–181**:1046–9. https://doi.org/10.1016/0022-3115(91)90271-8.

[8] Ding Q, Zhang Y, Chen X, et al. 2019 *Nature 2019 574:7777* **574**:223–7. https://doi.org/10.1038/s41586-019-1617-1.

[9] Karthik GM, Kim HS. 2021 *Metals and Materials International 2021 27:1* **27**:1–39. https://doi.org/10.1007/S12540-020-00931-2.

[10] Cui L, Deng D, Jiang F, et al. 2022 *J Mater Sci Technol* **111**:268–78. https://doi.org/10.1016/J.JMST.2021.10.006.

[11] Smith TR, Sugar JD, San Marchi C, et al. 2019 *Acta Mater* **164**:728–40. https://doi.org/10.1016/J.ACTAMAT.2018.11.021.

[12] Gorsse S, Hutchinson C, Gouné M, et al. 2017 *Sci Technol Adv Mater* **18**:584–610. https://doi.org/10.1080/14686996.2017.1361305.

[13] Yin H, Song M, Deng P, et al. 2021 *Addit Manuf* **41**:101981. https://doi.org/10.1016/J.ADDMA.2021.101981.

[14] Wang YM, Voisin T, McKeown JT, et al. 2017 *Nature Materials 2017 17:1* **17**:63–71. https://doi.org/10.1038/nmat5021.

[15] Barkia B, Aubry P, Haghi-Ashtiani P, et al. 2020 *J Mater Sci Technol* **41**:209–18. https://doi.org/10.1016/J.JMST.2019.09.017.

[16] Deng P, Karadge M, Rebak RB, et al. 2020 *Addit Manuf* **35**:101334. https://doi.org/10.1016/J.ADDMA.2020.101334.

[17] Zhou T, Zheng T, Yildiz AB, et al. 2022 *Addit Manuf* **58**:103047. https://doi.org/10.1016/J.ADDMA.2022.103047.

[18] Benoit MJ, Tabaie S, Waqar T, et al. 2023 *Addit Manuf* **72**:103615. https://doi.org/10.1016/J.ADDMA.2023.103615.

[19] Que Z, Chang L, Saario T, et al. 2022 *Addit Manuf* **60**:103205. https://doi.org/10.1016/J.ADDMA.2022.103205.

[20] Rebak RB, Terrani KA, Gassmann WP, et al. 2017 *MRS Adv* **2**:1217–24. https://doi.org/10.1557/ADV.2017.5/METRICS.

[21] Yamamoto Y, Pint BA, Terrani KA, et al. 2015 *Journal of Nuclear Materials* **467**:703–16. https://doi.org/10.1016/J.JNUCMAT.2015.10.019.

[22] Zinkle SJ. 2008 *NATO Science for Peace and Security Series B: Physics and Biophysics*:227–44. https://doi.org/10.1007/978-1-4020-8422-5_11.

[23] Briceño M, Fenske J, Dadfarnia M, et al. 2011 *Journal of Nuclear Materials* **409**:18–26. https://doi.org/10.1016/J.JNUCMAT.2010.12.026.

[24] Bhattacharyya D, Yamamoto T, Wells P, et al. 2019 *Journal of Nuclear Materials* **519**:274–86. https://doi.org/10.1016/J.JNUCMAT.2019.03.022.
[25] Yao B, Edwards DJ, Kurtz RJ. 2013 *Journal of Nuclear Materials* **434**:402–10. https://doi.org/10.1016/J.JNUCMAT.2012.12.002.
[26] Wakai E, Hishinuma A, Kato Y, et al. 1995 *Le Journal de Physique IV* **05**:C7-277. https://doi.org/10.1051/JP4:1995733.
[27] Aydogan E, Weaver JS, Maloy SA, et al. 2018 *Journal of Nuclear Materials* **503**:250–62. https://doi.org/10.1016/J.JNUCMAT.2018.03.002.
[28] Chen WY, Miao Y, Gan J, et al. 2016 *Acta Mater* **111**:407–16. https://doi.org/10.1016/J.ACTAMAT.2016.03.060.
[29] Edmondson PD, Briggs SA, Yamamoto Y, et al. 2016 *Scr Mater* **116**:112–6. https://doi.org/10.1016/J.SCRIPTAMAT.2016.02.002.
[30] Harrison RW, Carruthers AW, Hinks JA, et al. 2019 *Scr Mater* **172**:33–7. https://doi.org/10.1016/J.SCRIPTAMAT.2019.06.034.
[31] Vojtech V, Basu I, Wheeler JM, et al. 2023 *Acta Mater* **261**:119355. https://doi.org/10.1016/J.ACTAMAT.2023.119355.
[32] Liu X, Gigax JG, Poplawsky JD, et al. 2020 *Materialia (Oxf)* **9**:100542. https://doi.org/10.1016/J.MTLA.2019.100542.
[33] Fleischer RL. 1961 *Acta Metallurgica* **9**:996–1000. https://doi.org/10.1016/0001-6160(61)90242-5.
[34] Fleischer RL. 1963 *Acta Metallurgica* **11**:203–9. https://doi.org/10.1016/0001-6160(63)90213-X.
[35] Friedel J. Dislocations. Elsevier; 1964. https://doi.org/10.1016/C2013-0-02250-5.
[36] Suzuki H. 1979 *Strength of Metals and Alloys*:1595–614. https://doi.org/10.1016/B978-1-4832-8412-5.50247-2.
[37] Mott NF. 1952 *The London, Edinburgh, and Dublin Philosophical Magazine and Journal of Science* **43**:1151–78. https://doi.org/10.1080/14786441108521024.
[38] Mott NF, Nabarro FRN. 1948.
[39] Labusch R. 1970 *Physica Status Solidi (b)* **41**:659–69. https://doi.org/10.1002/PSSB.19700410221.
[40] Cahn JW. 1963 *Acta Metallurgica* **11**:1275–82. https://doi.org/10.1016/0001-6160(63)90022-1.
[41] Kato M, Horie S, Lee TC. 1986 *Physica Status Solidi (a)* **98**:209–14. https://doi.org/10.1002/PSSA.2210980123.
[42] Kato M. 1981 *Acta Metallurgica* **29**:79–87. https://doi.org/10.1016/0001-6160(81)90088-2.
[43] Ghista DN, Nix WD. 1969 *Materials Science and Engineering* **3**:293–8. https://doi.org/10.1016/0025-5416(69)90015-9.
[44] Bachmaier A, Rathmayr GB, Schmauch J, et al. 2019 *J Mater Res* **34**:58–68. https://doi.org/10.1557/JMR.2018.185/FIGURES/8.
[45] Pachaury Y, Kumagai T, Wharry JP, et al. 2022 *Acta Mater* **234**:118019. https://doi.org/10.1016/J.ACTAMAT.2022.118019.
[46] Kumagai T, Pachaury Y, Maccione R, et al. 2021 *Materialia (Oxf)* **18**:101165. https://doi.org/10.1016/J.MTLA.2021.101165.

Effect of microstructural heterogeneity and grain morphology on the annealing behavior of additively manufactured metastable β titanium alloy

J Q Chang, Y N Wu and Z B Zhang*

Centre for Adaptive System Engineering, ShanghaiTech University, Shanghai, China
*E-mail: Zhangzhb1@shanghaitech.edu.cn

Abstract. Titanium alloys fabricated by additive manufacturing usually exhibit significant heterogeneity in grain size and morphology. Recrystallization treatment is an effective approach to alter the microstructure and achieve a more homogeneous structure. In this study, recrystallization and grain growth behaviours of a metastable β titanium alloy fabricated by laser powder bed fusion (LPBF) and direct energy deposition (DED) were investigated. The LPBF sample developed a chessboard-like microstructure, with considerable variations in grain size and morphology across molten pools. Recrystallization started from the boundary of these pools, where grains were finer and the stored energy was higher. The DED sample had a heterogeneous structure with alternating layers of equiaxed grains and columnar grains oriented with <100> along the building direction. Results obtained from in-situ EBSD demonstrated that grain growth occurred preferentially in regions with fine equiaxed grains, and regions with columnar grains showed limited grain growth due to the relatively lower driving force and mobility.

1. Introduction

The texture and grain morphology of metastable β titanium alloys significantly influence their mechanical properties. Titanium alloys fabricated by additive manufacturing usually develop columnar grains and <100> texture of the β phase, leading to pronounced mechanical anisotropy [1–4]. Solution treatment above the β transus temperature (T_β) is employed to facilitate recrystallization and grain growth, eliminating columnar grains and reducing mechanical anisotropy by forming uniform equiaxed grains [5,6]. Most of the as-printed titanium alloys with a <100> texture and significant columnar grains transform into texture-free equiaxed structures after solution treatment [5–7]. However, these previous studies have primarily focused on α-β titanium alloys (e.g. Ti6Al4V) undergoing phase transitions, which lead to recrystallization [6,8], or β titanium alloys involving dislocation-based recrystallization [5,7]. This raises the question of how columnar grain and equiaxial grain of DED samples grow in the absence of dislocation-base or phase transformation-mediated recrystallization, and how heterogeneous grain morphology in LPBF samples affect recrystallization behaviour during annealing.

In this study, a metastable β titanium alloy, Ti-3Al-6Fe-6V-2Zr (Ti3662), was fabricated using laser powder bed fusion (LPBF) and direct energy deposition (DED). This resulted in the formation of three distinct types of structures: homogeneous equiaxed grains in the DED single track, a heterologous microstructure with alternating equiaxed and columnar grains produced by DED, and a chessboard-like microstructure with significant variations in grain size and morphology built through LPBF. The columnar to equiaxed transition (CET) phenomenon in Ti3662 with these microstructural characteristics was studied to reveal their microstructural dependence.

2. Experimental

The Ti-3Al-6Fe-6V-2Zr titanium alloy powders, abbreviated as Ti3662 hereafter, were customized for additive manufacturing using gas atomization to achieve a globular shape. The D50 mean diameter is 36.3 μm and 83.3 μm for LPBF and DED, respectively. Additive manufacturing was conducted using an SLM Solution 280 equipped with double 400 W fibre lasers and a DED instrument with a 6000 W flat-top fibre laser, both operating within an argon protective atmosphere.

Table 1. LPBF and DED process parameters.

Samples	*P* (W)	*t* (μm)	*v* (mm/s)	*h* (mm)	*d* (μm)	*R* (°)
Single track of DED	1200	——	15	——	3000	——
Bulk samples of DED	1200	520	15	2.5	3000	90
Bulk samples of LPBF	200	30	900	0.12	90	67

Here: *P*, laser power; *t*, layer thickness; *v*, scanning speed; *h*, hatch spacing; *d*, laser spot diameter and *R*, rotation angle between successive layers.

An off-site solution treatment was conducted using a heat treatment furnace. Bulk samples prepared by DED were subjected to treatment at 800 °C for 2 h, 3 h, 5 h, and 8 h, while bulk samples prepared by LPBF underwent treatment at 800 °C for 15 min, 30 min, and 45 min. All samples were water quenched to retain the full beta phase.

All samples were polished to #4000 SiC grit papers followed by electrochemical polishing using A3 electrolyte (6 vol% perchloric acid, 34 vol% butanol, and 60 vol% methanol) to remove subsurface damage. Electron backscattered diffraction (EBSD) characterizations were performed utilizing a Symmetry S2 detector at an operating voltage of 30 kV and current of 15 nA. Data analysis was carried out using AztecCrystal software, and the grain orientation spread (GOS) was calculated to determine the degree of orientation change between each pixel in the grain and the average orientation of the grain.

In-situ EBSD experiments were performed on both single track and bulk DED samples during in-situ heating tests. For the heating process, a Kammrath & Weiss GmbH mini heater, capable of reaching temperatures up to 1000 °C, was used, with sample temperature monitored indirectly via thermocouple contact with the heating substrate. Prior to heating, both the samples and pure zirconium sheets acting as oxygen absorbents were vacuumed for six hours to minimize oxygen presence [9]. All in-situ EBSD maps were collected at temperatures above 800 °C to prevent unexpected effects during cooling process, with the exact temperature specified in the figures.

3. Result

The figure 1(a) shows the typical microstructure of the as-built single track by DED, obtained from a transverse section perpendicular to the scanning direction (SD). The grain size of the β phase is approximately 30.2 μm, without appreciable texture. In contrast, bulk sample (figure 1(b)) exhibits a heterogeneous structure with alternating layers of columnar grains and equiaxed grains. The columnar grains exhibit a pronounced <100> texture along the building direction (BD), formed due to heating effect of subsequent re-melting. The grain size of the equiaxed layer is characterized by the equivalent circular diameter, measured at 87±29 μm. The grain size of the columnar layer is described using the fitted elliptic aspect ratio, with a length of 289±111 μm and a width of 94±34 μm. As seen from the GOS map, most grains have GOS values less than 1°, which can be categorized as recrystallized with very low stored energy [10, 11]. Annealing at 800 °C (above T_β) of the bulk samples reveals the different responses of the equiaxed and columnar regions during annealing. The grain morphology remained basically unchanged after ST for 2 hours. Following a solution treatment for 3 hours, equiaxed grains became coarser while columnar grains persisted. Continued annealing for longer time results in significant grain coarsening, and the columnar grains were replaced by equiaxed grains.

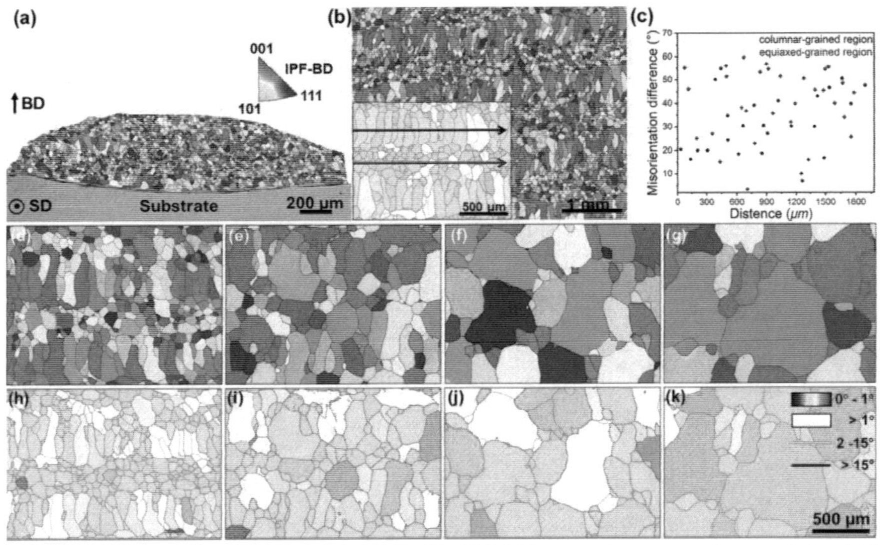

Figure 1. Microstructures of single track and bulk Ti3662 samples manufactured by DED. (a) EBSD orientation map of as-built single track on transverse section; (b) EBSD orientation map of as-built block sample on a section defined by the building direction BD and transverse X direction. The inset is a partial GOS map; (c) misorientation difference of adjacent grains as measured from the inset in figure 1(b); (d)-(g) EBSD orientation maps of block sample after solution treatment at 800 °C for 2h, 3h, 5h and 7h, respectively and (h)-(k) the corresponding GOS maps. The threshold value of misorientation angle to define a grain boundary is set to be 15°.

The figure 2(a) and 2(b) show the IPF orientation maps of the Ti3662 alloy produced by LPBF, acquired on the X-Y and BD-X planes, respectively. A chessboard-like structure is observed on the X-Y section, with alternating domains of fine grains and coarse grains. This checkerboard as-built microstructure is commonly observed in metals prepared by LPBF, typically exhibiting visible periodicity corresponding to the laser scanning strategies [10,11]. The regions with fine grains are oriented with <001> along BD. Such fine, <001>-oriented grains are in the overlapping area of two adjacent molten pools, which experienced more heating during LPBF process compared to the region with coarse grains. The temperature gradient in the overlapping area was altered due to the heat input from the adjacent molten pool. The repeated alterations in the direction of the temperature gradient resulted in increased nucleation, leading to finer grains. Consequently, the grains in the overlapping area is finer and maintain the initial <001>//BD orientation due to continuous competition between epitaxial growth and heterogeneous nucleation with the thermal gradients, and have higher stored energy, making them preferential sites for recrystallization.

Unlike the Ti3662 alloy fabricated by DED, most grains in the LPBF Ti3662 alloy have high GOS values, which indicates a higher density of low angle dislocation boundaries and dislocations stored during processing (figure2(c) and 2(g)). Grains with lower GOS values are mainly near the molten pool boundaries. After 30 minutes of solution treatment at 800 °C, the morphology of the as-printed structure vanished. The GOS map reveals that most grains have completely recrystallized, replaced by equiaxed and large grains with an average size of 114.6 ± 42.9 μm. The grain size remained unchanged after an additional 15 minutes of solution treatment. It is important to note that the grains of the as-printed structure prepared by LPBF are less thermally stable, which is a direct result of higher stored energy due to the smaller molten pool and higher cooling rate in the LPBF process.

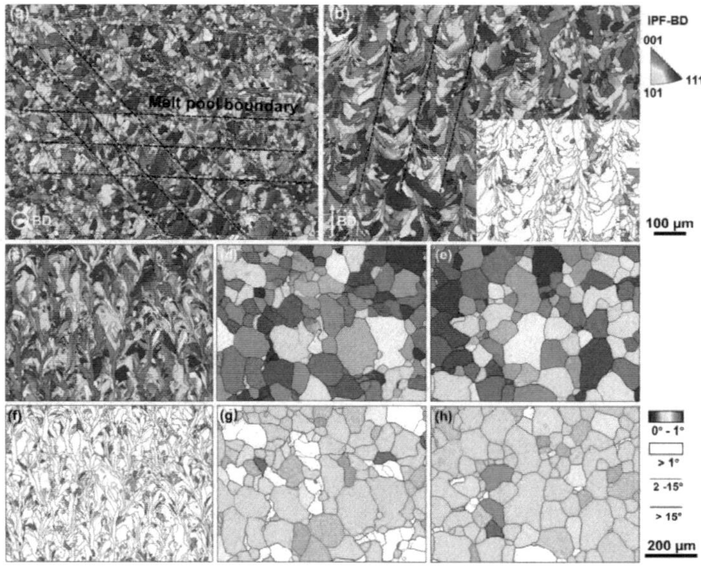

Figure 2. Microstructure of Ti3662 manufactured by LPBF. (a) EBSD orientation map of as-built bulk sample acquired on the X-Y section; (b) EBSD orientation map of as-built bulk sample acquired on the BD-X transverse section. The inset is partial GOS map; (c)-(e) EBSD orientation map of block sample after ST at 800 °C for 15min, 30min and 45min respectively and (f)-(h) corresponding GOS maps.

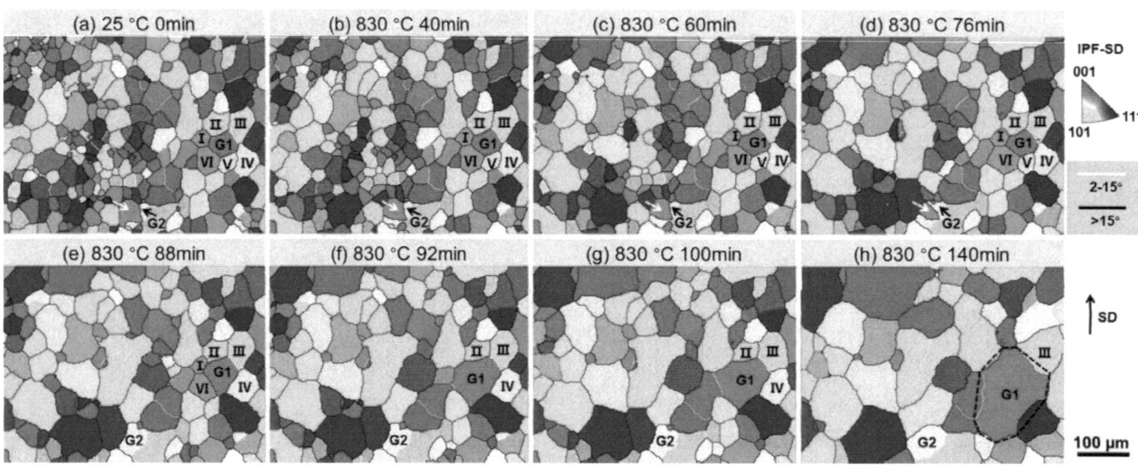

Figure 3. Orientation maps obtained by in-situ EBSD experiment illustrate the microstructure evolution at 830 °C for various time periods of a single-track sample deposited by DED.

The in-situ EBSD experiment at high temperatures was conducted on the DED single track sample, as illustrated in figure 3. The purpose of this experiment was to investigate the evolution of the microstructure of a single track when subjected to heating effects from adjacent molten pools. Each EBSD scanning acquisition lasted approximately 4 minutes, and the focal length was swiftly adjusted after each acquisition to mitigate the effects of sample drift. During the solution treatment process, it was observed that the grain size gradually increased from an initial value of 37.9 ± 13.5 μm to 87.1 ± 32.4 μm. Given the negligible fraction of low angle grain boundaries in the DED sample, it can be concluded that the heating process primarily led to grain growth without nucleation. The main driving force for this grain growth is attributed to grain boundary energy, with smaller grains being consumed by larger ones. Notably, no abnormal grain growth with preferred orientations was observed during this process, and ultimately, the final microstructure remained uniformly equiaxed. This suggests that with proper control over the heating effect from adjacent tracks, a homogeneous final microstructure for bulk samples could be achieved.

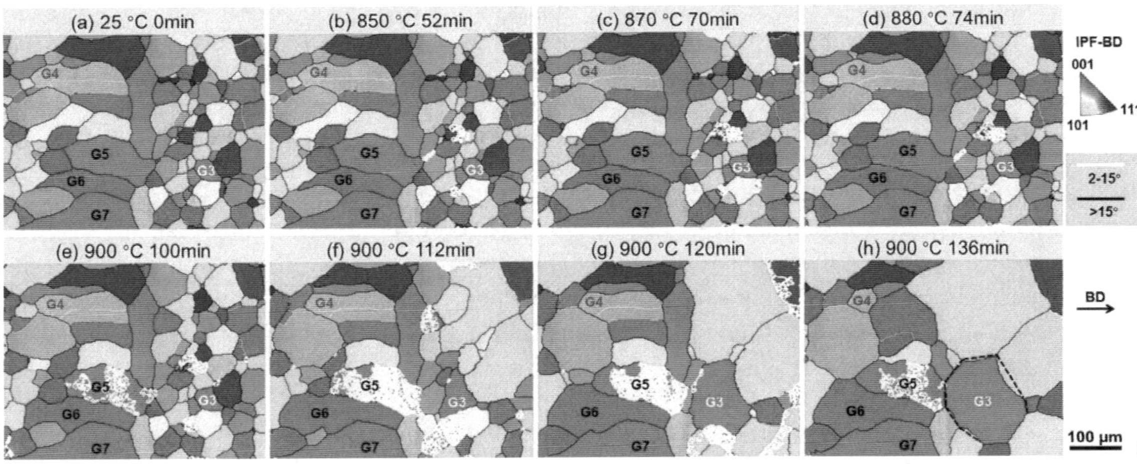

Figure 4. Orientation maps obtained by in-situ EBSD showing the microstructure evolution at temperatures between 850 °C to 900 °C for different time periods of bulk sample manufactured by DED.

For the bulk sample containing multiple tracks, higher annealing temperatures are necessary to induce grain growth. This requirement arises because the grains have already undergone coarsening due to the cyclic heating effect from adjacent tracks. The in-situ heating EBSD results of the Ti3662 sample, which exhibits a columnar/equiaxed heterogeneous structure, are depicted in figure 4. Upon annealing at temperatures exceeding 850 °C, the finer equiaxed grains on the right side of the image begin to coarsen first, while the columnar region on the left maintains its initial morphology at temperatures below 900 °C. With increasing temperature and duration of annealing, columnar grains start to grow, leading to a transformation in grain boundary morphology from ellipsoid to regular polygon. Ultimately, this process leads to the replacement of the heterogeneous microstructure, with alternating equiaxed and columnar regions, by a coarse equiaxed structure.

4. Discussion

In the heat-affected zones (HAZs), the microstructure tends to form a <100> texture and columnar grains [7], a phenomenon more pronounced in DED. When the grains in the equiaxed-grain layer are affected by heat input, they grow along the temperature gradient. The rate of grain boundary migration perpendicular to the temperature gradient is very low, resulting in the formation of columnar grains.

In contrast to the current understanding [5–8], the present research findings suggest that neither phase transformation nor recrystallization can be attributed to the formation of equiaxed grains of the β phase after ST. Due to the high content of β-stable elements in Ti3662, no α phase or α′ martensite precipitated in the as-printed samples, and there was no dislocation-based nucleation due to low dislocation density in as-built DED samples. On the contrary, LAGB composed of dislocations is more stable during ST, as exemplified by G1 with adjacent grains in figure 3 [12]. The CET in the columnar region occurs due to the gradual consumption of columnar grains by grain growth in the equiaxed region, as demonstrated by G4 in figure 4 [7]. Similar phenomena regarding grain growth behaviour in a mixture of columnar and equiaxed grains have also been reported in alloys processed using conventional approaches [10,13].

The coarsening of equiaxed and columnar grains is driven by curvature, as depicted in figure 3 and 4 [14,15], while mobility depends on boundary misorientation difference. In comparison to equiaxed grains, the growth of columnar grains is sluggish, indicating that columnar grains possess higher thermal stability. This can partly be attributed to the low grain boundary mobility of LAGBs between adjacent columnar grains. As illustrated in figure 1(c), the fraction of LAGBs is comparatively larger between columnar grains than between equiaxial grains. Additionally, the driving force of flat grain boundaries is minimal, resulting in G5, G6 and G7 in figure 4 maintaining their original morphology after annealing for more than two hours.

5. Conclusion

In conclusion, the CET during super-transus ST contributed to equiaxed grain growth and consumption of columnar grains. The grain growth process of additively manufactured Ti3662 is strongly influenced by grain morphology and surrounding structure. The highly curved grain boundaries in LPBF Ti3662 and equiaxed grains in DED provided sufficient driving force for facile grain growth and structural coarsening during annealing. A larger misorientation and curvature between adjacent equiaxed grains results in higher mobility than the grain boundary between columnar grains. These results highlight the importance of tailoring grain morphology through adjustments in additive manufacturing parameters for subsequent annealing processes.

Reference

[1] Yao M, Yao Z, Tao X, Huang C and Zhang S 2022 Alleviating plastic anisotropy of boron modified titanium alloy by constructing layered structure via electron beam directed energy deposition *Addit. Manuf.* **50** 102561

[2] Zhang G, Lu X, Li J, Chen J, Lin X, Wang M, Tan H and Huang W 2022 In-situ grain structure control in directed energy deposition of Ti6Al4V *Addit. Manuf.* **55** 102865

[3] Aitang Xue H D Xin Lin, Lilin Wang, Xufei Lu, Lukai Yuan and Huang W 2023 Achieving fully-equiaxed fine β-grains in titanium alloy produced by additive manufacturing *Mater. Res. Lett.* **11** 60–8

[4] Liu Z, Zhao Z, Liu J, Wang L, Yang G, Gong S, Wang Q and Yang R 2019 Effect of α texture on the tensile deformation behavior of Ti–6Al–4V alloy produced via electron beam rapid manufacturing *Mater. Sci. Eng. A* **742** 508–16

[5] Ng C H, Bermingham M J and Dargusch M S 2022 Controlling grain size, morphology and texture in additively manufactured β-titanium alloy with super transus hot isostatic pressing *Addit. Manuf.* **59** 103176

[6] Nandwana P, Lee Y, Ranger C, Rollett A D, Dehoff R R and Babu S S 2019 Post-processing to Modify the α Phase Micro-Texture and β Phase Grain Morphology in Ti-6Al-4V Fabricated by Powder Bed Electron Beam Melting *Metall. Mater. Trans. A* **50** 3429–39

[7] Qi M, Huang S, Ma Y, Youssef S S, Zhang R, Qiu J, Lei J and Yang R 2021 Columnar to equiaxed transition during β heat treatment in a near β alloy by laser additive manufacture *J. Mater. Res. Technol.* **13** 1159–68

[8] Chen J, Fabijanic D, Brandt M and Xu W 2023 Transformation-mediated in-situ β recrystallization in laser powder-bed fusion additively manufactured Ti-6Al-4V *Mater. Charact.* **206** 113452

[9] Davis A E, Donoghue J, Kennedy J R, Byres N and Prangnell P B 2021 In-situ observation of single variant α colony formation in Ti-6Al-4V *Acta Mater.* **220** 117315

[10] Park J M, Choe J, Park H K, Son S, Jung J, Kim T-S, Yu J-H, Kim J G and Kim H S 2020 Synergetic strengthening of additively manufactured (CoCrFeMnNi)99C1 high-entropy alloy by heterogeneous anisotropic microstructure *Addit. Manuf.* **35** 101333

[11] Yu Xuan R Y Jiaqiang Chang, Yao Ou and Zhang Z 2024 Heterogeneous structure architected by additive manufacturing: facile route towards strong and ductile steel *Mater. Res. Lett.* **12** 199–207

[12] Huang S, Ma Y, Zhang S, Youssef S S, Qiu J, Wang H, Zong B Y, Lei J and Yang R 2018 Nonuniform Recrystallization and Growth Behavior of β Grains Dominated by Grain Misorientation and Interfacial Energy in Metastable β Titanium Alloy *Metall. Mater. Trans. A* **49** 6390–400

[13] ang Y, Yang B, Gao M, Zhao E and Guan R 2022 Microstructure evolution, mechanical property response and strengthening mechanism induced by compositional effects in Al–6 Mg alloys *Mater. Des.* **220** 110849

[14] Ye X, Suo Z, Heng Z, Chen B, Wei Q, Umeda J, Kondoh K and Shen J 2024 An in-situ study of static recrystallization in Mg using high temperature EBSD *J. Magnes. Alloys* **12** 1419–30

[15] Luo Z, Yoshino M, Terano M and Yamanaka A 2018 Texture evolution in single crystal iron static recrystallization through in-situ EBSD observation *Procedia Manuf.* **15** 1565–72

IOP Conf. Series: Materials Science and Engineering **1310** (2024) 012034 doi:10.1088/1757-899X/1310/1/012034

Multiscale characterization of the additive-manufacturing-induced cell structure in 316L stainless steel: A comparative study

X Wang*, V K Nadimpalli, D Juul Jensen and T Yu

Department of Civil and Mechanical Engineering, Technical University of Denmark, Kgs. Lyngby DK-2800, Denmark

*E-mail: xiawnga@dtu.dk (X Wang)

Abstract. This study evaluates the potential of various electron microscopy-based methods for the characterization of the cell structures in additive manufactured stainless steel 316L. We focus on the capability of different techniques to characterize the multiscale microstructural features of the cell structures, including cell morphology, elemental segregation, particle distribution, and dislocation arrangement. Additionally, we suggest a manual method to quantify the very small misorientation angles across cell walls and within the cells. The advantages and limitations of different methods are discussed.

1. Introduction

In recent years, metal additive manufacturing (AM) has gained increasing attention [1]. Laser powder bed fusion (L-PBF) is a popular AM method [2]. The literature has demonstrated that L-PBF stainless steel components, printed using optimized printing conditions, typically exhibit superior mechanical properties compared to conventionally manufactured ones [3]. Given that almost fully dense samples can now be printed, the pursuit of an in-depth understanding of the microstructural characteristics is timely.

The L-PBF process introduces a distinctive microstructure. For example, a hierarchical microstructure is developed in L-PBF stainless steel 316L [4]. Notably, the additive-manufacturing-induced cell structure has gained significant interest in recent years [5].

The morphology of the cell structure appears to be equiaxed, extended, or columnar in the observation plane as found in 2D studies, matching well with the suggestion that in 3D the cells have a prism-like shape [6]. In the as-printed condition, the cell walls are usually decorated with a relatively high density of dislocations and segregated atoms [7]. Some studies attribute the enhanced tensile strength of L-PBF steel to this special cell structure [3].

However, due to technical challenges relating to the *combination* of small cell sizes and small misorientation angles, the microstructural features of the cell structure are still not fully understood. This study applies various electron microscopy-based methods to characterize the cell structure in L-PBF 316L, including cell morphology, segregation phenomenon, particle distribution, and dislocation arrangement. The aim is to evaluate the potential and limitations of each method for these characterizations.

IOP Conf. Series: Materials Science and Engineering **1310** (2024) 012034 doi:10.1088/1757-899X/1310/1/012034

2. Experimental

A cubic stainless steel 316L sample (10x10x10 mm^3) was 3D printed using a custom-built open-architecture L-PBF system [8]. To print dense samples (<0.2% porosity), pre-optimized parameters were used: laser power P = 250 W, scanning speed s = 650 mm/s, layer thickness t =50 μm, and hatch distance h = 100 μm. After each layer, the scanning direction was rotated 90°. The printing was done in a nitrogen atmosphere with a total oxygen level of <1000 ppm.

A 3-mm disk sample was prepared from the stable printing region. Electrolytic polishing was done by Struers TenuPol-5 at a temperature of -25 °C. The electrolyte contained 10% perchloric acid [9].

The sample was characterized by both scanning electron microscopy (SEM) and transmission electron microscopy (TEM). Two SEM modes, namely secondary electron (SE) and backscattered electron (BSE) modes, were used (i) to give an overview of the sample and select a region of interest (ROI); and (ii) to roughly characterize the cell structures. An accelerating voltage of 5 kV was used in both the SE and BSE modes. TEM was used to characterize the same ROI, including observing the dislocation structures and measuring misorientation angles across cell walls and within cells. Concerning the latter, a simple manual method based on TEM Kikuchi patterns is suggested. Finally, SEM-based and TEM-based energy dispersive X-ray spectroscopy (EDS) as well as high-angle annular dark-field scanning transmission electron microscopy (HAADF-STEM) were applied to characterize the segregation at cell walls.

3. Results and discussions

3.1 Overview of the sample

Figure 1 gives an overview of the sample as observed by SEM. By twin-jetting, a hole was produced, and the thin area around the hole is suitable for TEM observation. In the SE mode the detailed shape of the hole, including the rolled-up edges and cracks, is clearly visible. These features indicate high residual stress in the as-printed sample. The creation of a hole disturbs the stress equilibria causing relaxation of stresses, resulting in such complex shapes.

Orientation contrast is already clear in the SE mode, and this contrast becomes stronger in the BSE mode. Based on the overview images, one ROI was chosen for further study, as shown in figure 1(b).

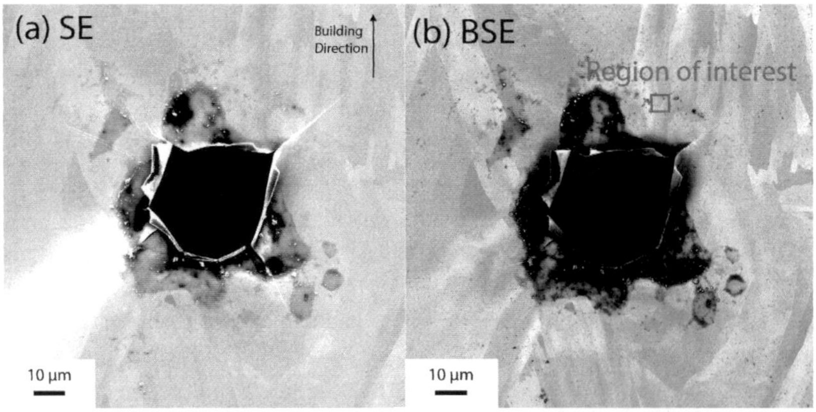

Figure 1. Overview of the sample: (a) SE mode and (b) BSE mode. The hole was produced by electrolytic polishing. One ROI is chosen for detailed characterization, as shown in the red box in (b).

3.2 Fine scale characterization

High resolution microstructural results are shown in figure 2. In both the SEM SE and BSE images, a columnar cell morphology is observed, in which the cell walls appear bright while the cell interior is darker. The contrast difference could be due to chemical segregation or a high density of dislocations at cell walls. In this ROI, the cells are aligned along the building direction and the average width (perpendicular to the building direction) of the cells is 0.8 µm. Many particles are observed. Some of these are large, typically >100 nm, and apparently randomly distributed, while others are relatively small, typically <50 nm. Most of the small particles are located at the cell walls.

The same area was also inspected by TEM. A bright-field image is shown in figure 2(c). Five columnar cells, named cell 1 (C1) to cell 5 (C5), are revealed when tilting the sample to an appropriate imaging condition. At the cell walls, dislocations are highly entangled. Loosely distributed dislocations, which are invisible in SEM images, are found in the cell interiors. Large particles are observed as dark spherical dots. The small particles can only be seen when the sample is tilted to certain angles. This is illustrated in figure 2(d)-(e): when the sample was tilted -2.5° about the holder axis, many small particles located at the cell walls are observed. When tilting the sample to 11.4°, the cell walls with entangled dislocations are clearly seen, but the small particles are hidden by the diffraction contrast from the dislocations.

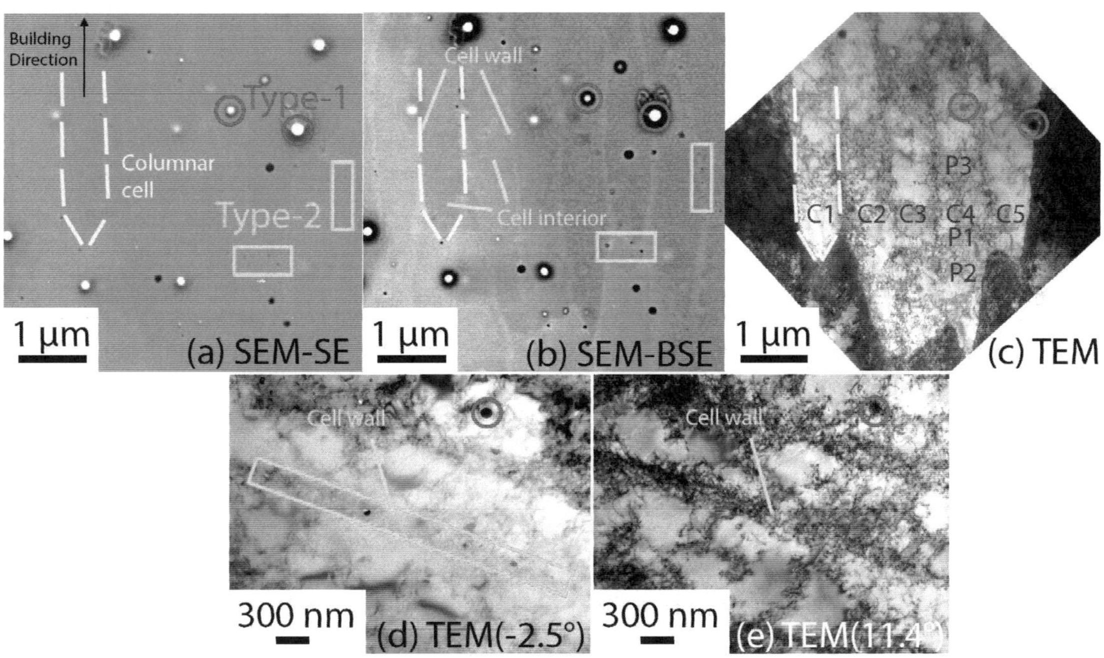

Figure 2. The same sample area characterized by (a) SEM-SE mode, (b) SEM-BSE mode, and (c) TEM-bright field. In (a)-(b), columnar cells are observed. In (c), dislocation networks within these columnar cells are observed. Two types of particles, large and small, are seen by SEM and TEM. (d) and (e) illustrate that tilting the sample is necessary to observe the small particles.

Segregation at cell walls is another interesting feature in AM samples. As shown in figure 3(a)-(d), segregation at cell walls is not detected by SEM-based EDS, which may be related to the low SEM spatial resolution and a weak segregation in the present sample. TEM-based EDS was

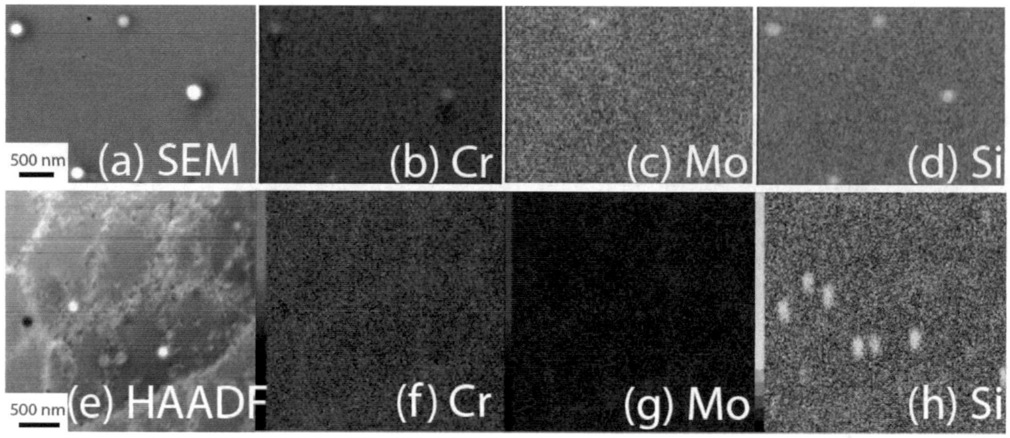

Figure 3. (a)-(d) Results from SEM-based EDS, showing no segregation at cell walls. (e)-(h) Results from TEM-based EDS, showing very weak Cr and Mo segregation at cell walls.

conducted at a thinner area, see figure 3(e)-(h). The TEM results show very weak Cr and Mo segregation at cell walls. To detect this weak segregation, a large beam current and a long scanning period of around 1 hour were used.

Concerning the large particles discussed above, the SEM-based and TEM-based EDS both reveal that these particles are rich in Si, which agrees with previous results [10].

3.3 Determination of small misorientation angles

To precisely characterize very small misorientation angles, for example across a cell wall or an interior dislocation boundary, both a high spatial resolution and a high angular resolution are required. This section introduces a simple manual method with that purpose. The method relies on measuring the translation and rotation of TEM Kikuchi patterns and converting the values into misorientation angles.

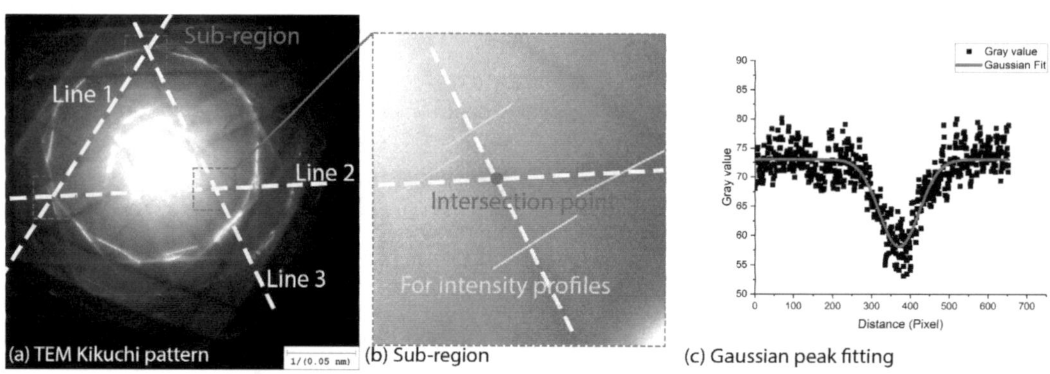

Figure 4. Determination of small misorientation angles by TEM Kikuchi diffraction. (a) Typical TEM Kikuchi pattern. Three sub-regions and three dark Kikuchi lines are chosen to measure the translation and in-plane rotation of Kikuchi patterns. (b) A magnification of the selected sub-region. The red dot shows the intersecting point between two dark Kikuchi lines which are marked in dash yellow. Cyan lines indicate where intensity profiles (gray values) are recorded. (c) Intensity profile across a dark Kikuchi line and Gaussian peak fitting to the intensity profile.

A series of Kikuchi patterns are recorded within the ROI by the TEM built-in camera in the convergent beam condition. Calculating local misorientation angles is possible by measuring the translation and rotation of the two relevant Kikuchi patterns. More specifically, the translation and rotation are acquired by: (1) selecting a sub-region—an example is shown in figure 4(a), where two dark lines intersect; (2) drawing four lines (marked in cyan) across the dark lines (marked by dashed yellow lines) and acquiring four intensity profiles—an example is shown in figure 4(b); (3) conducting Gaussian peak fitting to determine the peak positions, as shown in figure 4(c); (4) calculating the centre positions of the dark lines based on the peak fitting; (5) calculating the rotation angles based on the change of slope of dark lines, and (6) calculating the coordinates of the intersection point, which subsequently are used to calculate the relative translation of two Kikuchi patterns. The misorientation angles are finally converted from the translation distances of the intersecting points and the rotation angles of the dark lines. In this work, three sub-regions and three dark lines were selected in each Kikuchi pattern. Based on the known misorientation angle between two zone axes, it is found that each pixel corresponds to a misorientation angles in the Kikuchi pattern of 0.003°.

Table 1. Misorientation angles across five cell walls and across dislocation boundaries within a cell – see figure 2 where the cells C1-5 and the points P1-3 are marked.

	Cell 1&2	Cell 2&3	Cell 3&4	Cell 4&5	Point 1&2 in Cell 4	Point 1&3 in Cell 4
Translation angle (°)	0.26	0.13	0.18	0.13	0.21	0.19
Rotation angle (°)	0.18	0.35	0.39	0.70	0.60	0.20
Misorientation angle (°)	0.32	0.37	0.43	0.71	0.64	0.28

As an example, misorientation angle results for 4 cell boundaries and between 2 sets of interior points in a cell are shown in table 1. It is found that the misorientation angles across cell walls are very low, and the misorientation angles within one columnar cell are comparable with the values across cell walls. It is worth noting that in samples, as the present, with large residual stresses and a high density of interior dislocations such low misorientation angles cannot be detected precisely by standard electron backscatter diffraction (EBSD). Thus, manual measurements based on TEM Kikuchi patterns, as suggested here, may be a quick method to determine the misorientation angles of a few cell walls of interest.

3.4 Comparison of different methods

SEM-based methods give an excellent overview of the sample. Comparing the SE and BSE mode of operation, the SE mode provides more information about the sample areas near the hole. In BSE mode, the orientation contrast is clearer, which can be used for a quick characterization of the whole sample. Moreover, BSE mode also gives Z-contrast and dislocation contrast, enabling a clearer observation of cell walls in non-etched samples.

Both SEM-based and TEM-based EDS are commonly used to study chemical composition. For the present sample, only TEM-based EDS could detect the very weak segregation of Cr and Mo at the finely spaced cell walls due to the small TEM interaction volume (i.e. the high voltage electron beam penetrates the foil without much lateral broadening). This required a very long scanning time, which was possible without too severe contamination due to the good vacuum inside the TEM.

SEM-based EBSD is a commonly used technique to measure orientations and thus also misorientations, and a modern EBSD system can achieve an angular resolution of the order of 0.1°

[11]. However, for AM samples like the present one with large residual stresses and a high density of dislocations, EBSD is unable to reliably detect misorientation angles below 0.5° due to its small camera length and a large interacting volume. Instead, measuring small misorientation angles by TEM-based Kikuchi patterns can theoretically achieve a higher angular resolution because of the larger camera length. The simple method presented here illustrates the potential of using TEM-based Kikuchi patterns for the measurement of cell wall misorientation angles in AM samples.

4. Summary

To illustrate the potentials of a series of electron microscope methods, the cell structure in L-PBF stainless steel 316L was characterized by SEM (SE and BSE modes) and TEM. Also, EDS in SEM and TEM were applied for segregation mapping. For the cell structure, SEM-based methods are effective for characterizing the overall morphology, while TEM-based methods are suitable for examining dislocation structures within cells. In this study, only TEM-based EDS could detect the very weak segregation of Cr and Mo. The small misorientations between and within cells could be measured based on the TEM Kikuchi patterns, and a quick manual method is suggested.

Acknowledgements

The work of XW and TY was supported by the Independent Research Fund Denmark DFF-FTP [grant number 2035-00061B], while Villum Fonden [MicroAM - grant number VIL54495] supported the work of VKN and DJJ.

References

[1] Lewandowski J J and Seifi M 2016 *Annu. Rev. Mater. Res* **46** 151–186
[2] Sing S L and Yeong W Y 2020 *Virtual Phys. Prototy* **15** 359–370
[3] Liu L, Ding Q, Zhong Y, Zou J, Wu J, Chiu Y L, Li J, Zhang Z, Yu Q and Shen Z 2018 *Mater. Today* **21** 354–361
[4] Wang Y M et al. 2018 *Nat. Mater.* **17** 63–71
[5] Kong D, Dong C, Wei S, Ni X, Zhang L, Li R, Wang L, Man C and Li X 2021 *Addit. Manuf.* **38** 101804
[6] Wang X 2024 *Mater. Sci. Technol.* accepted
[7] Tekumalla S, Seita M and Zaefferer S 2024 *Acta Mater.* **262** 119413
[8] Kjer M B, Pan Z, Nadimpalli V K and Pedersen D B 2023 *J. Manuf. Mater. Process.* **7** 143
[9] Christiansen G, Bowen J R and Lindbo J 2002 *Mater. Charact.* **49** 331–335
[10] Funch C V, Grumsen F B, da Silva Fanta A B, Christiansen T L and Somers M A J 2023 *Heliyon* **9** e16555
[11] Wilkinson A J 2001 *Scr. Mater.* **44** 2379–2385

Neutron diffraction analysis of microstructural evolution and mechanical behavior in an additively manufactured multiphase alloy

K Yamanaka[1*], M Mori[2], Y Onuki[3] and S Sato[4]

[1] Institute for Materials Research, Tohoku University, Sendai, Japan
[2] Department of General Engineering, National Institute of Technology, Sendai College, Natori, Japan
[3] Department of Advanced Machinery Engineering, Tokyo Denki University, Tokyo, Japan
[4] Graduate School of Science and Engineering, Ibaraki University, Hitachi, Japan

*E-mail: kenta.yamanaka.c5@tohotoku.ac.jp

Abstract. Additive manufacturing (AM) involves an unprecedented thermal history during solidification and post-melt high-temperature exposure, leading to unique microstructural evolution. In this study, we employed neutron-diffraction-based microstructural analysis to better understand the microstructural evolution and mechanical behavior of AM alloys, with a particular focus on multiphase alloys. Samples of Ti–6Al–4V alloy used as a model material were prepared using electron beam powder bed fusion (EB-PBF) under varying building conditions. Time-of-flight neutron diffraction (TOF-ND) measurements were performed using an iMATERIA (BL20), J-PARC, Japan. Using Rietveld texture analysis (RTA), we revealed the textural evolution during hierarchical microstructural development from solidification to solid-state phase transformations in the EB-PBF process. The effects of building conditions on the textures in the as-built states and their evolution during subsequent tensile loading were analyzed.

1. Introduction

Additive manufacturing (AM) has emerged as a promising technology for fabricating complex geometries and difficult-to-process materials [1–4]. Electron beam powder bed fusion (EB-PBF) [5,6] is a representative metal AM process that uses an electron beam to selectively melt a thin powder layer (powder bed) in a vacuum. PBF processes involve unprecedented thermal histories characterized by rapid heating/cooling cycles upon electron-beam irradiation and significant thermal gradients in a confined melt pool. Notably, the use of an electron beam enables an extremely high scanning speed, facilitating preheating of the powder bed and non-equilibrium solidification at high solidification rates. Moreover, the post-melt high-temperature exposure, which is unique to the EB-PBF process, promotes solid-state phase transformations in nonequilibrium solidification microstructures. Therefore, EB-PBF provides unique opportunities for designing and developing novel alloys with desirable properties [7,8]. However, conventional characterization techniques, such as electron microscopy and X-ray diffraction, are often inadequate to fully understand the process–microstructure–property relationship for AM materials because the microstructural evolution in EB-PBF occurs hierarchically over a wide length scale.

Neutron diffraction (ND) is a powerful tool for acquiring bulk-averaged microstructural characteristics because neutron beams have high penetration capability [9] and can be used to analyze relatively large volumes (~ 10 cm^3). Therefore, the microstructural information obtained can be correlated with the macroscopic mechanical properties and plastic deformation behavior obtained from mechanical testing using bulk specimens. Although the ND technique has been utilized to evaluate the residual stress of AM components [10–12], recent ND studies have focused on characterizing the microstructures and mechanical behaviors of AM materials [13–15].

We have been devoted to ND-based microstructural analyses, which consist of line-profile analyses of dislocation structures and Rietveld texture analyses of crystallographic orientations and phase fractions, often combined with in situ measurements [16–19]. Our focus has been on multiphase alloys, where phase transformations, crystallographic orientations, phase fractions, and interactions between the constituent phases govern the mechanical behavior. In this paper, we report the application of ND-based microstructural characterization techniques for EB-PBF Ti−6Al−4V alloys to analyze the as-built textures that developed under different building conditions and their evolution upon subsequent tensile deformation.

2. Materials and method

In this study, we employed Ti−6Al−4V alloy, a representative duplex titanium alloy consisting of hexagonal close-packed (hcp) α- and body-centered cubic (bcc) β-phases. Samples were prepared using an EB-PBF instrument (Arcam Q20, GE Additives, USA) with varying printing parameters, that is, power (P) and scan speed (v). The line energy ($E_l = P/v$) for the low and high energy input conditions was 64.0 and 85.3 J mm^{-1}, respectively. The other parameters, such as the line offset, focus offset, and preheating temperature, were identical for both conditions.

Microstructural observations of the as-built specimens were conducted using field-emission scanning electron microscopy (FE-SEM; JSM-IT800, JEOL, Japan) at 15 kV. Electron backscatter diffraction (EBSD) measurements were performed using a FE-SEM (JSM-7100F, JEOL, Japan) operated at 25 kV. The samples for microstructural observation were prepared using a standard metallographic procedure and finished with a colloidal silica suspension (OP-S, Struers, Japan).

Time-of-flight (TOF) ND measurements were performed using an iMATERIA (BL20), J-PARC, Japan [20]. As illustrated in Figure 1 [21], the iMATERIA beamline comprises backscattering (BS), a special environment (SE), and low-angle (LA) detector banks, covering a wide range of scattering vectors. Tensile samples with a gauge portion 6 mm in diameter and 30 mm in length were prepared from the as-built specimens. Each sample was fixed to a tensile testing machine mounted on a goniometer and strained until failure during TOF-ND measurements. The beam direction was

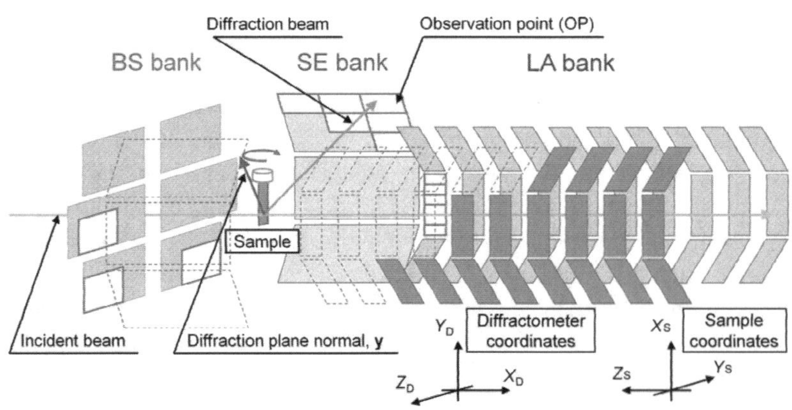

Figure 1. Schematic of iMATERIA diffractometer (BL20) at J-PARC, Japan [21].

perpendicular to the longitudinal direction (i.e., the building direction (BD)) of the tensile samples. The initial strain rate was set to 1×10^{-4} s^{-1}. The TOF-ND data were recorded continuously during tensile loading. Rietveld texture analysis (RTA) was performed to determine the textures and phase fractions [22,23] using Materials Analysis Using Diffraction (MAUD) software [24].

3. Results and discussion

Figure 2 shows the as-built microstructures of the Ti–6Al–4V alloy specimens prepared under two different building conditions. Both specimens exhibit acicular α + β microstructures, where the β-phase appears as a bright contrast. The acicular α + β microstructure is produced via the β → α' (HCP) martensitic transformation and subsequent decomposition of the α'-martensite [25]. Notably, a higher energy input (E_l) led to the coarsening of the microstructure.

Figure 3 shows the results of RTA for the as-built microstructures. Pole figures for both the α-matrix and nanoscale β-precipitates were successfully determined. For the low E_l condition, the *c*-axis of the α-phase was aligned at approximately 45° and 90° relative to the BD. This can be explained by the parent β-grains growing with the ⟨001⟩ direction being parallel to the BD and a Burgers orientation relationship

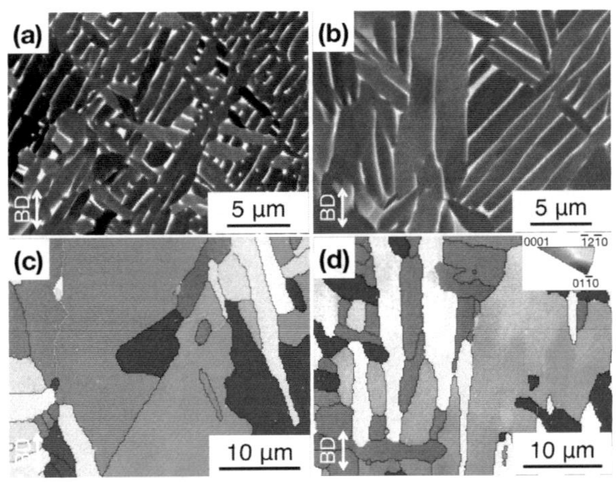

Figure 2. (a, b) SEM-BSE images and (c, d) IPF maps of as-built Ti–6Al–4V alloy specimens prepared under (a, c) low and (b, d) high E_l conditions.

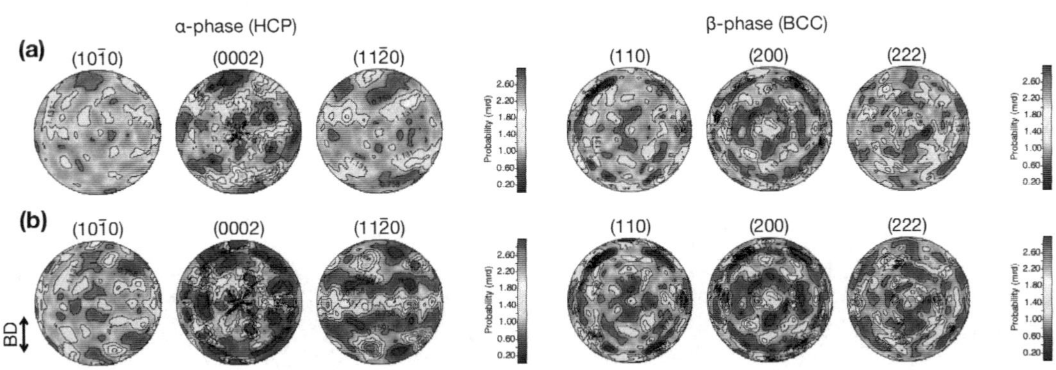

Figure 3. Pole figures of α-matrix and nanoscale β-precipitates for as-built Ti–6Al–4V alloy specimens prepared under (a) low and (b) high E_l conditions.

(OR): $(0001)_\alpha//\{\bar{1}10\}_\beta$, $\langle 11\bar{2}0\rangle_\alpha//\langle 111\rangle_\beta$ [26]. The presence of a Burgers OR is also observed in the α-matrix and nanoscale β-precipitates. Therefore, the RTA results indicate the existence of a multiple Burgers OR among the prior β-grains, α-matrix, and nanosized β-precipitates. A similar tendency was identified in a Ti–6Al–4V alloy prepared under standard conditions for the Arcam A2X system [27]. In contrast, the E_l condition resulted in a different α-texture where the *c*-axis of the α-phase is aligned parallel to the BD, indicating a notable difference in the prior β microstructures between the two building conditions. The β-phase fraction, determined by RTA, was nearly identical between the specimens at approximately 4 vol. %.

In contrast, our previous study [28] highlighted that the ND intensity for Ti–6Al–4V alloys is generally insufficient to measure the dislocation density in the nanoscale β-phase with line-profile analysis due to the negative and positive neutron scattering lengths of Ti and Al, respectively. Therefore, the direct detection of plastic deformation based on dislocation density in the nanoscale β-phase is challenging. However, as depicted in Figure 4, increasing tensile strain increased the pole density for both α- and β-textures in the elastic regime (point 1) and after yielding (point 2). This indicates that α-matrix and nanoscale β-phase deform concurrently under external loading, although lattice rotations were not significant.

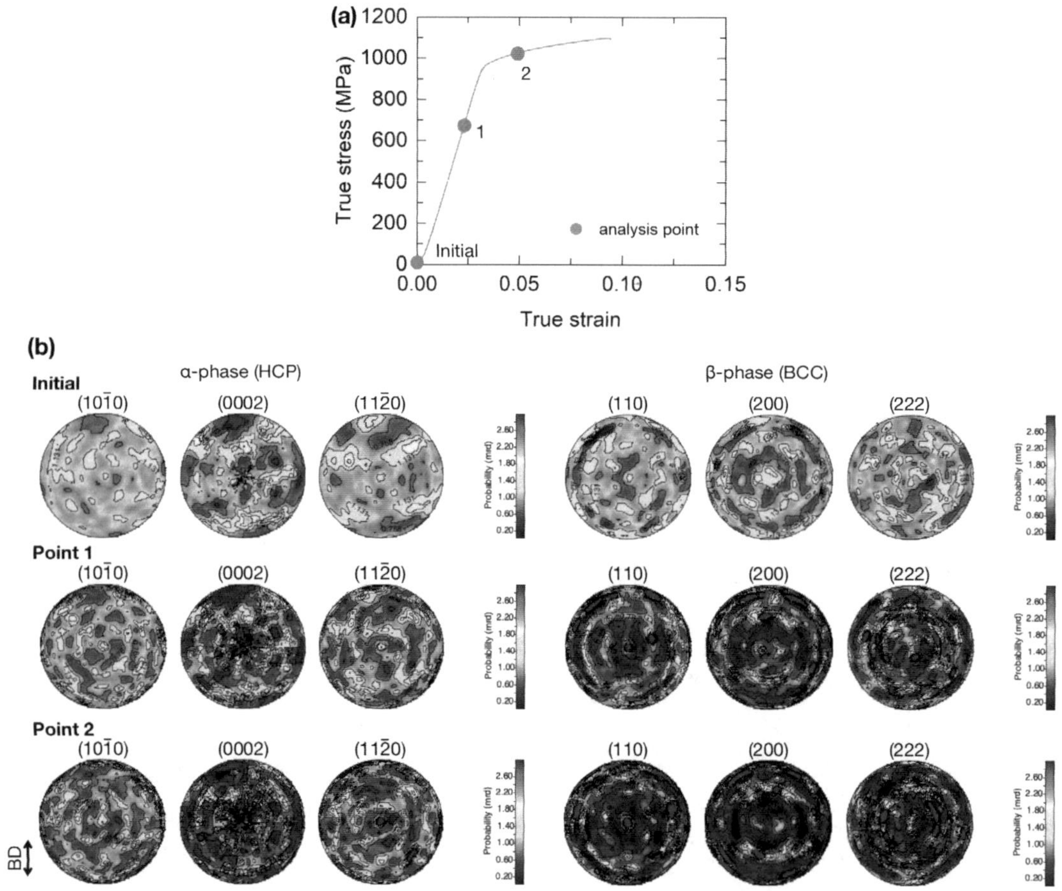

Figure 4. (a) True stress–true strain curve and (b) pole figures of α-matrix and nanoscale β-precipitates for Ti–6Al–4V alloy obtained under low E_l condition.

4. Conclusions

In this study, we conducted TOF-ND measurements on EB-PBF Ti–6Al–4V alloys prepared under different E_l conditions. Pole figures for both the α-matrix and nanoscale β-precipitates were determined through RTA based on the TOF-ND datasets obtained from a sufficient gauge volume. This allowed us to uncover the effect of the process parameters on the as-built textures and cooperative deformation behavior of the duplex microstructure.

Line profile analysis techniques, such as convolutional multiple whole profile (CMWP) fitting [29], can be integrated using high-angle-resolution datasets obtained from the BS bank of the iMATERIA beamline. These techniques are also applicable to various multiphase alloys such as Ni-based superalloys, where coarse columnar or single-crystalline matrices with nanoscale precipitation of intermetallic compounds and metastable alloys involve deformation-induced phase transformations.

Acknowledgements

This research was supported by the Japan Society for the Promotion of Science (JSPS [grant numbers 22H05274 and 24H01007]. Neutron diffraction experiments were performed at J-PARC according to proposal 2023PM2004.

References

[1] DebRoy *et al* 2018 *Prog. Mater. Sci.* **92** 112
[2] Herzog D, Seyda V, Wycisk E and Emmelmann C 2016 *Acta Mater.* **117** 371
[3] Sames W J, List F A, Pannala S, Dehoff R R and Babu S S 2016 *Int. Mater. Rev.* **61** 315
[4] Lewandowski J J and Seifi M 2016 *Annu. Rev. Mater. Res.* **46** 151
[5] Murr L E 2015 *Addit. Manuf.* **5** 40
[6] Körner C 2016 *Int. Mater. Rev.* **61** 361
[7] Aota S *et al* 2022 *Addit. Manuf.* **59** 103134
[8] Ellis E A I *et al* 2021 *J. Nucl. Mater.* **555** 153041
[9] Withers P J 2004 *J. Appl. Crystallogr.* **37** 607
[10] Wu A S, Brown D W, Kumar M, Gallegos G F and King W E 2014 *Metall. Mater. Trans. A* **45** 6260
[11] Wang Z, Denlinger E, Michaleris P, Stoica A D, Ma D and Beese A M 2017 *Mater. Des.* **113** 169
[12] Nadammal N *et al* 2021, *Addit. Manuf.* **38** 101792
[13] Wang Z, Stoica A D, Ma D and Beese A M 2018 *Mater. Sci. Eng. A* **714** 75
[14] Pokharel R, Patra A, Brown D W, Clausen B, Vogel S C and Gray G T 2019 *Int. J. Plast.* **121** 201
[15] Ghorbanpour S *et al* 2020 *Int. J. Plast.* **125** 63
[16] Sesha K S N, Yamanaka K, Mori M, Onuki Y, Sato S, Fabrègue D and Chiba A 2022 *Mater. Res. Lett.* **10** 754
[17] Mori M *et al* 2022 *J. Mech. Behav. Biomed. Mater.* **133** 105360
[18] Yamanaka K, Mori M, Yokosuka D, Yoshida K, Onuki Y, Sato S and Chiba A 2024 *Mater. Res. Lett.* **12** 1
[19] Onuki Y *et al* 2018 *Quantum Beam Sci.* **2** 11
[20] Ishigaki T *et al* 2009 *Nuc. Instrum. Methods Phys. A* **600** 189
[21] Onuki Y *et al* 2016, *J. Appl. Crystallogr.* **49** 1579
[22] Wenk H-R, Lutterotti L and Vogel S C 2010 *Powder Diffr.* **25** 283
[23] Onuki Y, Hoshikawa A, Nishino S, Sato S and Ishigaki T 2018 *Adv. Eng. Mater.* **20** 1700227
[24] Lutterotti L, Matthies S, Wenk H R, Schultz A S and Richardson J W 1997 *J. Appl. Phys.* **81** 594
[25] Yamanaka K *et al* 2021 *Addit. Manuf.* **37** 101678
[26] Al-Bermani S S, Blackmore M L, Zhang W and Todd I 2010 *Metall. Mater. Trans. A* **41** 3422
[27] Mori M, Yamanaka K, Onuki Y, Sato S and Chiba A 2022 *Addit. Manuf. Lett.* **3** 100053
[28] Yamanaka K, Mori M, Onuki Y, Sato S and Chiba A 2022 *Metals* **13** 86
[29] Ribárik G, Gubicza J and Ungár T 2004 *Mater. Sci. Eng. A* **387–389** 343

IOP Conf. Series: Materials Science and Engineering **1310** (2024) 012036 doi:10.1088/1757-899X/1310/1/012036

Synthesis and characterization of the dealloyed hierarchical structure of AlSi10Mg-Cu alloys prepared using laser-powder bed fusion

S Sun, C F Zhang, W Y Wang, A Godfrey*

Key Laboratory of Advanced Materials (MOE), School of Material Science and Engineering, Tsinghua University, Beijing 100084, P.R. China

*E-mail: awgodfrey@mail.tsinghua.edu.cn

Abstract. Nanoporous structures with high surface area can be obtained by 3D printing of lattice structures and subsequent dealloying, with potential applications in the field of catalysis and batteries. Here we prepared a AlSi10Mg-Cu (38wt%) alloy by laser-powder bed fusion of mixed powders. A metastable structure was obtained after 3D printing, with significant differences in solidification structures in different regions. The solidification structure is inherited after dealloying, in which dendrites form submicron ligaments, and nanoscale ligaments are formed inside the dendrites. Dealloying of a 3D-printed lattice structure allows the preparation of hierarchical nanoporous copper with only limited crack formation.

1. Introduction

A wide range of hierarchical porous materials can be found in nature, such as roots, stems, plant leaves, and bones, and researchers have used this as inspiration to design hierarchical porous materials [1]. Metal additive manufacturing (3D printing) by laser-powder bed fusion (L-PBF) is a process that can be used to create strong, dense and complex parts and offers the potential to prepare hierarchical lattice structures of various geometry [2,3]. It is not possible with this technique, however, to print directly samples with submicron or nanoscale internal open structure, due to the relatively large size of the meltpool.

As an alternative, dealloying can be used to selectively remove one or more components or phases of an alloy precursor, to allow the preparation of a bicontinuous 3D structure with a high surface area at the nanoscale [4]. Its potential applications include actuators [5], catalysts [6], and battery current collector materials [7]. Through appropriate precursor selection or coarsening and secondary dealloying processes [8,9], the method can be used to prepare submicron and nanoscale hierarchical porous structures. As such the combination of 3D printing and dealloying has been explored for a number of systems including, including AuAg [10] and CuMn [11] by ink-jet printing, and AlCu [12], CuMn [13] CuAg [14] and CuTi [15] by L-PBF. In this work we examine the potential for using L-PBF of a AlSi10Mg-Cu (38%) powder mixture to achieve a multi-scale hierarchical porous Cu structure. This is achieved firstly by investigating the relationship between the dealloyed microstructure and the as-printed structure in bulk samples, and then by investigating dealloying of 3D printed lattice structures, where the goal is to obtain a multi-scale hierarchical strut-ligament structure without excessive crack formation during dealloying.

2. Experimental

Spherical AlSi10Mg powder and pure copper powder (average diameters of 15-53μm) were used to prepare samples. The powders were mixed with a 38 wt% Cu ratio (equivalent to a 15 % volume ratio) using a high-efficiency resonance mixer at 60Hz for 10min (LABSEM-1, CASC, China). Samples were printed using this mixed powder using a SLM 280 instrument (SLM Soutions, Germany) on a AlSi10Mg substrate (build plate). The L-PBF process was carried out in an Ar atmosphere. Single track printing was first performed to screen appropriate process parameters. Parameters of 200 w/500 mm s^{-1}, 400 w/1000 mm s^{-1} and 500 w/1000 mm s^{-1} were used to print bulk samples, using hatch distances of 130, 150 and 150 μm respectively with a powder layer thickness of 50μm and layer rotation of 67° in each case. Parameters of 400 w/1000 mm s^{-1} were selected to prepare a BCC lattice structure, as these showed the least cracking in bulk samples. For the lattice structure a total of 7×5×5 lattice cells were printed (lattice cell strut size of 2 mm, and strut diameter of 500 μm), printed on a pre-printed AlSi10Mg-Cu substrate.

Bulk samples were ground and mechanically polished for subsequent scanning electron microscope (SEM) characterization (Mirca 3LMH, Tescan, Czekh) and composition was analyzed by energy dispersive spectroscopy (EDS). The bulk samples and lattice structure samples were dealloyed using 0.5 mol L^{-1} sodium hydroxide until bubbles no longer appeared. After dealloying, samples were soaked in clean water and subsequently rinsed with alcohol. The samples were stored in a vacuum environment. The microstructure and composition of the dealloyed samples were also characterized using a SEM and by EDS.

3. Results and discussion

Figure 1a shows the powder morphology after vibration mixing and the superimposed element distribution, where it is seen that the AlSi10Mg powder and Cu powder are spherical and well mixed. Since there are few reports on the printing of this type of alloy system, we first performed single-track printing on the AlSi10Mg substrate to explore the melt track width. For all three conditions examined a relatively continuous melting track without transverse or longitudinal cracks was obtained, see figure 1(b-d), in each case with a wide melt track width of about 300 μm. Analysis of the melt track (not shown here) showed copper mixing to at least 200 μm below the substrate (base plate) plane, indicating significant internal flow in the melt pool.

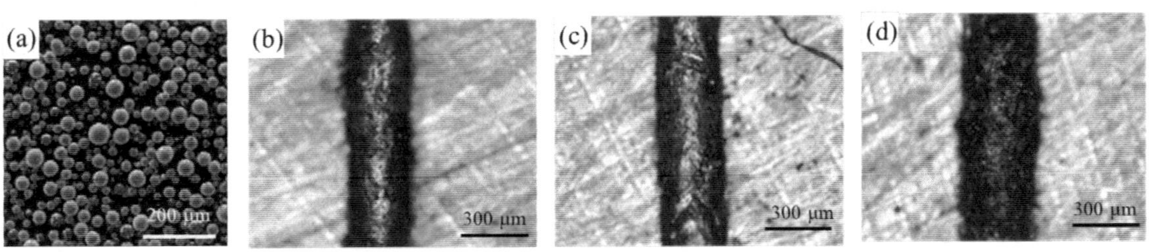

Figure 1. (a) SEM micrograph of the mixed AlSi10Mg-Cu (38wt%) powder superimposed by elemental distribution where green represents aluminium and red represents copper; (b-d) single track optical microscope images for 200 w/500 mm s^{-1}, 400 w/1000 mm s^{-1} and 500 w/1000 mm s^{-1}, respectively.

Figure 2 shows SEM images of bulk-printed samples sectioned in the plane perpendicular to the build direction (BD), defined here as the XY plane, or a in plane containing the build direction (defined here as the XZ plane, with Z||BD). For 200 w/500mm s^{-1} a bi-phase structure is obtained, figure 2(a1-a4), where the light-colored phase in these images was identified as Al_2Cu (based on

EDS measurements; table 1, point ①) and the dark-colored phase corresponds to Al supersaturated (due to the fast cooling rate in additive manufacturing) with copper and silicon (table 1, point ②). Long columnar dendrites are seen clearly in the XZ plane (figure 2a3), shown at higher magnification in figure 2a4. For the entire area shown in figure 2a1 the EDS-determined composition (table 1) is close to the proportion of each element in the initial powder. In contrast the composition for the area shown in figure 2a3 deviates from the powder composition.

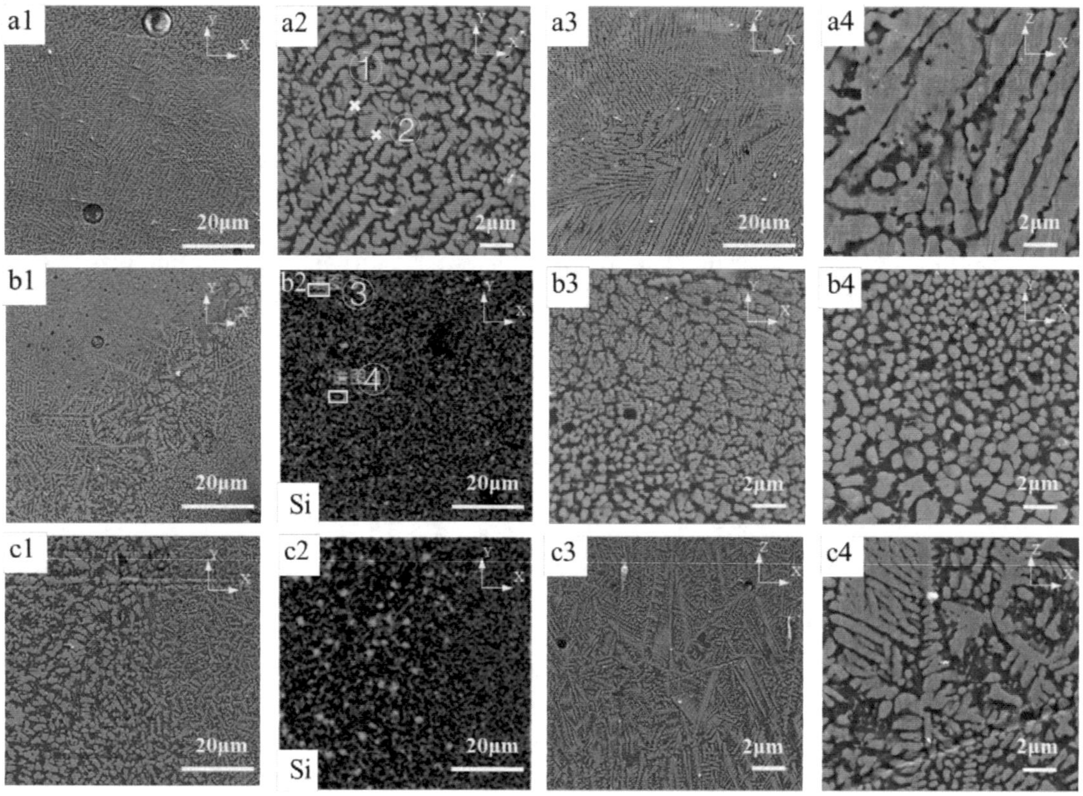

Figure 2. Microstructure as viewed in either the XY or XZ planes for samples printed at a) 200 w/ 500 mm s^{-1}, (b) 400 w/1000 mm s^{-1} and (c) 500 w/1000 mm s^{-1}. Sub-images b2 and c2 are EDS maps showing silicon distribution corresponding sub-images b1 and c1, respectively. EDS-determined compositions at locations 1-4 are given in table 1. The BD is parallel in these images to the Z direction.

Figure 2b(1-4) shows SEM-based images of a sample printed using 400 w/ 1000 mm s^{-1}. In figure 2b1 (XY plane) an obvious microstructural transition is seen from the upper left to the lower right of the image. The Si-distribution (EDS mapping) corresponding to the same region is shown in figure 2b2, which shows that Si is, however, relatively uniformly distributed across the entire area. EDS results from regions ③ and ④ also show no significant difference in the aluminium and copper content. Higher magnification images (figures 2b3,b4) show the structure in the upper left and lower right regions of figure 2b1, respectively. Dark spherical particles are also observed, identified from EDS results as Si particles. The different cooling rates at different locations in the melt pool may be a reason for the microstructural variations.

Figure 2c(1-4) shows SEM-based images for a sample printed using 500 w/ 1000 mm s^{-1}. Figure 2c1 shows a clear transition of microstructure from left to right in the image. The Si EDS map of the same area (figure 2c2) shows an obvious difference in silicon distribution in the two regions, suggesting that segregation of silicon may contribute to the difference in solidification microstructure. As shown in figures 2c3 and 2c4, an obvious dendritic structure is observed in the XZ plane under this high energy density, accompanied by the appearance of secondary dendrites and tertiary dendrites, resulting in a different microstructure compared to samples printed at lower laser energy density.

Table 1. EDS mass ratio of aluminium, copper and silicon corresponding to the indicated locations in figure 2.

Region/Point	Al	Cu	Si
Whole figure a1	54.6	39.2	6.2
Whole figure a3	47.2	47.1	5.7
Point ①	49.2	46.2	4.6
Point ②	67.7	25.7	6.6
Region ③	59.7	33.5	6.8
Region ④	61.2	30.7	8.1

The solid-liquid interface temperature, alloy composition, temperature gradient and solidification rate all have a significant impact on the final microstructure [16-18]. Considering our more complex alloy system and significantly different microstructures, we here just make a simple analysis of the above results. The melting point of AlSi10Mg is about 600°C, while the melting point of copper is 1084°C. The high energy density required for melting copper, may cause some loss by evaporation of aluminium during the printing process, and a keyhole laser mode may lead to pores inside the sample (see figure 2). Silicon has the highest melting point among the three main elements (1410°C) and as such will nucleate first during the solidification process, leading to some variation in the silicon segregation with laser power for the printed samples. A larger melt pool (from higher input power) will lead to more complex Marangoni convection, resulting in significantly different cooling rates, temperature gradients and alloy compositions at different locations, which will in turn affect the growth mode of the crystal and lead to a transition of the microstructure [19]. In addition, the powder layer thickness is only 50 microns, so a larger melt pool will also impact the microstructure of previously deposited layers to a greater depth. In some places there is a correlation between obvious segregation of silicon and differences in the solidification structure. However, in some areas the distribution of silicon is uniform, while different microstructures are also observed. This demonstrates the complexity of solidification microstructure formation. Only limited cracking was observed in the printed samples. We speculate the more plastic aluminium phase may help accommodate thermal stresses of the more brittle intermetallic compound Al_2Cu phase, thereby reducing crack susceptibility [16].

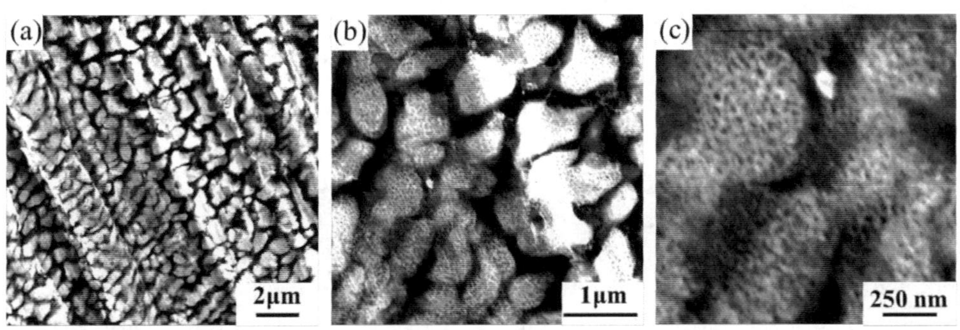

Figure 3. Morphology after dealloying of the 400 w/1000 mm s^{-1} sample as viewed in the XY plane.

IOP Conf. Series: Materials Science and Engineering **1310** (2024) 012036 doi:10.1088/1757-899X/1310/1/012036

Compared to AlSi10Mg, in our material where a large amount of copper is added, the Al-Si hypoeutectic structure transforms into an Al-Cu hypereutectic structure. Nevertheless the melt pool structure has certain similarities. For AlSi10Mg it has also been reported that due to variations in solidification conditions in the melt pool a transition from coarse dendrites to a finer scale structure can take place. An important difference though between the microstructures is that L-PBF of AlSi10Mg results in an obvious cellular structure formed by Si rich networks, whereas with the addition in our samples of a large amount of copper, a Si-rich segregation cell structure is no longer formed. The distribution of silicon in most areas is relatively uniform, but some areas have silicon-rich particles, which may be attributed to the complex compositional fluctuations and thermal gradients during solidification [20].

For the sample shown in figure 2c, we performed a sodium hydroxide dealloying treatment to remove the aluminium content in the L-PBF as-printed precursor. The resulting microstructure (as seen in the XY plane) is shown at various magnifications in figure 3. The dendritic characteristics of the precursor are preserved after dealloying. Although the supersaturated solid solution of aluminium contains a high concentration of copper, this phase is completely removed after dealloying as the percolation threshold for the formation of a nanoporous structure is not met. In contrast the morphology of the Al_2Cu phase is retained to form ligaments with a size of about 500 nm. As the sample retains structural rigidity, we can conclude that that these coarse ligaments form a three-dimensional bi-continuous structure. Moreover, due to the removal of Al from the Al_2Cu phase, the coarse ligaments are subdivided by nano-ligaments with a ligament size of less than 100 nm to form a nano-porous structure. Consequently, dealloying the bulk sample, results in a porous material with a hierarchical ligament structure.

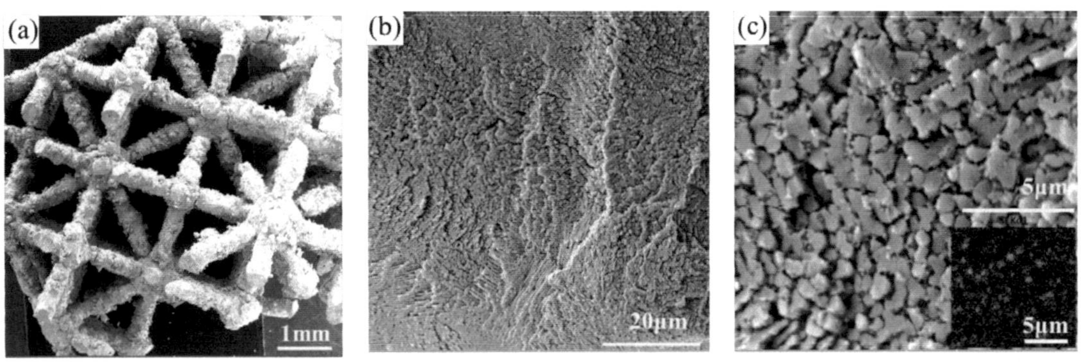

Figure 4. Lattice structure obtained after dealloying viewed at different magnifications. The inset in figure 4c shows the distribution of silicon in the same image.

In the lattice structure, due to the small size of the printed structs, the residual stress caused by rapid cooling can be effectively reduced. Additionally an open strut structure is conducive to rapid dealloying and therefore allows the realization of multi-level ligament/strut structure in a bulk sample. We used 400 w/1000 mm s^{-1} to print a lattice structure that was then dealloyed. As shown in figure 4a, in the printed-then-dealloyed lattice sample some cracks are present at the connection between the structs and nodes, especially for struts parallel to the build plane. This is because they are directly printed on powder due to the BCC-type lattice geometry. Such cracks should be eliminated by optimizing the lattice structure and printing conditions. The crack density inside the struts, however, is low. The strut fracture surface in a broken sample is shown in figure 4b and 4c. Similar to the results in figure 3, the complex solidification structure is inherited. According to the EDS results, only 1.3 wt% of aluminium remains, with the aluminium phase completely removed, and the original Al_2Cu phase forming a second-level sub-micron ligament structure. Under the dealloying conditions used, silicon is retained to a certain extent, as shown in the inset of figure 4c. The results confirm that 3D printing of a lattice structure from AlSi10Mg-Cu mixed powder allows a hierarchical porous copper structure to be achieved after dealloying, where this may have a potential application in copper-silicon batteries [7].

4. Summary and conclusions

Bulk samples and lattice structures of were prepared by L-PBF of a AlSi10Mg-Cu (38wt%) powder mixture. Due to differences in the melting points of the alloy elements, and differences in thermophysical conditions and compositions at different locations in the molten pool, a range of microstructures were obtained. In contrat to L-PBF of AlSi10Mg, a well defined segregation cell structure is not formed. Dealloying of the samples results in the formatation of a hierarchical bicontinuous porous structure, with finest scale ligaments of less than 100 nm. By L-PBF printing of a lattice structure a further level of porosity can be achieved, although some processing optimization and lattice design is required to avoid crack formation at lattice nodes. The ability to produce bulk samples with such nanoporous structure is expected to have good applications where excellent catalytic performance is required.

References

[1] Song T, Yan M and Qian M 2018 *Corros. Sci.* **134** 78-98
[2] Maconachie T, Leary M, Lozanovski B, Zhang X, Qian M, Faruque O and Brandt M 2019 *Mater. Design* **183** 108137
[3] Leary M, Mazur M, Elambasseril J, McMillan M, Chirent T, Sun Y, Qian M, Easton M and Brandt M 2016 *Mater. Design* **98** 344-57
[4] Erlebacher J, Aziz M J, Karma A, Dimitrov N and Sieradzki K 2001 *Nature* **410** 450-3
[5] Jin H J and Weissmuller J 2011 *Science* **332** 1179-82
[6] Wittstock A, Zielasek V, Biener J, Friend C M and Baumer M 2010 *Science* **327** 319-22
[7] Huang T, Sun D, Yang W, Wang H, Wu Q and Xiao R 2018 *Scr. Mater.* **146** 304-7
[8] Sun S et al 2022 *IOP Conf. Ser.: Mater. Sci. and Eng.* **1249** 012133
[9] Shi S, Li Y, Ngo-Dinh B N, Markmann J and Weissmuller J 2021 *Science* **371** 1026-33
[10] Zhu C, Qi Z, Beck V A, Luneau M, Lattimer J, Chen W, Worsley M A, Ye J, Duoss E B, Spadaccini C M, Friend C M and Biener J 2018 *Sci Adv* **4** eaas9459
[11] Mooraj S, Welborn S S, Jiang S, Peng S, Fu J, Baker S, Duoss E B, Zhu C, Detsi E and Chen W 2020 *Scr. Mater.* **177** 146-50
[12] Li C, Yuan S, Yao X, Yu X, Li B and Tu S-T 2023 *Fuel* **347** 128367
[13] Zhang Y, Sun X, Nomura N and Fujita T 2019 *Small* **15** e1805432
[14] Yan W Y, Zhang C and Liu L 2021 *ACS Appl. Mater. Inter.* **13** 45385-93
[15] Cai C, Guo S, Li B, Tian Y, Dong Qiu J C, Sun C N, Yan C, Qi H J and Zhou K 2021 *ACS Appl. Mater. Inter.* **13** 48709-19
[16] Hu L, Li Y, Luo G, Shi Q, Wang Y, Ma S, Cui Y, Wang H and Chen Z 2023 *Addit. Manuf.* **73** 103671
[17] Pauly S, Wang P, Kühn U and Kosiba K 2018 *Addit. Manuf.* **22** 753-7
[18] Rosa D M, Spinelli J E, Ferreira I L and Garcia A 2008 *Metall. Mater. Trans. A* **39** 2161-74
[19] Zhang T, Li H, Liu S, Shen S, Xie H, Shi W, Zhang G, Shen B, Chen L, Xiao B and Wei M 2019 *J. Phys. D: Appl. Phys.* **52** 055302
[20] Paul M J, Liu Q, Best J P, Li X, Kruzic J J, Ramamurty U and Gludovatz B 2021 *Acta Mater.* **211** 102914

Microstructural characterization of AISI 440C stainless tool steel fabricated by laser powder bed fusion

Z Pan[1*], Y Zhang[1], M Defer[1], X Xiao[2], D Juul Jensen[1], and V K Nadimpalli[1]

[1] Department of Civil and Mechanical Engineering, Technical University of Denmark, 2800 Kgs., Lyngby, Denmark
[2] National Synchrotron Light Source II, Brookhaven National Laboratory, Upton, 11973 New York, USA

*E-mail: zhipa@dtu.dk

Abstract. The microstructure of a stainless tool steel AISI 440C fabricated by laser powder bed fusion (L-PBF) without pre-heating of the build plate was characterized by multi-scale experimental methods. In combination with thermodynamic calculations, the solidification and cooling-down procedures were analyzed with the intention to understand the cracking behavior of high carbon tool steels processed by L-PBF. The results showed a fully austenitic structure in the as-built sample with sub-micro cellular structures and nano-sized carbides decorating the cell walls. Significant segregation exists merely at the intersection of cell walls while it is absent along high angle grain boundaries. Factors contributing to crack-free AISI 440C are discussed, providing guidelines for future L-PBF fabrication of high-carbon tool steels.

1. Introduction

L-PBF of tool steels is an attractive routine for the fabrication of intricate molds and tools. However, tool steels tend to exhibit relatively limited printability: cracks, warpage, and delamination often develop during the L-PBF process due to accumulated thermal stress [1,2]. One way to enhance printability is to reduce the carbon content in the alloys. Many studies have reported that maraging steels exhibit excellent printability due to its low carbon content (<0.03 wt.%) [3,4], and Deirmina et al. [5] found an enhanced printability in AISI H13 with ~0.25 wt.% C compared to that with ~0.4 wt.% C. Another approach to develop medium-carbon tool steels was proposed by Röttger et al. [6]: by tailoring the martensite start temperature to near room temperature, the tensile residual stresses developed in the austenite phase during cooling is counteracted. However, tool steels with a low or medium carbon content have low hardness (usually < 55 HRC) and additional carbon is necessary for further hardening.

It has been found that L-PBF of crack-free high carbon steels can be obtained by pre-heating of the build plate (200-500 °C). Then the thermal residual stresses are reduced due to a decrease in temperature gradient [2,7] as well as a reduction of retained austenite [8,9]. However, pre-heating the build plate increases the overall build time and energy consumption, reduces the recyclability of powders [10], and coarsens the microstructure in both as-built and heated-treated conditions [11]. Further, merely a few available L-PBF systems allows such a high degree of preheating, which severely limits the fabrication of high-carbon tool steels.

It is thus highly relevant to consider that it has been shown possible to print crack-free stainless steel AISI 440C, which contains ~1 wt.% C without pre-heating the build plate [12,13].

IOP Conf. Series: Materials Science and Engineering **1310** (2024) 012037 doi:10.1088/1757-899X/1310/1/012037

Nevertheless, the mechanisms behind this non-cracking behavior remain unclear. In the present work, we also managed to print crack-free AISI 440C without pre-heating the build plate. Based on an investigation of as-built microstructure and solidification path, the underlying mechanisms are discussed and a guide for successful L-PBF manufacturing of high-carbon steels is suggested.

2. Methods

AISI 440C powder was provided by Asgaard Metals. Its chemical composition is specified in table 1. The carbon, nitrogen, and oxygen concentrations were measured by combustion and gas fusion analysis using a LECO TC500 N/O determinator and a LECO CS230 C/S determinator, while other elements were measured using a Thermo Scientific ICAP 7200 ICP-OES Radial. The L-PBF process was performed in a custom-built open-architecture L-PBF system equipped with a fiber laser with a power output of up to 300W and a Gaussian beam with a focal diameter of 90 μm at 1080 nm [14]. The processing parameters were set as follows: 250 W power, 700 mm/s scanning speed, 50 μm layer thickness, 0.12 mm hatch distance, and a unidirectional scan with 67° rotation between layers. The printing was conducted in a N_2 atmosphere without pre-heating of the build plate, and cubes of 10x10x10 mm^3 were fabricated for microstructural analysis. The chemical composition of the as-printed AISI 440C part was measured using energy dispersive spectroscopy (EDS) in a Zeiss Sigma 300 scanning electron microscope (SEM) as well as combustion and gas fusion analysis. The results are given in table 1.

Table 1. Chemical composition of AISI 440C (wt.%, Fe balance).

	Cr	Mo	Si	Ni	Mn	V	C	N	O	S+P
Powder	16.3	0.4	0.2	0.3	0.3	0.2	0.998	0.114	0.056	<0.01
Part	16.5	0.5	0.4	0.3	0.2	0.2	0.987	0.110	0.016	-

The as-built AISI 440C parts were ground and polished finishing with a Struers oxide polishing suspensions. The polished sample was investigated using light optical microscopy (LOM, Keyence VHX-6000 and Zeiss Axiolab 5) and SEM. Electron backscattered diffraction (EBSD) and EDS were conducted at 20 kV, and data analysis was performed using AZtecCrystal software. X-ray diffraction (XRD) was performed for both the powder and the polished sample using a Bruker D8 Discovery diffractometer, Cu Kα radiation, with a Δ2θ step size of 0.02°, an exposure time of 1 s, and a 2θ range of 35°-100°. The polished sample was then etched using Kalling's No.1 reagent for 1 min and investigated using LOM. Hardness testing was performed using Struers DuraScan G5 tester with a load of 5 kg and a dwell time of 10 s. The reported hardness value was averaged over 8 indentations. Thermodynamic calculations were performed using Thermo-Calc software based on the TCFE12 database under Classic Scheil assumption with C and N as fast diffusers.

To characterize the 3D cellular structure, nano tomography transmission X-ray microscopy (TXM) was performed at the 18-ID Full-field X-ray Imaging beamline of the National Synchrotron Light Source II. The setup is detailed in [15]. A needle-shaped sample in the as-built condition was prepared by electropolishing a rod with a diameter of ~0.5 mm using Struers electrolyte A2 at 18 V. The X-ray beam energy was 7 keV. The sample was rotated 180° at 2°/s with an exposure time of 0.148 s. A total of 600 projections were recorded, and a voxel size of 22.49 nm was achieved

IOP Conf. Series: Materials Science and Engineering **1310** (2024) 012037 doi:10.1088/1757-899X/1310/1/012037

with a field of view of 57.70 µm. Image reconstruction was performed using a Python package TomoPy and Avizo was used for data analysis.

3. Results and discussion

On a macro-scale, no apparent lack of fusion pores or cracks were observed in the as-built AISI 440C samples, as shown in figure 1a. Its relative density is 99.7 % determined by Archimedes method with an absolute density of 7.8 g/cm^3. The X-ray diffractograms exhibit a primarily austenitic structure for the powder with some low intensity peaks, which may be attributed to micro-sized carbides. The printed part is fully austenitic, which relates to its high C+N content – both C and N are strong austenite stabilizers – as well as to the rapid cooling in L-PBF. This agrees well with the thermodynamic simulations, as shown in figure 1c, where austenite dominates with the formation of carbides at the end of solidification. The high cooling rate in L-PBF suppresses diffusion-controlled solid-phase transformation during the cooling down procedure, and the martensite start temperature is much lower than room temperature for AISI 440C [16]. As a result, the solidified austenite remains at room temperature. It is believed that the primary austenite solidification mode tends to develop liquid films along grain boundaries, leading to easy initiation of cracking [17]. Therefore, solidification mode here doesn't account for its non-cracking behavior.

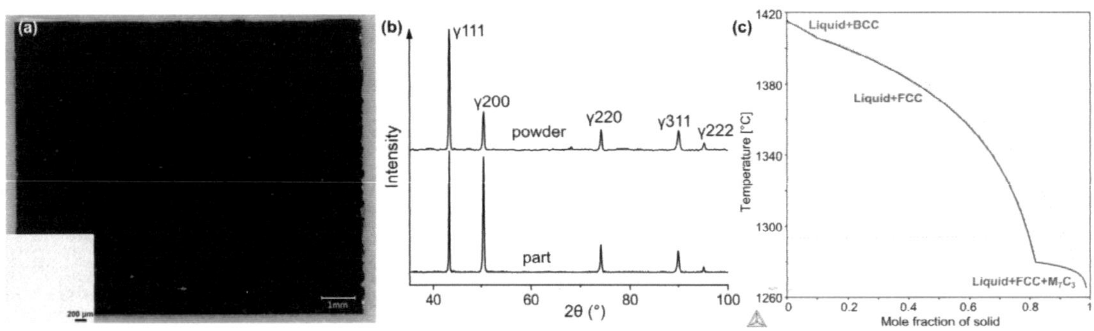

Figure 1. (a) A dark-field LOM image of as-polished AISI 440C sample with inset showing a magnified region in a bright-field mode; (b) X-ray diffractograms of powder and as-built part; and (c) predicted solidification path of AISI 440C by thermodynamic calculations.

On a meso-scale, figure 2 illustrates the columnar austenitic grains which are elongated along build direction (BD). A weak cube texture appears, where a <100> direction is aligned close to BD. The high angle grain boundaries (HAGBs, >15°) fraction here is ~63%, which is comparable to that found earlier in crack-free L-PBF AISI 316L (~59%) [18] but significantly higher than that in a austenitic steel with cracks (~43%) [19]. It has been suggested that cracks in L-PBFed austenitic steels tend to initiate from and propagate along HAGBs [20]. The high HAGB fraction in the present AISI 440C however indicates that HAGBs are not a dominant factor for the cracking behavior.

On a micro-scale, cell walls and cell interiors are clearly visible due to segregation of Cr and the cell walls are decorated with nano-sized carbides (~63 nm diameter), as shown in figure 3. The structure below melt pool boundaries tends to be cellular dendritic with more apparent segregation, whereas cellular structures with weaker segregation are observed above melt pool boundaries (see figure 3c). A 3D view of the cellular structure characterized with TXM is shown in figure 4, depicting two dominant grains distinguished by two sets of cellular structures with differently oriented cell walls. Cell walls weave into a perpendicular pattern and are normal to the

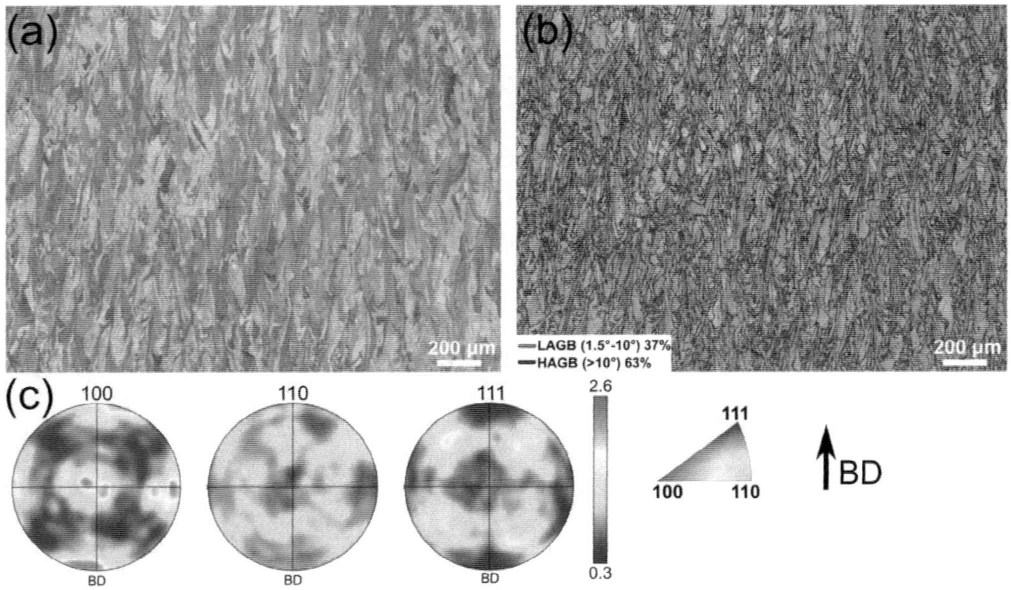

Figure 2. EBSD map showing (a) crystallographic orientations parallel to BD and (b) band contrast map with HAGB in blue and low angle grain boundaries (LAGB) in red; (b) the corresponding pole figures of austenitic grains in as-built AISI 440C.

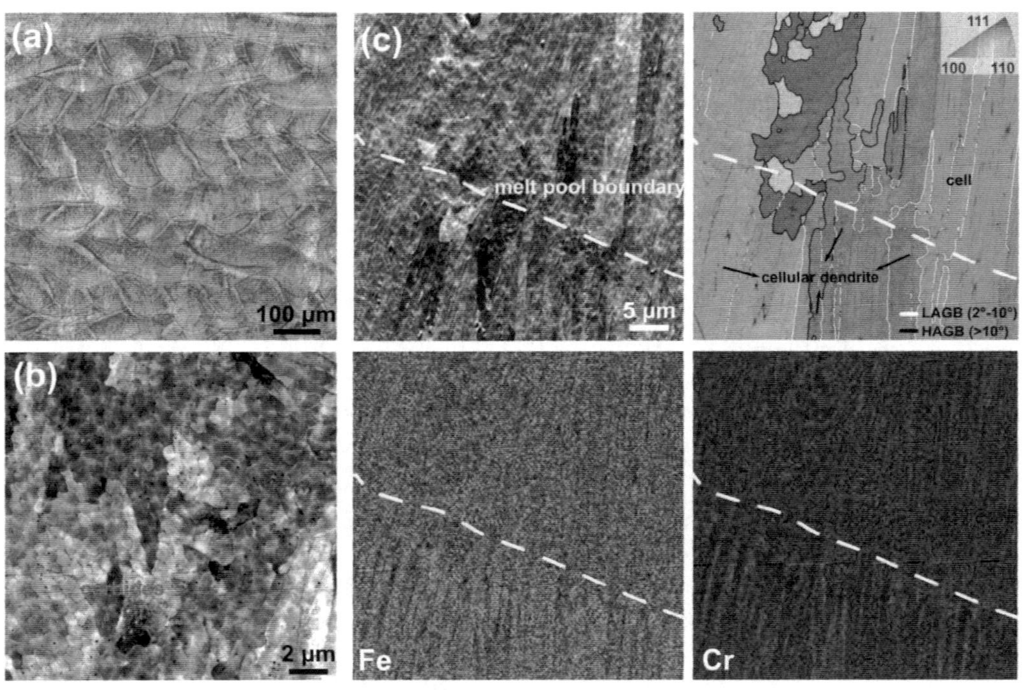

Figure 3. (a) LOM micrographics and (b) SEM images of as-built sample; (c) SEM image showing cellular structures with the corresponding EBSD and EDS maps showing crystallographic orientations parallel to BD and distribution of Fe and Cr.

cell growth direction, possibly orienting towards the three crystallographic <100> axes. The misorientation between the two grains is determined to be 48°. Significant segregation tends to

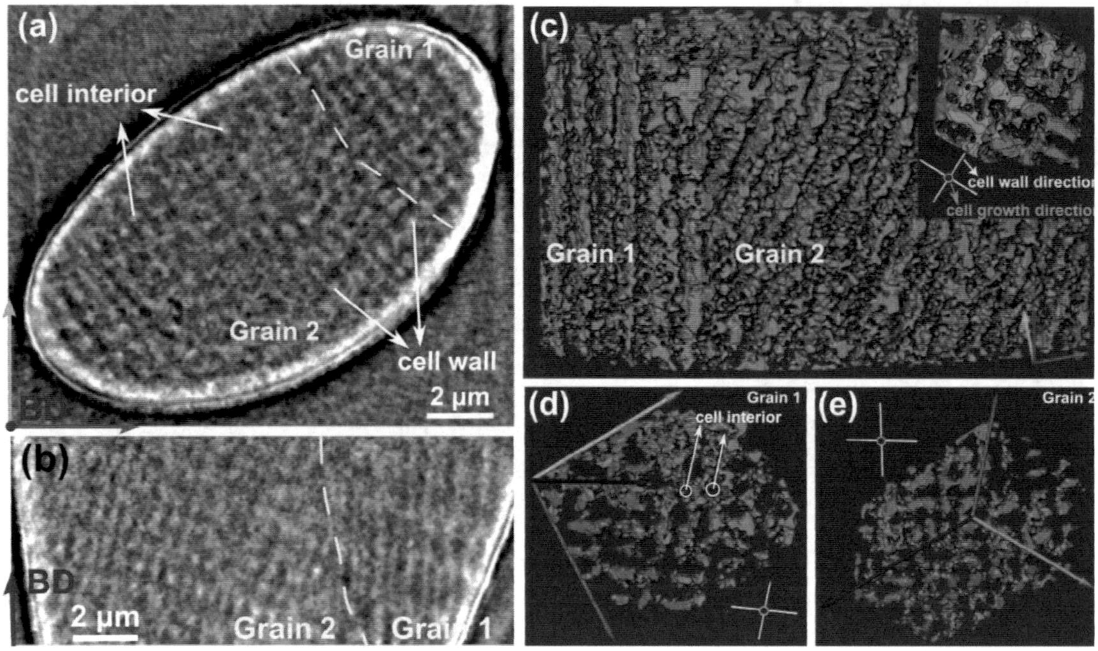

Figure 4. Typical processed slices of the as-built AISI 440C, (a) normal to BD and (b) parallel to BD. (c) A reconstructed 3D image of cell walls and inset showing a 4 × 4 × 4 μm^3 sub-volume with its cell growth direction aligned to the view normal direction; regions with high intensities (significant segregation) are marked in blue and the rest of cell walls are in grey. (d-e) Two sets of reconstructed cell walls in a 4 × 4 × 6.75 μm^3 sub-volume from grain 1 and 2 with cell growth directions and dominate cell wall directions indicated. Blue arrow indicates BD.

appear at the intersection of the cell walls within each grain (inset in figure 4c), rather than along HAGBs. This result is consistent with that seen in the EDS results (see figure 3c). The cell size is 716±141 nm determined by SEM images and 667±52 nm obtained from TXM results. Such sub-micron cellular structures, together with nano-sized carbides, promote uniform distribution of alloying elements at the grain scale, which may be beneficial for suppressing the formation of liquid film along HAGB during solidification and thereby prevent the formation of hot cracking.

The micro-hardness of the as-built AISI 440C is 441±2 HV_5, significantly lower than that of the heat-treated samples following a conventional austenization and quenching (up to 732 HV_5). The comparably soft condition is likely to help accommodate the thermal stresses developed during cooling, thereby preventing cold cracking.

4. Summary and outlook

The microstructure of crack-free AISI 440C stainless tool steel fabricated by L-PBF without pre-heating of the build plate has been characterized in detail and discussed with a view towards crack formation mechanisms in L-PBF high carbon tool steels. The as-built AISI 440C exhibits a fully austenitic structure, inherited from the solidified austenite, with columnar grains growing along the BD. Its HAGB fraction is high and comparable to other austenitic steels fabricated by L-PBF, suggesting HAGBs are not a dominating factor for cracking. Sub-micron cellular structures appear, with significant segregation at the intersection between cell walls and nano-sized carbides decorate the cell walls. Segregation at the intersection of the cell walls contributes to the uniform distribution of elements, thereby avoiding segregation along HAGBs. A relatively low hardness is obtained in the as-built condition, contributing to the lack of cold cracking. Future studies are encouraged that investigate the influence of process parameters as well as alloying elements (e.g., higher Cr and C within chemical specification, considering the current composition is just above the lower limit) on the cracking behavior.

Acknowledgement

ZP, YZ, MD, and DJJ acknowledge the financial support by the research grant (MicroAM-VIL54495) from VILLUM FONDEN. VKN acknowledges the Innovation Fund Project DECOR (0175-00043B). for financial support. We all thank Thomas Dahmen, Emilie H Valante, Magnus B Kjer and David B Pedersen for fruitful discussions and support during sample manufacturing.

References

[1] Kou S 2002 *Welding Metallurgy* (Hoboken, NJ, USA: John Wiley & Sons, Inc.)
[2] Liu Z, Chua C, Leong K, Kempen K, Thijs L, Yasa E, Van-Humbeeck J and Kruth J 2011 *5th International Conference on Advanced Research and Rapid Prototyping* (Leiria, Portugal) 339–46
[3] Kempen K, Yasa E, Thijs L, Kruth J P and Van Humbeeck J 2011 *Phys. Procedia* **12** 255–63
[4] Tan C, Zhou K, Ma W, Zhang P, Liu M and Kuang T 2017 *Mater. Des.* **134** 23–34
[5] Deirmina F, Quarzago L, Butcher D, Bettini E, Mehraban S, Hann J, Holländer Pettersson N, Lavery N, Röttger A and Pellizzari M 2024 *Mater. Sci. Eng. A* **901**
[6] Röttger A, Boes J, Großwendt F and Weber S 2023 *Addit. Manuf.* **61**
[7] Zumofen L, Beck C, Kirchheim A and Dennig H-J 2018 *Industrializing Additive Manufacturing - Proceedings of Additive Manufacturing in Products and Applications - AMPA2017* (Cham: Springer International Publishing) 210–9
[8] Saewe J, Benjamin W M, Jauer L and Schleifenbaum J 2019 *Metal Additive Manufacturing Conference* (Örebro, Sweden)
[9] Kosiba K et al. 2023 *J. Mater. Sci. Technol.* **156** 1–19
[10] Krist F 2021 *Metal AM* (vol 7) (Shrewsbury, Inovar Communications Ltd)
[11] Mertens R, Dadbakhsh S, Van Humbeeck J and Kruth J P 2018 *Procedia CIRP* **74** 5–11
[12] Bang G B, Hwang Y J, Kim W R, Song Y H, Kim G H, Hyun S K, Cha S C, Park H J, Lee T W and Kim H G 2022 *Mater. Sci. Eng. A* **860** 144259
[13] Gatões D, Alves R, Alves B and Vieira M T 2022 *Materials.* **15**
[14] Kjer M B, Pan Z, Nadimpalli V K and Pedersen D B 2023 *J. Manuf. Mater. Process.* **7** 143
[15] Coburn D S et al. 2019 *Rev. Sci. Instrum.* **90**
[16] Yang J R, Yu T H and Wang C H 2006 *Mater. Sci. Eng. A* **438–440** 276–80
[17] Kujanpää V P, David S A and White C L 1986 *Weld. J.* **65** 203–12
[18] Wang Y M et al. 2018 *Nat. Mater.* **17** 63–70
[19] Pan Z, Karthik V, Vase C, Aagaard S, Zhou L, Bolt M, Lundin T and Zhang Y 2024 *Mater. Charact.* **211** 113907
[20] Huang G, Wei K, Deng J, Liu M and Zeng X 2022 *J. Manuf. Process.* **83** 235–45

IOP Conf. Series: Materials Science and Engineering **1310** (2024) 012038 doi:10.1088/1757-899X/1310/1/012038

Effect of solution treatment temperature on recrystallisation behaviour of Haynes 282 manufactured through laser powder bed fusion

K Swaminathan* **and J Andersson**

Department of Engineering Science, University West, Trollhättan, Sweden

*E-mail: Kameshwaran.swaminathan@hv.se

Abstract. Increasing layer thickness in Laser Powder Bed Fusion (PBF-LB) process of metals enable increasing productivity and facilitate industrialisation of metal additive manufacturing (AM) process. Understanding of microstructure in as-built stage and possible post-processing steps to modify the microstructure is critical for metal AM components. Haynes 282 Nickel based superalloy, typically used in aerospace and energy industries, was manufactured using PBF-LB process at 60 microns layer thickness. Two different solution treatment temperatures were studied to analyse the recrystallization behaviour of the as-built material. The as built microstructure consisted high dislocation density given the rapid cooling in PBF-LB process. Solution treatment at 1150°C resulted in reduced dislocation density but similar morphology to grains in as built condition with visible residual melt pool boundaries. Solution treatment at 1250°C resulted in recrystallised grain structure. The recrystallisation behaviour is discussed with relation to manufacturing process and kinetic behaviour of alloying elements.

1. Introduction

Haynes 282 is a nickel-based superalloy developed in 2005 targeting applications with service temperature above 650°C. The alloy was developed to target ease of manufacture through lean γ' volume fraction of ~20% while providing good creep performance at targeted service temperatures on comparison with Rene 41 and Waspaloy[1]. Specific applications include turbine exhaust case in aeroengines [2] and hot section components of advanced ultra super critical boiler[3]. Aerospace industry can benefit from reduced weight of the components, and along with other industries with elevated service temperatures, design optimization and improved product performance are also targeted. Laser powder bed fusion is one such additive manufacturing technique, providing those benefits. The process involves layer by layer building of components comprising design complexity to a near net shape[4-5]. The adaptation and implementation of laser powder bed fusion in industries could be accelerated by improving productivity. One way to do so, is by increasing the layer thickness of build and optimize process parameters.

Laser powder bed fusion of Haynes 282 has been studied by few research groups usually between 20-40 microns layer thickness[6-14]. Process optimization through power, speed, hatch distance, scan strategy and heating of build plate has been reported[9-10, 12]. The current authors have utilized focus offset to optimize the process for higher layer thickness [15].

Post-processing of laser powder bed fusion-built Haynes 282 has been studied for conventional and alternate heat treatment including hot isostatic pressing[6, 8-11, 13, 14]. The vertical direction samples had better properties compared to horizontal direction in tensile and stress rupture properties. Some of the studies focussed on mechanical characterization of vertical direction specimens only[14, 16]. This anisotropy is usually attributed to the fact that more grain boundary area is available in horizontal direction samples rendering poorer mechanical properties compared to vertical direction samples[8]. Recrystallisation of the matrix to minimise the anisotropy can benefit the end application and typical service conditions.

This study aims at studying laser powder-built Haynes 282 with 60-micron layer thickness, for available options of solution treatment including the possibility of recrystallisation. The water quench study is aimed at understanding the rate of progression of the solution treatment while the air-cooled study is performed to reduce the cooling rate closer to industrial settings. Crystallographic analysis is performed on air cooled samples to understand the recrystallisation behaviour combined with thermodynamic calculations. Understanding the effect of post-process on PBF-LB microstructure could be used in design for life limiting damage mechanism of intended industrial applications.

2. Experimental methods

The test coupons for this study were manufactured in a SLM 280 2.0 Twin Laser Machine using Haynes 282 powders manufactured by Sandvik under special license. The chemistry of the powders and are presented in table 1. All the studies were carried out using a vertically built coupon measuring 15 mmx15 mmx110 mm. The process parameters were optimized in a previous study using focus offset of +4mm resulting in a spot size of 105µm, in a previous study[15] . The process parameters used for the build were 383 W for power, 683 mm/s for laser scan speed and 0.144 mm for hatch distance. There was a 67^0 rotation between each layer to break down the epitaxial growth and reducing texture. The heat treatment was carried out at 1150°C and 1250°C followed by air cooling or water quenching as specified. The specimens for heat treatment were at 5- 10 mm thick sections out of the aforementioned test coupon.

As built and heat-treated samples were sectioned to prepare the plane normal to the build direction. The metallographic preparation involved grinding and polishing with finishing steps of 3 µm diamond solution followed by 0.5 µm alumina solution. Kallings reagent was used to etch the samples and reveal the microstructure. Zeiss Axio optical microscope and Zeiss Gemini 450 field emission scanning electron microscope (SEM) were used for microstructural characterization. Electron back scatter diffraction (EBSD) detector from Oxford Instruments was used for data collection with a 2 µm step size, and Aztec software was used for data analysis. Struers Duramin–40 hardness tester was used for Vickers hardness measurement with 1kg load and dwell time of 10seconds with a minimum of 9 indents for every measurement. TCNI10 and MOBNI5 databases in Thermocalc software was used for segregation model for alloying elements during solidification and property diagram prediction for the given chemical composition.

3. Results and Discussion

Figure 1 shows the optical micrograph from top, middle and bottom of the 110 mm long test coupon. In all the images, the laser scan lines are visible approximately around the hatch distance (0.144mm) used during the build process. The different angles at top, middle and bottom could be due to scan rotation angle between each layer. Top section is within 5 mm from top, middle section is approximately at the middle of the test section, bottom section is approximately 5 mm away from build plate. The higher magnification optical images show similarity in morphology between top, middle and bottom of the test coupon, with cellular structure inside the grains. This indicates that the

IOP Conf. Series: Materials Science and Engineering **1310** (2024) 012038 doi:10.1088/1757-899X/1310/1/012038

repeated heating and cooling effects are similar at all heights of the coupon starting from 5 mm away from build plate.

Table 1. Chemical composition of the Haynes 282 powder used in this study

Element	Cr	Co	Mo	Ti	Al	C	B	O	N	Ni
Weight %	18.9	9.5	8.3	2.0	1.4	0.07	0.005	0.011	0.004	Bal.

Figure 1: Low and high magnification optical images of top, middle and bottom sections of the 15 mmx15 mmx110 mm vertically built coupon. Low magnification image scale bar is 50μm and high magnification image scale bar is 10μm.The high magnification images show cellular structure inside the grains.

Figure 2 shows the high magnification SEM image showing cellular structure of the order of few microns inside of each grain. This indicates the rapid cooling of the order of 10^5 to 10^6 K/s[6, 17]. The cells are decorated with inter-dendritic particles possibly primary carbides (Ti, Mo) C as indicated by line scans in figure 2 and elemental segregation profile in figure 3. Ti and Mo are rejected more into the liquid as solidification process nears completion.

Figure 3 also shows the property diagram of Haynes 282 indicating that phase fraction of primary carbides starts reducing around 1200 °C and at about 1250 °C almost a half order of magnitude. Hence 1150 °C, typical temperature for conventional solution treatment and 1250°C for recrystallisation of the matrix were chosen for further experimentation. The difference in morphology through solution treatment at 1150 °C and 1250 °C is due to the dissolution of (Ti, Mo) C as shown in figure 3 starting around 1200 °C. This dissolution helps the grain boundaries to move and recrystallise and relinquish the stored energy in the as-built grains. This can be seen by straighter grain boundaries in 1250 °C solution treatment compared to 1150 °C.

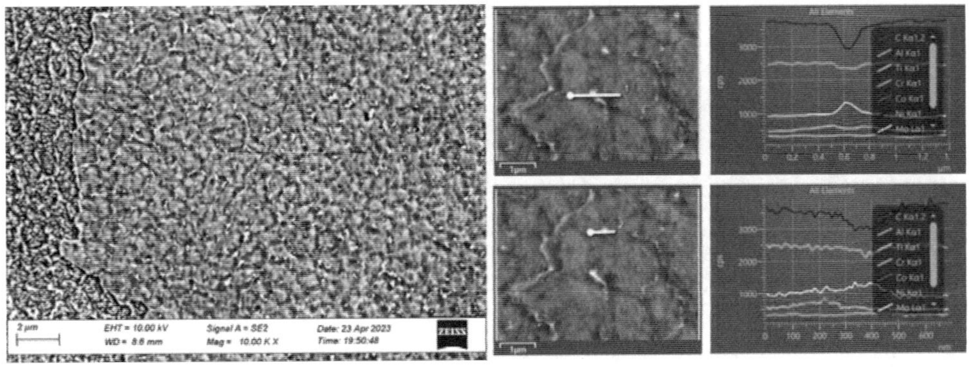

Figure 2: SEM image showing cellular structure present in the as-built structure. The line scans across inter-dendritic particles are shown to have increasing amounts of Mo, Ti, C compared to the matrix, possibly rejected through solidification process and forming primary carbides.

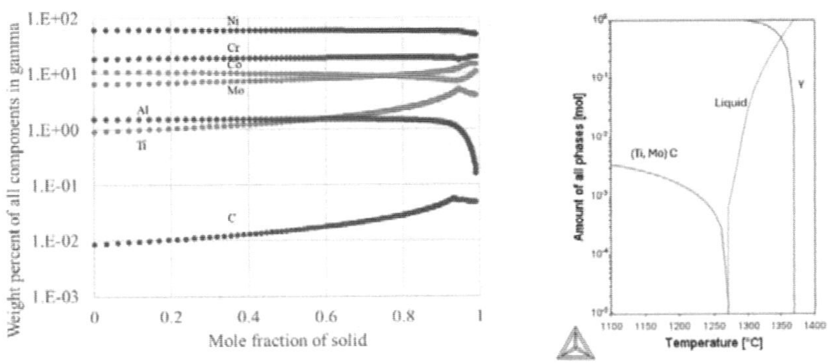

Figure 3: Solidification profile for Haynes 282 showing segregation of alloying elements starting from liquid to end of solidification process (left). Phase fraction diagram showing phase fractions of liquid and high temperature solid phases between 1100 °C and 1400 °C (right).

For further analysis, samples were solution treated at 1150 °C and 1250 °C for one hour, to simulate the time for components with thick sections, and cooled in air since water quenching might result in distortion for components with different section thickness. Figure 4 shows the inverse pole figure map and kernel average misorientation maps. The hardness of 1150°C/1h air cooled sample was 350±15 HV1, increase in hardness compared to as built (300±15 HV1) could be due to the cooling gamma prime. The hardness of 1250 °C/1h air cooled sample was lowest, at 250±12 HV1, indicating the effect of recrystallisation. Area weighted average grain size was 59±18 μm for as built and 52±17 μm for 1150 °C/1h indicating no recrystallisation. Area weighted average grain size was 107±29 μm for 1250 °C/1h study along with twin boundaries, indicate recrystallisation followed by partial grain growth. Similar effect of temperature on recrystallisation was recently reported for wire arc additively manufactured Haynes 282 as well[18].

The amount of local misorientation and stored energy is qualitatively assessed through kernel average misorientation (KAM) maps. Coincident site lattice boundaries are prevalent in recrystallised microstructure indicating strain free grains with twins generated in the microstructure. These crystallography-based details were analysed for as-built, 1150 °C/1h and 1250 °C/1h air cooled samples as shown in figure 4. The as-built microstructure possessed maximum stored energy due to the fast cooling and heating nature inherent to the process. 1150 °C/1h microstructure possessed residual stored energy, despite the heat treatment, indicating the recrystallisation has not initiated, as the melt pool boundaries were still partially visible.

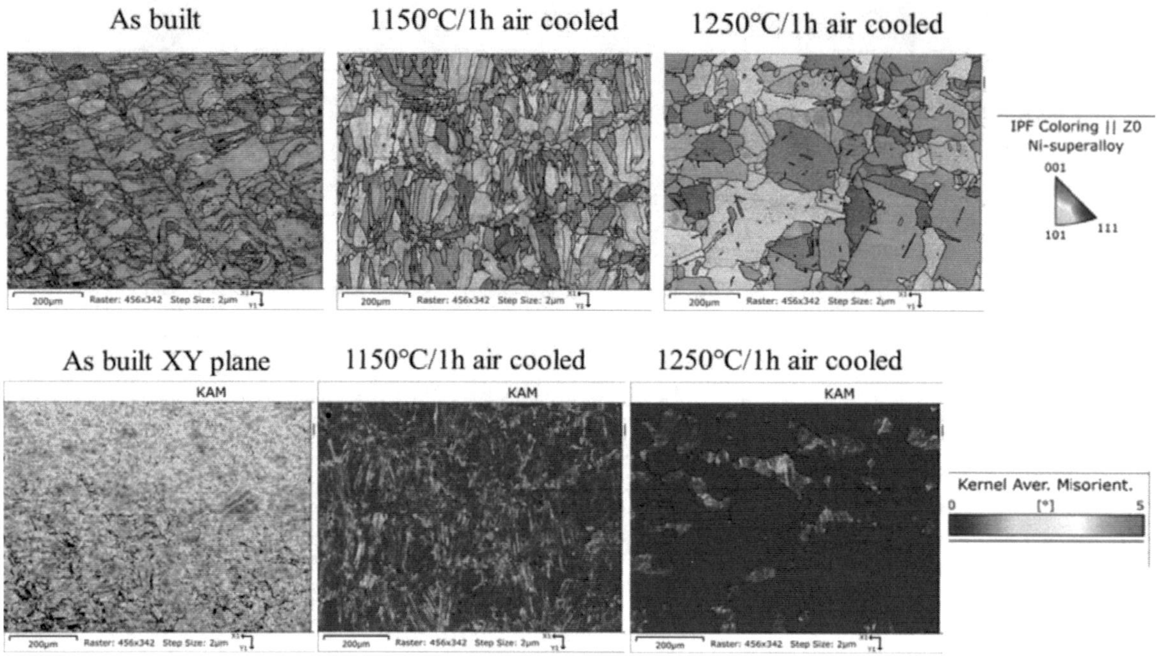

Figure 4: Inverse pole figure maps of as built and solution treated specimens showing the morphology and crystallographic orientation of the grains in the plane normal to the build direction (top). Kernel average misorientation maps of the same plane in as built and solution treated specimens, qualitatively indicating the local misorientation and the amount of stored energy present in the specimen.

The recrystallised microstructure is expected to reduce the anisotropy in mechanical properties. This could also alter the dynamic behaviour i.e. in fatigue and creep mechanisms. For example, this can facilitate similar behaviour of coupons and components in vertical and horizontal direction of additively manufactured sample, which is not yet fully explored in Haynes 282. Additional work is in process to determine the effect of recrystallization on the mechanical behaviour.

4. Conclusion

Different morphology of microstructure possible for AM built Haynes 282 (AM H282). The as-built microstructure consisted of cellular structure due to the rapid cooling rate. Two different solution temperatures resulted in different morphologies of the microstructure. 1150°C solution treatment was not sufficient to completely remove the melt pool boundaries. 1250°C solution treatment resulted in recrystallised matrix due to dissolution of (Ti, Mo) C at that temperature. Recrystallised microstructure could be beneficial to reduce the anisotropy in mechanical behaviour of AM H282.

References

[1] Haynes International, "HAYNES ® 282 ® alloy Material Brochure," 2020

[2] Borg M, Berglund J, Svahn A, Ockborn J, and Pettersson B, 2015, *High durability turbine exhaust case,* last viewed on 07.04.2024, available online: https://patents.google.com/patent/US20150107223A1/en

[3] Tortorelli P F, Unocic K A, Wang H, Santella M L, and Shingledecker J P, "Alloys for Advanced Ultrasupercritical (A-USC) Steam Boilers DOE Fossil Energy Crosscutting Research Program Review," 2016

[4] Gibson I, Rosen D, and Stucker B, 2015, doi: 10.1007/978-1-4939-2113-3

[5] Sames W J, List F A, Pannala S, Dehoff R R, and Babu S S, *International Materials Reviews*, vol. 61, no. 5. Taylor and Francis Ltd., pp. 315–360, 2016. doi: 10.1080/09506608.2015.1116649

[6] Hariharan V S, Pramod S, Kesavan D, Murty B S, and Phanikumar G, *J Mater Sci*, vol. 57, no. 21, pp. 9693–9713, Jun. 2022, doi: 10.1007/s10853-022-07170-3

[7] Tano I N, Rasouli E, Ziev T, Wu Z, Lamprinakos N, Seo J, Balhorn L S, Vaishnav P, Rollett A and Narayanan V *Solar Energy*, vol. 234, pp. 152, 2022, doi: 10.1016/j.solener.2022.01.056

[8] Shaikh A S, Schulz F, Minet-Lallemand K, and Hryha E, *Mater Today Commun*, vol. 26, Mar. 2021, doi: 10.1016/j.mtcomm.2021.102038

[9] Otto R, Brøtan V, Carvalho P A, Reiersen M, Graff S, Sunding M F, Berg O A, Diplas S and Azar A S *Mater Des*, vol. 204, Jun. 2021, doi: 10.1016/j.matdes.2021.109656

[10] Deshpande A, Nath S D, Atre S, and Hsu K, *Metals (Basel)*, vol. 10, no. 5, May 2020, doi: 10.3390/met10050629

[11] Christofidou K A, Pang H T, Li W, Pardhi Y, Jones C N, Jones N G and Stone H J, in *Minerals, Metals and Materials Series*, pp. 1014. doi: 10.1007/978-3-030-51834-9_99

[12] Islam Z, Agrawal A K, Rankouhi B, Magnin C, Anderson M H, Pfefferkorn F E and Thoma D J, *Metall Mater Trans A Phys Metall Mater Sci*, vol. 53, no. 1, pp. 250–263, Jan. 2022, doi: 10.1007/s11661-021-06517-w

[13] Boswell J, Jones J, Barnard N, Clark D, Whittaker M, and Lancaster R, *Mater Des*, vol. 205, Jul. 2021, doi: 10.1016/j.matdes.2021.109725

[14] Ghiaasiaan R, Ahmad N, Gradl P R, Shao S, and Shamsaei N, *Materials Science and Engineering A*, vol. 831, Jan. 2022, doi: 10.1016/j.msea.2021.142234

[15] Swaminathan K, Olsson K, Raza T, Harlin P, and Andersson J, *Proceedings of the 10th International Symposium on Superalloy 718 and Derivatives*, 2023, pp. 553–570. doi: 10.1007/978-3-031-27447-3_35

[16] Ahmad N, Ghiaasiaan R, Gradl, Shao S, and Shamsaei N, *Int J Fatigue*, vol. 176, Nov. 2023, doi: 10.1016/j.ijfatigue.2023.107880

[17] Kurz W, Rappaz M, and Trivedi R, *International Materials Reviews*, vol. 66, no. 1. Taylor and Francis Ltd., pp. 30–76, 2021. doi: 10.1080/09506608.2020.1757894

[18] Pizano L F L, Sridar S, De Vecchis R R, Wang X, Biddlecom J, Pataky G J, Sudbrack C and Xiong W *J Manuf Process*, vol. 119, pp. 781–789, Jun. 2024, doi: 10.1016/j.jmapro.2024.04.015

Effect of multi-pass shot peening on the microstructure of LPBF AlSi10Mg alloy

P Snopiński[1*], T Yu[2], X Zhang[2] and D Juul Jensen[2]

[1] Department of Engineering Materials and Biomaterials, Silesian University of Technology, Gliwice, Poland
[2] Department of Civil and Mechanical Engineering, Technical University of Denmark, Kgs. Lyngby, Denmark

*E-mail: przemyslaw.snopinski@polsl.pl

Abstract. In this study, we investigated the effect of multi-pass shot peening (SP) on the microstructural evolution of a Laser Powder Bed Fusion AlSi10Mg alloy. Characterization techniques included optical microscopy, scanning electron microscopy, and electron backscatter diffraction (EBSD). The results revealed that multi-pass SP induces a gradient microstructure with notable alterations in the grain structure extending to a depth of approximately 45 μm. EBSD analysis revealed that the SP treatment induced grain refinement and formation of subgrains (average distances between boundaries vary from 3.6 μm in the un-deformed area to 0.4 μm near the heavily deformed surface area). It is discussed how this refined microstructure following SP may lead to substantial improvements in mechanical properties and fatigue strength, and thus hold promises for increased performance of critical components across various industrial applications.

1. Introduction

Laser powder bed fusion (LPBF) has emerged as a promising additive manufacturing (AM) technique for fabricating complex metallic components with high precision and tailored material properties [1]. Among the alloys utilized in LPBF, AlSi10Mg alloy is particularly interesting due to its lightweight nature, excellent strength-to-weight ratio, and corrosion resistance, making it ideal for aerospace and automotive applications [2]. However, together with the advantages of the LPBF technology come inherent challenges that impact the reliability and mechanical performance of LPBF AlSi10Mg components. Residual stresses, microstructural anomalies, and surface irregularities often affect AM parts, potentially compromising their integrity during service [3].

To overcome these challenges, mechanical post-treatments can come into play. One promising method, shot peening (SP), can be used to induce beneficial compressive residual stresses along with various other surface modifications [4] Shot peening, a surface treatment process in which metal surfaces are bombarded with high-velocity spherical particles, has shown considerable potential for improving the properties of various AM metal components [5-7]. This relates to microstructure refinement and the introduction of compressive residual stresses which are helpful to restrain the crack initiation and its propagation [8].

The effects of post-built treatments, such as SP, on the mechanical properties of the AlSi10Mg alloy were investigated by several researchers. For example, Bagherifard et al. [9] observed an increase in the surface hardness up to 160 HV together with a 50% improvement in fatigue strength. Similar effects were observed by Uzan et al. [10] and Maleki et al. [11] These improvements must be related to the change in microstructure and residual stresses. However, to the best of our knowledge, the microstructural evolution of AlSi10Mg exposed to SP has not been extensively reported in the existing literature. This provided the impetus for the present investigation, in which an LPBF AlSI10Mg alloy was subjected to multi-pass SP. Through a combination of microscopy techniques including optical, scanning electron microscopy (SEM), and electron backscatter diffraction (EBSD), this study aims to provide insight into the microstructural modifications induced by multi-pass SP treatment.

2. Research methodology

The AM powder with an average particle size of ~40 μm was supplied by Sigma Aldrich. The main chemical composition, determined by energy dispersion spectroscopy (EDS) in wt%, was 87.8 Al, 10.5 Si, 0.5 Mg, 0.15 Ti, 0.15 Cu, 0.09 Fe, and the powder particles were spherical. The samples were printed in the Z-direction (vertical) by LPBF using a TruPrint 1000 system from Trumpf (Ditzingen, Germany). The 3D-printing procedure has been described in detail elsewhere [12].

SP was carried out in a commercial SP machine manufactured by Shanghai Carthing Machinery CO., LTD [13] by an air blasting system. The selected abrasive media was ZrO_2-based ceramic beads with a size of 0.6 mm and a hardness of ~700 HV. The SP process was carried out with a nozzle distance of 140 mm, an angle of 90 degrees and a pressure of 60 PSI (~4.13 bar). The nozzle velocity was set to 150 mm/min. The SP treatment was performed on both parallel surfaces of the 3D printed AlSi10Mg plates having a size of 10×5×50 mm. The SP procedure was repeated 5 times. A comparison of the samples before and after SP is shown in figure 1.

Samples for microstructural investigation were hot mounted in a conductive resin and subsequently prepared according to the standard metallographic procedure involving grinding with the 1200 grit SiC water paper, followed by polishing with 3 μm diamond suspension, and fine polishing with colloidal silica suspension. Microstructure detail was revealed by etching with Keller's reagent (distilled water 190 ml, nitric acid HNO_3 5 ml, hydrochloric acid HCL 3 ml, hydrofluoric acid HF 2 ml) for 30 s.

The microstructure was inspected using an Axio Observer Z1 light microscope and a high-resolution Zeiss Sigma 500 VP field-emission scanning electron microscope (FE-SEM, Carl Zeiss Microscopy GmbH, Jena, Germany) equipped with an EBSD camera. A large area of 700 × 500 μm was scanned with a 2 μm step size to analyze the microstructure after SP. A smaller area of 60 × 70 μm was additionally mapped with a step size of 50 nm to examine the impact of SP in detail. Oxford instruments AZtecHKL and ATEX software [14] were used to post-process and analyze the EBSD data. A two-step cleanup procedure, consisting of grain confidence index standardization followed by neighbor orientation correlation under the following conditions—a

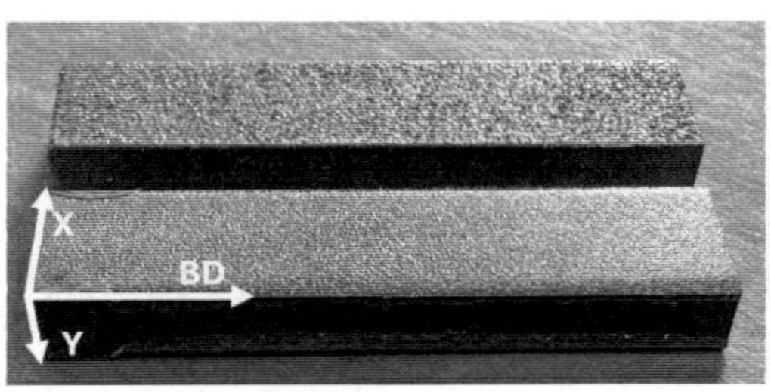

Figure 1. The AlSi10Mg alloy samples prior to (top) and after the SP process (bottom).

IOP Conf. Series: Materials Science and Engineering **1310** (2024) 012039 doi:10.1088/1757-899X/1310/1/012039

grain misorientation of 5°, a minimum grain size of 9 pixels and a minimum confidence index 0.1, was applied. The grain size analysis was conducted by using the linear intercept method.

3. Results

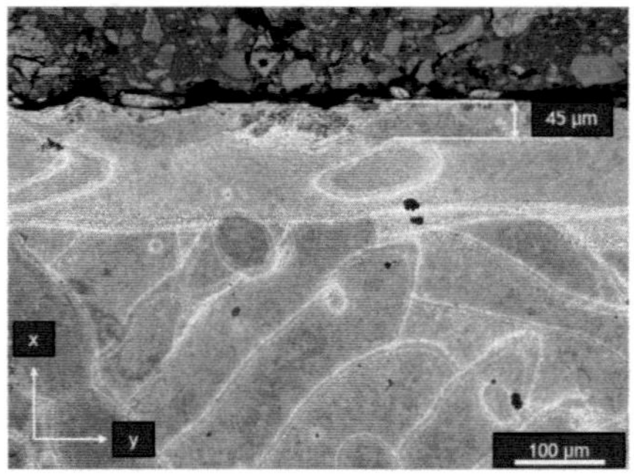

Figure 2. Bright field optical micrograph showing the cross-section of the SP AlSi10Mg specimen - the resin into which the sample was mounted is seen at the top.

Figure 2 presents a light optical micrograph of the sample cross-section (X-Y plane), highlighting key microstructural features. As can be seen, the chemical etching has exposed the melt pool tracks and the inhomogeneous microstructure, which is oriented according to the 67° rotation strategy applied between subsequent layers.

Additionally, it is evident that only the contour layer was significantly affected by SP, as the most substantial modifications are visible in the surface layer (close to the resin), within a depth of approximately 45 µm.

Figure 3 shows secondary electron micrographs of the AlSi10Mg samples before and after SP. The SEM image of the as-built sample reveals a microstructure consisting of a fine eutectic Al/Si network, which is characteristic of the LPBF AlSi10Mg alloy [15], see figure 3(a).

After SP, a clear change in the microstructure is observed. In particular, the eutectic Si network that originally surrounded the Al matrix grains in the as-built sample is affected. It can be seen that the multi-pass SP leads to a fragmentation of the eutectic Si network. After SP spherical Si particles are homogeneously distributed in the SP affected area, as shown in figure 3(b). The transition from an eutectic Si network to spherical Si particles after SP agrees well with the mechanisms reported in previous studies of cast and shot-peened Al-Si alloys, e.g. [16].

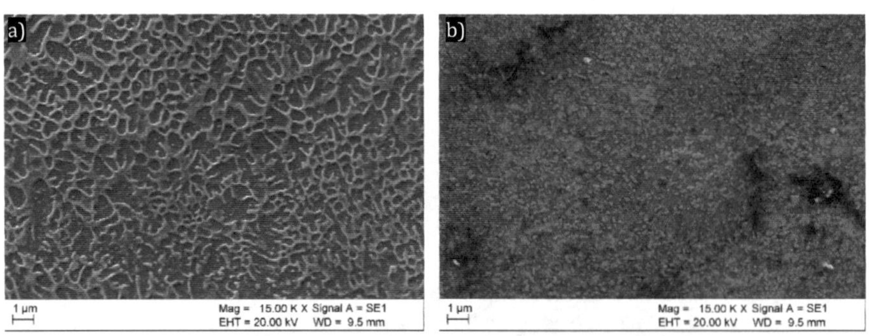

Figure 3. SEM images of the AlSi10Mg alloy a) as-built condition, b) heavily deformed surface area of the sample subjected to the multi-pass shot peening treatment (both images illustrate the xy plane).

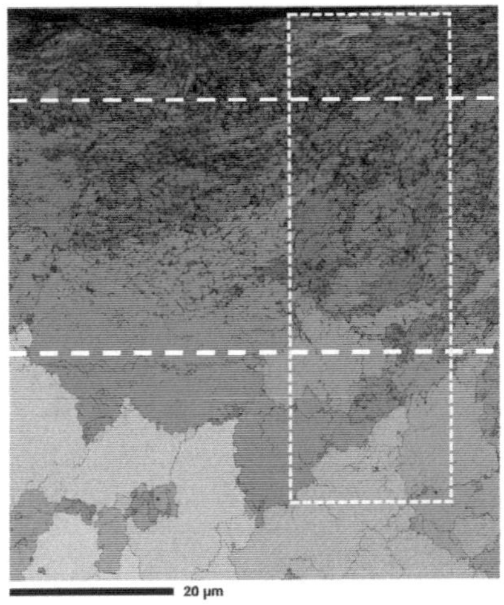

Figure 4. Band contrast map of the AlSi10Mg alloy subjected to SP treatment. The EBSD map was collected with a step size of 50 nm.

Figure 4 shows a band contrast map, which was acquired with a step size of 50 nm. As can be seen, a plastically deformed layer with a microstructure gradient was produced by the SP. According to the microstructure morphology, the observed section can be divided into three regions: an ~ 10 μm heavily-deformed surface region, a deformed sub-surface region and the non-deformed matrix (see white dashed lines). The deformed sub-surface region is approximately 30 μm thick and has a gradient microstructure.

Figure 5 clearly visualizes the degree of grain refinement. As can be observed, an ultrafine grain structure with an average grain size of ~0.4 μm has formed in the heavily-deformed surface region within the top 10 μm depth (figure 5(a)).

As the depth increases to 10–40 μm, the size of the grains is slightly larger (for example it is ~0.48 μm on average in the 10–20 μm layer). Both fine and coarser grains are mixed here (figure 5(b)). The grain refinement is not observed in deeper regions where the transferred energy [17] was too low (figures 5(e–f)). In this region, the EBSD analysis yielded an average grain size of ~3.6

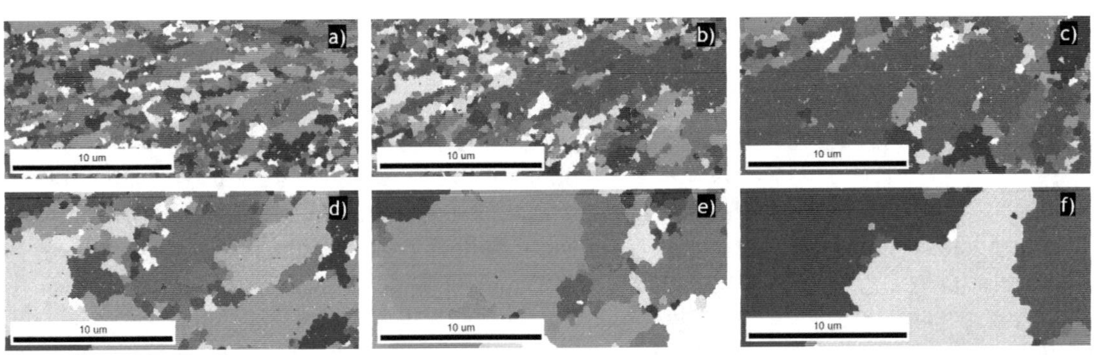

Figure 5. Unique grain colour maps illustrating grain structures in multi-pass shot peened AlSi10Mg sample. These maps (a) - (f) are segmented from the larger EBSD map presented in figure 4 (yellow dashed square box), providing a detailed visualization of the grain colour variations and their distribution at a finer scale. Each subfigure represents a consecutive 10 μm thick slice, offering insights into the microstructural evolution and heterogeneity within the material. It should be noted that in the unique grain colour map each grain is shaded with a random colour.

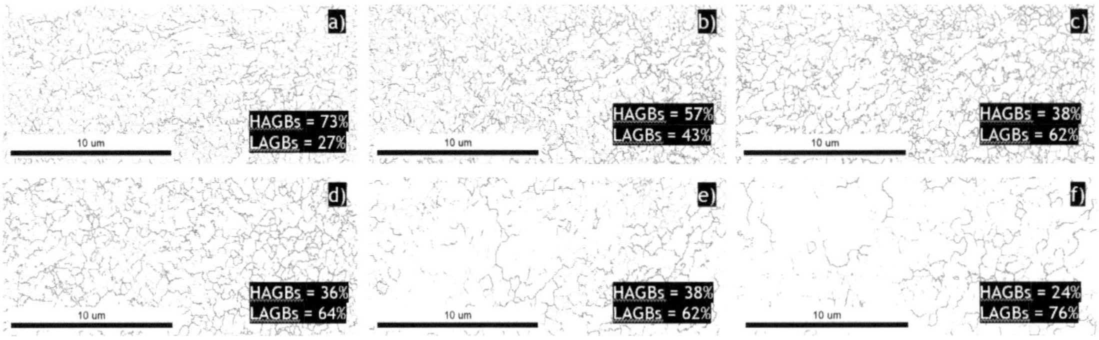

Figure 6. The EBSD grain boundary maps: a) plastic deformation area, b) transition area, c) un-deformed area. High-angle boundaries (>15°) in green and low-angle boundaries (2°–15°) in red.

μm with a large variation in grain size due to the complex solidification process during 3D printing.

Figure 6 illustrates the boundary misorientations obtained by the EBSD analysis. The fraction of high-angle grain boundaries (HABs) in the heavily deformed surface area with a depth of 10 μm is 73% (figure 6(a)). This high percentage of HABs is normally observed in the surface region of SP-treated metals, where the local high strain introduces a high density of dislocations, subdividing the original grains by gradually forming HABs during the SP process [18]. Figure 6 shows the interesting observation that the LABs are distributed fairly uniformly in the deformed layers while heterogeneously in the deep non-deformed layer. This may be directly related to the Si network at cell boundaries, which was believed to effectively stop dislocation movement during deformation and contribute to strain hardening and grain refinement [19]. However, the mechanisms behind this need further detailed investigations in the near future. At deeper depth away from the SP-treated surface, both the HAGB and LAGB spacings increase significantly, figure 6(e). This microstructural evolution is attributed to the presence of significantly lower cumulative plastic strains below the heavily-deformed surface area [20] In the un-deformed area, figure 6(f), the fraction of LABs is 76%. Such a high fraction of LABs can be attributed to the presence of pre-existing dislocation boundaries formed due to rapid cooling and thermal cycling [21], which is typical for the LPBF parts.

As shown above, the multi-pass SP treatment changed the material microstructure (by creating ultra-fine grains in the heavily-deformed surface area). It is thus expected that both the fatigue performance and the mechanical properties will be affected accordingly. It has been reported in the literature that SP significantly affects the fatigue performance by reducing the surface grain size [22]. This is because a smaller grain size increases the strength, thereby delaying crack initiation, and it also reduces the length a crack can grow to before it encounters a grain boundary, thereby SP contributes to resisting crack propagation [23]. Consequently, the endurance stress for a finer microstructure will be greater than that for a coarse-grained counterpart. In addition, the SP treatment can introduce considerable compressive stresses into the surface layer [24]. The positive effects of these compressive stresses are well documented and contribute to increased fatigue resistance and improved stress corrosion resistance [25, 26].

4. Conclusions and outlook

In this work the influence of multi-pass SP on the microstructure characteristics of LPBF AlSi10Mg was investigated. The main conclusions are as follows:

- The multi-pass SP affected the eutectic Si network of the as-built structure, resulting in the formation of nanoscale Si particles at the sample surface.
- The multi-pass SP formed a microstructural gradient in an approximately 45 μm think surface layer.
- EBSD micrographs confirmed the significant refinement of the grain structure. It is found that an ultra-fine grained layer with an average size of ~0.4 μm formed as a result of the SP treatment.

With this in-depth understanding of the SP microstructure, it would be of interest to analyse the direct correlations between SP microstructure and mechanical properties. These include tensile strength, nano-hardness, fatigue resistance and wear behaviour after SP treatment. Such a direct analysis would also have to include local residual stress quantifications. As a next step, such relationships between microstructure, stress and properties may guide *knowledge-based* further optimization of 3D printed and post-built processed Al10SiMg.

Acknowledgements

This work received financial support by the research grant (MicroAM-VIL54495) from VILLUM FONDEN. P Snopinski was funded by the National Science Centre, Poland based on the decision number 2021/43/D/ST8/01946.

References

[1] Narasimharaju SR, Zeng W, See TL, Zhu Z, Scott P, Jiang X and Lou S. *J Manuf Process* 2022; **75** 375–414
[2] Kan WH, Nadot Y, Foley M, Ridosz L, Proust G and Cairney JM. *Addit Manuf* 2019
[3] Polat E and Saklakoglu IE. Fatigue Fract Eng Mater Struct 2024 **47** 240–253
[4] Zine PU and Joshi PK. *Mater Today Proc* 2023 **72** 870–877
[5] Qin Z, Li B, Chen R, Zhang H, Xue H, Yao C and Tan L. *Int J Fatigue* 2023 **168** 107429
[6] Zhang H, Cao S, Li C, He N and Wang F. *Mater Sci Eng A* 2023 **883** 145477
[7] Önder S, Saklakoğlu N and Sever A. *Mater Charact* 2023 **196** 112571
[8] Zhang Q, Xu S, Wang J, Zhang X, Wang J and Si C. *Surf Topogr* 2022 **10** 015010
[9] Bagherifard S, Beretta N, Monti S, Riccio M, Bandini M and Guagliano M. *Mater Des* 2018 **145** 28–41
[10] Uzan NE, Ramati S, Shneck R, Frage N and Yeheskel O. *Addit Manuf* 2018 **21** 458–464
[11] Maleki E, Bagherifard S, Heydari Astaraee A, Sgarbazzini S, Bandini M and Guagliano M. *Mater Sci Eng A* 2023 **881** 145397
[12] Snopiński P, Matus K, Łagoda M, Appiah ANS and Hajnyš J. *J Alloys Compd* 2024 **970** 172576
[13] Angelou N, Zhang X, Sj̈oholm M, Lorentzen L and Huang X. DTUWind Energy. DTU Wind Energy E 2017 151
[14] Beausir, B and Fundenberger J, Analysis Tools for Electron and X-ray diffraction, ATEX - software, www.atexsoftware.eu, Université de Lorraine - Metz, 2017
[15] Prashanth KG and Eckert J. *J Alloys Compd* 2017 **707** 27–34
[16] Cho KT, Yoo S, Lim KM, Kim HS and Lee WB. *J Alloys Compd* 2011 **509** S265–S270
[17] Jamalian M and Field DP. *Mater Charact* 2019 **148** 9–16
[18] Wang Z, Gao C, Liu Z, Wang Z, Liu X, Wong K, Zhou Z and Xiao Z. *Mater Sci Eng A* 2020 **772** 138696
[19] Wang P, Rabori AS, Dong Q, Ravkov L, Balogh L and Fallah V. *Mater Charact* 2023 **198** 112730
[20] Alyousif OM, Engelberg DL and Marrow TJ. *J Mater Sci* 2008 **43** 1270–1277
[21] Dong Z et al. *Int. J. Extreme Manuf* 2024 **6** 045003
[22] Karimbaev R, Pyun YS, Maleki E, Unal O and Amanov A. *Mater Sci Eng A* 2020 **791** 139752
[23] Turnbull A and De Los Rios E.R. *Fatigue Fract Eng M* 1995 **18** 1455-1467
[24] Yang R, Zhang X, Mallipeddi D, Angelou N, Toftegaard HL, Li Y, Ahlström J, Lorentzen L, Wu G and Huang X. *IOP Conf. Ser.: Mater. Sci. Eng.* 2017 **219** 012046, 38th Risø International Symposium on Materials Science 4–8 September 2017, Risø, Denmark
[25] Wu J, Liu H, Wei P, Lin Q, Zhou S, *Int J Mech Sci* 2020 **183** 105785
[26] Sherafatnia K, Farrahi G, Mahmoudi A, Ghasemi A, Mater Sci Eng A 2016 **657** 309-321

IOP Conf. Series: Materials Science and Engineering **1310** (2024) 012040 doi:10.1088/1757-899X/1310/1/012040

Elucidating the impact of laser beam shape on the as-printed microstructure in 316L stainless steel

W E Alphonso[1*], M A Ribeiro[1], R Rothfelder[2], M Schmidt[2], J H Hattel[1], D Juul Jensen[1] and M Bayat[1]

[1] Department of Civil and Mechanical Engineering, Technical University of Denmark, Lyngby, Denmark
[2] Department of Mechanical Engineering, Institute of Photonic Technologies (FAU), Erlangen, Germany

*E-mail: wayeal@dtu.dk

Abstract. In Laser Powder Bed Fusion (L-PBF), the intrinsic inclination of the laser beam creates variability in the interaction between the laser and the melt pool, which along with thermal conditions within the pool, results in the formation of defects in the parts. To elucidate the impact of laser beam inclination on the melt pool, a deposition scale model using the finite volume method (FVM) is developed. The laser model demonstrates that as the laser beam inclines towards the periphery of the build plate, the projected laser spot size enlarges, resulting in lowered energy density. To assess how the laser beam inclination affects multi-layer printing, cubic specimens are printed at both the center and the far corner of a rectangular build plate of 280 mm x 280 mm size. X-ray computer tomography (X-CT) is used to investigate internal porosities and electron backscatter diffraction (EBSD) is used to characterize the microstructure of the printed cubes. The experimental results are discussed based on the FVM simulations.

1. Introduction

LPBF is a widely used metal additive manufacturing (MAM) technique that employs a laser to melt selectively and fuse layers of metal powder, creating intricate three-dimensional structures [1]. Despite its numerous advantages, the L-PBF process is influenced by several factors often understated during the optimization of process parameters to achieve maximal part density. One critical parameter is the intrinsic inclination angle of the laser beam, which can induce anisotropic elongation of the circular beam spot shape on the build plate surface [2]. The variation of the laser inclination angle across different regions of the powder bed significantly impacts the melt pool conditions, thereby influencing the microstructure and the final part quality. Moreover, the inclination angle or incidence angle indirectly contributes to the offset in lens focal distance, resulting in defocusing at the periphery of the build plate [3].

An experimental study was conducted in [4] where the effect of laser incidence angle on porosity and laser shape distortion was investigated. The study revealed that as the laser's incidence angle increases, the beam shifts to an elliptical shape, resulting in a decrease in energy density. This decrease subsequently impacts the melt pool geometry, leading to wider and shallower melt pools. Additionally, the effect of inclination angle on porosity and surface roughness was investigated which concluded a direct correlation wherein increased inclination angle leads to elevated levels of porosity and surface roughness. A numerical investigation explored the influence of incident angle on conduction melt pool dynamics and solidified microstructures [5]. It was observed that increasing the laser incidence angle from 60° to 120°

resulted in shallower melt pool depths, while maintaining consistent width due to constant power density, along with higher thermal gradient, reduced solidification rate, and the emergence of a more dominant columnar grain structure. Further numerical exploration was conducted with the inclination angle and scanning direction both aligned and oriented transversely [6]. This study revealed that inclining the laser beam resulted in shallower melt pools when the spot was elongated transversely, and deeper melt pools when elongated along the scanning direction. Such alterations induced variations in both the thermal gradient and the rate of solidification growth. These works [4–6] emphasize the significance of laser inclination angle, yet the resulting as-printed microstructures and detailed melt pool dynamics remain unexplored.

This study aims to investigate the impact of laser beam shape variation, resulting from the intrinsic inclination angle, on melt pool dynamics and the as-printed microstructure through numerical modeling and experimental analysis, respectively. Single-track investigations are conducted during the numerical modeling phase, as they are fundamental to the formation of a three-dimensional (3D) part. For the experimental investigations, a cube of 10mm x 10mm x 10mm is printed and subsequently investigated using X-CT and EBSD.

2. Numerical Model and Experiments

2.1 High fidelity deposition scale numerical model

A deposition-scale numerical model based on FVM has been developed to replicate the impact of laser inclination angle on the L-PBF process. This multi-physics computational fluid dynamics (CFD) model incorporates melting, solidification, evaporation, fluid flow, heat transfer, ray tracing with multiple reflections, and Marangoni convection [7]. The model features two phases: the metal, which can exist in solid or liquid states, and a void phase above the powder bed. A volume of fluid (VOF) method is implemented to track the interface between these two phases. This method is represented by the scalar α (-), which ranges between 0 and 1, allowing for precise interface tracking, and can be expressed by equation (1).

$$\frac{\partial}{\partial t}\alpha+\vec{\nabla}.\alpha\vec{V}=0 \qquad (1)$$

$\vec{V}$ in equation (1) represents the velocity of the fluid at the interface. More detailed information on the numerical model can be found in the previous work of the author [7]. In this study, a single track is consolidated on a powder bed with 50 μm layer thickness, using a laser beam directed perpendicularly and at an inclination of 17°.

2.2 Experiments

An array of 75 cubes was printed in 316L stainless steel using an SLM 280 commercial L-PBF system, with raw powder from Höganäs featuring particle sizes between 15 μm and 45 μm. A 150W laser with a 0.6 m/s scanning speed was used to build the cubes, with a 50 μm powder layer thickness and a 100 μm laser spot size. The cubes were evenly spaced across the build plate. This study focuses on two specimens; one on the center cube, with a theoretical inclination angle of 0°, and one in the corner, with a theoretical inclination angle of 17°, located at the outermost edge of the rectangular build plate. One-quarter of the cube sample underwent X-ray computed tomography for density and pore morphology analysis. Another quarter of the cube was hot embedded for microstructural analysis using LOM, and the remainder was used for EBSD characterization.

IOP Conf. Series: Materials Science and Engineering **1310** (2024) 012040 doi:10.1088/1757-899X/1310/1/012040

A Nikon XT H 225 metrology device was used for the X-ray CT inspection from which X-ray tomograms of the bulk sample were obtained. The instrument was operated with a tungsten reflection target using 200 kV, 13 W, and a 2 mm thick copper filter. For image acquisition, 1571 projections, each with 1s exposure time and a 2x2 binning were used. The reconstructed voxel size was 13.55 μm. The reconstruction of the tomographs was performed using the software "CT Pro 3D" (Nikon Metrology Inc.) and included a correction for beam hardening effects. The reconstructed tomographs were loaded into Avizo for subsequent processing, including alignment, cropping, denoising, thresholding, and segmentation. Each pore was labeled, and its volume was calculated to determine the porosity of the cube. EBSD analysis was performed utilizing a ZEISS Sigma field emission scanning electron microscope (FE-SEM). The SEM was operated at an acceleration voltage of 20kV and a step size of 0.5 μm was utilized.

3. Results and Discussion

3.1 Numerical Analysis

The boundary conditions, numerical domain size, and other details are illustrated in figure 1(a). To validate the numerical model, the experimentally obtained melt pool shape and size were compared to the numerical predictions, as shown in figure 1(b). The model accurately predicts the melt pool shape and size, with the predicted melt pool width being 240 μm for the non-inclined beam and 255 μm for the inclined beam. Once the melt pool reaches a steady state, the model also predicts the melt pool depth, which is 42 μm for the non-inclined beam and 35 μm for the inclined beam. The simulations omitted the powder bed to minimize uncertainty in the validation case, and experiments were conducted directly on the base plate without a powder layer.

Figures 1(d,e) show 3D temperature contours for the non-inclined and inclined beam cases, respectively. The melt pool shapes and dimensions are extracted from the numerical models once they have reached the steady state. The measurements reveal that inclining the laser beam reduces the melt pool depth and increases its width compared to the non-inclined case. This supports literature findings that the melt pool depth decreases due to laser beam spot elongation, resulting in a lower energy density. For a more detailed thermal analysis, a line probe was positioned just below the melt pool surface in both cases. The temperature distribution along the scanning direction, shown in figure 1(c), indicates that inclining the laser beam reduces both the domain temperature and the peak temperature just below the melt pool compared to the non-inclined case. With an inclined beam, the maximum temperature is reached slightly later along the scanning track because the front wall of the melt pool is more inclined, pushing the peak temperature point further back along the track.

3.2 Experimental characterization

LOM images captured at various locations on the cube, including three sections; the center, top, and side regions for both the center and corner cube are depicted in figure 2. Notably, the center cube showed small, randomly distributed gas pores in addition to the presence of lack of fusion pores between the contour scan and the core. The process parameters, including laser power and scanning speed, were optimized to achieve sufficient energy density, aiming to minimize porosity when the laser inclination angle is 0° (center cube).

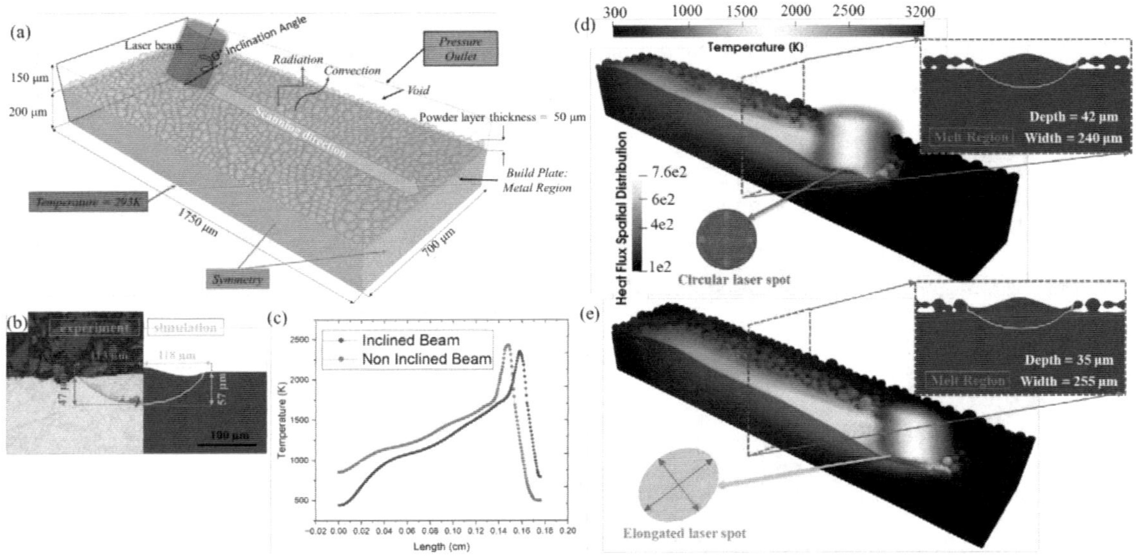

Figure 1 (a) Numerical domain and boundary conditions for the inclined and non-inclined single track simulations, (b) validation of the numerical model (laser power 250W and scanning speed 0.5 m/s), (c) temperature 75 μm beneath the powder bed in the scanning direction at t = 2.8 ms , and 3D temperature plots showing the melt pool shape and size at steady state for the (d) non-inclined and (e) inclined laser beam (laser power 150W and scanning speed 0.6 m/s).

The majority of sizable pores within the center cube were found between the core and contour scan, attributable to suboptimal scanning strategies and can be avoided by a 2-track contour scan. Conversely, lack of fusion pores was prevalent across all regions of the corner cube, including the top, center, and edge. This occurrence aligns with the numerical modeling for inclined beams, wherein reduced melt pool depth was simulated. Shallower and wider melt pools and can lead to the formation of lack of fusion pores within the bulk material also known as intralayer lack of sintering.

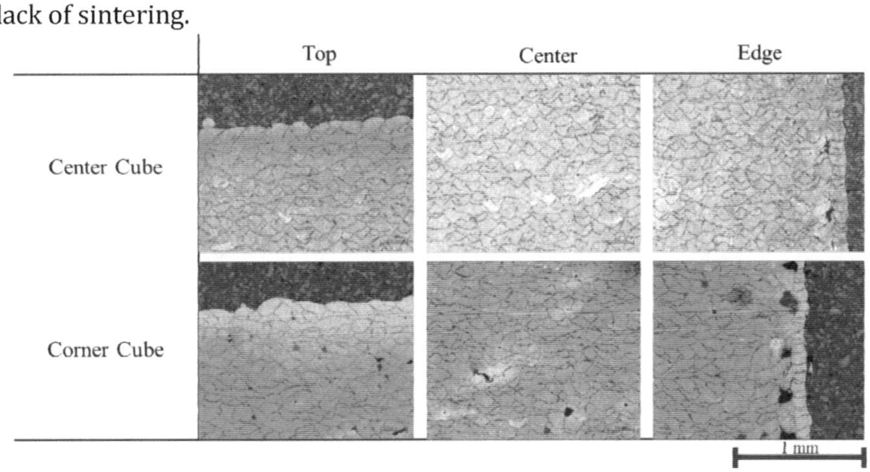

Figure 2 LOM images of the center and corner cubes with a focus on the top, center, and edge regions.

The X-ray CT results are shown in figure 3. Figure 3(a) reveals randomly distributed, spherical gas pores in the center cube, while figure 3(b) shows a significant increase in pores with sharp edges in the bulk, typical of lack of fusion porosities for the corner cube. The calculated porosity for the center and corner cubes is 0.19% and 5%, respectively, and their size distribution

IOP Conf. Series: Materials Science and Engineering **1310** (2024) 012040 doi:10.1088/1757-899X/1310/1/012040

is shown in figure 3(c) highlighting the significant impact of beam inclination angle on part density. Note that the volumes used for density measurement exclude the contour region, as the pores in this area can be avoided with the implementation of an optimized contour scanning strategy.

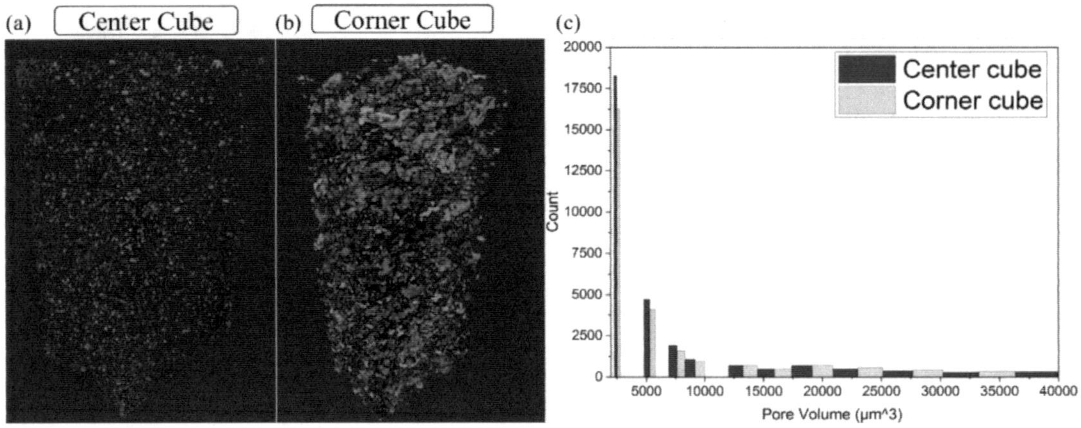

Figure 3 X-ray CT results of a cropped volume of (a) center cube and (b) corner cube with individual pores segmented (c) Pore volume (μm^3) measured in the central and corner cube.

EBSD analysis was performed within the middle bulk region of both the center and corner cubes. Figure 4(a,d) presents the Inverse Pole Figures (IPF) maps with the building direction oriented along the x-axis. Pole figures, shown in figure 4(b,e) indicate an insignificant difference in texture index which is around 0.15.

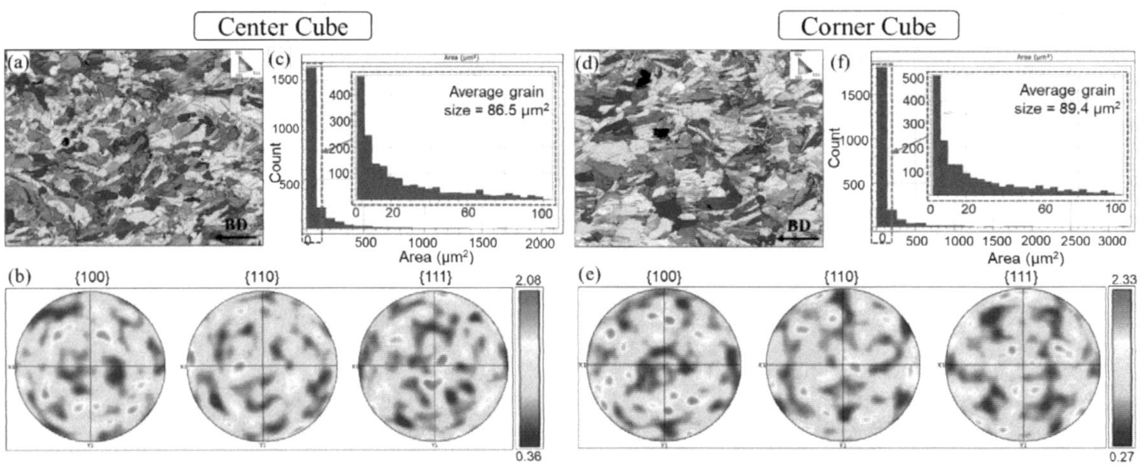

Figure 4 EBSD analysis of center and corner cube (a,d) IPF maps (b,e) Pole figures (c,f) grain size distribution based on area measurement with the in-set image focussing on grain areas below 100 μm.

It is observed that the corner cube has a larger number of small grains (see figure 4(c,f)) with an area of less than 100 μm^2. Here a grain is defined by a threshold misorientation angle of 10deg. The reduction in energy density due to the inclined laser beam can increase the solidification rate and reduce the thermal gradient as shown by the numerical model, which may explain the existence of the small grains. The grain size along the building direction was measured using the linear intercept method (LID). Three random lines were placed on the grain map, and the intersections with grain boundaries were counted. The results showed that the average number of intersections was higher in the center cube compared to the corner cube and the mean LID was 8 ± 0.2 μm for the center cube and 9 ± 0.6 μm for the corner cube, which suggests that the grains in the corner cube on average are smaller (see figure 4 (f)) but still slightly longer along the building direction compared to the center cube. Therefore, EBSD analysis concludes that the inclination angle affects solidification dynamics, leading to a change in grain shape in the corner cube. This may be associated with a reduction in nucleation events of equiaxed grains [8] caused by lower laser energy density, but further investigations are underway to clarify the dynamics.

Conclusion

This work investigates the impact of inclination angle on melt pool dynamics and thermal conditions in the L-PBF process using a high-fidelity multi-physics model validated with single-track experiments in 316L stainless steel. Two cases were simulated: one with a 0° inclination angle and another with a 17° inclination angle. The results suggest an inclined laser beam elongates the laser spot, reducing energy density and melt pool depth. Additionally, the inclination shifts peak temperatures by 100 μm in the scanning direction and lowers the peak temperature by approximately 273 K. Experimental studies involved analysing two 10mm x 10mm x 10mm cubes: one at the build plate center (0° inclination) and one at the edge (17° inclination). X-ray CT and LOM imaging revealed increased porosity and reduced melt pool depth in the corner cube compared to the center cube.

EBSD analysis revealed that the texture remains largely unaffected by the inclination angle. However, there is a notable change in average grain size and grain shape along the building direction. This change is attributed to altered solidification dynamics, where a decrease in laser energy density in the corner cube may affect the nucleation resulting in on average smaller grains many of which however are longer along the build direction.

Acknowledgments

WEA, MB, and DJJ acknowledge the financial support of the research grant (MicroAM-VIL54495) from VILLUM FONDEN. The authors gratefully acknowledge the printing work done by the Erlangen Graduate School in Advanced Optical Technologies (SAOT).

References

[1] Bayat M, Dong W, Thorborg J, To A C and Hattel J H 2021 A review of multi-scale and multi-physics simulations of metal additive manufacturing processes with focus on modeling strategies *Addit Manuf* **47** 102278.
[2] Ayoola W A, Suder W J and Williams S W 2019 Effect of beam shape and spatial energy distribution on weld bead geometry in conduction welding *Opt Laser Technol* **117** 280–7.
[3] Berez J, Dushaj E, Jost E, Saldana C and Fu K 2024 Measurement of focal plane error in laser powder bed fusion machines *Additive Manufacturing Letters* **9** 100196.
[4] Fathi-Hafshejani P, Soltani-Tehrani A, Shamsaei N and Mahjouri-Samani M 2022 Laser incidence angle influence on energy density variations, surface roughness, and porosity of additively manufactured parts *Addit Manuf* **50** 102572.
[5] Li E, Shen H, Wang L, Wang G and Zhou Z 2023 Laser shape variation influence on melt pool dynamics and solidification microstructure in laser powder bed fusion *Additive Manufacturing Letters* **6** 100141.
[6] Noori Rahim Abadi S M A, Mi Y, Sikstrom F, Ancona A and Choquet I 2021 Effect of shaped laser beam profiles on melt flow dynamics in conduction mode welding *International Journal of Thermal Sciences* **166** 106957.
[7] Alphonso W E, Baier M, Carmignato S, Hattel J H and Bayat M 2023 On the possibility of doing reduced order, thermo-fluid modelling of laser powder bed fusion (L-PBF) – Assessment of the importance of recoil pressure and surface tension *J Manuf Process* **94** 564–77
[8] Roehling T T, Shi R, Khairallah S A, Roehling J D, Guss G M, McKeown J T and Matthews M J 2020 Controlling grain nucleation and morphology by laser beam shaping in metal additive manufacturing *Mater Des* **195** 109071

IOP Conf. Series: Materials Science and Engineering **1310** (2024) 012041 doi:10.1088/1757-899X/1310/1/012041

Integration of spray-formed AISI H13 overspray powder in additive manufacturing to enable a circular ecosystem

M R L Zwicker[1], N S Tiedje[1], T Dahmen[2] and V K Nadimpalli[1]

[1] Department of Civil and Mechanical Engineering, Technical University of Denmark, DK-2800 Kgs. Lyngby, Denmark
[2] Asgaard Metals Aps, Industrivej 25-27, 3300 Frederiksværk, Denmark

E-mail: mrlzw@dtu.dk

Abstract. Overspray powder is a bi-product of the spray forming process and is commonly considered as scrap material. This study, however, assessed the feasibility of the use of spray-formed AISI H13 overspray as powder feedstock for LPBF as part of a circular economy concept. The overspray powder presented a suitable particle size distribution for LPBF applications and showed the capability of being printed crack free. X-ray diffraction was used to show that ferritic/austenitic overspray powder transformed into a martensitic austenitic microstructure in the as-printed conditions. Furthermore, texture analysis revealed a preferred crystallographic orientation in <100> parallel to the build direction for martensite while the retained austenite presents a preferred orientation in <110> along the build direction.

1. Introduction

The growing and simultaneously unsustainable global demand for industrial goods demonstrates how the industrial sector is contributing to the ongoing degradation of the environment. In addition to that, geopolitical disruptions such as the COVID-19 pandemic have shown weaknesses in global supply chains which in turn demonstrate the need for structural changes, for instance, by moving from a linear global economy to a local circular economy. Such changes are essential to build robust logistics while driving efficient resource allocation, waste reduction, and reuse in order to minimize the environmental impact of manufacturing companies [1-4].

To enable a closed loop system for recycling worn-out injection molds made from hot working tool steel AISI H13 (DIN 1.2344), the study at hand utilizes so-called overspray powder from the spray-forming process as powder feedstock for Laser Powder Bed Fusion (LPBF).

Spray forming generally consists of an atomizer whose droplets deposit onto a rotating plate yielding in the formation of an ingot. However, due to the characteristics of the process, the highest achievable sticking efficiency of the droplets onto the deposition surface amounts to approximately 80% with overspray powder as a bi-product [5].

Rodriguez et al. demonstrated that the powder characteristics of martensitic stainless steel 440C overspray powder as suitable for LPBF applications [6]. Hence, this study aims to show the feasibility of printing AISI H13 overspray powder and incorporating it within a circular ecosystem. The present phases of a spray formed ingot, its corresponding overspray powder, and the as printed material should be determined as well as the microstructure of the as printed material. Additionally, the general printability and cracking behaviour should be assessed, since H13 is considered as hot cracking susceptible due to its relatively large solidification range of approximately 85°C [7].

2. Experimental Section

2.1 Material

AISI H13 overspray powder, which originated from the spray forming (SF) process, was used for this investigation. For the printing experiments, the powder was sieved to reach a particle size range between 20μm and 45μm. The SF process, conducted at Asgaard Metals A/S under a nitrogen atmosphere, involved recycling scrap material with a known chemical composition. The resulting chemical composition of the AIS H13 powder and the corresponding spray-formed ingot are displayed in table 1.

Table 1. Nominal chemical composition of the spray formed ingot and measured carbon content of the corresponding powder

Name	C [wt.%]	Si [wt.%]	Mn [wt.%]	Cr [wt.%]	Mo [wt.%]	V [wt.%]	Fe
AISI H13 (nominal)	0.35 – 0.42	0.80 – 1.20	0.25 – 0.50	4.80 – 5.50	1.20 – 1.50	0.85 – 1.15	Bal.
Spray formed ingot	0.36	1.10	0.42	5.25	1.48	0.79	Bal.
Over-spray powder	0.42			Not measured			

During the SF process, the overspray is gathered both from underneath the spray chamber in the so-called 'Kammer' and within the spray chamber in the so-called 'Cyclon'.
As intended for commercial applications, powder from both locations was blended with a mass ratio of 50/50 and used for all LPBF builds. The particle size distribution before and after sieving for the two different locations and the powder blend can be found in table 2.
Around 12.4% volume fraction of the 'Cyclon' powder, and 4.8% volume fraction of the 'Kammer' powder are within a particle size range of 20μm – 45μm, making them suitable for LPBF applications.

Table 2. D_{10}, D_{50} and D_{90} values before and after sieving for 'Cyclon' and 'Kammer' powder, as well as the used 50/50 blend for a sieving fraction of 20μm – 45μm.

	Before Sieving		After sieving to 20μm – 45μm		
	Kammer	Cyclon	Kammer	Cyclon	Kammer-Cyclon blend
D_{10} [μm]	53.3	44.0	20.2	22.5	21.7
D_{50} [μm]	89.0	144.0	32.7	34.1	32.8
D_{90} [μm]	351.7	446.0	50.4	49.6	49.9

2.2 LPBF process

The printing experiment was conducted using the open architecture LPBF system developed by DTU [8], equipped with a 300W laser, 1080nm fiber laser, and a digital scanner with a 6μm resolution and featuring a total build envelope of ⌀250mm x 150mm in height [9]. Printing was performed under a nitrogen atmosphere (Oxygen levels <1000 ppm) and a preheated build plate at 80°C. Furthermore, a parallel scanning strategy was chosen with a rotation of 90° in each layer. No post-print heat treatment was conducted. Moreover, three different parameter sets were chosen to reach a volumetric energy density (VED) of 37.50, 55.55, and 71.40 $J \cdot mm^{-3}$. For each parameter set one cuboidal sample of 15mm x 15mm x 8mm in dimension was printed. While the hatch spacing and layer thickness were kept at 100 μm and 50 μm, the laser power and scan speed combinations were (150 W, 700 mm/s), (250 W, 900 mm/s) and (250 W, 700 mm/s) respectively.

2.3 Material Characterization

The alloy content of Fe, Si, Mn, Cr, Mo, and V in both the ingot and the 3D printed samples was measured by Glow Discharge Optical Emission Spectroscopy (GD-OES) according to ISO

IOP Conf. Series: Materials Science and Engineering **1310** (2024) 012041 doi:10.1088/1757-899X/1310/1/012041

11505:2012 [10] using a SPECTROMAXx Optical Emission Spectrometer by Ametek with an FE-10 calibration for low alloyed steels. The carbon content was determined by combustion and infrared absorption using a LECO CS230 and following ASTM 1019-18 [11]. Phase analysis of the ingot, 'Cyclon' and 'Kammer' powders, as well as printed samples fabricated with different energy densities, were conducted by X-ray diffraction using a D8 Discover diffractometer by Bruker running with copper radiation and an operating current and voltage of 40mA and 40kV respectively. Hereby, a parallel beam set-up was used with a 2θ range of 35-85 [deg. 2θ] and a step time of 5s. To analyze the data, EVA software by Bruker was used. Additionally, EBSD characterization was conducted with a Zeiss Sigma 300 with a detector system from Oxford Instruments. The chosen step size was 0.1 μm.

3. Results and Discussion

3.1 Powder Characterization

Figure 1(a) exhibits SEM images of the AISI H13 blend consisting of Cyclon and Kammer powder (D_{10}=21.7μm, D_{50}=32.8μm D_{90}=49.9μm) which shows a mainly spherical powder morphology with a dendritic solidification microstructure (1(b)). Additionally, fragmented particles can be observed (1(c-e)), such as splat caps, satellites, and nonspherical/irregularly shaped particles.

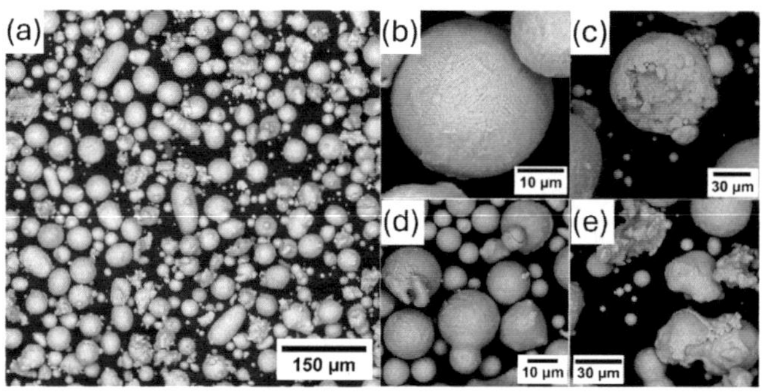

Figure 1. SEM images of (a) sieved and blended powder between 20μm and 45μm, (b)dendritic particle structure and (c-e) defects of the observed powders

Figure 2 (a1-a3) shows the normalized diffractograms of the spray-formed ingot and the Kammer and Cyclon overflow powders with a sieved fraction of 20μm to 45μm respectively between 35 and 85 °2θ. While the ingot consists of α-bcc peaks, both powders exhibit diffraction peaks for α-bcc and γ-fcc with a comparable fraction of the present phases in both powders whereas α-bcc indicates ferrite and γ-fcc indicates austenite. Due to a relativley slow cooling rate of the deposition surface of 0.1 to 10 Ks−1 it is assumed that the α-bcc peaks of the spray-formed ingot exhibit a purely ferritic microstructure.

Despite the rapid solidification of the powder during the atomization stage with a cooling rate ranging between 10^4 Ks^{-1} and 10^5 Ks^{-1} it is assumed that the α-bcc peaks represent ferrite. It is anticipated that the temperatures at the powder collection points 'Cyclon' and 'Kammer' are likley to show storing temperatures above the martensite start temperature of AISI H13 (390 °C), as determined by thermodynamic calculations. After the completion of the SF process, the powder was left to cool down relatively slowly to prevent martensite formation.

These results stand in contrast to those of Lei et al. [12] and Yan et al. [13] who reported a martensitic powder feedstock for gas-atomized AISI H13 powder with a comparable D50 fraction size. Additionally, Yan et al. [13] emphasize the presence of low-intensity peaks of the

(111) plane indicating a rather small fraction of austenite. The present intensities for the austenite phase exceed the reported data in the literature by far.

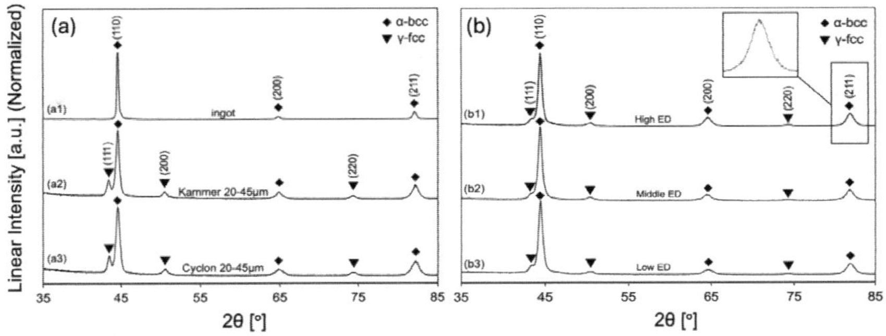

Figure 2. Diffractorgram within 35 and 85 °2θ of (a) the spray forming products (a1) ingot (a2) Kammer powder, (a3) Cyclon powder and (b) printed H13 with different energy densities (b1-b3); Normalized between 0 and 1 by the highest intensity

This suggests that spray-forming yields a mainly ferritic overspray powder for a size distribution between 20µm and 45µm instead of a martensitic microstructure compared to conventional gas atomization and additionally to a higher austenite content that was reported for gas atomized powder. The increased austenite content may be due to the austenite stabilizing nitrogen within the spray chamber or due to the reheating of solid particles upon impingement on the deposition surface during the SF process. The latter may lead to a reverse transformation of martensite to austenite within the already solidified martensitic particles. Since the powder was allowed to cool down slowly within the collection points a repeated martensitic transformation was prevented.

3.2 As-printed microstructure

Hot cracking phenomena during printing of H13 across many different process parameter combinations have been reported frequently e.g., by Quin et al. [7] or Köhler et al. [14], and is generally attributed to a comparably large solidification range and the rupture of liquid films in the final part of solidification. Contrary to that, no hot cracking was observed in the examined samples which might be due to reduced residual stresses from built plate pre-heating and/or due to a limited built height of 8 mm. Additionally, depending on the VED's 37.50, 55.55 and 71.50 $J \cdot mm^{-3}$ the average hardness values were 363.2, 393.8, and 379.0, HV_{10}, while the obtained relative densities were 91.4%, 97.8%, and 98.0% respectively.

Figure 2(b) shows the normalized diffractogram for the AISI H13 prints for the VED's of 37.50, 55.55, and 71.50 $J \cdot mm^{-3}$. Accordingly, the samples primarily consist of α-bcc peaks regardless of the energy density and show a soft γ-fcc (200) peak while the (110) bcc peak exhibits a notable (111) fcc shoulder. The rapid cooling rates in LPBF along with the comparable results by Yan et al. [11] and He et al. [15], [16] mainly indicate the presence of tempered martensite with low fractions of retained austenite. Moreover, it is noteworthy that all martensitic peaks exhibit no noticeable asymmetry nor skewness as it is exemplarily shown for the (211) peak, suggesting that all tetragonal martensite transformed into tempered martensite due to thermal cycling during the LPBF process.

Figure 3 shows (a) an overlay of an EBSD inverse pole figuring map (IPF) along the build direction (BD) and a band contrast image for a print with a VED of 71.50 $J \cdot mm^{-3}$. The corresponding phase map can be seen in figure 3(b). As indicated by the dotted white lines, the melt pool boundary, and the heat affected zone I and II (HAZI, HAZII) of two 90° rotated scan tracks can be observed. Furthermore, no noise reduction was applied to the dataset leading to a phase fraction of 16% fcc, 59.4% bcc, and 24.9% zero solutions.

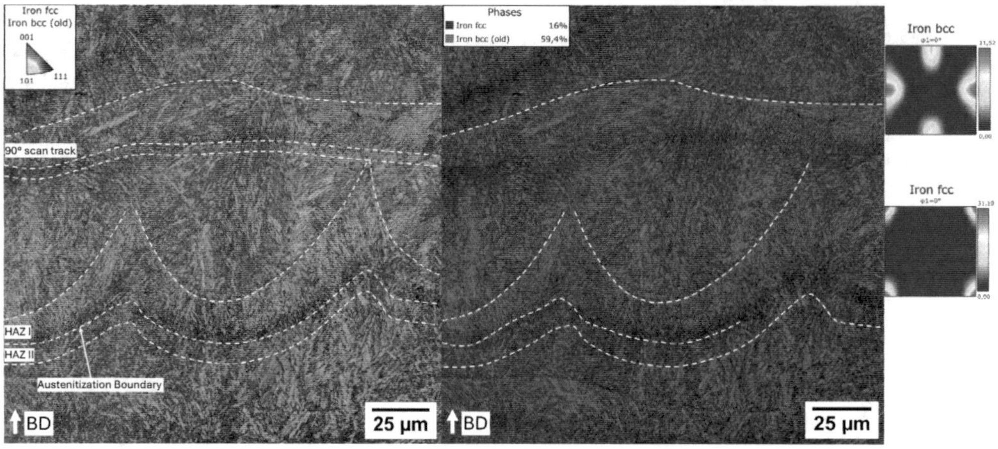

Figure 3. (a) EBSD pole figuring map (IPF) along build direction (BD) overlaid with the corresponding band contrast image, (b) phase map and (c) ODF's for a VED of 71.4 $\mathrm{J \cdot mm^{-3}}$

Consistent with the diffractograms, the microstructure primarily consists of a martensitic matrix with a small fraction of retained austenite. However, contrary to e.g. Naravan et al. [17], who used gas atomized powder feedstock and a pre-heated build plate of 200°C, or Trinh et al. [18], who also used gas atomized powder feedstock without a pre-heated build plate, the retained austenite does not appear to be homogeneously distributed along the cross-section. Regarding the phase identification, it should be noted that low indexing rates at the boundary between HAZI and HAZII can be observed likely due to grain refinement during austenitization within that area[19].

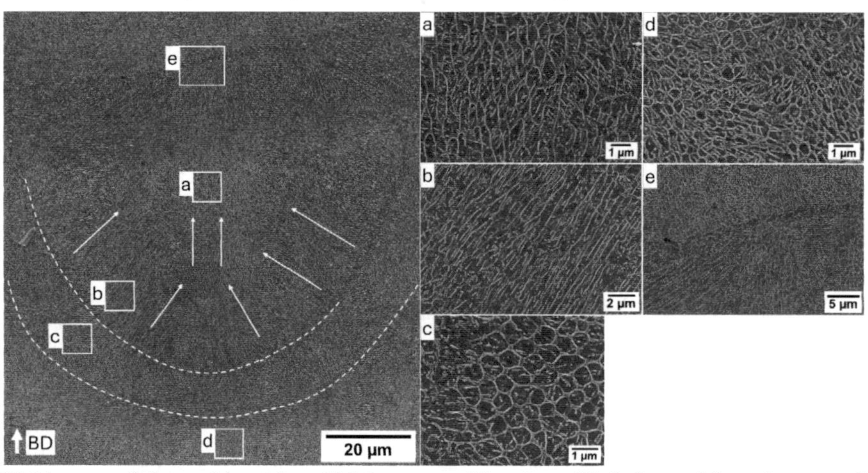

Figure 4. SEM image of the weld pool and higher magnification of (a) the weld pool center, (b) the weld pool boundary, (c) HAZI, (d) HAZII, and the next laser track which is 90° rotated

Moreover, it can be qualitatively observed in figure 3(b) that the retained austenite content within the martensite matrix of HAZI is notably smaller compared to the retained austenite content of the melt pool. This may be due to a micro in-situ heat treatment when the observed track was melted, leading to localized temperatures closely below the AC1 temperature, leading to a transformation of tetragonal martensite to tempered martensite with a bcc structure. This localized heat treatment appears to have a greater tempering effect on HAZI compared to the overall tempering cycles during the process on the rest of the microstructure. This may be related to the rather slow scanning speed of 700 mm/s at 250 W.

In contrast to HAZI, HAZII seems to show a higher austenite fraction than the melt pool indicating partial re-austenization. Further in-depth characterization is required to confirm this. Including the ODF's for bcc and fcc of figure 2(c) it becomes clear that bcc shows a preferred orientation in <100> parallel to the build direction while fcc presents a preferred orientation in <110> along the build direction.

Eventually, figure 4 represents SEM images of a weld pool in the center of the sample with VED of VED of 71.4 $J \cdot mm^{-3}$. It can be seen that the solidification front exhibits epitaxial grain growth with a columnar structure pointing radially into the melt pool (figure 4 (b)) and elongated grain growth with a cellular structure that appears stretched into the building direction (figure 4 (a)). Additionally, HAZI and HAZII show cellular networks as shown in figure 4 (c) and (d).

4. Summary and Outlook

Overspray powder is commonly considered as a scrap product of the spray forming process. To incorporate this scrap into a circular economy, this study investigated the feasibility of printing AISI H13 overspray. The as printed state presented a martensitic/austenitic microstructure showing elongated grains with a cellular structure in the core of the melt pool while the melt pool boundary showed epitaxial growth with a columnar grain structure.

Contrary to the existing literature, no hot cracking was observed, with the drawback of a poor part density of less than 98%. These findings show that spray formed H13 powders proved to be printable and further analysis is needed to understand how to optimize them and take advantage of the absence of hot cracking in this material. Concluding, the study generally shows the potential of the use of overspray powder to increase sustainability in additive manufacturing.

References

[1] Nygaard A 2023 *Circ. Econ. Sust.* **3(2)** 1099–1126
[2] Sosnowski P C and Cyplik P 2022 *Logforum* **18(4)** 413–420
[3] Lieder L and Rashid A 2016 *J Clean Prod* **115** 36–51
[4] Andrews D 2015 *Local Economy* **30(3)** 305–315
[5] Mathur P, Annavarapu S, Apelian D and Lawley A 1991 *Mater. Sci. Eng. A* **142(2)** 261–276
[6] Rodríguez V M, López-Morelos V H, Nadimpalli V K, Bue Pedersen D, Ruiz A and Somers M A J 2023 *Steel Res Int* **94(6)** 2200749
[7] Qin S, Bo Y, Herzog S, Hallstedt B, Kaletsch A and Broeckmann C 2022 *Powders* **1** 184–193
[8] *github DTU-OpenAM/OpenAM-PBF* last viewed on 13.06.2024 available online: https://github.com/DTU-OpenAM/OpenAM-PBF
[9] Kjer M B 2023 Systems Engineering in the Pursuit of Open Architecture Additive Manufacturing a study *PhD Thesis* DTU
[10] ISO 11505:2012 *Surface chemical analysis — General procedures for quantitative compositional depth profiling by glow discharge optical emission spectrometry.*
[11] ASTM E1019-18 *Standard Test Methods for Determination of Carbon, Sulfur, Nitrogen, and Oxygen in Steel, Iron, Nickel, and Cobalt Alloys by Various Combustion and Inert Gas Fusion Techniques.*
[12] Lei F, Wen T, Yang F, Wang J, Fu J, Yang H, Wang J, Ruan J and Ji S 2022 *Materials* **15(7)** 2686
[13] Yan J, Zhou Y, Gu R, Zhang X, Quach W-M and Yan M 2019 *Metals (Basel)* **9(1)** 86
[14] Köhler M L, Ali F, Herzog S, Suwanpinij P, Kaletsch A, and Broeckmann C 2022 *Steel Res Int* **93(1)** 202100454
[15] Yan J J, Zheng D L, Li H X, Jia X, Sun J F, Li Y L, Qian M and Yan 2017 M *J Mater Sci* **52(20)** 12476–12485
[16] He Y, Zhong M, Beuth J, and Webler B 2020 *J Mater Process Technol* **286** 116802
[17] Narvan M, Al-Rubaie K S, and Elbestawi M *Materials* **12(14)** 2284
[18] Van Trinh T, Nguyen S A, Pham K G, Seidel C, Pham A H and Phung C N 2023 *Acta Metall. Slovaca* **29(2)** 82–87
[19] de F. Silveira A C, Fechte-Heinen R and Epp J 2023 *Addit. Manuf.* **63** 103408

IOP Conf. Series: Materials Science and Engineering **1310** (2024) 012042 doi:10.1088/1757-899X/1310/1/012042

Recrystallization kinetics in 3D printed 316L stainless steel

C Zhang[1,2*], S Ahmed[2], V K Nadimpalli[2], T Yu[2], D Juul Jensen[2]

[1] Suzhou Lab. 215215 Suzhou, China
[2] Department of Civil and Mechanical Engineering, Technical University of Denmark, DK-2800 Kgs. Lyngby, Denmark

*E-mail: zhangcl@szlab.ac.cn

Abstract. During operation, many 3D printed components are likely to be exposed to external strains and/or high temperatures. However, the thermomechanical behaviours, e.g. plastic deformation and recrystallization, have not been thoroughly analysed, and are interesting to study because of the complex microstructure introduced by 3D printing. In the present work, the microstructural evolution and change in hardness during plastic deformation and recrystallization have been characterized in austenitic stainless steel 316L samples manufactured by laser powder bed fusion. The focus is on the recrystallization kinetics in both the as-printed and the plastically deformed state. It is discussed how these results compare to results for conventionally manufactured samples. The present study suggests that the inhomogeneity of 3D printed samples significantly affects the recrystallization behaviour.

1. Introduction

3D printing has become an emerging manufacturing technique, which is widely utilized to produce metallic components of stainless steels, Inconel 718, titanium alloys etc. [1–3]. The fast local solidification and thermal cycles during 3D printing result in microstructures very different from those after conventional manufacturing, often with an irregular grain (domain) morphology, and elemental segregation on cell/grain boundaries [2]. These structures are further away from equilibrium structures than the conventional ones and may thus exhibit different thermomechanical behaviour. It has for example been demonstrated that the as-printed cellular dislocation structure affects the twinning and dislocation behaviour during subsequent plastic deformation of 3D printed stainless steel (316L). As a consequence, certain mechanical properties are improved compared to conventional manufacturing [4,5].

Heat treatment is an important approach to manipulate the microstructure and thus the properties [6]. Post-print heat treatments are often required for 3D printed samples to homogenize the microstructure and reduce residual stress as well as anisotropy. Taking 316L as an example, extensive research has shown that 3D printed samples exhibit a sluggish recrystallization behaviour, which only happens during high (~1100 °C) temperature annealing [7–10]. However, most of these studies focus on the properties after post-print annealing, while rather little attention has been devoted to investigating the recrystallization kinetics and mechanisms. Aota et. al. [8] suggested that oxide particles in as-printed 316L provide significant dragging forces via Zenner pinning, explaining the sluggish recrystallization, whereas Pinto et al.

[11] attributed it to an effect of the particular texture in the as-printed state and thus a possible non-random distribution of nuclei. Recently, Sonis et al. [12] suggested that the many low angle boundaries in as-printed 316L, to some extent, affect both the nucleation and growth of recrystallizing grains. All these possibilities make it hard to optimize the heat treatment and thus the properties; a proper physical understanding is needed. The aim of the present work, is to take a step in that direction by quantifying the recrystallization kinetics of 3D printed 316L in both the as-printed as well as in a plastically deformed condition, and compare the results to those after conventional manufacturing.

2. Experimental

Two series of as-printed 316L samples with sizes of 3 mm × 4 mm × 14 mm and 10 mm × 10 mm × 10 mm were produced by laser powder bed fusion (L-PBF). The laser power and scanning speed were set to 370 W and 1.35 m/s, respectively for the first series, and 200 W and 0.75 m/s for the second series. A 67° and 45° rotation were applied between neighbouring layers for the first and second series, respectively.

To analyse the recrystallization kinetics, the first series of samples were annealed at 850 °C for up to 5 h and at 900 °C for some cases, whereas the second series of samples were annealed at 1175 °C for various times up to 21 h. Furthermore, to investigate how a possible plastic deformation of a 3D printed component would affect subsequent recrystallization, some of the first series as-printed samples were cold rolled to 30% reduction and subsequently annealed. The same procedure was applied to the conventionally manufactured 316L samples for comparison. Details on how the additional 30% cold rolling affects the deformation microstructure can be found in [5,7]. Four sample conditions are included in the present study, namely P1-0(as-printed, series 1), P2-0(as- printed, series 2), P1-30% (as-printed, series 1, plus 30% cold rolled) and C-30% (conventionally manufactured, plus 30% cold rolling).

A Zeiss Supra 35 scanning electron microscope (SEM) equipped with an HKL electron backscatter diffraction (EBSD) system was used to characterize the microstructure. The samples for EBSD/SEM characterization were prepared by grinding followed by mechanical polishing. The grain reconstruction was carried out with a critical misorientation of 10° using the MTEX toolbox. Recrystallizing grains/nuclei were identified by the following criteria: a maximum grain orientation spread (GOS) of 0.5° and a minimum size of 2 μm. The volume fraction of recrystallized grains V_v was calculated by $f = \frac{V_{rex}}{V_{ebsd}}$,where V_{rex} is the area of recrystallizing grains and V_{ebsd} is the total area of relevant EBSD maps. The micro-hardness of all the samples was measured by a Struers DuraScan micro hardness machine.

The classic KJMA model (as discussed in [13]) was used to analyse the recrystallization kinetics, where the recrystalized fraction is expressed as:

$$Vv = 1 - \exp\left(-kt^n\right) \quad (1)$$

$$n = \frac{\log(-\ln(1 - Vv)) - logk}{logt} \quad (2)$$

where t is the annealing time, and k and n are two parameters describing the kinetics. If (-ln(1-V_v)) is plotted versus t on double logarithm scale, a straight line should thus be obtained with n as the slope (see equation (2)). In the present work we shall use equations (1-2) to analyse the recrystallization kinetics based on V_v values experimentally determined by EBSD for the series 1 samples.

3. Results

3.1 Recrystallizing microstructure

Micrographs of the recrystallizing microstructure are shown in figure 1. For P1-0, there is no obvious recrystallization, even after annealing at 900 °C for 5 h (see figure 1). This sluggish recrystallization consents with earlier studies showing that as-printed 316L only recrystallizes at 1050 °C annealing because of the high thermal stability of the dislocation network [14]. For P1-30%, the recrystallization nucleation is inhomogeneous as compared to C-30%, as reported in [8]. This is likely inherited from the inhomogeneous as-printed microstructures and thus the deformed microstructure. In P1-30% recrystallization is observed already after 15 min annealing and the recrystallized fraction reaches 80% after 1 h, whereas the recrystallization just started in C-30% (i.e C-30% has an incubation time of about 1 h at 850 °C). This is reasonable because the high density of dislocations in P1-30% leads to a higher stored energy than in C-30%, which facilitates the nucleation and growth of recrystallizing grains. Nearly complete recrystallization was observed after annealing at 900 °C for 5 h in P1-30%.

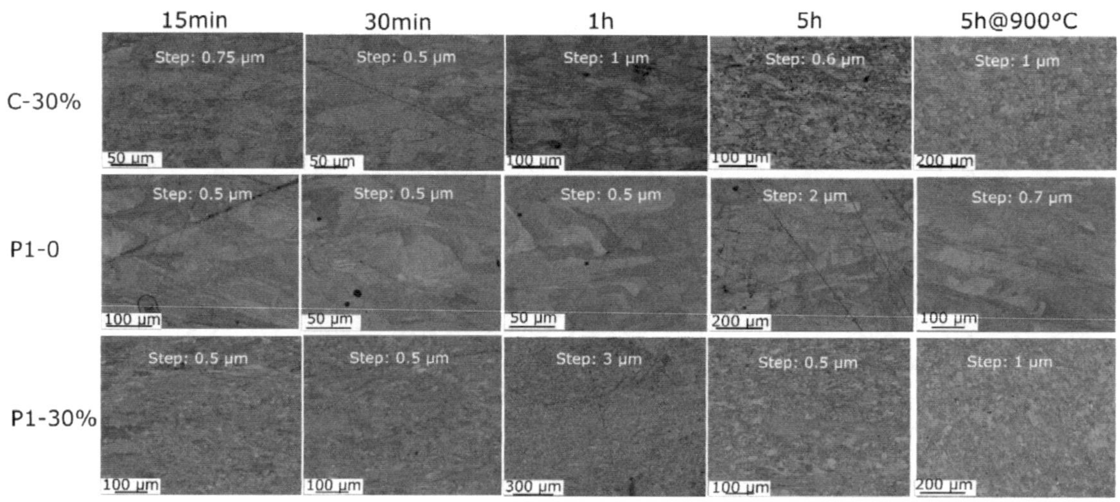

Figure 1. Band contrast maps for C-30%, P1-0, and P1-30% samples annealed at 850 °C for 15 min, 30 min, 1 h, and 5 h. Note: the last column is for annealing at 900 °C for 5 h.

During annealing, the low angle boundaries (LABs) in P1-0 appear to be relatively stable (figure 2a), which is a common observation, and has been documented in e.g. [14]. It is more interesting to observe that when the printed sample is slightly deformed, like in P1-30%, the microstructure becomes highly sensitive to annealing and the fraction of LABs decreases rapidly during annealing (figure 2b). This is even more remarkable because the conventionally manufactured C-30% sample is more stable, and the LAB fraction only decreases after an incubation period of 1 h (figure 2c). These data correspond very well with the recrystallization evolution in P1-30% and C-30%, where it is observed that P1-30% recrystallizes readily, while the onset of recrystallization is delayed in C-30% (figure 2d).

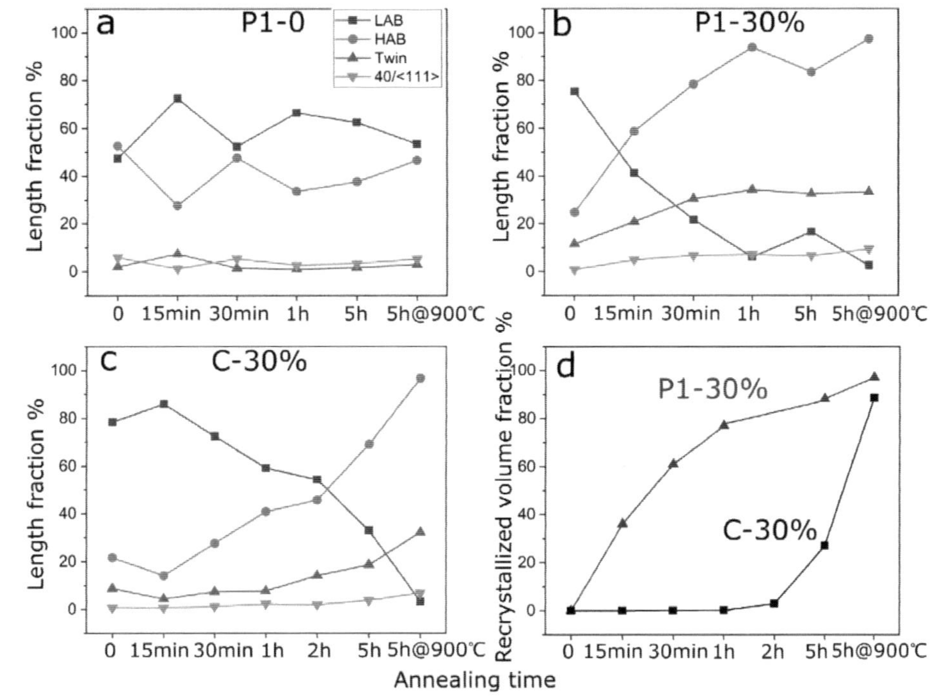

Figure 2. Analysis based on the EBSD characterization. (a-c), Length fraction of boundaries of various types in P1-0, P1-30%, and C-30%, respectively; (d), recrystallized volume fraction for P1-30% and C-30%. Note subfigures (a-c) share the same legend, and the last data points of (a-d) correspond to 900 °C for 5 h (other cases correspond to annealing at 850°C).

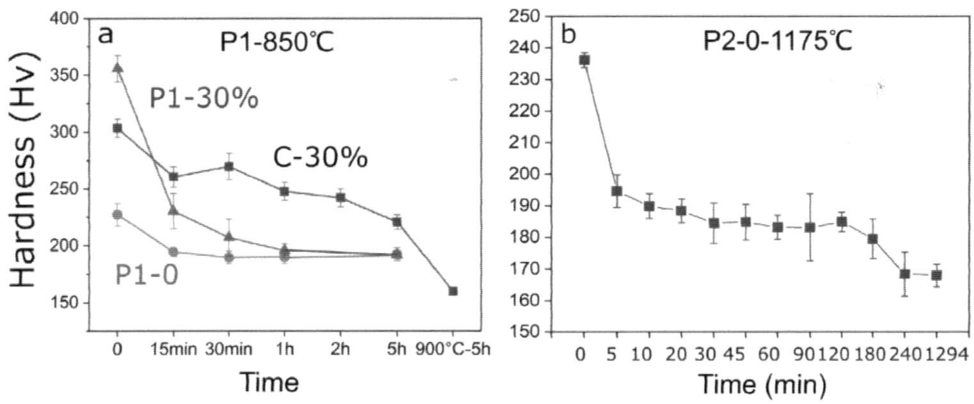

Figure 3. Hardness values as a function of annealing time. (a): P1-0, P1-30% and C-30% annealed at 850 °C (Note the last data point of C-30% corresponds to 900 °C for 5 h); (b): P2-0 annealed at 1175 °C. Please note the different hardness scales used in (a) and (b).

3.2 Mechanical properties

There is a significant drop in hardness in both P1-0 and P2-0 initially during annealing at 850 °C as well as at 1175 °C, while for somewhat longer annealing times the hardness remains constant at this lower level (figure 3). The initial drop in hardness appears to be dominated by recovery. For the high temperature anneal, it is observed that the hardness drops further after 2 h. This softening may be a result of further microstructural coarsening during recrystallization and grain growth. For P1-30% and C-30%, more traditional hardness reductions are observed as a function of time at 850 °C (see figure 3a), which corresponds to the recrystallization curves obtained by EBSD (see figure 2d).

3.3 Recrystallization kinetics
As P1-0 and P2-0 are observed to respectively only recover or undergo a complex process where recovery, recrystallization and probably also grain growth occur more or less simultaneously, in the following only the recrystallization kinetics will be analysed for P1-30% and C-30%. Figure 4 presents the KJMA plots. It is found that the Avrami exponent n (see equation 1) is rather low; only 0.5 for P1-30% compared to $n = 3.28$ for C-30%. An n-value around three, is according to the KJMA theory, fairly standard, and would align with a random distribution of nuclei growing three dimensionally. The low n-value for P1-30%, is more interesting. Zener pinning may be of importance here, as suggested in [8], but other reasons related both to the nucleation and growth may be of similar importance: It could relate to clustered nucleation (as the investigations indicate, see figure 1) and/or a low dimensional or otherwise not standard growth of the nuclei, as discussed in e.g. [15].

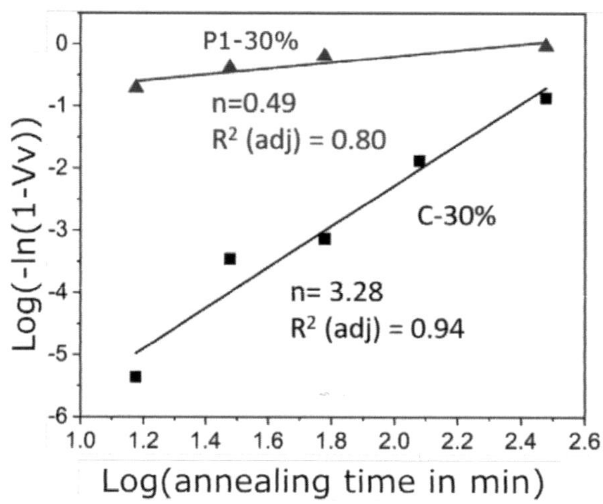

Figure 4. KJMA plot for P1-30% and C-30%, reporting the Avrami exponent n and the statistical parameter R^2.

4. Summary and outlook

The annealing response of both 3D printed samples and conventionally manufactured samples were analysed based on hardness and EBSD measurements. Additionally, some of the printed and the conventionally manufactured samples were cold rolled to 30% reduction in thickness to investigate effects of the additional deformation on the subsequent annealing behaviour. Recovery dominates the annealing process in the as-printed samples, even at the very high annealing temperature of 1175 °C. The rolled samples recrystallize normally. The printed sample recrystallize more rapidly than the conventional one, which is related to the higher stored energy in the printed sample. However, when the recrystallization kinetics are analysed in terms of the classic KJMA model, it is observed that the kinetics exponent n is much higher in the conventional than in the printed sample. The very low n exponent in the printed and rolled sample may be related to clustered non-random distribution of the nuclei, as well as a complex growth pattern.

Acknowledgments

The main part of this work was carried out within M4D project supported by the European Research Council (ERC) under the European Union's Horizon 2020 research and innovation program (Grant agreement No. 788567), while the latter part was done in the MicroAM project supported by VILLUM FONDEN (VIL54495). CZ further acknowledges the financial support by the China Scholarship Council (CSC) (N0. 201706120026).

References

[1] W.E. Frazier 2014 Metal additive manufacturing: A review *J. Mater. Eng. Perform.* **23** 1917–1928.

[2] Y.M. Wang, T. Voisin, J.T. McKeown, J. Ye, N.P. Calta, Z. Li, Z. Zeng, Y. Zhang, W. Chen, T.T. Roehling, R.T. Ott, M.K. Santala, P.J. Depond, M.J. Matthews, A. V. Hamza and T. Zhu 2018 Additively manufactured hierarchical stainless steels with high strength and ductility *Nat. Mater.* **17** 63–70.

[3] D. Herzog, V. Seyda, E. Wycisk and C. Emmelmann 2016 Additive manufacturing of metals *Acta Mater.* **117** 371–392.

[4] L. Liu, Q. Ding, Y. Zhong, J. Zou, J. Wu, Y.L. Chiu, J. Li, Z. Zhang, Q. Yu and Z. Shen 2018 Dislocation network in additive manufactured steel breaks strength–ductility trade-off *Mater. Today.* **21** 354–361.

[5] C. Zhang, D. Juul Jensen and T. Yu 2023 Effects of initial 3D printed microstructures on subsequent microstructural evolution in 316L stainless steel *Acta Mater.* **242** 118481.

[6] A. Rollett, G.S. Rohrer and J. Humphreys 2017 Recrystallization and Related Annealing Phenomena.

[7] R.W. Fonda, D.J. Rowenhorst, C.R. Feng, A.J. Levinson, K.E. Knipling, S. Olig, A. Ntiros, B. Stiles and R. Rayne 2020 The Effects of Post-processing in Additively Manufactured 316L Stainless Steels *Met. Mater Trans A.* **51** 6560–6573.

[8] L.S. Aota, P. Bajaj, K.D. Zilnyk, E.A. Jägle, D. Ponge, H.R.Z. Sandim and D. Raabe 2021 Recrystallization kinetics, mechanisms, and topology in alloys processed by laser powder-bed fusion: AISI 316L stainless steel as example *Materialia.* **20** 101236.

[9] M.R. Jandaghi, A. Saboori, L. Iuliano and M. Pavese 2021 On the effect of rapid annealing on the microstructure and mechanical behavior of additively manufactured stainless steel by Laser Powder Bed Fusion *Mater. Sci. Eng. A.* **828** 142109.

[10] D. Kong, C. Dong, X. Ni, L. Zhang, J. Yao, C. Man, X. Cheng, K. Xiao and X. Li 2019 Mechanical properties and corrosion behavior of selective laser melted 316L stainless steel after different heat treatment processes *J. Mater. Sci. Technol.* **35** 1499–1507.

[11] F.C. Pinto, L.S. Aota, I.R. Souza Filho, D. Raabe and H.R.Z. Sandim 2022 Recrystallization in non-conventional microstructures of 316L stainless steel produced via laser powder-bed fusion: effect of particle coarsening kinetics *J. Mater. Sci.* **57** 9576–9598.

[12] E. de Sonis, S. Dépinoy, P.F. Giroux, H. Maskrot, L. Lemarquis, O. Hercher, F. Villaret and A.F. Gourgues-Lorenzon 2022 Dependency of recrystallization kinetics on the solidification microstructure of 316L stainless steel processed by laser powder bed fusion (LPBF) *Mater. Charact.* **194** 112370.

[13] R.A. Vandermeer and D. Juul Jensen 2023 Microstructural path modeling of primary recrystallization, in: *J. Phys. Conf. Ser. IOP Publishing*, 012027.

[14] T. Voisin, J.B. Forien, A. Perron, S. Aubry, N. Bertin, A. Samanta, A. Baker and Y.M. Wang 2021 New insights on cellular structures strengthening mechanisms and thermal stability of an austenitic stainless steel fabricated by laser powder-bed-fusion *Acta Mater.* **203** 116476.

[15] F. Lin, Y. Zhang, W. Pantleon and D. Juul Jensen 2018 Importance of non-uniform boundary migration for recrystallization kinetics *Metall. Mater. Trans. A* **49** 5246–5258.

IOP Conf. Series: Materials Science and Engineering **1310** (2024) 012043 doi:10.1088/1757-899X/1310/1/012043

Exploring the feasibility of preparing Ti/Ti6Al4V composites by laser powder bed fusion

J Shen*, **Z Pan, V K Nadimpalli and T Yu**

Department of Civil and Mechanical Engineering, Technical University of Denmark, Kgs. Lyngby, DK-2800, Denmark

*E-mail: jinshe@dtu.dk

Abstract. Ti-based alloys and composites are advanced lightweight materials that are indispensable for many critical applications. This study presents a novel strategy of fabricating laminated plates by printing Ti6Al4V on the surface of pure Ti substrate to obtain a composite structure combining coarse equiaxial and fine acicular grains. The molten pool layer is well bonded to the substrate surface and the chemical composition is rather uniform, due to melting of the substrate surface together with the powders and subsequent solidification. These findings provide a new path for material design, which will broaden the alternatives of structural materials applied in extreme service enviroment.

1. Introduction

Titanium (Ti) alloys and composites are very promising candidates for structural components in the aerospace field, due to the high specific strength, excellent corrosion resistance and good cryo-toughness [1,2]. Given the moderate strength of pure Ti, the way to harden it is usually by grain refinement, hetero-structure design (lamellar or gradient structure) and adding alloying elements (e.g. Al, Cu, Si, V, Fe, Mo) [3-6]. Introducing ultrafine grains can significantly increase the strength but severely deteriorates the ductility. Besides, the low thermal stability of the ultrafine grains prevents them from serving in medium-to-high temperature environments. Components with lamellar structure (e.g. by hot compression or accumulative roll bonding) and gradient structure (e.g. by surface mechanical grinding treatment) are difficult to industrialize because of the low production efficiency and possible inherent defects [7,8]. Therefore, better fabrication methods are needed in order to utilize the advantage of hetero-structures.

In recent years, three-dimensional (3D) printing or additive manufacturing (AM) is getting more attention due to its high design freedom and low material waste, which is especially significant for Ti and its alloys due to their high price and poor processability in traditional manufacturing [9]. Laser powder bed fusion (L-PBF) is a popular AM technique used to create components layer by layer from metal powders [4,10-12]. The built components show complex microstructures and often anisotropy in mechanical properties due to directional solidification under a high cooling rate and a high thermal gradient during L-PBF [13].

Different from the common AM components that are fully built layer by layer from bottom to top, here we report a new design strategy, i.e., integrating AM with conventionally manufactured parts by printing Ti6Al4V on a Ti plate, which is fixed to the building plate of the L-PBF machine. The resulting Ti/Ti6Al4V plate contains both the fully recrystallized microstructure with coarse equiaxed grains and the printed microstructure with acicular martensite. This work opens a new door for designing composite microstructure and composite materials, enabling property optimization and new component design strategies.

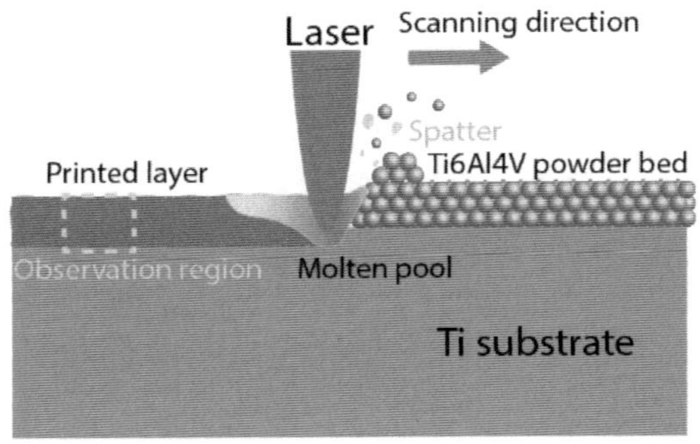

Figure 1. Schematic diagram of fabricating a Ti/Ti6Al4V composite sample via L-PBF.

2. Experimental materials and methods

2.1 Preparation of layered materials

Grade 2 commercial purity titanium (TA2) plates and Ti6Al4V powders were used in this work to prepare the composite plates. The as-received TA2 plates were prepared by multi-pass rolling and annealing, and have an equiaxed microstructure with an average grain diameter of ~23 μm. The average diameter of gas atomized Ti6Al4V powders was measured to be 27 μm.

The surface of the TA2 plate was thoroughly cleaned before printing to minimize the effect of the oxidation layer and surface contamination on the subsequent printing process. The plate was first mechanically cleaned using different grades of sandpapers, and then it was placed into deionized water and anhydrous ethanol for ultrasonic cleaning.

The subsequent procedure for fabricating the layered composite plate is illustrated in figure 1. The cleaned TA2 plate was fixed by screws in a groove of the building plate and then placed together into the building chamber, which was subsequent flushed with high purity argon. One layer of Ti6Al4V (120 mm × 30 mm) was printed on the TA2 plate, with a powder layer thickness of 50 μm, laser power 250 W, scanning speed 600 mm/s, and hatch spacing 0.12 mm.

2.2 Microstructure characterization

The microstructure observation was conducted in a Zeiss Sigma 300 scanning electron microscope (SEM) equipped with an electron backscatter diffraction (EBSD) system, operated at 20 kV. For the acquisition of EBSD data, the scanning step was set to 0.15 μm. AztecCrystal was utilized to analyze the crystallographic information.

3. Results and discussion

Figure 2 shows the microstructure covering the molten pool layer. The boundary between molten pool and Ti substrate is not straight, as indicated by the dashed line in figure 2(a). The average thickness of the molten pool layer is measured to be ~132 μm, which is much larger than that of the powder bed. This suggests that a suface layer, about 100 μm in thickness, of the TA2 plate was melted to form the molten pool during the printing process. No obvious cracking was found in the molten pool layer nor at its boundary, implying good bonding between the molten pool layer and the substrate. This is attributed to the fact that the melted powers are mixed with the melt zone on the surface of the TA2 plate under the effect of thermal fluid convection. Upon closer inspection,

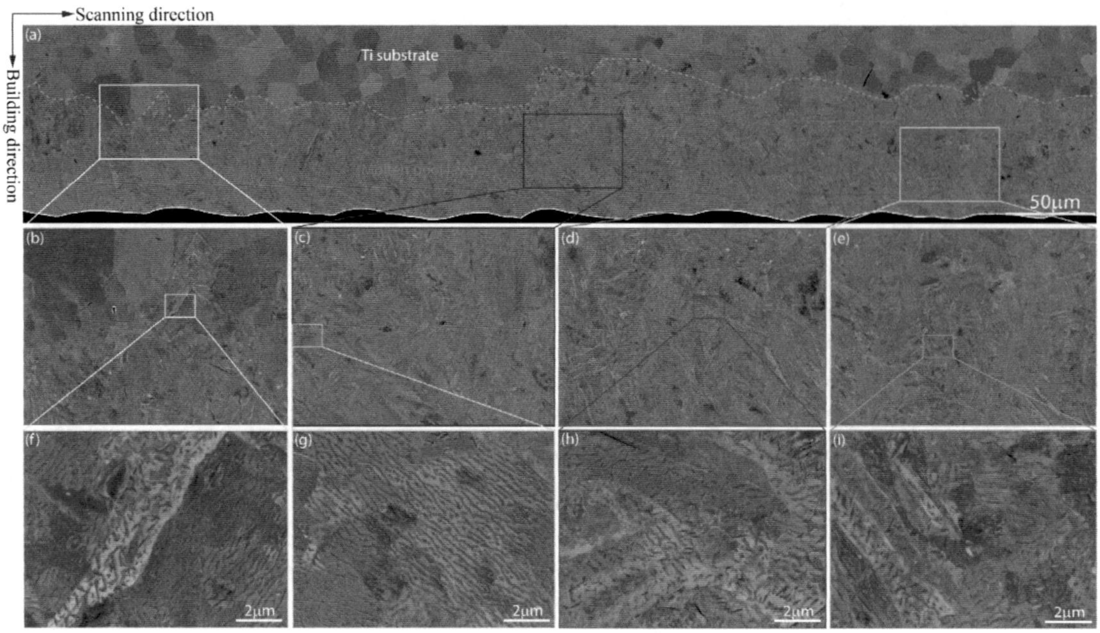

Figure 2. Multi-dimensional microstructure of the molten pool layer.

the microstructure near the pool boundary shows a lath-like shape, containing a large number of acicular α' martensite, as shown in figure 2(f). The formation of martensite is a result of the rapid cooling process, and similar results are also reported in the literature [11,14].

Figure 3 mainly shows the EBSD maps of the molten pool layer. The inverse pole figure (IPF) map indicates the weak crystallographic texture in the molten pool layer, as shown in figure 3(a). The average length of acicular grains in the molten pool layer is measured to be 17.2 μm, which is smaller than the average diameter of exquiaxed grains in the substrate. Meanwhile, the average length-to-width ratio of these acicular grains is as high as 9.3. Taken together, the grain size in the molten pool layer is much finer than that in the substrate. Particularly, the grain size at the top of the molten pool is larger than that at the bottom of the molten pool, due to the grain growth promoted by slower cooling here. The interface shows somehow a smooth transition, which might be due to the phase transformation occurred in the substrate immediately below the molten pool boundary. In the molten pool layer, the fraction of low angle grain boundaries (LAGB: 6.8%) are much lower than that of high angle grain boundaries (HAGB: 93.2%), as shown in figure 3(b). Moreover, a large number of special boundaries of $55{\sim}65^{\circ}\langle 1\bar{2}10\rangle$ are also detected here, with a percentage accounting for 27.4% of the total. These boundaries between different martensite variants are derived from the same prior β Ti grain [15]. Due to the thermal strain during cooling, the misorientation angles measured across mutiple acicular grains are in the the rage of 56°–62°, as shown in figure 3(c).

To evaluate the composition variation from the molten pool layer to the substrate, elemental analysis by energy dispersive spectroscopy (EDS) was performed. The results are shown in figure4 and table 1. The point analysis results show that there is only a small difference in the chemical composition between the molten pool layer and the substrate. This can be also seen from the line profiles shown in figure 4(b). Note that the mapping of element Al shows a slight segregation in the upper region of the molten pool layer, see figure 4(c). Nevertheless, the

IOP Conf. Series: Materials Science and Engineering **1310** (2024) 012043 doi:10.1088/1757-899X/1310/1/012043

elemental distribution in the molten pool layer can still be considered homogeneous, which indicates that the molten Ti6Al4V powders can be fully mixed with the Ti molten pool due to the strong convection effect.

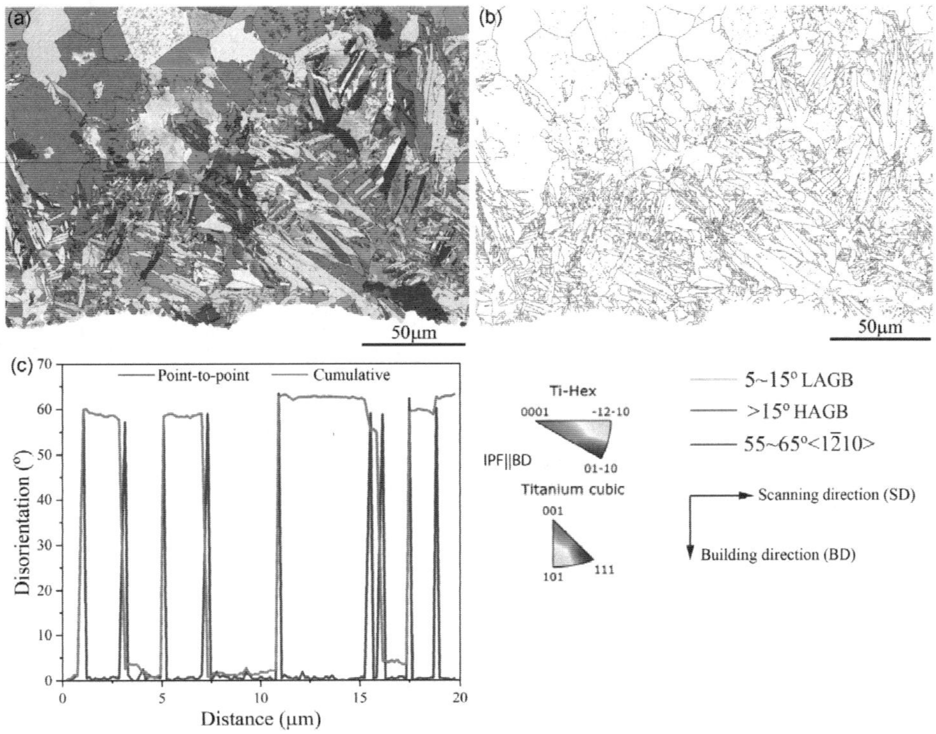

Figure 3. EBSD maps of the molten pool layer. (a) IPF map. (b) Grain boundary map. (c) Quantitative analysis of disorientation across multiple grain boundaries along the red arrow shown in (b).

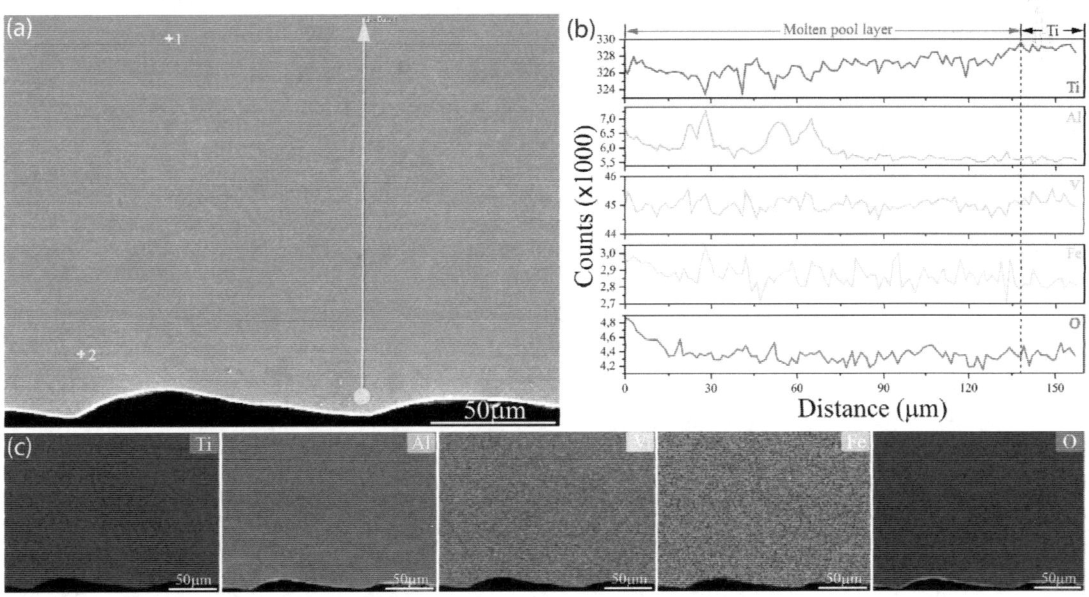

Figure 4. SEM-EDS analysis of the molten pool layer. (a) SEM image; (b) EDS line profile from the printed surface to the substrate; (c) EDS mapping of the same area.

Table 1. Chemical composition (wt.%) measured from two points shown in figure 4.

Spot	Ti	Al	V	Fe	O
1	99.68	0.18	0.01	0.04	0.08
2	99.35	0.42	0.08	0.06	0.09

4. Summary

In this study, we propose a novel strategy of preparing composite plate by additive manufacturing, combining coarse equiaxial and fine acicular grains in a plate. Based on the L-PBF technique, the molten pool layer is seamlessly integrated into the substrate plate by mixing the molten Ti6Al4V powders with the melt pool in the Ti substrate surface followed by solidification. A substantial fraction of specifical boundaries of $55\sim65^{\circ}\langle 1\bar{2}10\rangle$ are found in the molten pool layer due to martensitic transformation. These findings confirm the feasibility of fabricating layered materials by printing the intended materials directly on a substrate plate, which will broaden the design strategy of structural materials to meet specific performance requirements.

Acknowledgements

The work of JS and TY was supported by VILLUM FONDEN (VIL50197); while the work of ZP and VKN was supported by VILLUM FONDEN (VIL54495).

References

[1] Song T T et al. 2023 *Nature* **618** 63-71
[2] Qiu F C, Cheng T, Song Y C, Ivasishin O M, Savvakin D G, Ma G Y and Xu H Y 2024 *J. Mater. Sci. Technol.* **171** 24–36
[3] Meng L, Zhang Y Y, Zhao X L and Zhang D L 2021 *Mater Sci Eng A* **825** 141877
[4] Wang H, Chao Q, Chen H S, Chen Z B, Primig S, Xu W, Ringer S P and Liao X Z 2022 *Acta Mater.* **235** 118104
[5] Zhang D Y, Qiu D, Gibson M A, Zheng Y F, Fraser H L, St John D H and Easton M A 2019 *Nature* **576** 91-96
[6] Zhang J Q, Bermingham M J, Otte J, Liu Y G, Hou Z Y, Yang N, Yin Y, Bayat M, Lin W K, Huang X X, StJohn D H and Dargusch M S 2024 *Science* **383** 639–645
[7] Li D Y, Fan G H, Huang X X, Juul Jensen D, Miao K S, Xu C, Geng L, Zhang Y B and Yu T B 2021 *Acta Mater.* **206** 116627
[8] Yin Y F, Xu Wei, Sun Q Y, Xiao L and Sun J 2015 *T. Nonferr. Metal. Soc.* **25** 738-747
[9] Qu Z et al. 2024 *Nature* **626** 999-1006
[10] Tian X N, Zhu Y M, Lim C V S, Williams J, Boyer R, Wu X H, Zhang K and Huang A J 2021 *Addit. Manuf.* **46** 102151
[11] Xu W, Brandt M, Sun S, Elambasseril J, Liu Q, Latham K, Xia K and Qian M 2015 *Acta Mater.* **85** 74–84
[12] Lu S L, Zhang Z J, Liu R, Zhou X H, Wang X G, Zhang B N, Zhao X M, Eckert J and Zhang Z F 2022 *Addit. Manuf.* **59** 103092

[13] Bermingham M J, StJohn D H, Krynen J, Tedman-Jones S and Dargusch M S 2019 *Acta Mater.* **168** 261-274
[14] Xu W, Lui E W, Pateras A, Qian M and Brandt M 2017 *Acta Mater.* **125** 390-400
[15] Kazantseva N, Krakhmalev P, Thuvander M, Yadroitsev I, Vinogradova N and Ezhov I 2018 *Mater. Charact.* **146** 101-112

IOP Conf. Series: Materials Science and Engineering **1310** (2024) 012044 doi:10.1088/1757-899X/1310/1/012044

Cold spray - a solid-state additive manufacturing technology

Congcong Su[1], Yan Wang[1], Hanqing Che[1*], Stephen Yue[2], Xiaoxu Huang[1]

[1] International Joint Laboratory for Light Alloys (MOE), College of Materials Science and Engineering, Chongqing University, Chongqing 400044, China
[2] Department of Mining and Materials Engineering, McGill University, Montreal, Quebec H3A 0C5, Canada

*E-mail: hanqing.che@cqu.edu.cn

Abstract. Cold spray is a solid-state powder deposition technique and has evolved into an additive manufacturing process. Unlike conventional additive manufacturing technologies that rely on melting and solidification, cold spray additive manufacturing (CSAM) forms components at low temperatures at a relatively high build rate. This article introduces the technical principles, process parameters and typical microstructure of cold spray, as well as its applications in the fabrication of 3D components and damaged component repair. Current issues faced in cold spray research and future development directions in CSAM are also discussed.

1. Introduction

Cold spray was first noticed by Soviet scientist Anatolii Papyrin during supersonic wind tunnel experiments in the 1980s [1]. During the experimental process, when the velocity of tracer particles (Al) used in the supersonic wind tunnel exceeded a certain value, the resulting effect of those particles on the target material surface transformed from erosion to deposition. This discovery indicated the potential for high-speed solid metal particles to directly deposit on a material surface, forming a coating. Therefore, the fundamentals for cold spray technology were established and further developed as a coating technique [1]. The prominent feature of cold spray is its low processing temperatures and solid-state deposition. It can also deposit materials at relatively high build rates, normally 2-7 kg/h [2].

With recent developments, cold spray technology has evolved into a solid-state additive manufacturing method, which is also known as cold spray additive manufacturing (CSAM). CSAM has characteristics different from traditional melting and solidification-based additive manufacturing technologies, such as selective laser melting (SLM), electron beam melting (EBM), and wire arc additive manufacturing (WAAM). Components manufactured with CSAM experience lower thermal impact, shorter production time, and fewer limitations in product size. It can also circumvent common issues appearing in high-temperature processes, including excessive oxidation, high residual thermal stress, and undesired phase transformation. Thus, It is particularly suitable for preparing high-quality materials sensitive to oxidation, heat, and phase changes [3, 4].

This article offers an overview of the principles, process parameters and microstructure of cold spray technology and discusses its CSAM applications in the fabrication of 3D components and damaged component repair. The final summary highlights the advantages of CSAM and outlooks on the scientific and industrial issues it is currently facing.

IOP Conf. Series: Materials Science and Engineering **1310** (2024) 012044 doi:10.1088/1757-899X/1310/1/012044

2. Cold spray process

2.1 Technical principles

Cold spray is a low-temperature, solid-state powder consolidation process that forms components by depositing layers sequentially. In this process, compressed gases (He, N_2, and air), often heated, serve as propellants to accelerate the powder (diameters typically between 5-50 μm) through a de Laval nozzle to high velocities (300-1200 m/s) [5]. Upon impact with a substrate, the particles deform and bond to the substrate. The process is schematically shown in figure 1 [6]. The formation of the deposit primarily depends on the velocity of the particles. Deposition occurs only when the particle velocity is between two critical velocities, the threshold for deposition (often denoted v_{crit}) and that for erosion; otherwise, it will cause particle rebound or substrate erosion [7]. The deposition involves two stages: the initial layer is formed as particles impact on and bond to the substrate, followed by particles being deposited onto the previously deposited layer (i.e., the build-up).

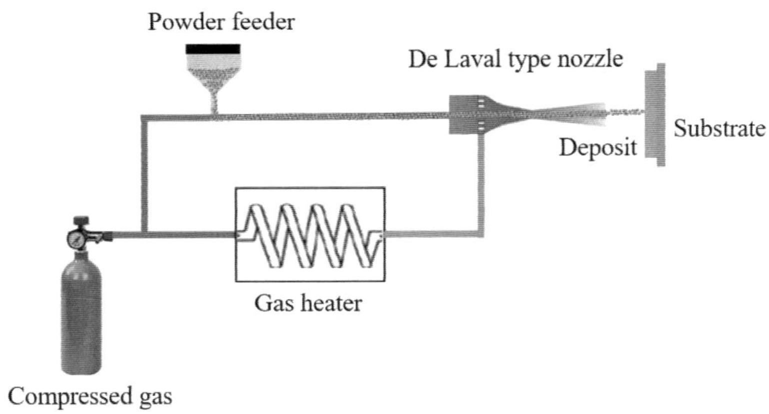

Figure 1. Schematic diagram of cold spray [6].

During the process of high-speed particle impacting on the substrate, a significant portion of the kinetic energy is converted into thermal energy and plastic deformation for both the particles and the substrate. Several bonding mechanisms have been proposed by researchers so far [8]. Among them, the adiabatic shear instability (ASI) mechanism is a widely accepted one [9, 10]. When a particle impacts a substrate, deformation in the contact area results in the fracture of oxide film on the particle and substrate surfaces, leading to intimate contact. Subsequently, the interface between the particle and substrate undergoes ASI in the direction tangential to the interface due to shear forces, resulting in material jetting. The ASI mechanism describes that the tangential deformation is attributed to an adiabatic thermal softening which can surpass the work-hardening of materials. Thus, metallurgical bonding forms as material jetting takes place at the particle-substrate interface through ASI [11].

Another widely acknowledged bonding mechanism is mechanical interlocking, which occurs due to the high-speed impact of particles, causing the particle-substrate interface to roll up and swirl [12]. In addition to the above two mechanisms, some scholars have proposed other mechanisms to explain bonding, such as hydrodynamic plasticity [13], heteroepitaxial growth (for ceramic substrates) [14], and local melting [15]. Although the bonding mechanism may require further clarification, it is apparent that the particles must undergo significant plastic deformation in order to achieve bonding to the substrate; this is in agreement with the critical velocity concept introduced above.

IOP Conf. Series: Materials Science and Engineering **1310** (2024) 012044 doi:10.1088/1757-899X/1310/1/012044

2.2 Process parameters

The properties of the component fabricated by cold spray depend on the process parameters. These include gas type (air, N_2, He, etc.), gas preheating temperature, gas pressure, nozzle geometry, powder feeding rate, spraying angle, and spraying distance. Among them, the carrier gas parameters are most important [16]. Commonly utilized gases include He, N_2, and air, with He being capable of rendering higher velocities than N_2 and air, thereby enabling high-quality bonding for many materials. However, the high cost of He constrains its broad application. While air is an economical option, it has poor acceleration effect and can cause excessive oxidation. The acceleration effect and economy of N_2 lie between He and air, making it more commonly used. Meanwhile, mixing He and N_2 has also been explored [16]. Additionally, gas temperature and pressure are crucial factors affecting the deposition and bonding condition of the particles. Within the deposition window, increasing gas pressure results in more kinetic energy being transferred to the particles, causing them to bombard the substrate at higher velocities, thereby improving the quality of particle bonding. The increase in gas temperature plays two roles. First, it can increase the inner energy of the gas in the converging part of the nozzle (before the throat), thereby leading to higher kinetic energy in the diverging part of the nozzle and thus increasing the particle velocity. Second, it can also soften the particle and effectively facilitate plastic deformation, therefore, improving deposition [5].

2.3 Cold sprayed microstructure

According to the ASI mechanism, significant plastic deformation takes place at the particle/substrate interface during the high-velocity impact. This has been confirmed by both numerical simulation (figure 2a) [17] as well as experimental observation (figure 2b) [18]. Such significant plastic deformation can generate a high density of dislocations in the deformed region, and even initiate dynamic recrystallization at the particle-substrate boundaries and result in ultra-fine grain structure. For example, 7075Al cold sprayed using helium gas showed elongated grains with 80-250 nm boundary spacing as well as nano gains with diameters of approximately 140 nm [19]. Moreover, if the particles experience limited plastic deformation upon impact, either due to insufficient kinetic energy or its intrinsic high strength, residual micro-defects such as pores can present in the deposit [20].

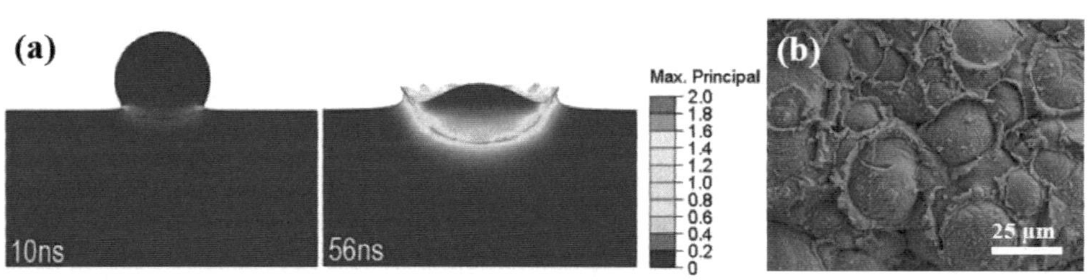

Figure 2. (a) Simulated plastic strain of the particle/substrate interface at an impact velocity of 850 m/s and a starting temperature of 20°C [17], and (b) top-surface SEM image of cold sprayed titanium coating exhibiting "jetting" [18].

3. Applications of cold spray in additive manufacturing

3.1 Fabrication of 3D components

Conventional additive manufacturing technologies, such as SLM, EBM, and WAAM, are mostly based on the melting and solidification process. They usually involve a laser or electron beam as the heat source to melt powders or wires into a liquid melt pool, which then rapidly solidify and bond with adjacent layers to produce components [21]. These technologies have some issues related to complex physical and chemical changes during melting/solidification, making the final microstructure difficult to control. Whereas cold spray as a solid-state process can avoid those melting-related issues. In addition, the cold spray possesses higher production rates than conventional additive manufacturing. The fabrication speed for EBM, SLM, and WAAM is usually less than 2 kg/h [22], while that of CSAM is usually 2-7 kg/h, and can even reach 40 kg/h [23].

Some engineering components prepared by CSAM are shown in figure 3. Figure 3a displays an aluminum bronze propeller made by the Australian company SPEE3D [24]. Figure 3b displays an aluminum 6061 automotive alternator shielding component made by the Australian Army, aimed at enhancing the strength and reliability of the vehicle's alternator cover [24]. Figure 3c illustrates a rocket nozzle fabricated by the German company, Impact Innovations, employed in the propulsion bell of rocket engines [25]. The geometry of nozzles is often intricate, and this intricacy has been the Achilles heel of CSAM, but developments in build design and process control have enabled the fabrication of rocket nozzles by CSAM. Some complex components can also be prepared by combining CSAM with subtractive manufacturing. As shown in figure 3d, Impact Innovations has manufactured a standalone turbojet aircraft engine fan shaft using Ti-6Al-4V [25]. The fan shaft was first deposited onto a pre-machined aluminum alloy mandrel, which was then removed by chemical dissolution, and finally, the preparation was completed with machining. In the same vein, the advancement of CSAM with structural design techniques like topology optimization can facilitate the creation of intricate lightweight structures with superior performance [26].

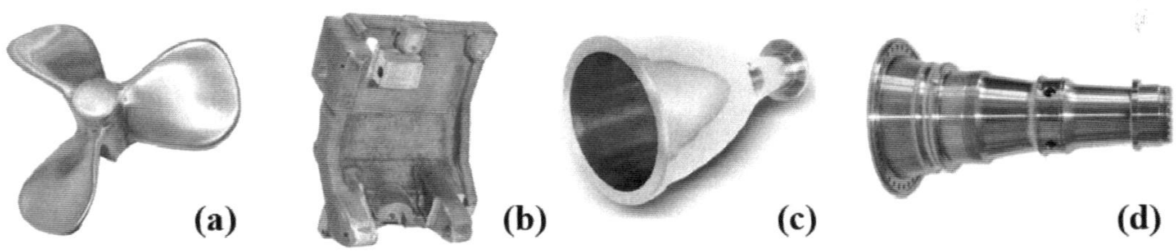

Figure 3. Components fabricated by CSAM: (a) aluminium bronze propeller [24], (b) automotive alternator shield [24], (c) rocket nozzle [25], (d) fan shaft [25].

Currently, CSAM has been applied to fabricate components of only a few materials, including Al, Cu, Ti, Ni, and steels. The fabrication of these materials can capitalize on the inherent advantage of low processing temperatures in CSAM, which can yield superior performance that other additive manufacturing processes cannot achieve. For example, Cu deposits fabricated by cold spray are almost twice as electrically conductive as SLM Cu [27].

3.2 Damaged component repair

Components, equipment, and machinery utilized in domains including aviation, automotive, and nuclear energy, are frequently exposed to harsh environments, resulting in wear, overload, corrosion, and mechanical damages. CSAM provides an economical and efficient solution for damage repairs. It can effectively restore damaged components to their original geometries. Compared to other high-temperature repair processes such as WAAM and laser cladding [28], CSAM operates at lower temperatures and can diminish the thermal impact on the substrates. This can help to prevent warping, cracking, and other melting-related issues. As technology advances, CSAM demonstrates increasing the potential for repair applications in a wider range of industries, especially for aerospace and military equipment [29].

CSAM has demonstrated its advantages in the repair of aircraft components. Figure 4 illustrates a repair example of a utility engine for a private jet conducted by CenterLine® (Canada)

[30]. This engine component is primarily composed of aluminum alloys and functions in the propulsion systems and auxiliary power units of commercial aircraft. This component frequently suffers from corrosion damage, as shown in figure 4a. CSAM allows for effective restoration of damaged areas (figure 4b), followed by completion of the repair work through machining (figures 4c and 4d) [30]. Moreover, it also plays a crucial role in the repair of military equipment. The U.S. Army Research Laboratory has utilized CSAM in the repair of the aluminum alloy mast bracket on Apache helicopters which often encounter corrosion and mechanical damage. After CSAM repair, the mast bracket retains identical tensile strength and fatigue performance as before repair [29].

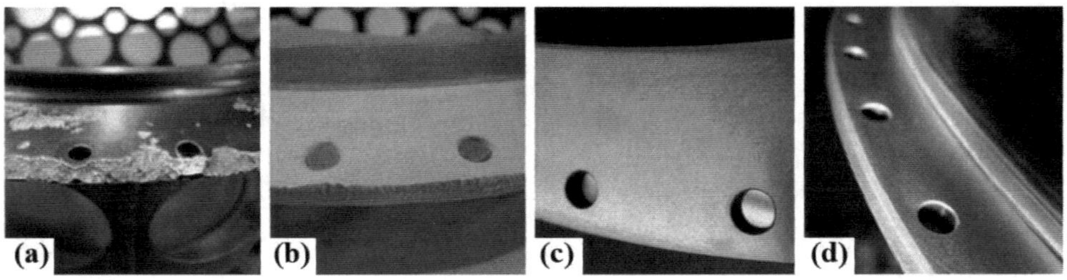

Figure 4. Repair of an aluminum component in a utility engine for a private jet: (a) before repair, (b) as sprayed with aluminum, (c) as machined, (d) finished component [30].

4. Summary and prospect

Cold spray, as a solid-state additive manufacturing technology, offers distinct advantages over conventional additive manufacturing, such as a short production time and fewer geometry limitations. It operates at low temperatures, thereby reducing thermal impact and degradation of materials. This technology is capable of producing large 3D components and has demonstrated advantages in damaged component repair, which can effectively restore the performance of components.

Meanwhile, the technology continues to encounter scientific and industrial obstacles. First, cold spray is a line-of-sight process, so during CSAM, as each layer of material is deposited, the geometric shape of the construction may obstruct the deposition path for subsequent materials. This is in particular an issue when fabricating components with complex geometry. It also demands a high level of precision in nozzle angle and path planning. Second, nozzle clogging is also a common issue in the CSAM process, since powders with low melting points and low hardnesses are prone to clogging. Although equipment manufacturers have spent much effort in nozzle design, the exact cause of nozzle clogging is not well understood. Lastly, CSAM of materials with low ductility may result in poor deposition efficiency and mechanical properties due to the tendency for brittle fracture rather than plastic deformation upon high-speed impact. Therefore, it is imperative to investigate hybrid technologies such as laser-assisted cold spray and refine post-processing treatments to improve the deposition quality of hard and brittle materials.

A significant amount of research and development efforts as well as scaling-up work is needed to expand the application of CSAM to more materials and more industries. Future studies are recommended to address the aforementioned issues while maintaining the distinct benefits of CSAM. Additionally, it is also important to establish performance assessment standards for components produced by CSAM to foster their wider industrial application.

References

[1] Alkhimov A P, Kosarev V F and Papyrin A N 1990 *Soviet Physics Doklady* **35** 1047
[2] Kindermann P, Strasser M, Wunderer M, Uensal I, Horn M and Seidel C 2023 *Progress in Additive Manufacturing* **22** 413
[3] Prashar G and Vasudev H 2021 *Journal of Cleaner Production* **310** 127606
[4] Kosarev V F, Klinkov S V, Alkhimov A P and Papyrin A N 2003 *Journal of Thermal Spray Technology* **12**(2) 265-81
[5] Yin S, Meyer M, Li W, Liao H and Lupoi R 2016 *Journal of Thermal Spray Technology* **25**(5) 1-23
[6] Zhang Z, Su C, Shao Z, Liao H, Ding X, Zhang L and Che H 2024 *Materials China* **43**(02) 102-14
[7] Schmidt T, GäRtner F, Assadi H and Kreye H 2006 *Acta Materialia* **54**(3) 729-42
[8] Yin S, Cavaliere P, Aldwell B, Jenkins R, Liao H, Li W and Lupoi R 2018 *Additive Manufacturing* **21** 628-50
[9] Grujicic M, Zhao C L, DeRosset W S and Helfritch D 2004 *Materials & Design* **25**(8) 681-8
[10] Assadi H, Klassen T, Gärtner F and Klassen T 2016 *Acta Materialia* **116** 382-407
[11] Li W and Gao W 2009 *Applied Surface Science* **255**(18) 7878-92
[12] Grujicic M, Saylor J R, Beasley D E, DeRosset W S and Helfritch D 2003 *Applied Surface Science* **219**(3-4) 211-27
[13] Hassani-Gangaraj M, Veysset D, Champagne V K, Nelson K A and Schuh C A 2018 *Acta Materialia* **158** 430-9
[14] Drehmann R, Grund T, Lampke T, Wielage B, Manygoats K, Schucknecht T and Rafaja D 2013 *Journal of Thermal Spray Technology* **23**(1-2) 68-75
[15] Bae G, Kumar S, Yoon S, Kang K, Na H, Kim H J and Lee C 2009 *Acta Materialia* **57**(19) 5654-66
[16] Li W, Cao C and Yin S 2020 *Progress in Materials Science* **110** 100633
[17] Aleksieieva O, Dereviankina L, Breuninger P, Bozoglu M, Tretiakov P, Toporov A and Antonyuk S 2022 *Coatings* **12**(9) 1297
[18] Wong W, Irissou E, Ryabinin A N, Legoux J and Yue S 2011 *Journal of Thermal Spray Technology* **20**(1) 213-26
[19] Rokni M R, Widener C A, Champagne V K and Crawford G A 2015 *Surface and Coatings Technology* **276** 305-15
[20] Poza P and Garrido-Maneiro M Á 2022 *Progress in Materials Science* **123** 100839
[21] Cooke S, Ahmadi K, Willerth S and Herring R 2020 *Journal of Manufacturing Processes* **57** 978-1003
[22] Gisario A, Kazarian M, Martina F and Mehrpouya M 2019 *Journal of Manufacturing Systems* **53** 124-49
[23] Huang R, Sun W, Guo S, Xie Y and Liu M 2020 *China Surface Engineering* **33**(4) 16-25
[24] https://www.spee3d.com/resources/case-studies/ June 9, 2024
[25] https://impact-innovations.com/en/applications/ June 9, 2024
[26] Lynch M E, Gu W, El-Wardany T, Hsu A, Viens D, Nardi A and Klecka M 2013 *Virtual & Physical Prototyping* **8**(3) 213-31
[27] Huang J, Yan X, Chang C, Xie Y, Ma W, Huang R, Zhao R, Li S, Liu M and Liao H 2020 *Surface and Coatings Technology* **395** 125936
[28] Kanishka K and Acherjee B 2023 *Journal of Manufacturing Processes* **89** 220-83
[29] Champagne V and Helfritch D 2015 *Materials Science and Technology* **31**(6) 627-34
[30] Villafuerte J and Wright D 2010 *Advanced Materials & Processes* **168** S13+

AUTHOR INDEX

Institute of Physics
Dirac House, Temple Back
Bristol BS1 6BE UK

ISSN: 1757-8981
ISBN 979-8-3313-0267-2